S0-EJV-227

WORLD HEALTH ORGANIZATION

INTERNATIONAL AGENCY FOR RESEARCH ON CANCER

IARC MONOGRAPHS

ON THE

EVALUATION OF THE CARCINOGENIC RISK OF CHEMICALS TO HUMANS

Polynuclear Aromatic Compounds,
Part 1,
Chemical, Environmental and Experimental Data

VOLUME 32

This publication represents the views and expert opinions of an IARC Working Group on the Evaluation of the Carcinogenic Risk of Chemicals to Humans which met in Lyon,

1-8 February 1983

December 1983

IARC MONOGRAPHS

In 1971, the International Agency for Research on Cancer (IARC) initiated a programme on the evaluation of the carcinogenic risk of chemicals to humans involving the production of critically evaluated monographs on individual chemicals. In 1980, the programme was expanded to include the evaluation of the carcinogenic risk associated with employment in specific occupations.

The objective of the programme is to elaborate and publish in the form of monographs critical reviews of data on carcinogenicity for chemicals and complex mixtures to which humans are known to be exposed, and on specific occupational exposures, to evaluate these data in terms of human risk with the help of international working groups of experts in chemical carcinogenesis and related fields, and to indicate where additional research efforts are needed.

This project was supported by PHS Grant No. 1 UO1 CA33193-01 awarded by the US National Cancer Institute, DHHS.

ISBN 92 832 12320 (soft-cover edition)

ISBN 92 832 1532X (hard-cover edition)

PRINTED IN FRANCE

CONTENTS

NOTE TO THE READER

The term 'carcinogenic risk' in the *IARC Monographs* series is taken to mean the probability that exposure to the chemical will lead to cancer in humans.

Inclusion of a chemical in the monographs does not imply that it is a carcinogen, only that the published data have been examined. Equally, the fact that a chemical has not yet been evaluated in a monograph does not mean that it is not carcinogenic.

Anyone who is aware of published data that may alter the evaluation of the carcinogenic risk of a chemical for humans is encouraged to make this information available to the Unit of Carcinogen Identification and Evaluation, International Agency for Research on Cancer, Lyon, France, in order that the chemical may be considered for re-evaluation by a future Working Group.

Although every effort is made to prepare the monographs as accurately as possible, mistakes may occur. Readers are requested to communicate any errors to the Unit of Carcinogen Identification and Evaluation, so that corrections can be reported in future volumes.

IARC WORKING GROUP ON THE EVALUATION OF THE CARCINOGENIC RISK OF CHEMICALS TO HUMANS: POLYNUCLEAR AROMATIC COMPOUNDS, PART 1, CHEMICAL, ENVIRONMENTAL AND EXPERIMENTAL DATA

Lyon, 1-8 February 1983

Members

P. Bogovski, Director, Institute of Experimental & Clinical Medicine, 42 Hiiu Street, Tallinn, Estonia 200016, USSR (*Chairman*)

V.J. Feron, Institute CIVO, Toxicology & Nutrition, TNO, PO Box 360, 3700 AJ Zeist, The Netherlands (*Co-rapporteur section 3.1*)

L. Fishbein, Acting Deputy Director, National Center for Toxicological Research, Jefferson, AR 72079, USA (*Co-rapporteur sections 1 and 2*)

C. Garner, Cancer Research Unit, University of York, Heslington, York YO1 5DD, UK (*Co-rapporteur section 3.2*)

G. Grimmer, Biochemisches Institut für Umweltcarcinogene, Sieker Landstrasse 19, 2070 Ahrensburg, Federal Republic of Germany

P. Grover, Institute of Cancer Research, Royal Cancer Hospital, Chester Beatty Research Institute, Fulham Road, London SW3 6JB, UK (*Co-rapporteur section 3.2*)

D. Hoffmann, Associate Director, Naylor Dana Institute for Disease Prevention, American Health Foundation, Valhalla, NY 10595, USA

N. Ito, Chairman & Professor, First Department of Pathology, Nagoya City University Medical School, 1 Kawasumi, Mizuho-cho, Mizuho-ku, Nagoya 467, Japan

J. Jacob, Biochemisches Institut für Umweltcarcinogene, Sieker Landstrasse 19, 2070 Ahrensburg, Federal Republic of Germany

K. Kurppa, Assistant Chief, Department of Epidemiology & Biometry, Institute of Occupational Health, Haartmaninkatu 1, Helsinki, Finland

L.S. Levy, Head, Industrial Toxicology Research & Advisory Unit, Department of Occupational Health & Safety, The University of Aston in Birmingham, Birmingham B4 7ET, UK

H. Liber, Department of Nutrition & Food Science, Massachusetts Institute of Technology, Cambridge, MA 02139, USA

J.K. Selkirk, Biology Division, Oak Ridge National Laboratory, PO Box Y, Oak Ridge, TN 37830, USA

B.W. Stewart, Senior Research Fellow, NSW Cancer Council, School of Pathology, The University of New South Wales, PO Box 1, Kensington, NSW, Australia 2033 (*Vice-chairman*)

F.M. Sullivan, Department of Pharmacology, Guy's Hospital Medical School, London SE1 9RT, UK (*Co-rapporteur section 3.2*)

B. Teichmann, Akademie der Wissenschaften der DDR, Zentralinstitut für Krebsforschung, Lindenberger Weg 80, 1115 Berlin-Buch, German Democratic Republic

W.G. Thilly, Department of Nutrition & Food Science, Massachusetts Institute of Technology, Cambridge, MA 02139, USA

A.W. Wood, Department of Biochemistry & Drug Metabolism, Hoffmann-La Roche Inc., Nutley, NJ 07110, USA

F. Zajdela, Director, Unité de Physiologie Cellulaire, INSERM (U22), Bâtiment 110, Centre Universitaire, 91405 Orsay Cédex, France

Representative from the National Cancer Institute

E.K. Weisburger, Assistant Director for Chemical Carcinogenesis, Division of Cancer Cause and Prevention, Building 31, Room 11A04, National Cancer Institute, Bethesda, MD 20205, USA

Representative from SRI International[1]

K.E. McCaleb, Director, Chemical-Environmental Program, SRI International, 333 Ravenswood Avenue, Menlo Park, CA 94025, USA

Secretariat

H. Bartsch, Division of Environmental Carcinogenesis (*Co-rapporteur section 3.2*)
J.R.P. Cabral, Division of Environmental Carcinogenesis
M. Friesen, Division of Environmental Carcinogenesis
L. Haroun, Division of Environmental Carcinogenesis (*Co-secretary*)
E. Heseltine, Editorial and Publications Services (*Editor*)

---

[1] Unable to attend

A. Likhachev, Division of Environmental Carcinogenesis
G. Mahon, Division of Epidemiology and Biostatistics
D. Mietton, Division of Environmental Carcinogenesis (*Library assistant*)
R. Montesano, Division of Environmental Carcinogenesis (*Co-rapporteur section 3.1*)
I. O'Neill, Division of Environmental Carcinogenesis (*Co-rapporteur sections 1 and 2*)
C. Partensky, Division of Environmental Carcinogenesis (*Technical officer*)
I. Peterschmitt, Division of Environmental Carcinogenesis, Geneva, Switzerland (*Bibliographic researcher*)
S. Poole, Birmingham, UK (*Technical editor*)
R. Saracci, Division of Epidemiology and Biostatistics
L. Simonato, Division of Epidemiology and Biostatistics
L. Tomatis, Director
H. Vainio, Division of Environmental Carcinogenesis (*Head of the Programme*)
J. Wahrendorf, Division of Epidemiology and Biostatistics
J.D. Wilbourn, Division of Environmental Carcinogenesis (*Co-secretary*)
H. Yamasaki, Division of Environmental Carcinogenesis

Secretarial assistance

M.-J. Ghess
D. Marcou
K. Masters
J. Mitchell
S. Reynaud

IN MEMORIAM

LEON M. SHABAD

Professor L.M. Shabad, Head of the Section of Chemical Carcinogenesis of the All-Union Cancer Research Centre in Moscow for the last 20 years, died on 29 August 1982.

Leon M. Shabad was born on 19 January 1902 in Minsk. In 1924 he graduated from the I Medical Institute in Leningrad, where he then started his research work in the pathology department. He was one of the first to observe, as part of a study on the carcinogenic activities of various tars, that the offspring of mice that had been painted with coal-tar were more susceptible to the carcinogenicity of this product than their dams.

In 1935, Professor Shabad set up an experimental cancer research laboratory at the All-Union Institute of Experimental Medicine in Leningrad. He was head of a laboratory at the Institute of Morphology from 1944 to 1952, and at the Institute of Oncology of the Academy of Medical Sciences from 1952 to 1958. In 1958, Professor Shabad joined the Institute of Experimental and Clinical Oncology in Moscow, which later became the All-Union Cancer Research Centre, to head the Section of Chemical Carcinogenesis. Beginning in 1935, he was one of the first scientists to start animal experiments with the newly-synthesized polynuclear aromatic hydrocarbons; and his early animal studies and work with clinical material enabled him to formulate morphological criteria for precancerous states and to describe consecutive stages in the development of neoplasia. He also put forward ideas concerning the endogenous formation of chemical carcinogens (1935, 1936) that were widely discussed, and he opened up new areas of research in chemical carcinogenesis, mainly in the field of industrial and environmental carcinogenesis.

During the last two decades of his career, Professor Shabad concentrated his attention on the prevention of cancer by limiting human exposure to carcinogens. Among his numerous publications (some 400, including eight books), one of the most important was a book entitled *Circulation of Carcinogens in the Environment*, which was published in 1973.

Professor Shabad was a member of many committees and editorial boards. Twenty-five years ago he organized the Committee on Carcinogenic Compounds and Cancer Prevention of the USSR Ministry of Health, and he implemented, through this committee, important practical measures aimed at decreasing human exposure to carcinogens. A member of the Academy of Medical Sciences of the USSR, he was also elected a Foreign Member of the Polish Academy of Sciences and given an honorary Doctorate by Charles University in Prague. He was a member of cancer research associations in many countries and in 1962, he was awarded a United Nations Organization prize for 'Outstanding Research on Cancer Causation and Prevention'. Professor Shabad served as vice-chairman of the scientific committee that met in Lyon in 1965 to make recommendations to the Director-General of WHO for the creation of an International Agency for Research on Cancer.

IN MEMORIAM

CHARLES HEIDELBERGER

Professor C. Heidelberger, Professor of Biochemistry and Director for Basic Research at the University of Southern California Cancer Centre in Los Angeles, died on 18 January, 1983.

Charles Heidelberger was born on 23 December 1920 in New York City. In 1942 he graduated with an SB degree in chemistry from Harvard University where he also obtained his MSc and PhD degrees in organic chemistry in 1944 and 1946, respectively. After two years of post-doctoral work at the University of California at Berkeley, where he synthesized ^{14}C-labelled dibenz[*a*,*h*]anthracene, the first ^{14}C-labelled carcinogen, Professor Heidelberger accepted a position as Assistant Professor of Oncology in the McArdle Laboratory in Madison, Wisconsin. He was promoted to Associate Professor in 1952, Professor in 1958 and became American Cancer Society Professor of Oncology in 1960. He also became Associate Director for Basic Research in the Wisconsin Clinical Cancer Center in 1973.

During his 28 years at the University of Wisconsin, Professor Heidelberger devoted himself to three main lines of research. The first was concerned with the mechanisms by which polynuclear aromatic hydrocarbons induce malignancy, and he successfully used ^{14}C-labelled compounds to demonstrate that the hydrocarbons become bound covalently to both protein and DNA when they are applied to mouse skin. The second major area was the early development of quantitative in-vitro systems in which malignant transformation of mammalian cells could be induced by addition of chemicals to the media. The third, and perhaps most important area in terms of human cancer was concerned with the preparation and study of the properties of fluorinated pyrimidines as chemotherapeutic agents. This work developed from his synthesis, in the 1950s, of 5-fluorouracil, which is still used extensively in clinical medicine.

Professor Heidelberger moved to the University of Southern California Cancer Center in 1976, and he became a Distinguished Professor of that University in 1981. He was the author of over 300 research publications and was a member of the editorial boards of several journals. He also served on many national and international committees, which included the Board of Directors of the American Association for Cancer Research, the Council of the UICC and, at one time, the Fellowship Committees of both the UICC and the IARC. He became a Member of the National Academy of Sciences in 1978 and was the recipient of numerous international awards and honours, including, in 1982, the first Athayde International Cancer Prize.

IARC MONOGRAPH PROGRAMME ON THE EVALUATION OF THE CARCINOGENIC RISK OF CHEMICALS TO HUMANS[1]

PREAMBLE

1. BACKGROUND

In 1971, the International Agency for Research on Cancer (IARC) initiated a programme on the evaluation of the carcinogenic risk of chemicals to humans with the object of producing monographs on individual chemicals. The criteria established at that time to evaluate carcinogenic risk to humans were adopted by all the working groups whose deliberations resulted in the first 16 volumes of the *IARC Monographs* series. In October 1977, a joint IARC/WHO ad-hoc Working Group met to re-evaluate these guiding criteria; this preamble reflects the results of their deliberations(1) and those of subsequent IARC ad-hoc Working Groups which met in April 1978(2) and February 1982(3).

An ad-hoc Working Group, which met in Lyon in April 1979 to prepare criteria to select chemicals for *IARC Monographs*(4), recommended that the *Monographs* programme be expanded to include consideration of human exposures to complex mixtures which occur, for example, in selected occupations. The Working Group which met in June 1980 therefore considered occupational exposures in the wood, leather and some associated industries; their deliberations resulted in Volume 25 of the *Monographs* series. A further Working Group which met in June 1981 evaluated the carcinogenic risks associated with occupations in the rubber manufacturing industry, and their conclusions were published as Volume 28 of the *Monographs* series.

2. OBJECTIVE AND SCOPE

The objective of the programme is to elaborate and publish in the form of monographs critical reviews of data on carcinogenicity for chemicals, groups of chemicals or industrial processes to which humans are known to be exposed, to evaluate the data in terms of human risk with the help of international working groups of experts in chemical carcinogenesis and related fields, and to indicate where additional research efforts are needed.

---

[1] This project was supported by PHS Grant No. 1 UO1 CA33193-01 awarded by the US National Cancer Institute, DHHS.

The critical analyses of the data are intended to assist national and international authorities in formulating decisions concerning preventive measures. No recommendations are given concerning legislation, since this depends on risk-benefit evaluations, which seem best made by individual governments and/or other international agencies. In this connection, WHO recommendations on food additives(5), drugs(6), pesticides and contaminants(7) and occupational carcinogens(8) are particularly informative.

In February 1982, a special ad-hoc Working Group met in Lyon to re-evaluate all chemicals, groups of chemicals, industrial processes or occupational exposures evaluated in volumes 1-29 of the *IARC Monographs*(3) and for which some data on carcinogenicity to humans (case reports or epidemiological studies) had been published in the literature. Re-evaluations of the data on humans, the data on experimental animals and the data on short-term tests were made for each agent and an overall evaluation of the carcinogenicity to humans was then made. The Working Group concluded that seven industrial processes or occupational exposures were causally associated with cancer in humans, as were 23 chemicals or groups of chemicals (Group 1). Sixty-one chemicals, groups of chemicals or industrial processes were considered to be probably carcinogenic to humans (Group 2; 14 in Group 2A, with a higher degree of evidence, and 47 in Group 2B, with a lower degree of evidence; compounds for which no evaluation could be made were classified as Group 3). In addition, for 175 chemicals there is *sufficient evidence* of their carcinogenicity in experimental animals.

The *IARC Monographs* are recognized as an authoritative source of information on the carcinogenicity of environmental chemicals. The first users' survey, made in 1976, indicates that the monographs are consulted routinely by various agencies in 24 countries. Up to December 1983, 32 volumes of *Monographs* had been published or were in press; and a cross index of synonyms and trade names in volumes 1 to 26 has been published as Supplement 3 to the *Monographs*(9). Each volume of monographs is printed in 4000 copies for distribution to governments, regulatory agencies and interested scientists. The monographs are also available *via* the WHO Distribution and Sales Service.

3. SELECTION OF CHEMICALS FOR MONOGRAPHS

The chemicals (natural and synthetic, including those which occur as mixtures and in manufacturing processes) are selected for evaluation on the basis of two main criteria: (a) there is evidence of human exposure, and (b) there is some experimental evidence of carcinogenicity and/or there is some evidence or suspicion of a risk to humans. In certain instances, chemical analogues are also considered. The scientific literature is surveyed for published data relevant to the monograph programme. In addition, the IARC *Survey of Chemicals Being Tested for Carcinogenicity*(10) often indicates those chemicals that may be scheduled for future meetings.

Inclusion of a chemical in a volume does not imply that it is carcinogenic, only that the published data have been examined. The evaluations must be consulted to ascertain the conclusions of the Working Group. Equally, the fact that a chemical has not appeared in a monograph does not mean that it is without carcinogenic hazard.

As new data on chemicals for which monographs have already been prepared and new principles for evaluating carcinogenic risk receive acceptance, re-evaluations will be made at subsequent meetings, and revised monographs will be published as necessary.

4. WORKING PROCEDURES

Approximately one year in advance of a meeting of a working group, a list of the substances to be considered is prepared by IARC staff in consultation with other experts. Subsequently, all relevant biological data are collected by IARC; in addition to the published literature, US Public Health Service Publication No. 149(11) has been particularly valuable and has been used in conjunction with other recognized sources of information on chemical carcinogenesis and systems such as CANCERLINE, MEDLINE and TOXLINE. The major collection of data and the preparation of first drafts for the sections on chemical and physical properties, on production, use, occurrence and on analysis are carried out by SRI International, Stanford, CA, USA under a separate contract with the US National Cancer Institute. Most of the data so obtained on production, use and occurrence refer to the United States and Japan; SRI International and IARC supplement this information with that from other sources in Europe. Bibliographical sources for data on mutagenicity and teratogenicity are the Environmental Mutagen Information Center and the Environmental Teratology Information Center, both located at the Oak Ridge National Laboratory, TN, USA.

Six months before the meeting, reprints of articles containing relevant biological data are sent to an expert(s), or are used by the IARC staff, for the preparation of first draft monographs. These drafts are compiled by IARC staff and are sent prior to the meeting to all participants of the Working Group for their comments. The Working Group then meets in Lyon for seven to eight days to discuss and finalize the texts of the monographs and to formulate the evaluations. After the meeting, the master copy of each monograph is verified by consulting the original literature, then edited by a professional editor and prepared for reproduction. The monographs are published within six to nine months after the Working Group meeting.

5. DATA FOR EVALUATIONS

With regard to biological data, only reports that have been published or accepted for publication are reviewed by the working groups, although a few ad-hoc exceptions have been made; in certain instances, reports from government agencies that have undergone peer review and are widely available are considered. The monographs do not cite all of the literature on a particular chemical: only those data considered by the Working Group to be relevant to the evaluation of the carcinogenic risk of the chemical to humans are included.

Anyone who is aware of data that have been published or are in press which are relevant to the evaluations of the carcinogenic risk to humans of chemicals for which monographs have appeared is urged to make them available to the Unit of Carcinogen Identification and Evaluation, Division of Environmental Carcinogenesis, International Agency for Research on Cancer, Lyon, France.

6. THE WORKING GROUP

The tasks of the Working Group are five-fold: (a) to ascertain that all data have been collected; (b) to select the data relevant for the evaluation; (c) to ensure that the summaries of the data enable the reader to follow the reasoning of the committee; (d) to judge the significance of the results of experimental and epidemiological studies; and (e) to make an evaluation of the carcinogenic risk of the chemical.

Working Group participants who contributed to the consideration and evaluation of chemicals within a particular volume are listed, with their addresses, at the beginning of each publication. Each member serves as an individual scientist and not as a representative of any organization or government. In addition, observers are often invited from national and international agencies, organizations and industrial associations.

7. GENERAL PRINCIPLES FOR EVALUATING THE CARCINOGENIC RISK OF CHEMICALS

The widely accepted meaning of the term 'chemical carcinogenesis', and that used in these monographs, is the induction by chemicals of neoplasms that are not usually observed, the earlier induction by chemicals of neoplasms that are commonly observed, and/or the induction by chemicals of more neoplasms than are usually found - although fundamentally different mechanisms may be involved in these three situations. Etymologically, the term 'carcinogenesis' means the induction of cancer, that is, of malignant neoplasms; however, the commonly accepted meaning is the induction of various types of neoplasms or of a combination of malignant and benign tumours. In the monographs, the words 'tumour' and 'neoplasm' are used interchangeably. (In scientific literature the terms 'tumourigen', 'oncogen' and 'blastomogen' have all been used synonymously with 'carcinogen', although occasionally 'tumourigen' has been used specifically to denote a substance that induces benign tumours.)

(*a*) *Experimental Evidence*

(*i*) *Qualitative aspects*

Both the interpretation and evaluation of a particular study as well as the overall assessment of the carcinogenic activity of a chemical involve several qualitatively important considerations, including: (a) the experimental parameters under which the chemical was tested, including route of administration and exposure, species, strain, sex, age, etc.; (b) the consistency with which the chemical has been shown to be carcinogenic, e.g., in how many species and at which target organ(s); (c) the spectrum of neoplastic response, from benign neoplasm to multiple malignant tumours; (d) the stage of tumour formation in which a chemical may be involved: some chemicals act as complete carcinogens and have initiating and promoting activity, while others are promoters only; and (e) the possible role of modifying factors.

There are problems not only of differential survival but of differential toxicity, which may be manifested by unequal growth and weight gain in treated and control animals. These complexities are also considered in the interpretation of data.

Many chemicals induce both benign and malignant tumours. Few instances are recorded in which only benign neoplasms are induced by chemicals that have been studied extensively. Benign tumours may represent a stage in the evolution of a malignant neoplasm or they may be 'end-points' that do not readily undergo transition to malignancy. If a substance is found to induce only benign tumours in experimental animals, it should nevertheless be suspected of being a carcinogen and requires further investigation.

(*ii*) *Hormonal carcinogenesis*

Hormonal carcinogenesis present certain distinctive features: the chemicals involved occur both endogenously and exogenously; in many instances, long exposure is required; tumours occur in the target tissue in association with a stimulation of non-neoplastic growth, but in some cases hormones promote the proliferation of tumour cells in a target organ. Hormones that occur in excessive amounts, hormone-mimetic agents and agents that cause hyperactivity or imbalance in the endocrine system may require evaluative methods comparable with those used to identify chemical carcinogens; particular emphasis must be laid on quantitative aspects and duration of exposure. Some chemical carcinogens have significant side effects on the endocrine system, which may also result in hormonal carcinogenesis. Synthetic hormones and anti-hormones can be expected to possess other pharmacological and toxicological actions in addition to those on the endocrine system, and in this respect they must be treated like any other chemical with regard to intrinsic carcinogenic potential.

(*iii*) *Quantitative aspects*

Dose-response studies are important in the evaluation of carcinogenesis: the confidence with which a carcinogenic effect can be established is strengthened by the observation of an increasing incidence of neoplasms with increasing exposure.

The assessment of carcinogenicity in animals is frequently complicated by recognized differences among the test animals (species, strain, sex, age), route(s) of administration and in dose-duration of exposure; often, target organs at which a cancer occurs and its histological type may vary with these parameters. Nevertheless, indices of carcinogenic potency in particular experimental systems [for instance, the dose-rate required under continuous exposure to halve the probability of the animals remaining tumourless(12)] have been formulated in the hope that, at least among categories of fairly similar agents, such indices may be of some predictive value in other systems, including humans.

Chemical carcinogens differ widely in the dose required to produce a given level of tumour induction, although many of them share common biological properties, which include metabolism to reactive [electrophilic(13-15)] intermediates capable of interacting with DNA. The reason for this variation in dose-response is not understood, but it may be due either to differences within a common metabolic process or to the operation of qualitatively distinct mechanisms.

(*iv*) *Statistical analysis of animal studies*

Tumours which would have arisen had an animal lived longer may not be observed because of the death of the animal from unrelated causes, and this possibility must be allowed for. Various analytical techniques have been developed which use the assumption of independence of competing risks to allow for the effects of intercurrent mortality on the final numbers of tumour-bearing animals in particular treatment groups.

For externally visible tumours and for neoplasms that cause death, methods such as Kaplan-Meier (i.e., 'life-table', 'product-limit' or 'actuarial') estimates(12), with associated significance tests(16,17), have been recommended.

For internal neoplasms which are discovered 'incidentally'(16) at autopsy but which did not cause the death of the host, different estimates (18) and significance tests(16,17) may be necessary for the unbiased study of the numbers of tumour-bearing animals.

All of these methods(12,16-18) can be used to analyse the numbers of animals bearing particular tumour types, but they do not distinguish between animals with one or many such tumours. In experiments which end at a particular fixed time, with the simultaneous sacrifice of many animals, analysis of the total numbers of internal neoplasms per animal found at autopsy at the end of the experiment is straightforward. However, there are no adequate statistical methods for analysing the numbers of particular neoplasms that kill an animal. The design and statistical analysis of long-term carcinogenicity experiments were recently reviewed, in Supplement 2 to the *Monograph* series(19).

(*b*) *Evidence of Carcinogenicity in Humans*

Evidence of carcinogenicity in humans can be derived from three types of study, the first two of which usually provide only suggestive evidence: (i) reports concerning individual cancer patients (case reports), including a history of exposure to the supposed carcinogenic agent; (ii) descriptive epidemiological studies in which the incidence of cancer in human population is found to vary (spatially or temporally) with exposure to the agent; and (iii) analytical epidemiological studies (e.g., case-control or cohort studies) in which individual exposure to the agent is found to be associated with an increased risk of cancer.

An analytical study that shows a positive association between an agent and a cancer may be interpreted as implying causality to a greater or lesser extent, on the basis of the following criteria: (a) There is no identifiable positive bias. (By 'positive bias' is meant the operation of factors in study design or execution which lead erroneously to a more strongly positive association between an agent and disease than in fact exists. Examples of positive bias include, in case-control studies, better documentation of exposure to the agent for cases than for controls, and, in cohort studies, the use of better means of detecting cancer in individuals exposed to the agent than in individuals not exposed.) (b) The possibility of positive confounding has been considered. (By 'positive confounding' is meant a situation in which the relationship between an agent and a disease is rendered more strongly positive than it truly is as a result of an association between that agent and another agent which either causes or prevents the disease. An example of positive confounding is the association between coffee consumption and lung cancer, which results from their joint association with cigarette smoking.) (c) The association is unlikely to be due to chance alone. (d) The association is strong. (e) There is a dose-response relationship.

In some instances, a single epidemiological study may be strongly indicative of a cause-effect relationship; however, the most convincing evidence of causality comes when several independent studies done under different circumstances result in 'positive' findings.

Analytical epidemiological studies that show no association between an agent and a cancer ('negative' studies) should be interpreted according to criteria analogous to those listed above: (a) there is no identifiable negative bias; (b) the possibility of negative confounding has been considered; and (c) the possible effects of misclassification of exposure or outcome have been weighed. In addition, it must be recognized that in any study there are confidence limits around the estimate of association or relative risk. In a study regarded as 'negative', the upper confidence limit may indicate a relative risk substantially greater than unity; in that case, the study excludes only relative risks that are above the upper limit. This usually means that a 'negative' study must be large to be convincing. Confidence in a 'negative' result is increased when several independent studies carried out under different circumstances are in agreement. Finally, a 'negative' study may be considered to be relevant only to dose levels within or below the range of those observed in the study and is pertinent only if sufficient time has elapsed since first human exposure to the agent. Experience with human cancers of known etiology suggests that the period from first exposure to a chemical carcinogen to development of clinically observed cancer is usually measured in decades and may be in excess of 30 years.

The Working Group whose deliberations resulted in Supplement 4 to the *Monographs*(3) defined the degrees of evidence for carcinogenicity from studies in humans as:

i. *Sufficient evidence* of carcinogenicity, which indicates that there is a causal relationship between the agent and human cancer.

ii. *Limited evidence* of carcinogenicity, which indicates that a causal interpretation is credible, but that alternative explanations, such as chance, bias or confounding, could not adequately be excluded.

iii. *Inadequate evidence*, which applies to both positive and negative evidence, indicates that one of two conditions prevailed: (a) there were few pertinent data; (b) the available studies, while showing evidence of association, did not exclude chance, bias or confounding.

iv. *No evidence*, applies when several adequate studies were available which do not show evidence of carcinogenicity.

(c) *Relevance of Experimental Data to the Evaluation of Carcinogenic Risk to Humans*

Information compiled from the first 29 volumes of the *IARC Monographs*(3,20,21) shows that of the chemicals or groups of chemicals now generally accepted to cause or probably to cause cancer in humans (Groups 1 and 2A), all (with the possible exception of arsenic) of those which have been tested appropriately produce cancer in at least one animal species. For several of the chemicals (e.g., aflatoxins, 4-aminobiphenyl, diethylstilboestrol, melphalan, mustard gas and vinyl chloride), evidence of carcinogenicity in experimental animals preceded evidence obtained from epidemiological studies or case reports.

Assessment of evidence of carcinogenicity from studies in experimental animals

Overall evidence of carcinogenicity in experimental animals was classified into four groups:

i. *Sufficient evidence* of carcinogenicity, which indicates that there is an increased incidence of malignant tumours: (a) in multiple species or strains; or (b) in multiple experiments (preferably with different routes of administration or using different dose levels); or (c) to an unusual degree with regard to incidence, site or type of tumour, or age at onset. Additional evidence may be provided by data on dose-response effects, as well as information from short-term tests or on chemical structure.

ii. *Limited evidence* of carcinogenicity, which means that the data suggest a carcinogenic effect but are limited because: (a) the studies involve a single species, strain or experiment; or (b) the experiments are restricted by inadequate dosage levels, inadequate duration of exposure to the agent, inadequate period of follow-up, poor survival, too few animals, or inadequate reporting; or (c) the neoplasms produced often occur spontaneously and, in the past, have been difficult to classify as malignant by histological criteria alone (e.g., lung adenomas and adenocarcinomas and liver tumours in certain strains of mice).

iii. *Inadequate evidence*, which indicates that because of major qualitative or quantitative limitations, the studies cannot be interpreted as showing either the presence or absence of a carcinogenic effect.

iv. *No evidence* applies when several adequate studies show, within the limits of the tests used, that the chemical is not carcinogenic. The number of negative studies is small, since, in general, studies that show no effect are less likely to be published than those suggesting carcinogenicity.

The categories *sufficient evidence* and *limited evidence* refer only to the strength of the experimental evidence that these chemicals are carcinogenic and not to the extent of their carcinogenic activity nor to the mechanism involved. The classification of any chemical may change as new information becomes available.

For many of the chemicals evaluated in the first 31 volumes of the *IARC Monographs* for which there is *sufficient evidence* of carcinogenicity in animals, data relating to carcinogenicity for humans are either insufficient or nonexistent. **In the absence of adequate data on humans, it is reasonable, for practical purposes, to regard chemicals for which there is *sufficient evidence* of carcinogenicity in animals as if they presented a carcinogenic risk to humans**. The use of the expressions 'for practical purposes' and 'as if they presented a carcinogenic risk' indicates that at the present time a correlation between carcinogenicity in animals and possible human risk cannot be made on a purely scientific basis, but only pragmatically. Such a pragmatical correlation may be useful to regulatory agencies in making decisions related to the primary prevention of cancer.

In the present state of knowledge, it would be difficult to define a predictable relationship between the dose (mg/kg bw/day) of a particular chemical required to produce cancer in test animals and the dose which would produce a similar incidence of cancer in humans. Some data, however, suggest that such a relationship may exist(22,23), at least for certain classes of carcinogenic chemicals, but no acceptable methods are currently available for quantifying the possible errors that may be involved in such an extrapolation procedure.

Assessment of data from short-term tests

In recent years, several short-term tests for the detection of potential carcinogens have been developed. When only inadequate experimental animal data are available, positive results in a variety of validated short-term tests (see section 8(*c*)(ii)) can be taken as an indication that the compound is a potential carcinogen and that it should be further tested in animals for an assessment of its carcinogenicity. Negative results from short-term tests cannot be considered as evidence to rule out carcinogenicity(3). Whether short-term tests will eventually be as reliable as long-term tests in predicting carcinogenicity in humans will depend on further demonstrations of consistency with long-term experiments and with data from humans.

In view of the limitations of current knowledge about mechanisms of carcinogenesis, certain cautions should be emphasized: (i) at present, these tests should not be used by themselves to conclude whether or not an agent is carcinogenic; (ii) even when positive results are obtained in one or more of these tests, it is not clear that they can be used reliably to predict the relative potencies of compounds as carcinogens in intact animals; (iii) since the currently available tests do not detect all classes of agents that are active in the carcinogenic process (e.g., hormones, promoters), one must be cautious in utilizing these tests as the sole criterion for setting priorities in carcinogenesis research and in selecting compounds for animal bioassays.

The Working Group which met in February 1982 to re-evaluate chemicals and industrial processes associated with cancer in humans(3) sometimes considered the results from short-term tests in making an overall evaluation of the carcinogenic risk of chemicals to humans. In some cases, the Working Group considered that the known chemical properties of a compound and the results from short-term tests warranted its transfer from Group 3 to 2B or from Group 2B to 2A.

Because of the large number and wide variety of short-term tests that may be relevant for the prediction of potential carcinogens, assessing the overall evidence of activity of a compound in short-term tests is difficult. The data relative to each compound can, however, be classified by grouping results under the type of test used and the biological complexity of the test system. *'DNA damage'* would include evidence for covalent binding to DNA, induction of DNA breakage or repair, induction of prophage in bacteria, and a positive response in tests of comparative survival in DNA repair-proficient and DNA repair-deficient bacteria. *'Mutagenicity'* refers to induction of mutations in cultured cells or in organisms (e.g., heritable alterations in phenotype, including forward or reverse point mutations, recombination, gene conversion, and specific-locus mutation). *'Chromosomal anomalies'* refers to the induction of chromosomal aberrations, including breaks, gaps, rearrangements and micronuclei, sister chromatid exchange and aneuploidy. (This classification does not imply that some chromosomal anomalies are not mutational events.) *'Other'* refers to various additional endpoints, including cell transformation, i.e., morphological transformation and colony formation in agar; dominant lethal tests; morphological abnormalities in sperm; and mitochondrial mutation. Biological systems include: *'Prokaryotes'*, i.e., bacteria, in the presence or absence of a cellular or subcellular metabolic activation system; *'Fungi and plants'*; *'Insects'*, usually *Drosophila melanogaster*; *'Mammalian cells* (in vitro)', either rodent or human somatic cells or cell lines in culture; *'Mammals* (in vivo)', studies in which the test compound was administered to intact experimental animals; and *'Humans* (in vivo)', studies of cells from groups of individuals drawn from a population exposed to the substance in question.

Overall evidence of activity in short-term tests is adjudged to fall into one of four categories, *sufficient*, *limited*, *inadequate* or *no evidence*. The criteria generally used are:

i. *Sufficient evidence*, when there were a total of at least three positive results in at least two of three test systems measuring DNA damage, mutagenicity or chromosomal anomalies. When two of the positive results were for the same biological endpoint, they had to be derived from systems of different biological complexity.

ii. *Limited evidence*, when there were at least two positive results, either for different endpoints or in systems representing two levels of biological complexity.

iii. *Inadequate evidence*, when there were too few data for an adequate evaluation, or when there were contradictory data.

iv. *No evidence*, when there were many negative results from a variety of short-term tests with different endpoints, and at different levels of biological complexity. If certain biological endpoints are not adequately covered this is indicated.

In establishing these categories greater weight may be given to the three primary endpoints - DNA damage, mutagenicity and chromosomal anomalies - and judgements made on the quality as well as on the quantity of the evidence. In a minority of cases, strict interpretation of these criteria may be affected by consideration of a variety of other factors (such as the purity of the test compound, problems of metabolic activation, appropriateness of the test system) such that, in the judgement of the Working Group, a compound may be placed in a lower or higher category.

Assignment of a chemical to one of these categories involves several arbitrary decisions, since many of the tests systems are still under validation. Thus, the selection of specific tests remains flexible and should reflect the most advanced state of knowledge in this field.

8. EXPLANATORY NOTES ON THE MONOGRAPH CONTENTS

(*a*) *Chemical and Physical Data (Section 1)*

The Chemical Abstracts Services Registry Number, the latest Chemical Abstracts Primary Name (9th Collective Index)(24) and the IUPAC Systematic Name(25) are recorded in section 1. Other synonyms and trade names are given, but no comprehensive list is provided. Further, some of the trade names are those of mixtures in which the compound being evaluated is only one of the ingredients.

The structural and molecular formulae, molecular weight and chemical and physical properties are given. The properties listed refer to the pure substance, unless otherwise specified, and include, in particular, data that might be relevant to carcinogenicity (e.g., lipid solubility) and those that concern identification.

A separate description of the composition of technical products includes available information on impurities and formulated products.

(*b*) *Production, Use, Occurrence and Analysis (Section 2)*

The purpose of section 2 is to provide indications of the extent of past and present human exposure to the chemical.

(*i*) *Synthesis*

Since cancer is a delayed toxic effect, the dates of first synthesis and of first commercial production of the chemical are provided. In addition, methods of synthesis used in past and present commercial production are described. This information allows a reasonable estimate to be made of the date before which no human exposure could have occurred.

(*ii*) *Production*

Since Europe, Japan and the United States are reasonably representative industrialized areas of the world, most data on production, foreign trade and uses are obtained from those countries. It should not, however, be inferred that those nations are the sole or even the major sources or users of any individual chemical.

Production and foreign trade data are obtained from both governmental and trade publications by chemical economists in the three geographical areas. In some cases, separate production data on chemicals manufactured in the United States are not available because their publication could disclose confidential information. In such cases, an indication of the minimum quantity produced can be inferred from the number of companies reporting commercial production. Each company is required to report on individual chemicals if the sales value or the weight of the annual production exceeds a specified minimum level. These levels vary for chemicals classified for different uses, e.g., medicinals and plastics; in fact, the minimal annual sales value is between $1000 and $50 000, and the minimal annual weight of production is between 450 and 22 700 kg. Data on production in some European countries are obtained by means of general questionnaires sent to companies thought to produce the compounds being evaluated. Information from the completed questionnaires is compiled by country, and the resulting estimates of production are included in the individual monographs.

(iii) Use

Information on uses is meant to serve as a guide only and is not complete. It is usually obtained from published data but is often complemented by direct contact with manufacturers of the chemical. In the case of drugs, mention of their therapeutic uses does not necessarily represent current practice nor does it imply judgement as to their clinical efficacy.

Statements concerning regulations and standards (e.g., pesticide registrations, maximum levels permitted in foods, occupational standards and allowable limits) in specific countries are mentioned as examples only. They may not reflect the most recent situation, since such legislation is in a constant state of change; nor should it be taken to imply that other countries do not have similar regulations.

(iv) Occurrence

Information on the occurrence of a chemical in the environment is obtained from published data, including that derived from the monitoring and surveillance of levels of the chemical in occupational environments, air, water, soil, foods and tissues of animals and humans. When available, data on the generation, persistence and bioaccumulation of a chemical are also included.

(v) Analysis

The purpose of the section on analysis is to give the reader an indication, rather than a complete review, of methods cited in the literature. No attempt is made to evaluate critically or to recommend any of the methods.

(c) Biological Data Relevant to the Evaluation of Carcinogenic Risk to Humans (Section 3)

In general, the data recorded in section 3 are summarized as given by the author; however, comments made by the Working Group on certain shortcomings of reporting, of statistical analysis or of experimental design are given in square brackets. The nature and extent of impurities/contaminants in the chemicals being tested are given when available.

(i) Carcinogenicity studies in animals

The monographs are not intended to cover all reported studies. Some studies are purposely omitted (a) because they are inadequate, as judged from previously described criteria(19,26-29) (e.g., too short a duration, too few animals, poor survival); (b) because they only confirm findings that have already been fully described; or (c) because they are judged irrelevant for the purpose of the evaluation. In certain cases, however, such studies are mentioned briefly, particularly when the information is considered to be a useful supplement to other reports or when it is the only data available. Their inclusion does not, however, imply acceptance of the adequacy of their experimental design or of the analysis and interpretation of their results.

Mention is made of all routes of administration by which the compound has been adequately tested and of all species in which relevant tests have been done(7,28). In most cases, animal strains are given. [General characteristics of mouse strains have been reviewed(30).] Quantitative data are given to indicate the order of magnitude of the effective carcinogenic doses. In general, the doses and schedules are indicated as they appear in the paper; sometimes units have been converted for easier comparison. Experiments in which the compound was administered in conjunction with known carcinogens and experiments on factors that modify the carcinogenic effect are also reported. Experiments on the carcinogenicity of known metabolites and derivatives are also included.

(*ii*) *Other relevant biological data*

Lethality data are given when available, and other data on toxicity are included when considered relevant. The metabolic data are restricted to studies that show the metabolic fate of the chemical in animals and humans, and comparisons of data from animals and humans are made when possible. Information is also given on absorption, distribution, excretion and placental transfer.

Effects on reproduction and prenatal toxicity. Data on effects on reproduction, teratogenicity and feto- and embryotoxicity from studies in experimental animals and from observations in humans are also included. There appears to be no causal relationship between teratogenicity(31) and carcinogenicity, but chemicals often have both properties. Evidence of prenatal toxicity suggests transplacental transfer, which is a prerequisite for transplacental carcinogenesis.

Indirect tests (mutagenicity and other short-term tests). Data from indirect tests are also included. Since most of these tests have the advantage of taking less time and being less expensive than mammalian carcinogenicity studies, they are generally known as 'short-term' tests. They comprise assay procedures which rely on the induction of biological and biochemical effects in in-vivo and/or in-vitro systems. The end-point of the majority of these tests is the production not of neoplasms in animals but of changes at the molecular, cellular or multicellular level, as described in section 7(c).

The induction of cancer is thought to proceed by a series of steps, some of which have been distinguished experimentally(32-36). The first step - 'initiation' - is thought to involve damage to DNA resulting in heritable modifications in, or rearrangements of, genetic information. Proliferation of cells whose properties have been permanently altered during initiation (which may involve somatic mutation) is thought to result in the formation of clones of cells whose further progress to malignancy is dependent on a series of events - 'promotion' and 'progression' - the underlying mechanisms of which are largely unknown. Although this is a useful model, it should be kept in mind that the carcinogenic process may not always proceed by such a multi-step mechanism.

The idea that damage to DNA is a critical event in the initiation of carcinogenesis is based on a large body of data which show that many carcinogens are reactive electrophiles *per se*, or can be readily converted to reactive electrophiles by enzymic pathways characteristic of eukaryotic metabolism(37). A variety of DNA-carcinogen adducts, formed by reaction of electrophilic moieties with nucleophilic centres in DNA, have been identified in DNA recovered from reactions performed with carcinogens *in vitro*, or from cultured cells or intact organisms treated with carcinogens(35,38,39). Moreover, the recognition that many classes of carcinogens (including ionizing and ultra-violet radiation and chemicals of a very wide range of structure and reactivity) are mutagenic(40) supports the idea that DNA is a critical target of carcinogenic agents. Assays for mutagenicity and related effects all exploit this characteristic ability of carcinogens to cause DNA damage or chromosomal anomalies either directly or indirectly. It should be noted, however, that some carcinogens may act by mechanisms that do not involve DNA damage(41) and thus would not cause such genetic effects.

In many of the short-term tests, the indicator organism may not possess or may have lost, following culture, the range of enzyme systems known in intact mammals to metabolize chemically unreactive carcinogens to reactive electrophiles. It is often necessary, therefore, to provide an exogenous source of such activity in the form of a tissue extract or cell feeder-layer or whole-cell systems prepared from mammalian sources(19). In-vitro metabolic systems may not accurately reflect the fate of a chemical subjected to the checks and balances afforded by absorption, distribution, metabolism and excretion in mammals(19), and this must be borne in mind when evaluating the results from short-term tests which employ in-vitro metabolic activation.

Tests have been devised which exploit the useful attributes of microbial or cellular genetic systems without compromising the integrity of mammalian pharmacodynamics and metabolism. Such 'host-mediated' assays involve the inoculation of indicator organisms into mammals (usually rodents) which are then dosed with the test chemical. There are limitations to both the numbers and types of organisms which can be introduced and recovered from dosed animals and to the access of indicator organisms to activated metabolites. Lack of sensitivity may therefore be a problem.

A group of short-term tests use 'transformation' of cultured mammalian cells, rather than manifestation of DNA damage or chromosomal anomalies, as an indicator of carcinogenic potential. Some of the assays also employ an exogenous metabolic activation system. Cell transformation is assessed by scoring characteristic changes in cellular and colonial morphology, or changes in growth characteristics (e.g., growth of colonies in soft agar) following treatment with the test compound. In some protocols, the ability of transformed cells to produce tumours is tested by injecting the cells into appropriate animals.

Studies may also be conducted on cells taken from people exposed to putative chemical carcinogens. The cells are examined for mutation and for chromosomal anomalies either directly or after short-term culture *in vitro*, or samples of sperm from such individuals may be analysed for morphological abnormalities. Evidence of absorption of putative carcinogens may be adduced from the assay of body fluids and excreta for DNA-damaging activity, using, for example, bacterial mutation assays.

The present state of knowledge does not permit the selection of a specific test(s) as the most appropriate for identifying all classes of potential carcinogens, although certain systems are more sensitive to some classes. Ideally, a compound should be tested in a battery of short-term tests. For optimum usefulness, data on purity must be given. For several recent reviews on the use of short-term tests see IARC(19), Montesano *et al.*(42), de Serres and Ashby(43), Sugimura *et al.*(44), Bartsch *et al.*(45) and Hollstein *et al.*(46).

(iii) *Case reports and epidemiological studies*

Observations in humans are summarized in this section. The criteria for including a study in this section are described above (section 7(*b*)).

(*d*) *Summary of Data Reported and Evaluation (Section 4)*

Section 4 summarizes the relevant data from animals and humans and gives the critical views of the Working Group on those data.

(*i*) *Experimental data*

Data relevant to the evaluation of the carcinogenicity of the chemical in animals are summarized in this section. The animal species mentioned are those in which the carcinogenicity of the substance was clearly demonstrated. Tumour sites are also indicated. If the substance has produced tumours after prenatal exposure or in single-dose experiments, this is indicated. Dose-response data are given when available.

Results from validated mutagenicity and other short-term tests and from tests for prenatal toxicity are reported if the Working Group considered the data to be relevant. The degree of evidence of activity in short-term tests is mentioned in this section.

(ii) Human data

Human exposure to the chemical is summarized on the basis of data on production, use and occurrence. Case reports and epidemiological studies that are considered to be pertinent to an assessment of human carcinogenicity are described. Other biological data which are considered to be relevant are also mentioned.

(iii) Evaluation

This section comprises the overall evaluation by the Working Group of the carcinogenic risk of the chemical, complex mixture or occupational exposure to humans. All of the data in the monograph, and particularly the summarized experimental and human data, are considered in order to make this evaluation. This section should also be read in conjunction with section 7c of this Preamble.

References

1. IARC (1977) IARC Monograph Programme on the Evaluation of the Carcinogenic Risk of Chemicals to Humans. Preamble. *IARC intern. tech. Rep. No. 77/002*

2. IARC (1978) Chemicals with *sufficient evidence* of carcinogenicity in experimental animals - *IARC Monographs* volumes 1-17. *IARC intern. tech. Rep. No. 78/003*

3. IARC (1982) *IARC Monographs on the Evaluation of the Carcinogenic Risk of Chemicals to Humans*, Supplement 4, *Chemicals and Industrial Processes Associated with Cancer in Humans*, 292 pages

4. IARC (1979) Criteria to select chemicals for *IARC Monographs. IARC intern. tech. Rep. No. 79/003*

5. WHO (1961) Fifth Report of the Joint FAO/WHO Expert Committee on Food Additives. Evaluation of carcinogenic hazard of food additives. *WHO tech. Rep. Ser., No. 220*, pp. 5, 18, 19

6. WHO (1969) Report of a WHO Scientific Group. Principles for the testing and evaluation of drugs for carcinogenicity. *WHO tech. Rep. Ser., No. 426*, pp. 19, 21, 22

7. WHO (1974) Report of a WHO Scientific Group. Assessment of the carcinogenicity and mutagenicity of chemicals. *WHO tech. Rep. Ser., No. 546*

8. WHO (1964) Report of a WHO Expert Committee. Prevention of cancer. *WHO tech. Rep. Ser., No. 276*, pp. 29, 30

9. IARC (1972-1983) *IARC Monographs on the Evaluation of the Carcinogenic Risk of Chemicals to Humans*, Volumes 1-32, Lyon, France

Volume 1 (1972) Some Inorganic Substances, Chlorinated Hydrocarbons, Aromatic Amines, *N*-Nitroso Compounds and Natural Products (19 monographs), 184 pages

Volume 2 (1973) Some Inorganic and Organometallic Compounds (7 monographs), 181 pages

Volume 3 (1973) Certain Polycyclic Aromatic Hydrocarbons and Heterocyclic Compounds (17 monographs), 271 pages

Volume 4 (1974) Some Aromatic Amines, Hydrazine and Related Substances, *N*-Nitroso Compounds and Miscellaneous Alkylating Agents (28 monographs), 286 pages

Volume 5 (1974) Some Organochlorine Pesticides (12 monographs), 241 pages

Volume 6 (1974) Sex Hormones (15 monographs), 243 pages

Volume 7 (1974) Some Anti-thyroid and Related Substances, Nitrofurans and Industrial Chemicals (23 monographs), 326 pages

Volume 8 (1975) Some Aromatic Azo Compounds (32 monographs), 357 pages

Volume 9 (1975) Some Aziridines, *N*-, *S*- and *O*-Mustards and Selenium (24 monographs), 268 pages

Volume 10 (1976) Some Naturally Occurring Substances (32 monographs), 353 pages

Volume 11 (1976) Cadmium, Nickel, Some Epoxides, Miscellaneous Industrial Chemicals, and General Considerations on Volatile Anaesthetics (24 monographs), 306 pages

Volume 12 (1976) Some Carbamates, Thiocarbamates and Carbazides (24 monographs), 282 pages

Volume 13 (1977) Some Miscellaneous Pharmaceutical Substances (17 monographs), 255 pages

Volume 14 (1977) Asbestos (1 monograph), 106 pages

Volume 15 (1977) Some Fumigants, the Herbicides 2,4-D and 2,4,5-T, Chlorinated Dibenzodioxins and Miscellaneous Industrial Chemicals (18 monographs), 354 pages

Volume 16 (1978) Some Aromatic Amines and Related Nitro Compounds - Hair Dyes, Colouring Agents and Miscellaneous Industrial Chemicals (32 monographs), 400 pages

Volume 17 (1978) Some *N*-Nitroso Compounds (17 monographs), 365 pages

Volume 18 (1978) Polychlorinated Biphenyls and Polybrominated Biphenyls (2 monographs), 140 pages

Volume 19 (1979) Some Monomers, Plastics and Synthetic Elastomers, and Acrolein (17 monographs), 513 pages

Volume 20 (1979) Some Halogenated Hydrocarbons (25 monographs), 609 pages

Volume 21 (1979) Sex Hormones (II) (22 monographs), 583 pages

Volume 22 (1980) Some Non-nutritive Sweetening Agents (2 monographs), 208 pages

Volume 23 (1980) Some Metals and Metallic Compounds (4 monographs), 438 pages

Volume 24 (1980) Some Pharmaceutical Drugs (16 monographs), 337 pages

Volume 25 (1981) Wood, Leather and Some Associated Industries (7 monographs), 412 pages

Volume 26 (1981) Some Antineoplastic and Immunosuppressive Agents (18 monographs), 411 pages

Volume 27 (1982) Some Aromatic Amines, Anthraquinones and Nitroso Compounds, and Inorganic Fluorides Used in Drinking-water and Dental Preparations (18 monographs), 344 pages

Volume 28 (1982) The Rubber Industry (1 monograph), 486 pages

Volume 29 (1982) Some Industrial Chemicals and Dyestuffs (18 monographs), 416 pages

Volume 30 (1983) Miscellaneous Pesticides (18 monographs), 424 pages

Volume 31 (1983) Some Food Additives, Feed Additives and Naturally Occurring Substances (22 monographs), 314 pages

Volume 32 (1983) Polynuclear Aromatic Compounds, Part 1, Chemical, Environmental and Experimental Data (42 monographs), 477 pages

10. IARC (1973-1979) *Information Bulletin on the Survey of Chemicals Being Tested for Carcinogenicity*, Numbers 1-8, Lyon, France

Number 1 (1973) 52 pages
Number 2 (1973) 77 pages
Number 3 (1974) 67 pages
Number 4 (1974) 97 pages
Number 5 (1975) 88 pages
Number 6 (1976) 360 pages
Number 7 (1978) 460 pages
Number 8 (1979) 604 pages
Number 9 (1981) 294 pages
Number 10 (1982) 326 pages

11. PHS 149 (1951-1980) Public Health Service Publication No. 149, *Survey of Compounds which have been Tested for Carcinogenic Activity*, Washington DC, US Government Printing Office

1951 Hartwell, J.L., 2nd ed., Literature up to 1947 on 1329 compounds, 583 pages

1957 Shubik, P. & Hartwell, J.L., Supplement 1, Literature for the years 1948-1953 on 981 compounds, 388 pages

1969 Shubik, P. & Hartwell, J.L., edited by Peters, J.A., Supplement 2, Literature for the years 1954-1960 on 1048 compounds, 655 pages

1971 National Cancer Institute, Literature for the years 1968-1969 on 882 compounds, 653 pages

1973 National Cancer Institute, Literature for the years 1961-1967 on 1632 compounds, 2343 pages

1974 National Cancer Institute, Literature for the years 1970-1971 on 750 compounds, 1667 pages

1976 National Cancer Institute, Literature for the years 1972-1973 on 966 compounds, 1638 pages

1980 National Cancer Institute, Literature for the year 1978 on 664 compounds, 1331 pages

12. Pike, M.C. & Roe, F.J.C. (1963) An actuarial method of analysis of an experiment in two-stage carcinogenesis. *Br. J. Cancer*, *17*, 605-610

13. Miller, E.C. & Miller, J.A. (1966) Mechanisms of chemical carcinogenesis: nature of proximate carcinogens and interactions with macromolecules. *Pharmacol. Rev.*, *18*, 805-838

14. Miller, J.A. (1970) Carcinogenesis by chemicals: an overview - G.H.A. Clowes Memorial Lecture. *Cancer Res.*, *30*, 559-576

15. Miller, J.A. & Miller, E.C. (1976) *The metabolic activation of chemical carcinogens to reactive electrophiles*. In: Yuhas, J.M., Tennant, R.W. & Reagon, J.D., eds, *Biology of Radiation Carcinogenesis*, New York, Raven Press

16. Peto, R. (1974) Guidelines on the analysis of tumour rates and death rates in experimental animals. *Br. J. Cancer*, *29*, 101-105

17. Peto, R. (1975) Letter to the editor. *Br. J. Cancer*, *31*, 697-699

18. Hoel, D.G. & Walburg, H.E., Jr (1972) Statistical analysis of survival experiments. *J. natl Cancer Inst.*, *49*, 361-372

19. IARC (1980) *IARC Monographs on the Evaluation of the Carcinogenic Risk of Chemicals to Humans*, Supplement 2, *Long-term and Short-term Screening Assays for Carcinogens: A Critical Appraisal*, Lyon

20. IARC Working Group (1980) An evaluation of chemicals and industrial processes associated with cancer in humans based on human and animal data: *IARC Monographs* Volumes 1 to 20. *Cancer Res.*, *40*, 1-12

21. IARC (1979) *IARC Monographs on the Evaluation of the Carcinogenic Risk of Chemicals to Humans*, Supplement 1, *Chemicals and Industrial Processes Associated with Cancer in Humans*, Lyon

22. Rall, D.P. (1977) *Species differences in carcinogenesis testing*. In: Hiatt, H.H., Watson, J.D. & Winsten, J.A., eds, *Origins of Human Cancer*, Book C, Cold Spring Harbor, NY, Cold Spring Harbor Laboratory, pp. 1383-1390

23. National Academy of Sciences (NAS) (1975) *Contemporary Pest Control Practices and Prospects: the Report of the Executive Committee*, Washington DC

24. Chemical Abstracts Services (1978) *Chemical Abstracts Ninth Collective Index (9CI), 1972-1976*, Vols 76-85, Columbus, OH

25. International Union of Pure & Applied Chemistry (1965) *Nomenclature of Organic Chemistry*, Section C, London, Butterworths

26. WHO (1958) Second Report of the Joint FAO/WHO Expert Committee on Food Additives. Procedures for the testing of intentional food additives to establish their safety and use. *WHO tech. Rep. Ser., No. 144*

27. WHO (1967) Scientific Group. Procedures for investigating intentional and unintentional food additives. *WHO tech. Rep. Ser., No. 348*

28. Berenblum, J., ed. (1969) Carcinogenicity testing. *UICC tech. Rep. Ser., 2*

29. Sontag, J.M., Page, N.P. & Saffiotti, U. (1976) Guidelines for carcinogen bioassay in small rodents. *Natl Cancer Inst. Carcinog. tech. Rep. Ser., No. 1*

30. Committee on Standardized Genetic Nomenclature for Mice (1972) Standardized nomenclature for inbred strains of mice. Fifth listing. *Cancer Res., 32*, 1609-1646

31. Wilson, J.G. & Fraser, F.C. (1977) *Handbook of Teratology*, New York, Plenum Press

32. Berenblum, J. (1975) *Sequential aspects of chemical carcinogenesis: Skin*. In: Becker, F.F., ed., *Cancer. A Comprehensive Treatise*, Vol. 1, New York, Plenum Press, pp. 323-344

33. Foulds, L. (1969) *Neoplastic Development*, Vol. 2, London, Academic Press

34. Farber, E. & Cameron, R. (1980) The sequential analysis of cancer development. *Adv. Cancer Res., 31*, 125-226

35. Weinstein, I.B. (1981) The scientific basis for carcinogen detection and primary cancer prevention. *Cancer, 47*, 1133-1141

36. Slaga, T.J., Sivak, A. & Boutwell, R.K., eds (1978) *Mechanisms of Tumor Promotion and Cocarcinogenesis*, Vol. 2, New York, Raven Press

37. Miller, E.C. & Miller, J.A. (1981) Mechanisms of chemical carcinogenesis. *Cancer, 47*, 1055-1064

38. Brookes, P. & Lawley, P.D. (1964) Evidence for the binding of polynuclear aromatic hydrocarbons to the nucleic acids of mouse skin: relation between carcinogenic power of hydrocarbons and their binding to deoxyribonucleic acid. *Nature*, *202*, 781-784

39. Lawley, P.D. (1976) *Carcinogenesis by alkylating agents*. In: Searle, C.E., ed., *Chemical Carcinogens* (*ACS Monograph 173*), Washington DC, American Chemical Society, pp. 83-244

40. McCann, J. & Ames, B.N. (1976) Detection of carcinogens as mutagens in the *Salmonella*/microsome test: Assay of 300 chemicals: Discussion. *Proc. natl Acad. Sci. USA*, *73*, 950-954

41. Weisburger, J.H. & Williams, G.M. (1980) *Chemical carcinogens*. In: Doull, J., Klaassen, C.D. & Amdur, M.O., eds, *Casarett and Doull's Toxicology: The Basic Science of Poisons*, 2nd ed., New York, MacMillan, pp. 84-138

42. Montesano, R., Bartsch, H. & Tomatis, L., eds (1980) *Molecular and Cellular Aspects of Carcinogen Screening Tests* (*IARC Scientific Publications No. 27*), Lyon

43. de Serres, F.J. & Ashby, J., eds (1981) *Evaluation of Short-Term Tests for Carcinogens. Report of the International Collaborative Program*, Amsterdam, Elsevier/North-Holland Biomedical Press

44. Sugimura, T., Sato, S., Nagao, M., Yahagi, T., Matsushima, T., Seino, Y., Takeuchi, M. & Kawachi, T. (1976) *Overlapping of carcinogens and mutagens*. In: Magee, P.N., Takayama, S., Sugimura, T. & Matsushima, T., eds, *Fundamentals in Cancer Prevention*, Tokyo/Baltimore, University of Tokyo/University Park Press, pp. 191-215

45. Bartsch, H., Tomatis, L. & Malaveille, C. (1982) *Qualitative and quantitative comparison between mutagenic and carcinogenic activities of chemicals*. In: Heddle, J.A., ed., *Mutagenicity: New Horizons in Genetic Toxicology*, New York, Academic Press, pp. 35-72

46. Hollstein, M., McCann, J., Angelosanto, F.A. & Nichols, W.W. (1979) Short-term tests for carcinogens and mutagens. *Mutat. Res.*, *65*, 133-226

EXPLANATORY NOTE TO VOLUME 32

Evaluation of the carcinogenic risk associated with polynuclear aromatic compounds is the subject of this and the ensuing three volumes of Monographs. In this volume, experimental studies on the carcinogenic activity of single compounds (48 polynuclear aromatic hydrocarbons and related heterocyclic compounds) are evaluated. Monographs evaluating the carcinogenic risk of certain complex mixtures (mineral oils and carbon black) will be contained in Volume 33. Carcinogenic risk associated with certain industries, in which human exposure to polynuclear aromatic compounds occurs, will be evaluated in Volumes 34 and 35.

Humans are exposed to complex mixtures containing polynuclear aromatic compounds rather than to the single substances often examined experimentally. Accordingly, evaluations of carcinogenic activity made in the present volume of monographs are restricted to statements regarding carcinogenic activity in experimental animals. For information regarding human exposure to polynuclear aromatic compounds in an environment, and particularly an occupational context, reference should be made, in the first instance, to *IARC Monographs* Volumes 33, 34 and 35.

GENERAL REMARKS ON THE SUBSTANCES CONSIDERED

1. INTRODUCTION

This thirty-second volume of the *IARC Monographs* series is concerned with the risk of 42 polynuclear aromatic hydrocarbons (PAHs) and six heterocyclic compounds that have been tested for carcinogenicity and that have been shown to occur in the environment. Only condensed aromatic hydrocarbons and aza arenes with three or more rings are considered. These include the PAHs, anthanthrene, anthracene, benzo[*ghi*]fluoranthene, benzo[*k*]fluoranthene, benzo[*a*]fluorene, benzo[*b*]fluorene, benzo[*c*]fluorene, benzo[*ghi*]perylene, benzo[*c*]phenanthrene, coronene, cyclopenta[*cd*]pyrene, dibenz[*a,c*]anthracene, dibenz[*a,j*]anthracene, dibenzo[*a,e*]fluoranthene, 1,4-dimethylphenanthrene, fluoranthene, fluorene, 1-, 2-, 3-, 4-, 5- and 6-methylchrysenes, 2- and 3-methylfluoranthenes, 1-methylphenanthrene, perylene, phenanthrene, pyrene and triphenylene and the aza arene compounds benz[*a*]acridine and carbazole.

Additionally, a number of polynuclear aromatic compounds (PACs) (including several PAHs and one aza arene) that were previously evaluated by IARC (1973) and for which, at that time, the evidence of carcinogenicity was not considered to be *sufficient* (IARC, 1979) are further evaluated in this monograph. These include: benzo[*j*]fluoranthene, benzo[*e*]pyrene, chrysene, dibenzo[*a,l*]pyrene and benz[*c*]acridine.

A number of PACs that were previously evaluated by the IARC (1973) as showing *sufficient evidence* (see IARC, 1979) of carcinogenicity are not further evaluated in this volume. These include: benz[*a*]anthracene, benzo[*b*]fluoranthene, benzo[*a*]pyrene, dibenz[*a,h*]anthracene, dibenzo[*a,e*]pyrene, dibenzo[*a,h*]pyrene, dibenzo[*a,i*]pyrene, indeno[1,2,3-*cd*]pyrene, dibenz[*a,h*]acridine, dibenz[*a,j*]acridine and 7*H*-dibenzo[*c,g*]carbazole. Information on these compounds can be found in IARC (1973) and other reviews such as that of Dipple (1976), Freudenthal and Jones (1976), Bingham *et al.* (1980) and Conney (1982). However, the chemical and physical properties of these compounds and summaries of mutagenicity data are included in the present volume of monographs.

The Working Group was aware of the extensive number of studies on the carcinogenicity and mutagenicity of derivatives and metabolites of many of the compounds considered, but a full evaluation of these data was considered beyond the scope of this volume. Brief summaries describing the known metabolites of PACs and the biological activity of those metabolites are included in the individual monographs. A general review of PAC metabolism is given on p. 53.

Nitropolynuclear aromatic hydrocarbons (nitroarenes) have also been detected in various environmental samples (e.g., effluents, airborne particulates) (Pitts, 1979; King *et al.*, 1980; Kubitscheck & Williams, 1980; Pitts *et al.*, 1982); these compounds are presumed to result from the ambient nitration of PAHs that originate from incomplete mobile and stationary combustion processes (Mermelstein *et al.*, 1981). Nitroarenes are not included in this volume of monographs , but some will be considered by the Working Group convening in June l983 to prepare Volume 33 of the *IARC Monographs*.

Since humans are not exposed to individual PACs but always to complex mixtures, in the present volume of monographs only experimental data are evaluated. These monographs will serve as a basis for future IARC Working Groups who will prepare monographs on mineral oils and carbon blacks (in June 1983 for Volume 33) and on several industries in which exposure to PACs occurs, i.e., the aluminium production industry, coal gasification plants, coke production plants (including coke ovens) and iron and steel foundries (in October 1983 for Volume 34). Occupations where creosote, tars, pitch and asphalt (bitumen) are used and other industries in which exposure to PACs also occurs, such as shale-oil extraction and refining industries, will be considered in February 1984 for Volume 35.

Exposure to PAH occurs principally by direct inhalation of tobacco smoke, polluted air and, possibly, ingestion of contaminated and processed food and water or by dermal contact with soot, tars and oils. The PAHs have long been of concern as a potential human health hazard, since many members of this class are tumour initiators or promoters or tumorigenic and/or mutagenic *in vitro* and *in vivo* (Cook *et al.*, 1958; Andelman & Suess, 1970; Shabad *et al.*, 1971; Committee on Biological Effects of Atmospheric Pollutants, 1972; IARC, 1973; Andelman & Snodgrass, 1974; Dipple, 1976; Freudenthal & Jones, 1976; Hoffmann & Wynder, 1977; Gelboin & Ts'O, 1978; Jones & Freudenthal, 1978; Egan *et al.*, 1979; Jones & Leber, 1979; Cooke & Dennis, 1981; Lee & Grant, 1981; Cook *et al.*, 1982).

1.1 Correlation of human exposure with PAH carcinogenesis

Combustion, pyrolysis and pyrosynthesis of organic matter result in the formation of PAHs. Such thermal degradation products have been found in trace amounts in tobacco smoke, marijuana smoke, polluted air, some food and in drinking-water.

Tobacco smoking causes cancer at several sites: lung, larynx, oral cavity, oesophagus, bladder, pancreas and kidney (US Department of Health and Human Services, 1982). Epidemiological evidence, though inconclusive, points to urban air pollution as a possible contributory agent in respiratory cancers in some cities. There is no epidemiological evidence to suggest that intake of food containing traces of PAHs makes any appreciable contribution to the risk of human cancer.

Although there is no epidemiological study that has unequivocally established a relation between the occurrence of PACs and lung cancer or other cancers, several studies of workers show an increased risk of skin and scrotal cancer after exposure to soots, tars and mineral oils and of lung cancer after exposure to coal-gas and coke-oven emissions. Since all these materials contain PAHs, the question arises: what portion of carcinogenicity may originate from PAHs?

These studies are not described in detail here, because epidemiological studies in which the carcinogenic risk of tobacco smoking and of certain occupational exposures in which PACs occur will be evaluated in future *IARC Monographs*.

1.2 Bioassays with organic particulates

PAH concentrates from cigarette smoke and urban air have been shown to induce benign and malignant tumours in the skin of mice (Wynder & Wright, 1957; Kotin & Falk, 1959; Hueper *et al.*, 1962; Whitehead, 1977), subcutaneous sarcomas in rats (Seelkopf *et al.*, 1963) and malignant tumours in the bronchi of rats (Davis *et al.*, 1975).

Studies in mice by skin painting (Hoffmann & Wynder, 1977) demonstrated that the carcinogenicity of automobile exhaust condensate is related to its content of PAHs. This finding was confirmed by Brune (1977), Brune *et al.* (1978) and Grimmer *et al.* (1982a). Studies using mouse skin painting and implantation into the lungs of rats (Grimmer *et al.*, 1982a,b,c, 1983b) indicate that PAHs containing more than three rings are the major contributors to the carcinogenicity of emission condensates from automobile exhausts and from flue gas from coal-burning stoves and of lubricating oil and crank-case oil.

A number of studies have demonstrated the mutagenicity of airborne particulates (Talcott & Wei, 1977; Tokiwa *et al.*, 1977; de Wiest *et al.*, 1982).

2. SOURCES AND OCCURRENCE OF POLYNUCLEAR AROMATIC COMPOUNDS

PACs are a group of highly lipophilic chemicals that are present ubiquitously in the environment as pollutants, often in very small quantities of the order of μg/kg or ng/m^3.

Organic matter in combustion system exhausts consists chiefly of unburned alkanes and a series of PACs. These PACs range in chemical nature from the nonpolar PAHs to more polar, oxygen-containing compounds, such as aldehyde- and carboxylic acid-substituted PAH derivatives. In general, the PAHs themselves represent a substantial fraction of all the PACs, but it should be noted that the oxygen-containing derivatives of the PAHs also contribute substantial mass to combustion effluents and have not yet been studied extensively in terms of biological activity.

The sources and occurrence, including the main sources of emissions, of PACs have been reviewed (Shabad, 1967; Shabad *et al.*, 1971; Egan *et al.*, 1979; Grimmer, 1983a). A summary of benzo[*a*]pyrene concentrations in areas polluted from various sources was made by Sawicki (1967, 1976). Hangebrauck *et al.* (1964) assembled data on the emission of benzo[*a*]pyrene and other PAHs from coal-fired residential stoves, intermediate sized coal-fired units and coal-fired plants, as well as from intermediate and smaller oil-fired and gas-fired units, incineration and opening burning. A review on sources of polynuclear matter has been published (Committee on Biological Effects of Atmospheric Pollutants, 1972).

Numerous studies have demonstrated the presence of many compounds of this class in cigarette smoke, air (industrial and ambient), water, food, soil, sediments, aquatic organisms, mineral oils and refined petroleum products. They have high-melting points, are virtually insoluble in water and have low vapour pressures. Many PACs are often associated with or adsorbed on particulate matter and a large mass fraction of airborne PAHs is associated with ultrafine particles (Albagli *et al.*, 1974; Starkey & Warpinski, 1974; Pierce & Katz, 1975). Although airborne PACs have been studied for almost 30 years, knowledge of their chemical lifetimes is based mainly on studies in model systems (Falk *et al.*, 1956; Thomas *et al.*, 1968; Tebbens *et al.*, 1971). Recent work would indicate that these lifetimes are of the order of days (Korfmacher *et al.*, 1980; Butler & Crossley, 1981). Monitoring data suggest that at least 20 PACs are stable enough to travel considerable distances in combination with airborne particulate matter (Lunde, 1976; Lunde & Bjørseth, 1977; Lee & Grant, 1981).

Many of the current combustion processes and certain industrial processes (e.g., coke production and petroleum refining) have led to the widespread presence of PACs in industrial and ambient atmospheres. The major source of these environmental chemical pollutants is coal combustion, encompassing industrial operations and power plants using fossil fuels, waste incinerators, domestic heaters and vehicles powered by gasoline or diesel fuels (Sawicki *et al.*, 1962; Diehl *et al.*, 1967; Committee on Biological Effects of Atmospheric Pollutants, 1972; IARC, 1973; US Environmental Protection Agency, 1975; Gelboin & Ts'O, 1978). Emissions of PAHs from fossil fuel combustion can vary over several orders of magnitude depending upon the particular fuel and combustion conditions. Emissions of PACs from the burning of coal or wood for residential space heating are several orders of magnitude greater than those from gas or oil burning (Committee on Biological Effects of Atmospheric Pollutants, 1972; Cooper, 1980). Urban and highly industrialized environments contain higher levels of PACs than do rural locations (Committee on Biological Effects of Atmospheric Pollutants, 1972).

During forest fires, PACs are released into the atmosphere. McMahon and Tsoukalas (1978) reported emission of benzo[*a*]pyrene at 238-3454 ng/g of wood burned in backing fires and 38-97 ng/g wood burned in heading fires. Ilnitsky *et al.* (1977) have shown that volcanic activity can be a source of PAC pollution; however, samples of volcanic ash and of soil and plants in an area near a volcano did not contain more benzo[*a*]pyrene than occurred in the background.

Of additional potential concern is the presence of PACs in fly ash (Eiceman *et al.*, 1979) and ash residues (Davies *et al.*, 1976) from incinerators combusting raw municipal refuse. There is a paucity of data concerning the presence of PACs and other trace organic compounds in ash resulting from the incineration of sewage sludge and municipal waste. In limited studies, sludge ash residues from four cities in the USA were found to contain 0.1-1 μg/g. The ash containing the highest amount also exhibited the greatest variety of PACs (Wszolek & Wachs, 1982).

PACs in water may originate from fallout of particulate matter transported through air, from absorption of gaseous compounds, as well as from polluted water (Kveseth *et al.*, 1982). Some of these PACs are only slowly degraded; hence PAC pollution of water can represent a potential health hazard to man *via* consumption from drinking-water (WHO, 1964) and, indirectly, from the aquatic environment *via* initial absorption and accumulation in marine organisms (Neff *et al.*, 1976; Stich *et al.*, 1976; Pancirov & Brown, 1977; Kveseth *et al.*, 1982).

As noted earlier, humans are exposed to PACs by three routes: respiration, ingestion and by skin contact. Since occupational exposure - the greatest source of skin contact - is not considered in this volume, only non-occupational exposures *via* the respiratory tract and through the gastrointestinal tract are discussed here.

2.1 Respiratory tract

Non-occupational respiratory exposure is mainly to tobacco smoke and urban air. Urban air pollution is from various sources, including vehicle exhausts and combustion products from residential and industrial heating.

(*a*) *Tobacco smoke*

Table 1 gives the levels of 34 parent PACs and some of their methyl derivatives that have been unambiguously identified in mainstream and sidestream smoke from cigarettes, cigars and pipes, marijuana smoke and smoke-polluted environments. Other PACs have been reported to occur in cigarette smoke constituents; however, their identification needs confirmation. Cigarette smoke probably contains a large spectrum of alkylated PACs in trace amounts (Lee *et al.*, 1976; Snook *et al.*, 1977, 1978).

Table 1. Concentrations of some polynuclear aromatic hydrocarbons and heterocyclic compounds in tobacco smoke (with references[a])

	Cigarette main stream smoke (μg/100 cigarettes)	Cigarette side stream smoke (μg/100 cigarettes)	Cigarette smoke-polluted environments (ng/m³)	Cigar smoke (μg/100 g)	Pipe smoke (μg/100 g)	Marijuana smoke (μg/100 cigarettes)
Polynuclear aromatic hydrocarbons						
Anthanthrene	0.2-2.2 (7,23)	3.9 (7)	0.5-3 (8,19)			0.5 (22)
Anthracene	2.3-23.5 (20,22,23)			11.9 (3)	110.0 (3)	3.3 (22)
Benz[*a*]anthracene	0.4-7.6 (1,20,22,23,28)			2.5-3.9 (13)		3.3-7.5 (18,22)
Benzo[*b*]fluoranthene	0.4-2.2 (11,20,23)					
Benzo[*j*]fluoranthene	0.6-2.1 (11,22)					3.0 (22)
Benzo[*k*]fluoranthene	0.6-1.2 (21,23)					1.1 (22)
Benzo[*ghi*]fluoranthene	0.1-0.4 (34,36)					present (22)
Benzo[*a*]fluorene	4.1-18.4 (1,7,22)	75 (7)	39 (8)			4.2 (22)
Benzo[*b*]fluorene	2 (11)					present (22)
Benzo[*ghi*]perylene	0.3-3.9 (7,11,22)	9.8 (7)	5.9-17 (8,19)			0.7 (22)
Benzo[*c*]phenanthrene	present (30)					
Benzo[*a*]pyrene	0.5-7.8 (11,14,17,20,23, 24,28,32,33)	2.5-19.9 (7,27)	2.8-760 (5,6,26)	1.8-5.1 (3,13,14)	8.5 (3)	2.9-3.1 (18,22)
Benzo[*e*]pyrene	0.2-2.5 (7,20,22,23)	13.5 (7)	3-18 (8,19)			1.8 (22)
Chrysene	0.6-9.6 (1,4,10,11,20,22,23)					5.5 (22)
Coronene	0.1 (30)		0.5-2.8 (19)			
Dibenz[*a,c*]anthracene	present (30)					present (22)
Dibenz[*a,h*]anthracene	0.4 (11)					present (22)
Dibenz[*a,j*]anthracene	1.1 (7)	4.1 (7)	6 (8)			
Dibenzo[*a,e*]pyrene	present (30)					
Dibenzo[*a,h*]pyrene	present (30)					
Dibenzo[*a,i*]pyrene	0.17-0.32 (25)					
Dibenzo[*a,l*]pyrene	present (33)					
Fluoranthene	1-27.2 (7,16,20,22,29,33)	126 (7)	99 (8)	20.1 (3)		8.9 (22)
Fluorene	present (9,31)					
Indeno[1,2,3-*cd*]pyrene	0.4-2.0 (1,38)					
1-Methylchrysene	0.3 (10)					present (22)
2-Methylchrysene	0.12 (10)					present (22)
3-Methylchrysene	0.61 (10)					present (22)
4-Methylchrysene	present (31)					present (22)
5-Methylchrysene	0.06 (10)					present (22)

	Cigarette main stream smoke (μg/100 cigarettes)	Cigarette side stream smoke (μg/100 cigarettes)	Cigarette smoke-polluted environments (ng/m³)	Cigar smoke (μg/100 g)	Pipe smoke (μg/100 g)	Marijuana smoke (μg/100 cigarettes)
6-Methylchrysene	0.7 (10)					present (22)
2-Methylfluoranthene	present (22)					present (22)
3-Methylfluoranthene	present (22)					present (22)
Perylene	0.3-0.5 (23,30)	3.9 (7)	0.1-11 (8,19)			0.9 (22)
Phenanthrene	8.5-62.4 (20,22,23)			115 (3)		8.9 (22)
Pyrene	5-27 (2,4,7,20,22, 23,25,29,33)	39-101 (7,21)	2-66 (8,19)	17.6 (3)	75.5 (3)	6.6 (22)
Triphenylene	present (30)					
Heterocyclic compounds						
Carbazole	100 (15)					6.5 (22)
Dibenz[*a*,*h*]acridine	0.01 (35)					
Dibenz[*a*,*j*]acridine	0.27 (35)					
7*H*-Dibenzo[*c*,*g*]carbazole	0.07 (35)					
Benzo[*c*]fluorene	present (12,30)					
1,4-Dimethyl-phenanthrene[b]	present (30)					
1-Methyl-phenanthrene	3.2 (22)					4.2 (22)

[a] References: (1) Ayres & Thornton (1965); (2) Bonnet & Neukomm (1956); (3) Campbell & Lindsey (1957); (4) Ellington *et al.* (1978); (5) Elliot & Rowe (1975); (6) Galuskinova (1964); (7) Grimmer *et al.* (1977a); (8) Grimmer *et al.* (1977b); (9) Grob & Voellmin (1970); (10) Hecht *et al.* (1974); (11) Hoffmann & Wynder (1960); (12) Hoffmann & Wynder (1971); (13) Hoffmann & Wynder (1972); (14) Hoffmann *et al.* (1963); (15) Hoffmann *et al.* (1968); (16) Hoffmann *et al.* (1972); (17) Hoffmann *et al.* (1974); (18) Hoffmann *et al.* (1975); (19) Just *et al.* (1972); (20) Kiryu & Kuratsune (1966); (21) Kotin & Falk (1960); (22) Lee *et al.* (1976); (23) Masuda & Kuratsune (1972); (24) Müller *et al.* (1964); (25) Müller *et al.* (1967); (26) Perry (1973); (27) Pyriki (1963); (28) Rathkamp *et al.* (1973); (29) Severson *et al.* (1979); (30) Snook *et al.* (1977); (31) Snook *et al.* (1978a); (32) US Department of Health & Human Services (1982); (33) Van Duuren (1958a); (34) Van Duuren (1958b); (35) Van Duuren *et al.* (1960); (36) Wynder & Hoffmann (1959); (37) Wynder & Hoffmann (1961); (38) Wynder & Hoffmann (1963)

[b] Unspecified isomer

(*b*) *Urban atmospheres*

Only benzo[*a*]pyrene has been determined in most investigations. Its presence has been used as an indicator that other PAHs occur; however, in some cases, no constant ratio to other PAHs can be observed. Unfortunately, most published data on PAH concentrations in the atmosphere relate only to benzo[*a*]pyrene; such data exist for Australia, Belgium, Canada, Czechoslovakia, Denmark, the Federal Republic of Germany, Finland, France, Hungary, Iceland, India, Iran, Italy, Japan, The Netherlands, Norway, Poland, South Africa, Spain, Sweden, Switzerland, the United Kingdom, the USA and the USSR (compiled by Sawicki, 1976). In several cases, benzo[*a*]pyrene concentrations were determined in summer as well as in winter: generally, the concentration during winter was higher, indicating the contribution of coal-fired residential heating, which emits levels of benzo[*a*]pyrene of the order of mg/kg coal, whereas oil heating produces only about 10-100 ng/kg fuel.

Temporal variations in the concentration of several PACs during a day have been observed in an area heated predominantly by coal-fired stoves. Variations in PACs are also seen over longer periods and appear to be due to differences in vulnerability to photochemical and chemical decomposition and the different vapour pressures of various PAC at ambient temperatures (Grimmer *et al.*, 1981b; Grimmer, 1983b).

Similar profiles were found for 14 PAHs in four different cities in the Federal Republic of Germany (Funcke *et al.*, 1982). However, varying profiles of PAH have been found even within one city (Grimmer *et al.* 1981b,d).

High PAH concentrations in cities in the Federal Republic of Germany during the winter (e.g., for benzo[*a*]pyrene up to 333 ng/m^3) can be attributed to the wide use of coal for domestic heating especially in the period before 1975 (Grimmer, 1983b). Lower concentrations have been measured in the atmosphere of US cities (Hoffmann & Wynder, 1976).

Atmospheric polynuclear content is subject to regulation in many nations.

Combustion effluents from stationary sources

(i) *Coal-fired residential stoves*

No data on the total profile of flue gas particles emitted by hard-coal-fired stoves are available. PAC concentrations in the flue gas emitted from various coals, such as anthracite nuts, coal briquets, broken coke and other special coals, vary widely (Beine, 1970; Brockhaus & Tomingas, 1976). Furthermore, PAC mass emission depends on the type of the residential stove (e.g., slow combustion stove) and on the operating conditions (full, medium or low load) regulated by the air input (Grimmer *et al.*, 1983a).

Table 2 gives data on PACs emitted from two types of stove burning various types of fuel under low-load conditions - those commonly used in households. The actual ranges found are cited in the monographs on the individual compounds.

Table 2. Polynuclear aromatic compound emissions from different stationary sources (mg/kg)[a]

Polynuclear aromatic compound	Hard-coal briquet[b]		Anthracite nut[b]		Brown-coal briquet[b]		Oil-fired stove/vaporizing pot burner		Oil-fired heating/high pressure jet burner[c]	
	Range	Av.	Range	Av.	Range	Av.	Range	Av.	Range	Av.
Anthanthrene	0.11-0.90	0,51	0.00-0.09	0.045	0.03-0.70	0.37	0.001	0.001	-	-
Benz[*a*]anthracene	0.88-6.50	3.69	0.04-0.09	0.065	0.86-9.40	5.13	0.044-0.263	0.167	0.000030-0.000080	0.000047
Benzo[*ghi*]fluoranthene	0.50-3.00	1.75	0.01-0.02	0.015	0.70-8.90	4.80	0.150-0.470	0.301	-	-
Benzo[*b*]fluoranthene + benzo[*j*]fluoranthene + benzo[*k*]fluoranthene	1.97-17.20	9.59	0.05-1.29	0.67	0.38-7.60	3.99	0.143-0.405	0.252	0.000028-0.000076	0.000060
Benzo[*ghi*]perylene	0.23-2.50	1.37	0.02-0.19	0.105	0.30-2.70	1.50	0.010-0.061	0.034	0.000003-0.000022	0.000008
Benzo[*a*]pyrene	0.39-5.20	2.80	0.00-0.07	0.035	0.21-5.80	3.01	0.003-0.024	0.010	0.000002-0.000010	0.000007
Benzo[*e*]pyrene	1.28-10.30	5.79	0.02-0.34	0.18	0.27-3.80	2.04	0.035-0.139	0.081	0.000008-0.000030	0.000020
Chrysene + triphenylene	2.25-21.80	12.03	0.07-0.93	0.50	1.11-13.50	7.31	0.223-0.468	0.343	0.000376-0.000562	0.000472
Coronene	0.03-0.40	0.22	0.01-0.04	0.025	0.03-1.10	0.57	0.009-0.031	0.018	-	-
Cyclopenta[*cd*]pyrene	0.09-0.90	0.35	0.00-0.01	0.005	0.04-3.80	1.92	0.000	-	-	-
Fluoranthene	2.37-19.28	9.81	0.13-0.32	0.23	2.99-19.40	16.20	2.148-2.797	2.466	0.000668-0.001748	0.001274
Indeno[1,2,3-*cd*]pyrene	0.14-2.30	1.22	0.02-0.18	0.10	0.20-1.60	0.90	0.020-0.085	0.046	-	-
Perylene	0.09-0.70	0.40	0.00	-	0.21-1.90	1.06	0.003-0.005	0.004	-	-
Pyrene	1.78-11.20	6.49	0.09-0.31	0.20	2.68-31.00	16.84	0.107-1.604	1.006	0.000244-0.000530	0.000354

[a] Adapted from Brockhaus & Tomingas (1976) and Grimmer *et al.* (1983a)

[b] Hard-coal briquet, hard-coal fired residential stove (range includes two different types of stoves); anthracite nut, anthracite-fired residential stove (range includes two different types of stoves); brown-coal briquet, brown-coal fired residential stove (range includes two different types of stoves); av., average

[c] From Behn & Grimmer (1981)

(ii) *Brown-coal (lignite)-fired residential stoves*

Brown coal (or lignite) is an intermediate between peat and true coals, with a high moisture content.

PAHs and their sulphur and oxygen analogues have been determined in emissions from brown-coal-briquet-fired residential stoves: over 170 components were characterized and quantified at concentrations above 0.01 mg/kg combusted fuel, and 52 PACs were identified by comparison with reference substances. Only a few thiophene derivatives were present at low concentrations. This is contrary to the situation with emissions from hard coal and oil combustion. Picenes and several methylpicenes were present in mass concentrations comparable to that of benzo[*e*]pyrene and may serve to distinguish brown-coal emissions from those of other combustion sources. The range of PAC mass emission per kilogram is shown in Table 2 (Grimmer *et al.*, 1983a).

(iii) *Oil-fired heating units*

No inventory of the PACs emitted by oil-fired heating systems with high-pressure jet burners is available; however, some PACs emitted by a commercial oil-fired heating unit with high-pressure jet burners in combination with a steel-heating boiler have been determined (Table 2) (Behn & Grimmer, 1981).

(iv) *Oil-fired stoves with vaporizing pot burners*

The PAC profile of the flue gas of oil-fired vaporizing pot burners is shown in Table 2. At low load conditions, benzo[*a*]pyrene was emitted at levels of about 0.003-0.025 mg/kg.

(v) *Coal-fired steam power plants*

Investigations on the total PAC content of fly ash are still lacking, although some PAH concentrations have been reported (Cuffe *et al.*, 1964; Hangebrauck *et al.*, 1964; Gerstle *et al.*, 1965; Griest & Guerin, 1979; Sonnichsen *et al.*, 1980; Zelenski *et al.*, 1980; Guggenberger *et al.*, 1981). The reliability of PAC recovery from stack gas and fly ash is questionable (Sonnichsen *et al.*, 1980).

(vi) *Wood stoves*

Concentrations of up to 1550 μg/m^3 total PAHs have been measured in old-fashioned smoke-saunas in Finland immediately after heating. In more modern wood-heated saunas, with continuously burning stoves, the level of total PAHs measured was about 25 μg/m^3 (Häsänen *et al.*, 1983).

Exhaust emissions from gasoline engines and diesel engines

Although the exhaust gases from vehicles with internal combustion engines are only one source of urban air pollutants, they can contribute 80% and more of the collective polynuclear organic matter in the air in some cities (Goldsmith & Friberg, 1977; Schmidt *et al.*, 1977). In terms of total benzo[*a*]pyrene emissions in the USA, however, only a small portion derives from motor vehicles. It has been estimated that, of 894 tons/year emitted in the period from 1971-1973, only 11 tons (1.2%) originated from gasoline engine exhaust (Schmidt *et al.*, 1977). The importance of the PAHs as a major group of carcinogens in the organic particulate matter of emissions from gasoline and diesel engines has been demonstrated by a number of bioassays, including those in which PAH concentrates or fractions from the particulate matter have been tested on mouse skin or in the lungs of rats (Kotin *et al.*, 1954, 1955; Wynder & Hoffmann, 1962; Grimmer & Böhnke, 1978; National Research Council, 1981; Grimmer *et al.*, 1983b). As a result of the introduction of emission control devices, e.g., in Canada, France, Sweden, the United Kingdom and the USA, around 1970, release of organic particulate matter from gasoline engines has been diminished drastically (Schmidt *et al.*, 1977). This is an important factor in a comparison of PAH emissions from gasoline engine effluents. Whereas gasoline engines without emission control devices often release significantly more benzo[*a*]pyrene and other PAHs than diesel engines per volume of fuel burned, diesel engines release more PAHs than emission-controlled gasoline engines when compared on the basis of fuel consumption (Williams & Swarin, 1979; National Research Council, 1981).

(i) *Gasoline engine exhaust*

Table 3 lists PAH levels in exhaust effluents from gasoline engines (Grimmer *et al.*, 1977c). Although there are many other such reports in the literature, they cannot be compared with one another owing primarily to the use of different methods of PAH collection and differences in the reporting of unit measures of isolated amounts (e.g., μg/l fuel, μg/travel distance [km or miles], μg/exhaust volume or μg/g particulate matter). It can be seen from the table that, in comparison with tobacco smoke and coal tar, gasoline and diesel engine emissions contain relatively lower concentrations of alkylated PAHs than of the parent PAH.

Table 3. Polynuclear aromatic compounds in exhaust emissions from gasoline engines (no emission control devices): Isolated amounts[a]

Polynuclear aromatic compound	μg/l fuel[b]	
	Air-cooled engine	Water-cooled engine
Anthanthrene	17	26
Anthracene	534	642
Benz[*a*]anthracene	83	50
Benzo[*a*]fluorene	82	136
Benzo[b]fluorene + benzo[*c*]fluorene	65	112
Benzo[*b*]fluoranthene	48	29
Benzo[*ghi*]fluoranthene	244	112
Benzo[*j*]fluoranthene	27	11
Benzo[*k*]fluoranthene	17	7
Benzo[*ghi*]perylene	333	115
Benzo[*c*]phenanthrene	+	+
Benzo[*a*]pyrene	81	50
Benzo[*e*]pyrene	59	37
Chrysene	123	85
2- + 5-Methyl-	5	5
3-Methyl-	+	+
4- + 6-Methyl-	5	5
Coronene	271	106
Cyclopenta[*cd*]pyrene	987	750
Dibenzo[*a,h*]anthracene	+	+
Dibenzo[*a,j*]anthracene	+	+

Polynuclear aromatic compound	μg/l fuel[b]	
	Air-cooled engine	Water-cooled engine
Fluoranthene	1060	1662
Fluorene	+	+
Indeno[1,2,3-*cd*]pyrene	86	32
Perylene	14	7
Phenanthrene	2930	2356
1-Methyl-	256	404
Dimethyl-	+	+
Pyrene	2150	2884
Triphenylene	60	40

[a] Data from Grimmer *et al.* (1977c)

[b] +, Present but not quantitated

During accelerating or decelerating, or during cruising at increased speeds, for engines operated with or without emission control devices, higher amounts of PAHs are emitted into the environment. Furthermore, benzo[*a*]pyrene emission increases with ascending percentage of the aromatic portion of the fuel. It has also been shown that the PAH emission rate is significantly influenced by the air:fuel ratio, being highest with rich carburation. Benzo[*a*]pyrene emissions also rise with increasing oil consumption (Begeman & Colucci, 1970).

Begeman and Colucci (1970) as well as Müller and Meyer (1974) have shown that gasoline itself contains PAHs. Begeman and Colucci (1968) assumed that the PAH in fuel may contribute to PAH emissions from the engines. The contributory role of PAHs derived directly from fuel was demonstrated by operating an engine with gasoline to which ^{14}C-benzo[*a*]pyrene had been added and by quantifying the radioactivity in the exhaust effluents: 36% of the benzo[*a*]pyrene in the exhaust effluents was attributed to nondecomposed benzo[*a*]pyrene from the fuel.

(ii) *Diesel engine exhaust*

A number of studies, beginning in 1957, have established that diesel engines emit a large spectrum of PAHs similar to that generated by gasoline engines (National Research Council, 1981). Generally, the quantities of PAHs from the exhaust of a light-duty diesel engine (0.3-51 μg/km) (benzo[*a*]pyrene) appear to be in the same range as those from a gasoline engine without an exhaust control device (0.5-85 μg/km). However, a catalytic converter reduces the benzo[*a*]pyrene emissions of a gasoline engine significantly (0.05-0.3 μg/km) (Williams & Swarin; 1979; National Research Council, 1981).

(iii) *Aircraft engine exhaust*

In a study by Shabad and Smirnov (1972) aircraft engines released benzo[*a*]pyrene at 2-10 mg/min of operation, and the aromatic portion of the fuel was the major precursor for the pyrosynthesis of PAHs. At high speeds (>14 000 rev/min), the benzo[*a*]pyrene emission increased drastically.

Figure 1 depicts the relative amounts of PAHs in various combustion effluents.

Fig. 1. Relative emissions of polynuclear aromatic hydrocarbons from: 1 coal stove; 2, diesel engine; 3, gasoline engine; 4, coke plant; 5, oil burner (atomizer) taking benzo[*e*]pyrene as reference (1.0). The heights of the signals correspond to areas on the flame-ionization detector (i.e., quantities).

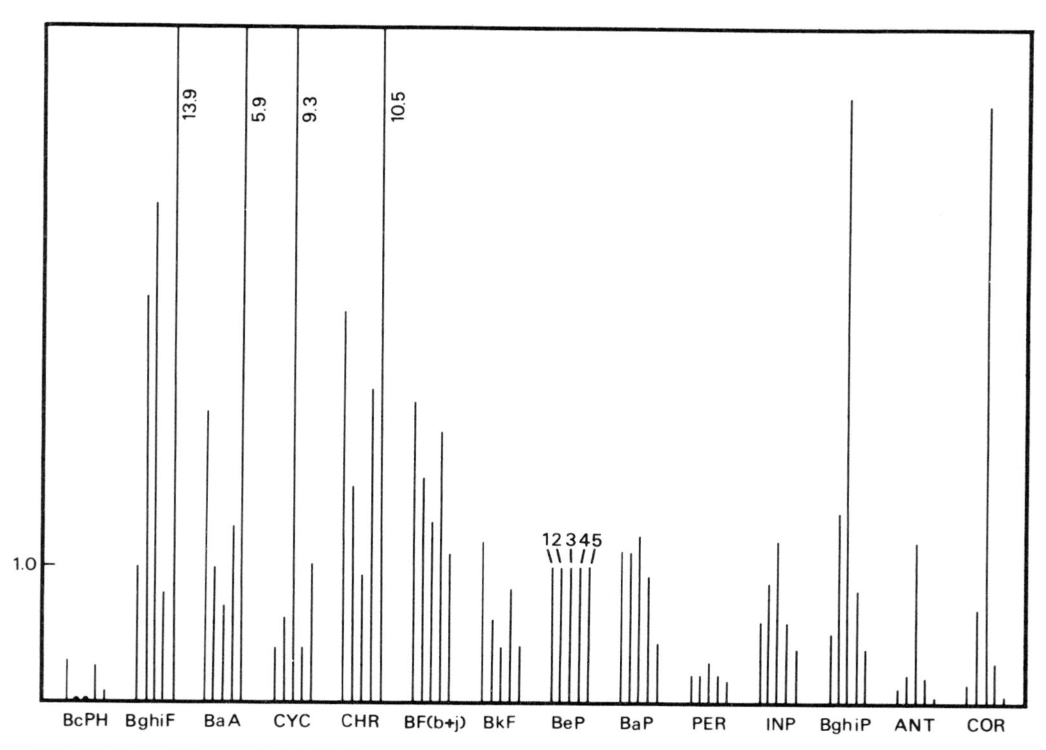

[a] Modified from Grimmer *et al.* (1980a)

[b] Abbreviations: BcPH, benzo[*c*]phenanthrene; BghiF, benzo[*ghi*]fluoranthene; BaA, benz[*a*]anthracene; CYC, cyclopenta[*cd*]pyrene; CHR, chrysene; BF(b+j), benzo[*b*]fluoranthene + benzo[*j*]fluoranthene; BkF, benzo[*k*]fluoranthene; BeP, benzo[*e*]pyrene; BaP, benzo[*a*]pyrene; PER, perylene; INP, indeno[1,2,3-*cd*]pyrene; BghiP, benzo[*ghi*]perylene; ANT, anthanthrene; COR, coronene

Legislation

The Clean Air Act of 1956 in the United Kingdom brought about a significant decrease in the concentration of PAHs in urban air (Lawther & Waller, 1976).

2.2 Ingestion

The major source of PAHs in drinking-water is pollution; other sources are effluents from asphalt pavements and used motor oil. The major sources of PAHs in food are curing smokes, contaminated soil, polluted air and water and food preparation methods, such as cooking (National Research Council, 1982).

(*a*) *Aquatic environment - water, sludge, sediments*

PAHs have been found in soil samples in many countries, and concentrations vary depending on the distance from the source of pollution. Air pollution is considered to be the main source; however, microorganisms in the soil may either metabolize or accumulate benzo[*a*]pyrene.

In non-industrial areas (forest, wood and sand samples in Czechoslovakia, the Federal Republic of Germany, France and the USA and lava and humus soil in Iceland), concentrations ranging from 0-390 μg/kg have been found (Blumer, 1961; Borneff & Fischer, 1962; Mallet & Héros, 1962; Borneff & Kunte, 1963; Zdrazil & Pícha, 1966; Fritz & Engst, 1971; Grimmer *et al.*, 1972). Benzo[*a*]pyrene was found at concentrations of up to 1300 μg/kg in forest soil samples (Blumer, 1961).

Although the concentrations of PAHs in water are usually very low owing to the low solubility of this group of compounds (Lee *et al.*, 1981), interest has focused on this matrix because of its general distribution and its importance as a source of consumption. The occurrence of benzo[*a*]pyrene in fresh and sea-water in the USSR has been reviewed by Weldre *et al.* (1977). Bottom sediments, algae, higher water plants, zooplankton and fish accumulate benzo[*a*]pyrene to concentrations about 1000-100 000 times that in water (Weldre *et al.*, 1977).

Atmospheric washout by rain and extraction of soil, pavements, coke plants and other industrial areas, represents an accumulation of anthropogenous wastes which are not transferred to refuses in sewage sludges, and are another source of PACs in the aquatic environment. The subject has recently been summarized by Neff (1979). The distribution and fate of benzo[*a*]pyrene in soil was reviewed by Shabad (1968) and Shabad *et al.* (1971), and in water by Weldre *et al.* (1977).

(i) *Surface, rain, subterranean and tap water*

Because there is legislation (WHO, 1971; Commission of the European Communities, 1980) restricting the sum of six PAHs (benzo[*b*]fluoranthene, benzo[*k*]fluoranthene, benzo[*ghi*]perylene, benzo[*a*]pyrene, fluoranthene, and indeno[1,2,3-*cd*]pyrene) to less than 200 ng/l, most studies have dealt with these compounds. In recent investigations, total PAH profiles were recorded for drinking-water and river water (Olufsen, 1980; Grimmer *et al.*, 1981a,c) Table 4 presents concentration ranges of various PAHs in surface, tap, rain and subterranean water.

Table 4. Concentration range of some polycyclic aromatic hydrocarbons in water[a]

Polynuclear aromatic hydrocarbon	Occurrence	Concentration/range (ng/l)
Anthanthrene	surface water	0.2[b]-10.9[b]
Anthracene	surface water	1000
	tap water	1.1-59.7
Benz[*a*]anthracene	surface water	1.9-30.6[b]
	tap water	0.4-10.7
	rainfall	3.2-12.3
	subterranean water	0-1.3

Polynuclear aromatic hydrocarbon	Occurrence	Concentration/range (ng/l)
Benzo[*b*]fluoranthene	surface water	0-320
	tap water	0.6-45
	rainfall	4.4-840
	subterranean water	0.5-9.0
Benzo[*j*]fluoranthene	surface water	0.6-1.2
	rainfall	2.6-11.1
	subterranean water	0.6-1.3
Benzo[*k*]fluoranthene	surface water	0-400
	tap water	0.9-8.0
	rainfall	1.6-450
	subterranean water	0.2-3.5
Benzo[*ghi*]fluoranthene	surface water	1.0[b]-11.2[b]
Benzo[*ghi*]perylene	surface water	0-390
	tap water	0.8-130
	rainfall	0-275
	subterranean water	0.3-5
Benzo[*c*]phenanthrene	surface water	1.0[b]-9.1[b]
Benzo[*a*]pyrene	surface water	0-13 000
	tap water	0-1000
	rainfall	10-1000
	subterranean water	0.1-6
Benzo[*e*]pyrene	surface water	3.4[b]-30.8[b]
Chrysene	surface water	7.6[b]-62.0[b]
Fluoranthene	surface water	4.7-1200
	tap water	7.2-132.6
	rainfall	5.6-1460
	subterranean water	3.5-100.0
Fluorene	surface water	300
	tap water	4-16
Indeno[1,2,3-*cd*]pyrene	surface water	0-350
	tap water	0.3-75
	rainfall	0-1020
	subterranean water	0.2-5.0
Perylene	surface water	0.2-520
	tap water	0.1-1.4
	rainfall	0.0-1.0
	subterranean water	0-0.2
Phenanthrene	surface water	0-1300
	tap water	24-90
Pyrene	surface water	2.0-530
	rainfall	5.8-27.8
	subterranean water	1.6-2.5

[a] From Commission of the European Communities (1979), unless otherwise specified

[b] From Grimmer *et al.* (1981c)

(ii) *Effluent discharges and sewage sludges*

In contrast to surface water, in which PAC concentrations are in the nanogram range, effluent discharges contain PACs in the microgram range (Table 5). PAHs have been found in the mg/kg range in freeze-dried samples (Grimmer *et al.*, 1978; Grimmer & Naujack, 1979; Grimmer *et al.*, 1980b); the original sludges contained about 95% water, so that these figures (given in the individual monographs) must be corrected in order to compare them with those for other aquatic matrices.

Table 5. Concentration range of some polynuclear aromatic hydrocarbons in effluent discharge and sludge[a]

Polynuclear aromatic hydrocarbon	Occurrence	Concentration/range
Anthanthrene	effluent discharge	0.04[b]-0.6[b] μg/l
Anthracene	effluent discharge	1.6-7.0 mg/l
Benz[*a*]anthracene	effluent discharge sludge	0.05-27 000 μg/l 230-1760 μg/kg
Benzo(*b*)fluoranthene	effluent discharge sludge	0.04-40.0 μg/l 510-2160 μg/kg
Benzo[*j*]fluoranthene	effluent discharge sludge	0.02-29.6 μg/l 230-2060 μg/kg
Benzo[*k*]fluoranthene	effluent discharge sludge	0.01-15 μg/l 150-1270 μg/kg
Benzo[*ghi*]fluoranthene	effluent discharge	0.042[b]-0.663[b] μg/l
Benzo[*ghi*]perylene	effluent discharge sludge	0.02-40 μg/l 200-1220 μg/kg
Benzo[*c*]phenanthrene	effluent discharge	0.042[b]-0.699[b] μg/l
Benzo[*a*]pyrene	effluent discharge sludge	0.001-10 000 μg/l 3-1330 μg/kg
Benzo[*e*]pyrene	effluent discharge	0.323[b]-2.928[b] μg/l
Chrysene	effluent discharge	0.732[b]-6.44[b] μg/l
Fluoranthene	effluent discharge sludge	0.01-45 μg/l 580-4090 μg/kg
Fluorene	effluent discharge	170 μg/l
Indeno[1,2,3-*cd*]pyrene	effluent discharge sludge	0.01-30.0 μg/l 470-1200 μg/kg
Perylene	effluent discharge	0.03-3.0 μg/l
Phenanthrene	effluent discharge	70-1400 μg/l
Pyrene	effluent discharge sludge	0.00023-11.8 μg/l 570-3080 μg/kg

[a] From Commission of the European Communities (1979), unless otherwise specified

[b] From Grimmer *et al.* (1981c)

(iii) *Sediments*

PAH profiles of fresh water and marine sediments differ from those observed in other matrices in so far as they are of two origins: anthropogenic origin and early fossilation or diagenetic origin. Terpenoid-derived hydrocarbons such as substituted tetrahydrophenanthrenes, substituted tetra- and octahydrochrysenes and substituted tetra- and octahydropicenes, indicate a diagenetic origin (Laflamme & Hites, 1979; Wakeham *et al.*, 1980); some have also been found in brown coal (Grimmer *et al.*, 1983a). PAH concentrations in sediments reflect the industrial situation of the surrounding area, since sediments may be regarded as an environmental sink (Grimmer & Böhnke, 1975, 1977; Müller *et al.*, 1977).

(iv) *Food*

It was reported in 1964 that the outer layers of charcoal-broiled steaks contain measurable amounts of PAHs (Lijinsky & Shubik, 1964; Table 6). Since that time, many global studies have documented that contamination of foods with PAHs is widespread. PAHs have frequently been found in processed fish and meat (Table 7). Steinig (1976) has shown that the PAH content of six different fish types is greatly dependent upon the degree of smoke exposure, concentrations of benzo[*a*]pyrene being in the range 0.17-1.8 μg/kg in the edible portion.

Table 6. Polynuclear aromatic hydrocarbons in charcoal-broiled steaks[a]

Compound	μg/kg
Anthanthrene	2
Anthracene	4.5
Benz[*a*]anthracene	4.5
Benzo[*ghi*]perylene	4.5
Benzo[*a*]pyrene	8
Benzo[*e*]pyrene	6
Chrysene	1.4
Coronene	2.3
Dibenz[*a*,*h*]anthracene	0.2
Fluoranthene	20
Perylene	2
Phenanthrene	11
Pyrene	18

[a] From Lijinsky & Shubik (1964)

Table 7. Concentration of chrysene, benz[*a*]anthracene and benzo[*a*]pyrene in processed food[a]

Compound	Refined oils or fats		Broiled meat or fish		Smoked fish	Smoked meat	
Benz[*a*]anthracene	coconut oil	0.5-13.7 (10)	meat and sausages	0.2-1.1 (9)	0.2-189 (12,15)	ham	1.3-12 (14,18)
	coconut fat	2-98 (1,10)		3.2-31[b](12)		surya	15.5 ± 2.5 (4)
		90-125 (2)	charcoal-broiled fish	0.6-2.9 (16)		mutton	0.05-0.3 (7)
						sausages	0.04-0.55 (7)
Benzo[*a*]pyrene	margarine	0.9-36 (10,2)	meat and sausages	0.17-0.63 (9)	1.0-78.0 (12)	ham	<0.5-14.6 (9,14,18)
	coconut oil	0.3-8.2		3.7-50.4[b](12)			
	butter	(5,6,17)	charcoal-broiled fish	0.2-0.9 (16)		surya	8.5 ± 2.0 (4)
	coconut fat	0.9-43.7 (1)				bacon	0.16-0.25 (13)
	plant oil	0.2-0.5 (18)				various meats	<33.5 (8)
	sunflower oil	29-62 (2)				sausages	0.05-0.08 (7)
		0.2 (18)				mutton	00.07-0.15 (7)
		5.2 (18)					
Chrysene		0.05-20 (3,10)	fish	0.4-4.3 (16)	0.3-173 (9,12,15)	ham	0.5-21.2 (14,18)
	coconut oil	2-129 (1,10)	meat and sausages	0.5-25.4 (9,11,14)			
	coconut fat	115-200 (2)					

[a] References: (1) Biernoth & Rost (1967); (2) Biernoth & Rost (1968); (3) Ciusa *et al.* (1965); (4) Emerole (1980); (5) Fábián (1968); (6) Fábián (1969); (7) Fretheim (1976); (8) Gray & Morton (1981); (9) Grimmer & Hildebrandt (1967a); (10) Grimmer & Hildebrandt (1967b); (11) Lijinsky & Ross (1967); (12) Lijinsky & Shubik (1965); (13) Lintas *et al.* (1979); (14) Malanoski *et al.* (1968); (15) Masuda & Kuratsune, 1971); (16) Masuda *et al.* (1966b); (17) Siegfried (1975); (18) Toth (1971)

[b] Charcoal-broiled T-bone steak

PAHs are present not only in broiled and smoked fish and meat and in roasted coffee (Fritz, 1972), but also in fresh meats, seafood, vegetables, oils, grains, fruits, whisky, etc. (Table 8). It can be assumed that, in the latter cases, the PAH contamination derives primarily from air and water pollution or from food or beverage containers. These data indicate that food is contaminated with trace amounts of PAHs. An attempt has been made to calculate the daily benzo[*a*]pyrene exposure of an 'average smoker' who resides in a city in the USA (Santodonato *et al.*, 1980).

Table 8. Concentrations of benz[*a*]anthracene, benzo[*a*]pyrene and chrysene in some food products (μg/kg)

Food	Benz[*a*]anthracene	Benzo[*a*]pyrene	Chrysene	Reference
Cereal	0.4-6.8	0.19-4.13	0.8-14.15	Grimmer & Hildebrandt (1965)
Potatoes (peelings) (tubers)		0.36 0.09		Archer *et al.* (1979)
Grain		0.73-2.3		Engst & Fritz (1975)
Flour, untreated Flour, dried		0.73 4.4		Fritz (1972) Fritz (1972)
Bread Bread, toasted		0.23 0.39-0.56		Lintas *et al.* (1979) Lintas *et al.* (1979)
Lettuce	6.1-15.4	2.8-12.8	5.7-26.5	Grimmer (1968)
Tomatoes	0.3	0.2-0.22	0.5	Grimmer (1968)
Spinach	16.1	7.4	28.0	Grimmer (1968)
Fruits		0.5-30		Engst & Fritz (1975)
Coffee, roasted	0.5-42.7	0.3-15.8	0.6-19.1	Kuratsune & Hueper (1960); Grimmer & Hildebrandt (1966); Fritz (1969)
Tea	2.9-36	3.9-21.3	4.6-63	Grimmer & Hildebrandt (1966)
Whisky	0.04-0.08	0.04	0.04-0.06	Masuda *et al.* (1966a)

(v) *Cosmetic and medicinal products*

Mineral oils and refined petroleum products used in cosmetics and medicinal products have also been shown to contain PAHs, a number of which are known carcinogens (Cook *et al.*, 1958; Lijinsky *et al.*, 1963; Helberg, 1964; Shabad *et al.*, 1970; IARC, 1973; McKay & Latham, 1973; Bingham *et al.*, 1980; Monarca *et al.*, 1981).

Fully-refined petroleum products with potential trace levels of PAHs are ingredients of many cosmetic preparations (e.g., cold creams, cleansing creams, suntan oils, baby lotions or creams and lipsticks) (Jellinek, 1970; Kraft *et al.*, 1972; Poucher, 1974; McCarthy, 1976; Monarca *et al.*, 1982). Some cosmetics prepared from vegetable oils may also contain trace amounts of benzo[*a*]pyrene and other PAH (Siegfried, 1975; Monarca *et al.*, 1982).

3. ANALYSIS

Methods for the analysis of PACs are described in detail in an IARC manual (Egan *et al.*, 1979), by Lee *et al.* (1981), and in two further monographs - a *Handbook of Polycyclic Aromatic Hydrocarbons* (Bjørseth, 1983) and *Environmental Carcinogens: Polycyclic Aromatic Hydrocarbons* (Grimmer, 1983a).

The matrices to be analysed vary so widely in their physical form that the analytical methodologies must be described separately.

Internal standards are widely used for quantitative analysis, e.g., in the form of picene or indeno[1,2,3-*cd*]pyrene (Grimmer *et al.*, 1982d) or as ^{14}C-labelled individual PAHs.

3.1 Collection and sampling

Adequate methods of collection and sampling are essential in order to obtain meaningful analytical results. Earlier measurements may not have reflected the levels of PACs actually present in the environment.

(*a*) *Air and combustion effluents*

The criterion for adequate collection is that the mixture of PACs found on the collection apparatus, e.g., on the filter be identical to that in ambient air or in flue gases.

In the atmosphere, PACs are adsorbed predominantly on suspended particulate matter. Some PACs with high vapour pressure at room temperature, such as anthracene, fluoranthene, fluorene, phenanthrene, pyrene and their methyl derivatives, although predominantly associated with particles, are also present in the gas phase, with an equilibrium between bound and free PACs. Therefore, complete trapping on particle filters (e.g., glass-fibre filters) cannot be expected. Extremely high- flow velocities allow the transfer even of compounds with very low vapour pressure from the filter to a subsequent sorbent.

The problem of re-evaporation of PACs from a filter (the 'blowing-off effect') occurs mainly with PACs with boiling-points below 400°C, such as the benzofluorenes, fluoranthene, phenanthrene, pyrene and their methyl derivatives (Rondia, 1965; Pupp *et al.*, 1974; König *et al.*, 1980; Lao & Thomas, 1980). A further problem is that sensitive PACs, such as anthanthrene, benz[*a*]anthracene and benzo[*a*]pyrene, are destroyed by light, oxidizing agents (SO_3, O_3) and nitration (NO_x) (Pitts, 1979; Hughes *et al.*, 1980; Lee *et al.*, 1980a; Nielsen, 1981). This degradation may be potentiated by the type of filter material used, e.g., quartz > glass > Teflon > Fluoropore (Lee *et al.*, 1980b).

As stated above, the PAC profiles of the gaseous sample and of the collection system must be identical. Since PAC mixtures consist of compounds possessing different boiling-points and/or different chemical reactivities, the time-dependent kinetic of the profiles (ratio of the PACs in the mixture) permits extrapolation to estimate the amounts present at the beginning of the sampling procedure. Both effects - chemical destruction of PACs on the particle filter and the re-evaporation of already collected PACs - can readily be recognized during the collection: with an invariable mass-stream, a correct collection process results in a linear correlation between the time of collection and the loading of the filter with PACs, independent of the molecular size or of the sensitivity of the different PACs to oxidation, nitration, etc. (Grimmer *et al.*, 1982d).

Standardization of sample collection can be assured by adhering to the following criteria:

(1) *Reproducibility of the sampling process*. With identical mass-stream per hour or simultaneous sampling using two different arrangements at the same place, the collection of PACs must result in identical profiles and in the same masses.

(2) *Invariability of the PAC profiles during sampling*. Variability of the PAC profile is a criterion for destruction and/or blow-off of the PACs already trapped on the filter owing to instability or volatility of certain individual compounds. There are significant differences in the vapour pressure of PACs containing 3-7 rings (Neff, 1979).

These criteria have rarely been discussed in investigations concerning collection apparatus for ambient air or emissions from fuel combustion. Thus, the US Environmental Protection

Agency (EPA) (1971) method 5 for in-stack sampling suggests use of a heated filter, followed by a series of impingers; however, with heated filters, destruction of various sensitive PACs and blow-off of low-boiling PACs cannot be excluded. A more efficient technique, recommended by Battelle Laboratories, uses a glass-cooler, followed by sorbents such as Tenax GC (Jones *et al.*, 1976) or XAD (Adams *et al.*, 1977), located between the filter and impingers. This arrangement yielded more than twice the amount of various PAC obtained with the method of the Environmental Protection Agency.

The criteria of repeatability of the sampling process and invariability of PAC profiles during sampling periods have been discussed with regard to apparatus for collecting automobile exhausts (Grimmer *et al.*, 1973a,b; Egan *et al.*, 1979), flue gases of oil-fired heating units (Behn *et al.*, 1980; Behn & Grimmer, 1981), flue gases of oil-fired vaporizing pot burners (Ratajczak *et al.*, 1982a,b) and emissions from brown-coal-fired residential stoves (Grimmer *et al.*, 1983a). Swarin and Williams (1980) investigated the variations of sampling rates, sampling times and effects of prolonged exposure in the case of diesel exhaust.

(*b*) *Tobacco smoke*

Analytical data for PAHs and heterocyclic compounds in tobacco smoke should be accepted only when certain criteria are met. In addition to avoidance of artefacts and contaminations and unambiguous identification of individual aromatic hydrocarbons by at least two reproducible physicochemical methods, there must be (1) meaningful smoking conditions and (2) appropriate collection of smoke particulates.

The advantages and shortcomings of automatic smoking devices for reproducing mainstream and sidestream smoke conditions have been reviewed by Dube and Green (1982). Standard smoking parameters for cigarettes, cigars and pipes have also been described (Elmenhorst & Stadler, 1967; International Committee for Cigar Smoke Study, 1974; Brunnemann *et al.*, 1976).

(*c*) *Water*

The solubility of PACs in water is low (Lee *et al.*, 1981). PACs in surface water or effluent discharge waste-water are mostly adsorbed onto particles; therefore, these samples must be carefully homogenized prior to analysis (Egan *et al.*, 1979).

(*d*) *Solid samples*

Most solid samples, such as food, soil or sediments, are not homogeneous and must also be carefully homogenized prior to analysis. The number of samples required to establish a statistically representative PAC profile in a large amount of material or for regulatory purposes represents a special problem; several ASTM (American Society for Testing and Materials) and ISO (International Standard Organization) standards have been proposed (Egan *et al.*, 1979).

3.2 *Extraction*

(*a*) *Air and combustion effluents*

The efficiency of extraction depends on (i) the properties of the particles (air-suspended matter of combustion effluent) that bind PAHs of different molecular weights to a varying degree, and (ii) the solvent. Generally, extraction with boiling solvents is more effective than Soxlhet extraction, or extraction under ultrasound conditions at room temperature.

Various solvents have been used to extract particles from combustion effluents or carbon blacks. Aromatic solvents are the most efficient (Cautreels & Van Cauwenberghe, 1977; Daisey & Leyko, 1979; Taylor *et al.*, 1980). Complete extraction of PACs from particles emitted by diesel engines (Swarin & Williams, 1980; Grimmer *et al.*, 1982c), carbon black (Locati *et al.*, 1979; Taylor *et al.*, 1980) and fly ash (Griest & Guerin, 1979; Zelenski *et al.*, 1980) is difficult.

The results of a comparison with several solvents demonstrated (Grimmer *et al.*, 1982d) that (i) the highest recovery is achieved with toluene, the lowest with methanol; (ii) the yield of different PAHs extracted by the same solvent decreases with increasing number of aromatic rings; (iii) there is a clear-cut equilibrium between the various PAHs more or less strongly bound to the 'active centres' of the surface and those in the solvent; and (iv) extraction by boiling cyclohexane (twice for three hours, using fresh solvent each time) is more effective than ultrasound (twice for three hours) at room temperature.

Since airborne particulate matter originates from a variety of sources, solvents such as cyclohexane or dichloromethane are in general not suitable for total extraction.

(*b*) *Water*

PACs can be extracted from drinking- or industrial-water by liquid-liquid extraction (Egan *et al.*, 1979). Various solvents have been found to be suitable (Woidich *et al.*, 1976; Hagenmaier *et al.*, 1977; Borneff & Kunte, 1979; Egan *et al.*, 1979; Saxena *et al.*, 1980; Grimmer *et al.*, 1981a), e.g., 1,1,2-trichlorotrifluoroethane, dichloromethane; however, the method may not be suitable when the PAHs are adsorbed on particulate matter. A short review including alternative methods, such as adsorption on resins, is given by Josefsson (1982). An analytical method for six PAHs, recommended by WHO (1971) and Borneff & Kunte (1979), involves extraction with cyclohexane.

(*c*) *Solid samples*

A general procedure for the extraction of PACs from solid materials, such as high-protein foods (fresh or smoked meat and fish) and plant materials rich in carbohydrates (vegetables, leaves, seeds, fruits, etc.), has been described (Egan *et al.*, 1979).

Samples of soil, aquatic sediments and sewage sludge contain organic, as well as inorganic, matter. Although it is generally sufficient to extract these materials with boiling acetone (Grimmer & Böhnke, 1975), some matrices retain PACs strongly, and a further extraction can be carried out using toluene. As described by Giger & Blumer (1974), PAHs can be extracted with boiling methanol followed by a benzene-methanol mixture.

(*d*) *Mineral and vegetable oils*

Since such samples are soluble in cyclohexane, isooctane and other lipophilic solvents, the resulting solutions can be extracted with methanol-water mixtures to remove nonaromatic compounds. Aromatic compounds may be enriched selectively with dimethylformamide:water (9:1), dimethyl sulphoxide or nitromethane (Egan *et al.*, 1979).

3.3 Clean up

This step is used to ensure complete separation of all impurities from PACs before instrumental analysis. The most frequently used procedures are liquid-liquid partition followed by adsorption chromatography using inorganic or organic supports, such as silica gel, aluminium oxide, Sephadex LH20 or LH60, Biobeads and acetyl cellulose, or partition chromatographic methods (Egan *et al.*, 1979; Lee *et al.*, 1981).

Secondary contamination from the laboratory environment and photo-oxidation should be avoided. A closed system has been proposed. The exclusion of organic compounds, such as phthalic esters, is essential, since flame-ionization detectors, which are preferred in gas chromatography, detect carbon-containing substances at levels of 0.1-0.01 ng. Sources of such contamination have been identified as surfaces of glassware and solvents that were not distilled in a closed apparatus (Grimmer *et al.*, 1982d).

3.4 Quantification and identification of PAHs in isolated PAC mixtures

Since the PAC mixture isolated from most sample types consists of more than 100 individual compounds, a procedure with high separation capacity (number of separation stages) is required.

(*a*) *Capillary-gas chromatography*

Presently, the most efficient method for the separation of complex mixtures of sufficiently volatile PAC, such as the PAH fraction isolated from air-suspended particulate matter, is capillary-gas chromatography (capillary GC), the number of separation stages of which exceeds 70 000 HETP (height equivalent of theoretical plates).

This method was already being used nearly 20 years ago to analyse PAHs in air-suspended particulate matter (Liberti *et al.*, 1964) and has been used widely since (Cantuti *et al.*, 1965; Grimmer & Böhnke, 1972; Lao *et al.*, 1973; Lee *et al.*, 1975; Cautreels & Van Cauwenberghe, 1977, 1978; König *et al.*, 1980; Strand & Andren, 1980; Grimmer *et al.*, 1981a,b, 1982d; Romanowski *et al.*, 1982). Numerous stationary phases have recently become available and qualified for PAC analysis, e.g., polydimethylsiloxanes (OV 101; SE 30; SP 2100; CP sil 5; DB 1), polymethylphenylsiloxanes (SE 52; SE 54; OV 3, OV 17; SP 2250; CP sil 19-CB), polyphenylether (Polysev), and carboran-siloxane (Dexsil 300, Dexsil 400). Stationary phases bound chemically to glass or fused silica permit gas chromatography at high temperatures without bleeding. This subject has been reviewed (Lee *et al.*, 1981).

Flame-ionization detection (FID) is used mainly for recording. With splitless or on-column injection, 5-10 ng benzo[*a*]pyrene are usually required to obtain an FID signal. The detection limit, defined as three times the noise level, is about 0.01-0.1 ng, depending on the retention time of the PAH. The detector response corresponds to the carbon mass of the compound being measured; thus, with optimum conditions, the areas of the FID signals are related linearly to the mass of the compound recorded, and the response factor is equal to 1.00. FID is highly sensitive and records identified and unidentified PAHs in proportion to their quantities. Thus, the chromatogram represents directly the quantitative composition of the PAC mixture without requiring correction factors.

The chemical structure of individual PACs inferred by retention time must be confirmed by mass spectrometric analysis and other high-resolution techniques.

(*b*) *High-performance liquid chromatography*

High-performance liquid chromatography (HPLC) reaches a maximum separation number of 20 000, corresponding to the separation power achieved with packed columns of 7-10 m in length. The separation power of HPLC is thus still limited, and this method is not presently widely used for analysis of complex mixtures originating from pyrolysis of organic materials or from incomplete combustion.

Reviews on the use of HPLC for the analysis of PACs are given by Egan *et al.* (1979) and Lee *et al.* (1981).

(*c*) *Thin-layer chromatography*

Thin-layer chromatography (TLC) requires only low qualification of the operator, but separates only a few substances: only 10-20 completely separated spots can be arranged between the starting point and the solvent front. Apart from cellulose acetate foils, mixed plates with alumina-acetylated cellulose coatings are used predominantly. Mixtures of less polar solvents, such as hexane, benzene and dichloromethane, as well as mixtures of these solvents with methanol, ether, water, etc., have been recommended for use as the mobile phase. A two-dimensional procedure for the analysis of airborne samples, and for the identification of six selected PAHs in water samples has been recommended (Egan *et al.*, 1979).

(*d*) *Quasi-linear fluorescence at low temperatures*

Airborne particulates can be extracted and separated by thin-layer chromatography and the fine-structure PAH spectra obtained at low temperature in frozen polycrystalline solutions. The method is given by Egan *et al.* (1979).

4. ABSORPTION, DISTRIBUTION, METABOLISM AND EXCRETION OF POLYNUCLEAR AROMATIC COMPOUNDS IN EXPERIMENTAL ANIMALS

The biological effects of xenobiotics are determined as much by the dynamics of their absorption, distribution, metabolism and excretion as by their intrinsic chemical properties. This assessment is especially true for PACs, since the parent hydrocarbons are chemically unreactive, and ultimate biological activity as well as detoxification is conferred by metabolic processes. Although many PAHs have long been recognized as ubiquitious environmental pollutants and potent carcinogens, their absorption, distribution and excretion have not been extensively evaluated. In contrast, metabolic studies of PAHs comprise a vast and exponentially growing literature. Investigators have naturally tended to focus their efforts on the more carcinogenic PAHs, which were evaluated previously by the IARC (1973) and are not re-evaluated for carcinogenicity in the present monograph. Benzo[*a*]pyrene has been the PAH most commonly studied. While many observations made with this hydrocarbon appear valid for the PAHs as a class, significant differences between compounds in terms of their bio-availability and metabolic fate have been noted. There is therefore a need for detailed studies with a number of other hydrocarbons and particulates.

4.1 Absorption

PACs are essentially devoid of polar and ionizable functional groups and would therefore be expected to dissolve readily in and cross the lipoprotein membranes of mammalian cells. The demonstrated toxicity of many PAHs in organs remote from the site of their administration confirms this expectation. Furthermore, the fact that isolated cells and tissues metabolize PAHs by means of intracellular enzymes, and that some of these metabolites react with intracellular constituents suggests that uptake across cellular membranes is an easy process.

The major routes of environmental exposure to PAHs are through the gut after ingestion of contaminated food or water, through the lungs by inhalation of aerosols and/or hydrocarbon-absorbed particles and through the skin by direct contact with the hydrocarbons.

Rees *et al.* (1971) demonstrated in rats that intragastric administration of benzo[*a*]pyrene is followed by rapid absorption, with peak levels of the hydrocarbon in the thoracic lymph duct being observed three to four hours after treatment. No more than 20% of the administered dose was accounted for in the lymph. Since the levels of benzo[*a*]pyrene in blood and other tissues were not reported, it is not clear whether the modest recovery is a reflection of incomplete absorption or uptake into the portal circulation. The authors utilized inverted intestinal sacs as well as intact rats to study the mechanism of absorption and suggested that physical adsorption of benzo[*a*]pyrene to the intestinal mucosa precedes passive diffusion through the intestinal wall, and that the whole process is described by first-order (exponential) kinetics. Carcinogenicity studies in which PAHs are administered intragastrically to animals and tumours are induced in the mammary tissue and lungs provide further evidence for significant enteral absorption of PAH (Huggins & Yang, 1962; Rigdon & Neal, 1966).

Absorption of PAHs across the pulmonary endothelium has been studied following inhalation of pure aerosols and after intratracheal administration of hydrocarbons absorbed on particles of various sizes and chemical compositions. Unfortunately, those studies that have examined the clearance of PAHs from the lungs have not generally differentiated between systemic absorption through pulmonary tissues and removal of the hydrocarbon by mucociliary clearance. The latter mechanism can result in high gastrointestinal levels and enteral absorption, since the cleared material is swallowed. Mitchell (1982) showed that 1-2-μm aerosol particles of benzo[*a*]pyrene are cleared from the lungs of rats and transported to internal organs biphasically; a rapid phase with a two-hour half-life precedes a slower phase with an approximately two-day half-life. Benzo[*a*]pyrene adsorbed on particles may take 20 times as long to be cleared from the lungs of mice as free benzo[*a*]pyrene (Creasia *et al.*, 1976). Furthermore, regional deposition and clearance rates from the respiratory tract are dependent on the size and chemical composition of the particle, as well as the structure of the hydrocarbon (Henry & Kaufman, 1973; Creasia *et al.*, 1976; Nagel *et al.*, 1976). Vainio *et al.* (1976) have demonstrated unequivocally the pulmonary absorption of benzo[*a*]pyrene by administering the hydrocarbon intratracheally to the isolated perfused lungs of rats and measuring the appearance of the parent hydrocarbon and its metabolites in the perfusion fluid. Mitchell and Tu (1979) detected significant levels of pyrene in the liver, kidney and muscle of rats following inhalation, but not after administration of the hydrocarbon by gavage.

Topical application of a benzene solution of ^{14}C-benzo[*a*]pyrene to the shaved backs of mice is followed by a biphasic disappearance of radioactivity, with half-lives of 40 and 104 h for the fast and slow phases, respectively (Heidelberger & Weiss, 1951). Since essentially all of the radioactivity was recovered in the faeces within 16 days, quantitative percutaneous absorption of this hydrocarbon is apparent. In contrast, these investigators reported that topical application of an equivalent dose of dibenzo[*a*,*h*]anthracene (0.2 μmol) dissolved in benzene was associated with little, if any, systemic absorption, and elimination was dependent on epidermal sloughing.

4.2 Distribution

Once PAHs are absorbed or injected into the bloodstream they are rapidly and widely distributed. Kotin *et al.* (1959) reported that intravenously-administered ^{14}C-benzo[*a*]pyrene (11 μg) was cleared from the blood of 200-g rats with a half-life of less than one minute, and no detectable blood level was observed ten minutes after injection. Schlede *et al.* (1970a) noted a similar rapid clearance of intravenously administered ^{3}H-benzo[*a*]pyrene from the blood of rats and reported that pretreatment of the animals with unlabelled PAHs for several days markedly stimulated the disappearance of the labelled benzo[*a*]pyrene from the blood. Since levels of benzo[*a*]pyrene were diminished in various tissues at the same time that biliary levels of labelled metabolites were increased, it was concluded that the unlabelled hydrocarbons induced the metabolism of benzo[*a*]pyrene (Schlede *et al.*, 1970a,b). Levels of PAHs observed in any particular tissue are dependent on the PAH administered, the route and vehicle of administration, the post-administration sampling times and the presence of inducers of PAH metabolism. Nevertheless, results of a number of studies (Heidelberger & Weiss, 1951; Kotin *et al.*, 1959; Bock & Dao, 1961; Mitchell, 1982) indicate that (i) detectable levels of hydrocarbon can be observed in most internal organs from minutes to hours after administration; (ii) mammary and other fat tissues are significant storage depots where hydrocarbons may accumulate and be slowly released; and (iii) the gut contains relatively high levels of hydrocarbons or hydrocarbon metabolites as the result of hepatobiliary excretion of metabolites or of swallowing of unmetabolized hydrocarbon following mucocillary clearance after inhalation exposure.

As discussed above, absorption and clearance of PAHs deposited in the respiratory tract are markedly dependent on the physical size and state of administered hydrocarbon. After intratracheal instillation, the pattern of distribution essentially parallels that observed after subcutaneous and intravenous administration of the carcinogen, except for the high local pulmonary concentration (Kotin *et al.*, 1959). Thus, Kotin *et al.* (1959) observed much higher and prolonged lung levels of intratracheally instilled benzo[*a*]pyrene than were observed after intravenous or subcutaneous administration of the hydrocarbon. Most recently, Mitchell (1982) has described hydrocarbon concentration-time curves for the major organs of the body following inhalation of ^{3}H-benzo[*a*]pyrene. Respiratory-tract tissues were cleared of the hydrocarbon with an initial half-life of two to three hours. Peak levels of radioactivity in the liver were observed one-half hour after exposure and declined with a half-life of six hours. Peak values in the kidney (almost twice those of liver) were observed approximately six hours after inhalation was terminated. From one-half to one hour after the administration of benzo[*a*]pyrene was terminated, the stomach and small intestine contained the highest levels of radioactivity, but these values declined rapidly with a concomitant increase in radioactivity in the large intestine and caecum.

PAHs cause significant toxicity and carcinogenicity in the fetuses and offspring of maternally-exposed animals (Tomatis, 1973). Consistent with these toxicological observations is the finding that benzo[*a*]pyrene and 7,12-dimethylbenz[*a*]anthracene can readily cross the placenta of intragastrically-dosed rats (Shendrikova & Aleksandrov 1974) and of mice treated intravenously with these hydrocarbons (Shendrikova *et al.*, 1973, 1974). More recently, Kelman and Springer (1981) monitored the appearance of radioactivity in the umbilical vein of guinea-pigs given an intravenous dose of ^{14}C-benzo[*a*]pyrene and concluded that the hydrocarbon readily gains access to the fetus, at rates comparable to those observed with tritiated water.

4.3 Excretion

Once metabolized (see next section), hepatobiliary excretion and elimination through the faeces is the major route by which PAHs are removed from the body, regardless of the original route of administration. Kotin *et al.* (1959) observed that 4-12% of a subcutaneously injected dose of benzo[*a*]pyrene was eliminated in the urine of mice within six days of injection, while 70-75% of the dose was recovered in the faeces. Similarly, in the rat, 4-40 times as much intravenously or intratracheally administered benzo[*a*]pyrene could be recovered in the intestine and faeces as in the urine. Other major findings of their study were (i) no evidence for elimination of benzo[*a*]pyrene *via* expired air; (ii) less than 1% of benzo[*a*]pyrene recovered in the bile was unmetabolized; (iii) the rate of biliary excretion of benzo[*a*]pyrene became saturated at an intravenous dose of 750 μg/kg; and (iv) cannulation of the bile duct reduced by half the urinary excretion of benzo[*a*]pyrene metabolites. This latter observation suggested the possibility of enterohepatic circulation of hydrocarbon metabolites. Chipman *et al.* (1982) demonstrated this pathway directly by collecting the biliary metabolites of benzo[*a*]pyrene, readministering them into the duodena of bile duct-cannulated rats, and observing the re-excretion of the metabolites in the bile and urine. These investigations and those of Hecht *et al.* (1979) have identified and quantified the biliary and faecal metabolites of benzo[*a*]pyrene.

Schlede *et al.* (1970b) demonstrated that metabolism rather than biliary transport was the rate-determining step in the biliary excretion of intravenously administered benzo[*a*]pyrene. Rats were pretreated with unlabelled benzo[*a*]pyrene for two days prior to an intravenous dose of radioactive benzo[*a*]pyrene in order to induce the oxidative enzyme system that metabolizes hydrocarbons. In comparison to unpretreated rats, there was a 20-fold higher concentration of radioactive metabolites in the bile seven minutes after injection, and peak biliary levels of metabolites were observed 15 min after injection, as compared to peak levels at 45 min post-injection in un-pretreated animals. Pretreatment of the rats had no effect on the rate of biliary excretion when the isolated metabolites were reinjected into rats.

Xenobiotics absorbed from the gastrointestinal tract enter the liver *via* the portal circulation prior to reaching the systemic circulation. The importance of hepatobiliary metabolism and excretion suggests that orally administered hydrocarbons would exhibit enhanced rates of excretion and lower tissue levels relative to routes of administration in which the hydrocarbon reached the systemic circulation without a first-pass through the liver. Consistent with this expectation, Aitio (1974) observed that 82% of an orally administered dose of 3-methylcholanthrene was excreted within 24 h, while only 30% of an intraperitoneally administered dose was excreted in 72 h. Intragastric administration of 3-methylcholanthrene was also associated with a rapid induction of the hydrocarbon-metabolizing monooxygenase system in the intestinal mucosa as well as in the liver.

4.4 Metabolism

(*a*) *General considerations*

Since lipophilic xenobiotics readily cross cellular membranes, they can be efficiently reabsorbed from the renal-tubular urine, or, if they enter the bile, through the bile canaliculi or intestinal mucosa. This is generally not the case for polar compounds, especially if they are ionizable, and organisms have thus evolved enzyme systems that convert lipophilic compounds to readily-excretable polar metabolites. The enzymes that metabolize xenobiotics have been classified into two broad categories on the basis of the types of reactions they catalyse.

Phase 1 enzymes catalyse oxidative, reductive or hydrolytic reactions that introduce or uncover functional groups on the xenobiotic agent. The NADPH-dependent cytochrome P-450 monooxygenase system is the most important of the phase 1 enzymes because it has a broad substrate specificity, catalyses a variety of oxidative reactions and can be induced in response to exposure to xenobiotics (Conney, 1967; La Du *et al.*, 1971; Testa & Jenner, 1976; Sato & Omura, 1978). The broad metabolic capacity of the monooxygenase system is due in part to the existence of multiple forms of cytochrome P-450 with different overlapping substrate specificities (Conney, 1982).

Phase 2 enzymes catalyse conjugative or synthetic reactions between the newly-formed functional groups and small-molecular-weight endogenous compounds derived from carbohydrates or amino acids (La Du *et al.*, 1971; Testa & Jenner, 1976; Caldwell, 1982). Many of these conjugates, such as the glucuronides, sulphate esters and mercapturic acids (*N*-acetylcysteine derivatives resulting from metabolic alteration of glutathione conjugates) are ionized at physiological pH and are readily excreted.

Many PAHs that are metabolized by the cytochrome P-450 monooxygenases are inducers of the enzyme system as well. As noted in the preceding section on excretion, the ability of the hydrocarbons to induce their own metabolism results in lower tissue levels and more rapid excretion of the hydrocarbon (Schlede *et al.*, 1970b; Aitio, 1974). Generally, pretreatment of animals with inducers of the monooxygenase systems is associated with a decreased tumour incidence (Wattenberg, 1978). However, studies with strains of mice that differ genetically in the capacity of their monooxygenase system to be induced by PAHs indicate that inducibility can be associated with an increased tumorigenic or toxicological response (Nebert, 1980). Induction of the monooxygenase system by PAHs or phenobarbital (another type of monooxygenase inducer) can result in different profiles of hydrocarbon metabolites, although the effect appears to be variable in extent (Holder *et al.*, 1974; Jacob *et al.*, 1981; Schmoldt *et al.*, 1981).

(*b*) *Benzo[a]pyrene as a model of PAH metabolism*

Analytical studies initiated over 25 years ago, principally by Sims and Boyland and their associates (for reviews see Sims & Grover, 1974; Dipple, 1976; Weinstein *et al.*, 1978) revealed that PAH metabolites isolated from experimental animals consist of hydroxylated derivatives, generally in the form of glucuronide, sulphate and mercapturic acid conjugates. Thus, the general scheme of drug and xenobiotic metabolism outlined above was applicable to PAHs. However, the principal interest in hydrocarbon metabolism arose from the realization that hydrocarbons, like many environmental carcinogens, were chemically unreactive and that their adverse biological effects were probably mediated by electrophilic metabolites capable of covalent interaction with critical macromolecules such as DNA (Miller, 1970; Jerina & Daly, 1974; Sims & Grover, 1974; Heidelberger, 1976; Weinstein *et al.*, 1978; Miller & Miller, 1982).

The identification of the biologically active metabolites of PAHs, coupled with advances in both the synthesis of known and potential hydrocarbon metabolites and metabolite analyses by high-performance liquid chromatography, has led in the last decade to a greatly enhanced appreciation of the complexity and diversity of hydrocarbon metabolism. These metabolic interrelationships are illustrated for benzo[*a*]pyrene in Figure 2; the structures of representative metabolites are given in Figure 3. A number of recent reviews have focused on the metabolism of benzo[*a*]pyrene and other PAHs, and particularly on their metabolic activation; the reader is referred to them for access to additional information and the primary literature that forms the basis of this overview (Sims & Grover, 1974; Selkirk, 1977; DePierre & Ernster, 1978; Yang *et al.*, 1978; Jerina *et al.*, 1980; Sims & Grover, 1981; Conney, 1982; Levin *et al.*, 1982; Pelkonen & Nebert, 1982; Sims, 1982; Cooper *et al.*, 1983).

Fig. 2. Metabolic fate of benzo[*a*]pyrene

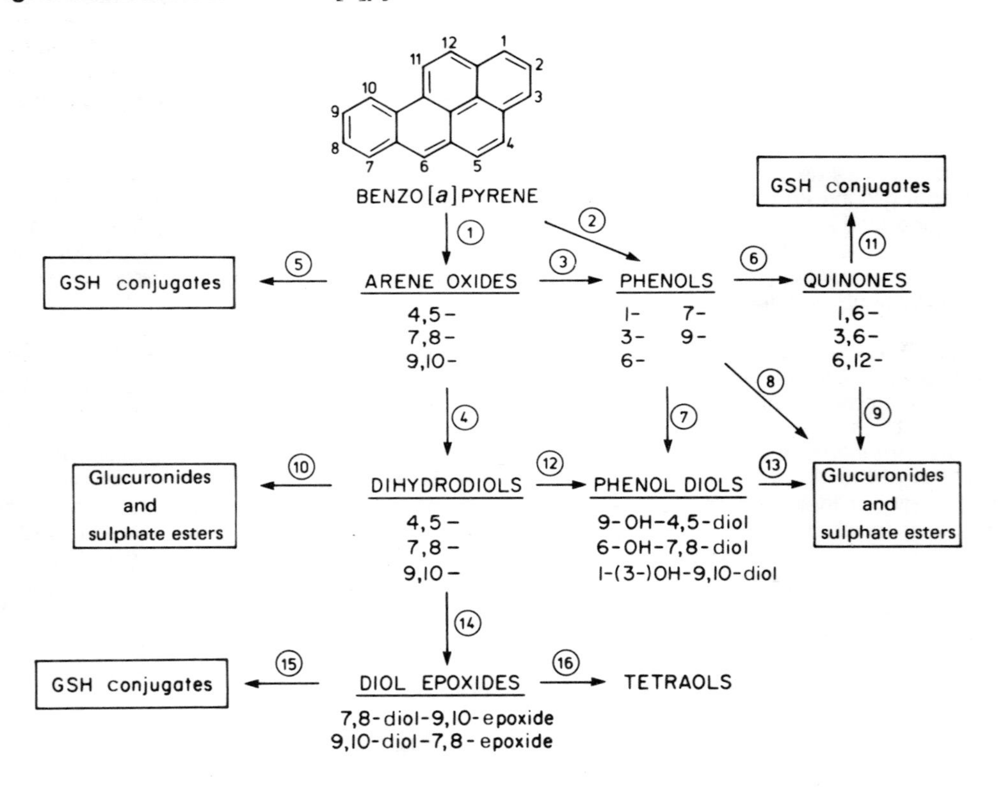

Benzo[*a*]pyrene is metabolized initially by the microsomal cytochrome P-450 monooxygenase system to several arene oxides (reaction 1, Fig. 2). Once formed, these arene oxides may rearrange spontaneously to phenols (reaction 3), undergo hydration to the corresponding *trans*-dihydrodiols in a reaction catalysed by microsomal epoxide hydrolase (reaction 4), or react covalently with glutathione, either spontaneously or in a reaction catalysed by cytosolic glutathione S-transferases (reaction 5). Phenols may also be formed by the cytochrome P-450 monooxygenase system by direct oxygen insertion (reaction 2), although unequivocal proof for

Fig. 3. Structures of types of polynuclear hydrocarbon metabolites referred to in the text[a]

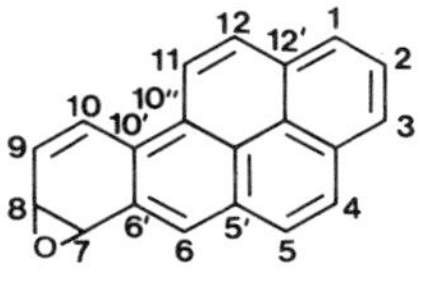

Arene oxide
7,8-epoxy B(*a*)P

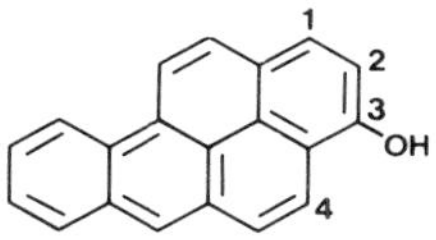

Phenol
3-OH B(*a*)P

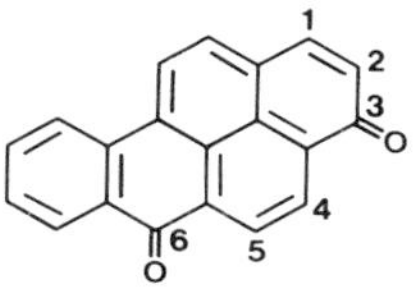

Quinone
B(*a*)P 3,6-dione

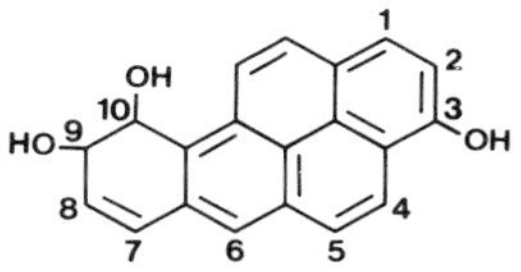

Phenol diol
3,9,10-trihydroxy 9,10-dihydro B(*a*)P

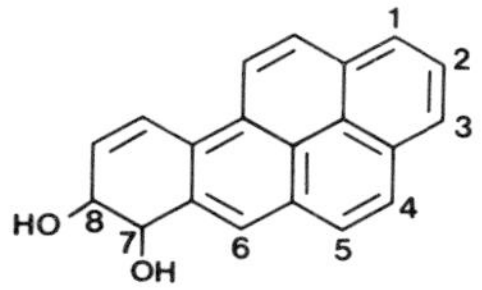

Dihydrodiol
7,8-dihydroxy 7,8-dihydro B(*a*)P

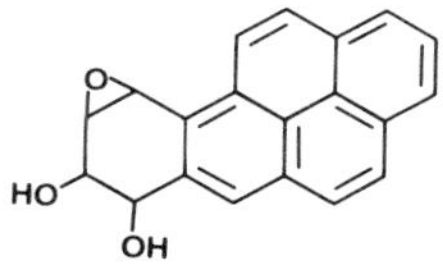

Diol epoxide
9,10-epoxy, 7,8-dihydroxy 7,8-dihydro B(*a*)P

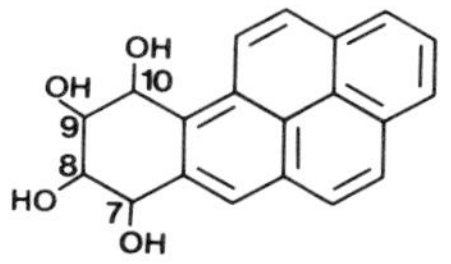

Tetraol
7,8,9,10-tetrahydroxy 7,8,9,10-tetrahydro B(*a*)P

[a] Abbreviation: B(*a*)P, benzo[*a*]pyrene

this mechanism is lacking. 6-Hydroxybenzo[*a*]pyrene is further oxidized either spontaneously or metabolically to the 1,6-, 3,6- or 6,12-quinones (reaction 6), and this phenol is also a presumed intermediate in the oxidation of benzo[*a*]pyrene to the three quinones catalysed by prostaglandin endoperoxide synthetase (Marnett *et al.*, 1977, 1979). Evidence exists for the further oxidative metabolism of two additional phenols: 3-hydroxybenzo[*a*]pyrene is metabolized to the 3,6-quinone (reaction 6), and 9-hydroxybenzo[*a*]pyrene is oxidized to the K-region 4,5-oxide, which is hydrated to the corresponding 4,5-dihydrodiol (reaction 7). The phenols, quinones and dihydrodiols can all be conjugated to glucuronides and sulphate esters (reactions 8-10), and the quinones also form glutathione conjugates (reaction 11) (for structures see Fig. 4).

Fig. 4. Types of conjugates that may be formed from polynuclear hydrocarbon metabolites

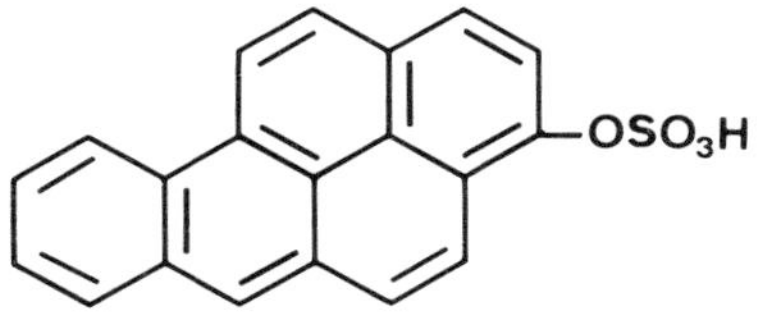

Sulphate ester

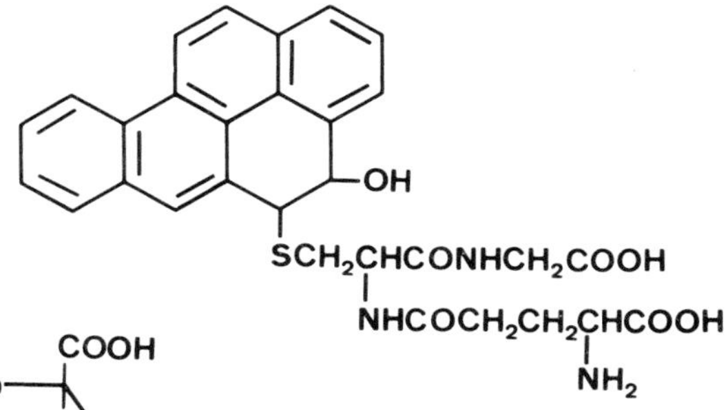

Gluthathione conjugate

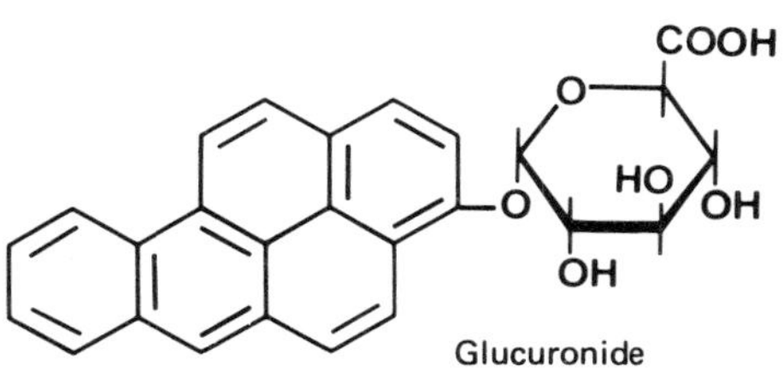

Glucuronide

In addition to being conjugated, the dihydrodiols undergo further oxidative metabolism. The cytochrome P-450 monooxygenase system metabolizes benzo[*a*]pyrene 4,5-dihydrodiol to a number of uncharacterized metabolites, while the 9,10-dihydrodiol is metabolized predominantly to its 1- and/or 3-phenol derivative (reaction 12), with only minor quantities of a 9,10-diol-7,8-epoxide being formed (reaction 14). In contrast to 9,10-dihydrodiol metabolism, the principal route of oxidative metabolism of benzo[*a*]pyrene 7,8-dihydrodiol is to a 7,8-diol-9,10-epoxide (reaction 14), and phenol-diol formation is a relatively minor pathway. The diol epoxides can be conjugated with glutathione either spontaneously or by a glutathione S-transferase-catalysed reaction (reaction 15). They may also hydrolyse spontaneously to tetraols (reaction 16, although epoxide hydrolase does not catalyse the hydration). Taken together, these reactions illustrate that benzo[*a*]pyrene, and hydrocarbons in general, can undergo a multitude of simultaneous and sequential transformations; they underscore the difficulties in determining which metabolites are responsible for the biological effects of the parent hydrocarbon.

An additional complexity in the evaluation of hydrocarbon metabolism is the fact that the compounds are metabolized to optically active products. Figure 5 illustrates the stereoselective metabolism of benzo[*a*]pyrene to its four metabolically possible 7,8-diol-9,10-epoxides. A total of four isomers are possible, since each diastereoisomer can be resolved into two enantiomers. With rat liver microsomes, the (+)-[7R,8S]-oxide of benzo[*a*]pyrene is formed in a 20-fold excess relative to the (-)-[7S,8R]-isomer and is metabolized stereospecifically by epoxide

Fig. 5. Metabolic formation of the 7,8-diol-9,10-epoxides of benzo[*a*]pyrene[a]

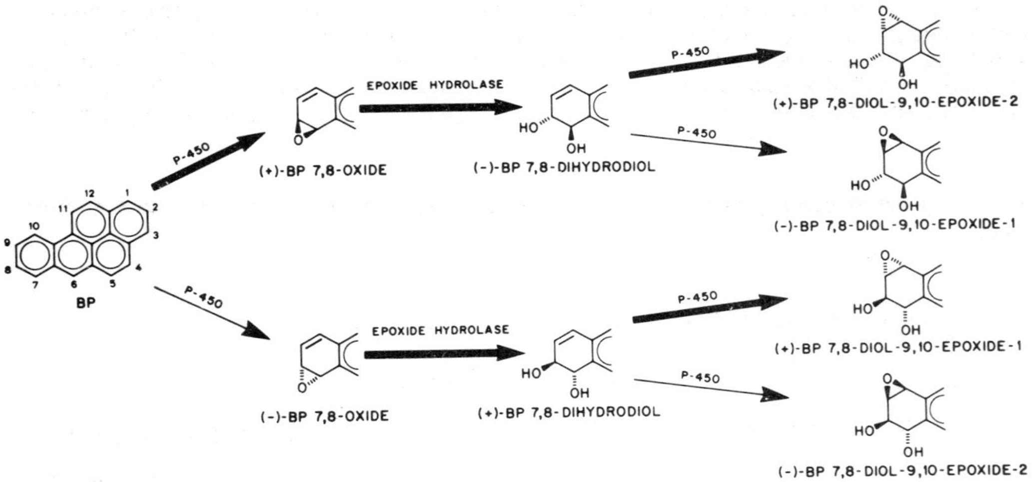

[a] Adapted from Levin *et al.* (1980)

Absolute stereochemistry of all metabolites is as shown. Heavy arrows indicate the predominant pathways. Diol epoxides exist as diastereoisomeric pairs in which the benzylic hydroxyl group and the epoxide oxygen are either *cis* (variously called 'diol epoxide I', 'diol epoxide II', 'syn-diol epoxide') or *trans* (variously called 'diol epoxide 2', 'diol epoxide I' or 'anti-diol epoxide').

I (+)-(7R,8S,9S,10R)-7,8-dihydroxy-9,10-epoxy-7,8,9,10-tetrahydro-benzo[*a*]pyrene
II (-)-(7R,8S,9R,10S)-7,8-dihydroxy-9,10-epoxy-7,8,9,10-tetrahydro-benzo[*a*]pyrene
III (+)-(7S,8R,9S,10R)-7,8-dihydroxy-9,10-epoxy-7,8,9,10-tetrahydro-benzo[*a*]pyrene
IV (-)-(7S,8R,9R,10S)-7,8-dihydroxy-9,10-epoxy-7,8,9,10-tetrahydro-benzo[*a*]pyrene

hydrolase to the (-)-[7R,8R]-dihydrodiol (Thakker *et al.*, 1977). More than 90% of the benzo[*a*]pyrene 7,8-oxide formed consists of the (+)-[7R,8S]-enantiomer (Levin *et al.*, 1982). This metabolically predominant dihydrodiol is metabolized primarily to a single diol epoxide isomer, (+) benzo[*a*]pyrene 7,8-diol-9,10-epoxide-2[7R,8S,9S,10R]. The biological significance of stereoselective formation of the 7,8-diol-9,10 epoxide isomers is that the metabolically predominant isomer is also the only isomer with high tumorigenic activity, and it is the predominant isomer found covalently bound to DNA in a variety of mammalian cells and organs exposed to benzo[*a*]pyrene.

(*c*) *Diol epoxides as ultimate carcinogens of PAH*

The current interest in diol epoxides as ultimate carcinogens stems from the initial observation (Sims *et al.*, 1974) that a 7,8-diol-9,10-epoxide is the intrinsically active metabolite responsible for the covalent binding of benzo[*a*]pyrene to DNA. It next remained to be determined whether diol epoxides were the general metabolically activated forms of PAH. If this were the case, it was also important to determine which diol epoxide(s) mediated the biological activity, since many PAH can form more than one diol epoxide. The unique structural features of the diol epoxide metabolite were proposed on the basis of structure-activity relationships of benzo[*a*]pyrene epoxides, a re-evaluation of published tumorigenicity data and quantum mechanical calculations (Jerina *et al.*, 1976). It was predicted that benzo-ring diol epoxides, in which the epoxide forms part of the bay region of the hydrocarbon molecule, should have high chemical reactivity, high biological activity and, if metabolically formed, would be likely candidates as ultimate carcinogenic metabolites of PAHs. A bay region occurs in a PAH when an angularly-fused benzo-ring is present; for example, the sterically-hindered region between the 10- and 11-positions constitutes the bay region of benzo[*a*]pyrene. To date, mutagenicity, DNA binding, cell transformation, metabolism and tumorigenicity studies have indicated that bay-region diol epoxides of about a dozen PAHs are ultimate carcinogenic metabolites. Furthermore, metabolic studies with hydrocarbons such as benzo[*e*]pyrene (MacLeod *et al.*, 1979; Wood *et al.*, 1979; MacLeod *et al.*, 1980; Thakker *et al.*, 1981) have indicated that minimal metabolic formation of the bay-region diol epoxide is consistent with the weak biological activity of the parent hydrocarbon.

It should be emphasized that the demonstration of high biological activity of a metabolically formed bay-region diol epoxide does not preclude the existence of other ultimately carcinogenic metabolites of PAHs, and that not all carcinogenic hydrocarbons have formal bay regions.

(*d*) *Metabolism of other PACs*

Summaries of the metabolism of other PACs are given in the individual monographs. In addition to the classes of metabolites formed from unsubstituted hydrocarbons, such as benzo[*a*]pyrene, alkyl-substituted hydrocarbons undergo oxidation of their side chains to form hydroxy-alkyls, which may be further oxidized to the corresponding acids and/or conjugates. Aza-arenes can also potentially undergo *N*-oxidation, although the limited studies performed to date do not indicate that this is a major pathway.

Information on the biological activity of known in-vivo or in-vitro metabolites is also summarized in the individual monographs for each hydrocarbon. A number of potential polynuclear hydrocarbon metabolites have also been synthesized and tested for biological activity, including tumorigenicity; but, although these compounds have provided useful insights into the mechanism of hydrocarbon activation and detoxification, no attempt has been made to include information on their biological effects.

(*e*) *Prenatal metabolism*

In general it seems that, in animals, the development of metabolizing enzymes occurs rather late in gestation. They can be induced to develop earlier by exposure to appropriate substrates, although usually to a lesser degree than in adults. Treatment of rats on day 18 of gestation with benzo[*a*]pyrene can induce fetal liver aryl hydrocarbon hydroxylase, and treatment from day 13 can induce it in the placenta. Similar findings have been made in mice (reviewed by Tomatis, 1974). Diastereoisomers of the 3,4-diol-1,2-epoxide of 7,12-dimethylbenz[*a*]anthracene formed adducts with maternal and fetal DNA following intravenous administration of the parent compound to pregnant rats (Doerjer *et al.*, 1978).

5. METABOLISM AND DISTRIBUTION OF POLYNUCLEAR AROMATIC COMPOUNDS IN HUMAN TISSUES

Benzo[*a*]pyrene has been the prototype polynuclear aromatic hydrocarbon for studies of the mechanism of chemical carcinogenesis in human tissue. The ready availability of synthetic benzo[*a*]pyrene metabolites has allowed a burgeoning literature on all aspects of in-vitro and in-vivo metabolism, mutagenesis and carcinogenesis. However, the metabolism of the majority of compounds included in this monograph has not been studied in human systems.

The metabolism of benz[*a*]anthracene has been studied in homogenates of human lung (Grover *et al.*, 1973) and bronchial tissue (Pal *et al.*, 1975). Analysis, carried out by thin-layer chromatography, showed that dihydrodiol and phenol metabolites were the major products. This is indicative of the same biochemical activation *via* oxide intermediates.

The metabolism of benzo[*a*]pyrene has been examined extensively in human cells, explant cultures, tissue homogenates and with microsomal preparations. Table 9 is a representative list of studies of the metabolism of benzo[*a*]pyrene in human tissues, which included organic solvent-soluble metabolites and water-soluble conjugates. The results illustrate the qualitative similarities of the metabolites produced by different human tissues. Furthermore, the metabolites are the same as those formed in a variety of animal tissues. However, there are quantitative variations among tissues, and relative resistance or susceptibility may be influenced by the rate at which a specific cell or tissue forms and detoxifies the reactive intermediates.

Table 9. Metabolism of benzo[*a*]pyrene by human tissues

Metabolizing system[a]	Dihydrodiols	Phenols	Quinones	Tetraols	Types of water-soluble derivative	References
Bladder (T)	4,5-; 7,8-; 9,10-	3-; 9-	Quinones	+	Glucuronides	Selkirk *et al.* (1983)
Bronchus (T)	4,5-; 7,8-; 9,10-	3-				Pal *et al.* (1975)
Bronchus epithelium (T)	4,5-; 7,8-; 9,10-[b]	3-		+	Sulphates (little)	Cohen *et al.* (1976)

Metabolizing system[a]	Dihydrodiols	Phenols	Quinones	Tetraols	Types of water-soluble derivative	References
Bronchus (T)	4,5-; 7,8-; 9,10-[b]	3-; 9-	1,6-; 3,6-; 6,12-	+		Harris *et al.* (1976)
Bronchus epithelium (T)	7,8-; 9,10-[b]	3-	Quinones	+		Autrup *et al.* (1978a)
Bronchus (T)	4,5-; 7,8-; 9,10-[b]	3-; 9-	Quinones	+	Sulphates, glucuronides, glutathione conjugates	Autrup *et al.* (1980a)
Bronchus (T)	4,5-; 7,8-; 9,10-	3-; 9-	Quinones	+	Glucuronides	Selkirk *et al.* (1983)
Colon (T)	4,5-; 7,8-; 9,10-[b]	3-	1,6-; 3,6-; 6,12-	+		Autrup *et al.* (1978b)
Colon (T)	4,5-; 7,8-; 9,10-[b]	3-; 9-	Quinones	+	Sulphates, glucuronides, glutathione conjugates	Autrup (1979)
Colon (T)	4,5-; 7,8-; 9,10-[b]	3-	Quinones	+	Sulphates, glucuronides, glutathione conjugates	Autrup *et al.* (1978b, 1980b)
Endometrium (T)	4,5-; 7,8-; 9,10-	3-; 9-	1,6-; 3,6-; 6,12-	+	Sulphates	Mass *et al.* (1981)
Oesophagus (T)	4,5-; 7,8-; 9,10-	3-; 9-	Quinones	+	Glucuronides	Selkirk *et al.* (1983)
Oesophagus	4,5-; 7,8-; 9,10-[b]	3-; 9-	Quinones	+	Sulphates, glucuronides, glutathione conjugates	Autrup (1980b)
Oesophagus (T)	4,5-; 7,8-; 9,10-[b]	3-; 9-	1,6-; 3,6-; 6,12-	+	Sulphates, glucuronides, glutathione conjugates	Harris *et al.* (1979)
Fibroblasts (C)	7,8-; 9,10-					Baird & Diamond (1978)
Hepatoma (C)	7,8-; 9,10-	3-	Quinones			Diamond *et al.* (1980)
Keratinocytes (epidermal) (C)	4,5- (negligible amount); 7,8-; 9,10-	3-; 9-	Quinones	+	Glucuronides	Kuroki *et al.* (1980)
Kidney (M)	4,5-; 7,8-; 9,10-	3-; 9-	Quinones			Prough *et al.* (1979)
Liver (M)	4,5-; 7,8-; 9,10-	3-; 9-	1,6-; 3,6-; 6,12-			Selkirk *et al.* (1975)
Liver (M)	4,5-; 7,8-; 9,10-	3-; 9-	Quinones			Prough *et al.* (1979)
Liver (H)	4,5-; 7,8-; 9,10-	3-; 9-	3,6-			Pelkonen *et al.* (1977)
Lung (T)	4,5- (small quantity); 7,8-; 9,10-	3- (little)			Sulphates	Cohen *et al.* (1976)
Lung (T)	7,8-; 9,10-	3-	3,6-; 6,12-	+		Stoner *et al.* (1978)
Lung (T)	4,5- & 7,8- (traces); 9,10-[b]	3-; 9-		+	Sulphates	Mehta & Cohen (1979)
Lung (T)	7,8-; 9,10-[b]	3-; 9- (only in tissue, not in medium)	Quinones (small amounts)	+	Sulphates	Mehta *et al.* (1979)
Lung (M)	4,5-; 7,8-; 9,10-	3-; 9-	Quinones			Prough *et al.* (1979)
Lung (M)	4,5-; 7,8-; 9,10-	Phenols	Quinones			Šipal *et al.* (1979)
Lung (S)	4,5-; 7,8-; 9,10-	3-; 9-	Quinones			Sabadie *et al.* (1981)
Lymphocytes (C)	4,5-; 7,8-; 9,10-	3-				Booth *et al.* (1974)

Metabolizing system[a]	Dihydrodiols	Phenols	Quinones	Tetraols	Types of water-soluble derivative	References
Lymphocytes (C)	4,5-; 7,8-; 9,10-	3-; 9-	1,6-; 3,6-; 6,12-			Selkirk *et al.* (1975)
Lymphocytes (C)	7,8-; 9,10- (probably)	3-; 9-	Quinones			Vaught *et al.* (1978)
Lymphocytes (C)	4,5-; 7,8-; 9,10- (traces)	3-; 7-; 9-	1,6-; 3,6-			Okano *et al.* (1979)
Lymphocytes (C)	7,8-; 9,10- (not detected with certitude)	3-; 9-	Quinones			Gurtoo *et al.* (1980)
Macrophages (C)	7,8-; 9,10-[b]	3- (traces)	Quinone (very low levels)	+	Glucuronides, sulphates	Autrup *et al.* (1978a)
Macrophages (C)	7,8-; 9,10-[b]	3-; 9- (traces; low levels)	Quinones (low levels)	+		Harris *et al.* (1978)
Macrophages (C)	4,5-; 7,8-; 9,10-	3-; 9-	1,6-; 3,6-; 6,12-	+		Autrup *et al.* (1979)
Macrophages (C)	4,5-; 7,8-; 9,10-	3-; 9-	Quinones		Glucuronides, sulphates	Marshall *et al.* (1979)
Mammary epithelium (C)	4,5-; 7,8-; 9,10-					Grover *et al.* (1980)
Mammary epithelium (C)	7,8-					MacNicoll *et al.* (1980)
Mammary epithelium (C)	4,5-; 7,8-; 9,10-	Phenols		+	Sulphates, glucuronides, glutathione conjugates	Bartley *et al.* (1982)
Mammary fibroblasts (C)	4,5-; 7,8-; 9,10-	Phenols		+	Sulphates, glucuronides, glutathione conjugates	Bartley *et al.* (1982)
Monocytes (C)	7,8-; 9,10- (probably)	3-; 9-	Quinones			Vaught *et al.* (1978)
Monocytes (C)	4,5-; 7,8-; 9,10- (traces)	3-; 7-; 9-	1,6-; 3,6-	+		Okano *et al.* (1979)
Placenta (M)	4,5-; 7,8-; 9,10-	Phenols				Namkung & Juchau (1980)
Placenta (M)	7,8-; 4,5- & 9,10- (very low levels)	3-; 9-	Quinones			Pelkonen & Saarni (1980)
Placenta (M)	7,8-	Phenols				Gurtoo *et al.* (1983)
Scalp hair follicles						Vermorken *et al.* (1979)
(T)	7,8-; 9,10-	3-				
(C)	7,8-; 9,10-	3-				
Skin epithelium (C)	7,8-; 9,10-	3-; 9-	1,6-; 3,6-; 6,12-			Fox *et al.* (1975)
Skin fibroblasts (C)	7,8-; 9,10-	3-; 9- (small amounts)				Fox *et al.* (1975)
Skin (T)	4,5-; 7,8-; 9,10-	3-; 9-	Quinones	+	Glucuronides	Selkirk *et al.* (1983)
Tracheobronchial tissue or trachea + bronchus (T)	4,5-; 7,8-; 9,10-[b]	3-; 9-	Quinones	+	Sulphates, glucuronides (low levels), glutathione conjugate	Autrup *et al.* (1980a)
Trachea (T)	4,5-; 7,8-; 9,10-[b]	3-; 9-	Quinones	+	Sulphates, glucuronides, glutathione conjugates	Autrup *et al.* (1980b)

[a] Abbreviations: C, cells in culture; H, homogenate; M, microsomes; T, tissue explants in organ culture; S, supernatant fractions

[b] Benzo[*a*]pyrene-9,10-diol was known to co-elute with benzo[*a*]pyrene-triol and/or tetraol derivatives.

The metabolic profiles reported from work with human tissues are in almost all cases identical to those obtained with other eukaryotes, indicating the involvement of similar enzyme systems. The same profile of reactive electrophilic intermediates found in other experimental systems appears to be formed in human tissues.

The hydrocarbon-nucleic acid adducts found in human tissues exposed to PAHs are essentially the same as those that have been found in animal tissues. It is also important to note that quantitative variations in metabolic activation and detoxification occur between genetically heterogeneous populations. For example, human bronchial explant cultures show wide inter-individual variations in the amount of benzo[*a*]pyrene that becomes bound to DNA (Harris *et al.*, 1976). As another example, benzo[*a*]pyrene in blood is bound to serum lipoproteins (Busbee *et al.*, 1982). Individual variations in carcinogen transport by blood lipoproteins may, in turn influence the rate of tissue deposition and therefore the overall metabolism. These results and the evidence (Table 9) that a large number of human tissues can metabolize benzo[*a*]pyrene indicate that all human tissues are capable of producing carcinogenic, reactive electrophiles.

In addition to these studies of metabolism, there have been several attempts to determine the levels of PAHs in both normal human tissues and human tumours. Gräf (1970) and Gräf *et al.* (1975) made extensive measurements of benzo[*a*]pyrene levels in many normal tissues, including liver, spleen, kidney, heart and skeletal muscle. Samples were taken at autopsy from people of ages ranging from birth to the fifth decade. Measurement was by ultraviolet spectroscopy, and results were reported as μg/100 g dry-tissue weight. The overall tissue average in these organs was 0.32 μg/100 g except in lungs and in tissues with high cellular proliferative activity, such as exocrine and endocrine glands and bone marrow, where it was 0.2 μg/100 g.

A second study of PAH distribution was performed by Obana *et al.* (1981), in which cancer-free liver and fat from six individuals were assayed for anthracene, benz[*a*]anthracene, benzo[*b*]fluoranthene, benzo[*k*]fluoranthene, benzo[*ghi*]perylene, benzo[*a*]pyrene, benzo[*e*]pyrene, dibenz[*a*,*h*]anthracene and pyrene. Identification and quantification were made by co-chromatography with authentic standards. Positive identification was made for pyrene, anthracene, benzo[*b*]fluoranthene, benzo[*ghi*]perylene, benzo[*k*]fluoranthene, and benzo[*a*]pyrene (which are listed in the order of their abundance). The results were reported, however, as ng/kg wet weight of tissue and are not directly comparable to the results of Gräf (1970) and Gräf *et al.* (1975).

Tomingas *et al.* (1976) analysed 24 bronchial carcinoma samples, taken during surgery or autopsy, for the presence of benz[*a*]anthracene, benzo[*b*]fluoranthene, benzo[*k*]fluoranthene, benzo[*ghi*]perylene, benzo[*a*]pyrene, benzo[*e*]pyrene, chrysene, coronene, dibenz[*a*,*h*]anthracene, fluoranthene, perylene and pyrene. Analysis was by thin-layer chromatography and fluorescence spectroscopy. Only benzo[*b*]fluoranthene, benzo[*a*]pyrene, fluoranthene and perylene were detected; the levels of the other eight PAHs were below the limits of detection of the analytical system. Benzo[*a*]pyrene was present in all samples, but the other three polycyclics were found in only some of the samples. Tissue averages of benzo[*a*]pyrene were 3.5 μg/g carcinoma and 0.09 μg/g tumour-free tissue (no dry or wet weight given).

In humans fetal enzymes seem to appear at a relatively earlier stage of development than in experimental animals. Significant levels of xenobiotic oxidizing enzymes appear in the liver and adrenal glands of the human fetus during the late embryonic period - from about 6-7 weeks gestation - reaching constant levels by 12-14 weeks (Pelkonen, 1979). The fetal levels of benzo[*a*]pyrene hydroxylase are only a low percentage of the adult values and do not appear to be inducible; however, the human placenta can show marked induction if the mother smokes, and in heavy smokers the levels of aryl hydrocarbon hydroxylase are close to the levels in adult liver (Pelkonen, 1979), although marked variations have been observed (Conney, 1982).

6. MOUSE SKIN ASSAY FOR CARCINOGENICITY

PAHs were the first class of compounds demonstrated to be carcinogenic in experimental animals. Benzo[*a*]pyrene and dibenz[*a,h*]anthracene were shown to produce tumours following repeated application to the skin of mice (IARC, 1973).

Although a variety of routes of administration (subcutaneous injection, intratracheal instillation, inhalation and oral) have since been used for the study of PAHs in experimental animals, an overwhelming number of studies have been performed using the induction of skin tumours as the biological end-point. Recommended protocols for carcinogenicity studies in experimental animals are outlined in Supplement 2 to the *IARC Monographs* (IARC, 1980); however, further details of the mouse-skin bioassay systems are given below. It will be remarked that virtually all mouse-skin application studies involve only female mice. This is due to the fact that males tend to fight more than females and thus to damage each others' skin.

Two types of mouse-skin bioassay are used to test the carcinogenicity of individual PAHs. In one, the test compound is applied repeatedly to the mouse skin for a long period (often for life). In the second form of the test, the compound is applied as a single (or a limited number of) dose(s); this is followed by a longer period of chemical or physical treatment with another non-carcinogenic factor, such as croton oil or phorbol esters. The latter test is known as the mouse-skin initiation-promotion model. In the first form of the test, the production of significant numbers of skin tumours indicates that a compound is a carcinogen. In the second form of the test, the production of significant numbers of tumours indicates that the compound is a tumour-initiating or tumour-promoting agent.

Testing for carcinogenicity on mouse skin typically involves application of the compound in an organic solvent ('vehicle') to the shaved skin of the backs of mice. Use of a brush for this purpose ('skin painting') does not permit quantitation of the dose and has been surplanted by more precise dispensing procedures. Generally, doses of the test compound are applied during a long period, which may extend to the animals' life-time.

The mouse-skin assay system allows continual observation of tissue exposed immediately to the test compound. However, criteria for observing and reporting results may vary. Tumours may be deemed to be lesions having certain macroscopic appearance, having reached a particular size and/or exhibiting particular histological characteristics. Often, tumour production is reported in the absence of histological confirmation. Some ambiguity may exist when lesions are inadequately described ('skin tumours') and especially for those with short observation times. Some papillomas are reversible lesions. In most experiments, examination of test animals has been restricted to the exposed skin; necropsy and examination of internal organs were rarely undertaken.

Considerations germane to assessment of carcinogen bioassay data in general (see 'Preamble') apply to mouse skin studies. Thus, the period for which animals are observed and the number of animals surviving treatment are critical to assessment of the bioassay. 'Vehicle' controls are often used, although controls may be subjected to skin shaving only .

A range of parameters is known to influence the outcome of mouse-skin assays. The selection of mouse strain may alter tumour production and latency because of differences in strain sensitivities and variations in cutaneous anatomy. The efficiency of carcinogenic induction is dependent on the stage of the hair-growth cycle. Sex and age of animals affect tumour yield. In respect of the test compound, significant variables include the purity, the dose and frequency of application as well as the solvent employed. Complete details regarding the foregoing parameters and procedures may be lacking in individual reports (particularly in early studies) and, where appropriate, such deficiencies are noted in the respective monographs.

Tumours produced by PAHs in mouse skin are benign; some may regress, and some progress to invasive carcinomas. Data from mouse-skin assays may contribute to *'sufficient'* evidence of carcinogenicity (see 'Preamble'). Certain PAHs, initially established as carcinogenic by application to mouse skin, have been shown to produce malignant tumours at other sites (forestomach, lung) following administration to experimental animals by other routes. Available data are consistent in showing that the mechanisms of carcinogenesis in mouse skin are common to those understood to be operative for tumour induction in other mammalian tissues.

Mouse-skin assays have also been performed in which PAHs have been administered in combination. PAHs used simultaneously are often referred to as 'cocarcinogens' if a positive effect on tumour yield is observed. The role of cocarcinogens, in, for example, mediating penetration of the epidermis, metabolic activation or degradation complicates evaluation of this type of study.

7. SHORT-TERM TESTS

In order to evaluate the activity of a chemical in short-term tests, a number of parameters, including toxicity to the indicator cells, should be taken into consideration. Most studies reviewed in this volume, using assays in yeast, mammalian cells and bacteria, scored for mutation, recombination or morphological transformation. The results were reported as frequency of the event among surviving cells. In the plate incorporation assay in *Salmonella typhimurium* his^-/his^+, the numbers of surviving mutant colonies were usually reported without accounting specifically for possible killing of the cells by the agent tested. Negative results obtained using this protocol may contrast with positive results obtained in the same organism when cell killing is taken into account.

Reports considered deficient in reporting or design were not cited, except when no other report was available; in such cases, critical comments were included in the summary table. Data generated with the baby hamster kidney (BHK) transformation assay (Styles, 1977; Purchase *et al.* 1978) and in an assay measuring the suppression of sebaceous glands on mouse skin were not considered (Bock & Mund, 1958; Lazar *et al.*, 1963).

8. REFERENCES

Adams, J.D., Menzies, K. & Levins, P. (1977) *Selection and Evaluation of Sorbent Resins for the Collection of Organic Compounds* (*EPA-600/7-77-044*), Washington DC, US Environmental Protection Agency, Office of Research and Development

Aitio, A. (1974) Different elimination and effect on mixed function oxidase of 20-methylcholanthrene after intragastric and intraperitoneal administration. *Res. Commun. chem. Pathol. Pharmacol.*, *9*, 701-710

Albagli, A., Oja, H. & Dubois, L. (1974) Size-distribution pattern of polycyclic aromatic hydrocarbons in airborne particulates. *Environ. Lett.*, *6*, 241-251

Andelman, J.B. & Suess, M.J. (1970) Polynuclear aromatic hydrocarbons in the water environment. *Bull. World Health Org.*, *43*, 479-508

Andelman, J.B. & Snodgrass, J.E. (1974) Incidence and significance of polycyclic aromatic hydrocarbons in the water environment. *CRC crit. Rev. environ. Control*, *4*, 68-83

Archer, S.R., Blackwood, T.R. & Wilkins, G.E. (1979) *Status Assessment of Toxic Chemicals: Polynuclear Aromatic Hydrocarbons* (*EPA-600/2-79-210L*), Cincinnati, OH, US Environmental Protection Agency

Autrup, H., Harris, C.C., Stoner, G.D., Selkirk, J.K., Schafer, P.W. & Trump, B. (1978a) Metabolism of [^{3}H]benzo[*a*]pyrene by cultured human bronchus and cultured human pulmonary alveolar macrophages. *Lab. Invest.*, *38*, 217-224

Autrup, H., Harris, C.C., Trump, B.F. & Jeffrey, A.M. (1978b) Metabolism of benzo(*a*)pyrene and identification of the major benzo(*a*)pyrene-DNA adducts in cultured human colon. *Cancer Res.*, *38*, 3689-3696

Autrup, H., Harris, C.C., Schafer, P.W., Trump, B.F., Stoner, G.D. & Hsu, I.C. (1979) Uptake of benzo[*a*]pyrene-ferric oxide particulates by human pulmonary macrophages and release of benzo[*a*]pyrene and its metabolites. *Proc. Soc. exp. Biol. Med.*, *161*, 280-284

Autrup, H., Wefald, F.C., Jeffrey, A.M., Tate, H., Schwartz, R.D., Trump, B.F. & Harris, C.C. (1980a) Metabolism of benzo[*a*]pyrene by cultured tracheobronchial tissues from mice, rats, hamsters, bovines and humans. *Int. J. Cancer*, *25*, 293-300

Autrup, H., Jeffrey, A.M. & Harris, C.C. (1980b) *Metabolism of benzo(*a*)pyrene in cultured human bronchus, trachea, colon, and esophagus*. In: Bjørseth, A. & Dennis, A.J., eds, *Polynuclear Aromatic Hydrocarbons: Chemistry and Biological Effects, 4th Int. Symposium*, Colombus, OH, Battelle Press, pp. 89-105

Ayres, C.I. & Thornton, R.E. (1965) Determination of benzo(a)pyrene and related compounds in cigarette smoke. *Beitr. Tabakforsch.*, *3*, 285-290

Baird, W.M. & Diamond, L. (1978) Metabolism and DNA binding of polycyclic aromatic hydrocarbons by human diploid fibroblasts. *Int. J. Cancer*, *22*, 189-195

Bartley, J., Bartholomew, J.C. & Stampfer, M.B. (1982) Metabolism of benzo[*a*]pyrene by human epithelial and fibroblastic cells: Metabolite patterns and DNA adduct formation. *J. cell. Biochem.*, *18*, 135-148

Begeman, C.R. & Colucci, J.M. (1968) Benzo[*a*]pyrene in gasoline partially persists in automobile exhaust. *Science*, *161*, 271

Begeman, C.R. & Colucci, J.M. (1970) *Polynuclear Aromatic Hydrocarbon Emissions from Automotive Engines* (*No. 700469*), New York, Society of Automotive Engineers

Behn, U. & Grimmer, G. (1981) *Bestimmung polycyclischer aromatischer Kohlenwasserstoffe (PAH) im Abgas einer Zentralheizungsanlage auf Heizöl EL-Basis* [Determination of polycyclic aromatic hydrocarbons (PAH) in the flue gas of an oil-fired heating unit] (*BMI-DGMK-Projekt 111-02*), Hamburg, Deutsche Gesellschaft für Mineralölwissenschaft und Kohlechemie

Behn, U., Meyer, J.-P. & Grimmer, G. (1980) *Entwicklung und Erprobung einer Methode zur Sammlung und Analyse von polycyclischen aromatischen Kohlenwasserstoffen (PAH) aus dem Abgas ölbefeuerter Zentralheizungsanlagen* [Development and testing of a method for collection and analysis of polycyclic aromatic hydrocarbons (PAH) from flue gas of oil-fired heating units] (*BMI-DGMK-Projekt 111-01*), Hamburg, Deutsche Gesellschaft für Mineralölwissenschaft und Kohlechemie

Beine, H. (1970) About the content of 3,4-benzopyrene in the flue gas of domestic stoves with solid fuel (Ger.). *Staub-Reinhalt. Luft*, *30*, 334-336

Bentley, H.R. & Burgan, J.G. (1960) Polynuclear hydrocarbons in tobacco and tobacco smoke. II. The origin of 3:4 benzopyrene found in tobacco and tobacco smoke. *Analyst*, *85*, 723-727

Bickers, D.R. & Kappas, A. (1978) Human skin aryl hydrocarbon hydroxylase. *J. clin. Invest.*, *62*, 1061-1068

Biernoth, G. & Rost, H.E. (1967) The occurrence of polycyclic aromatic hydrocarbons in coconut oil and their removal. *Chem. Ind.*, 25 November, 2002-2003

Biernoth, G. & Rost, H.E. (1968) Occurrence of polycyclic aromatic hydrocarbons in edible oils and their range (Ger.). *Arch. Hyg. (München), 152,* 238-250

Bingham, E., Trosset, R.P. & Warshawsky, D. (1980) Carcinogenic potential of petroleum hydrocarbons. A critical review of the literature. *J. environ. Pathol. Toxicol.*, *3*, 483-563

Bjørseth, A. (1983) *Handbook of Polycyclic Aromatic Hydrocarbons*, New York, Marcel Dekker

Blumer, M. (1961) Benzpyrenes in soil. *Science*, *134*, 474-475

Bock, F.G. & Dao, T.L. (1961) Factors affecting the polynuclear hydrocarbon level in rat mammary glands. *Cancer Res.*, *21*, 1024-1029

Bock, F.G. & Mund, R. (1958) A survey of compounds for activity in the suppression of mouse sebaceous glands. *Cancer Res.*, *18*, 887-892

Bonnet, J. & Neukomm, S. (1956) On the chemical composition of tobacco smoke. I. Analysis of the neutral fraction (Fr.). *Helv. chim. Acta*, *39*, 1724-1733

Booth, J., Keysell, G.R., Pal, K. & Sims, P. (1974) The metabolism of polycyclic hydrocarbons by cultured human lymphocytes. *FEBS Lett.*, *43*, 341-344

Borneff, J. & Fischer, R. (1962) Carcinogenic substances in water and soil. IX. Investigation on filter deposits of a sea water supply plant for polycyclic aromatic hydrocarbons (Ger.). *Arch. Hyg. Bakt.*, *146*, 183-197

Borneff, J. & Kunte, H. (1963) Carcinogenic substances in water and soil. XIV. New studies on polycyclic aromatic hydrocarbons in soils (Ger.). *Arch. Hyg. Bakt.*, *147*, 401-409

Borneff, J. & Kunte, H. (1979) *Method 1. Analysis of polycyclic aromatic hydrocarbons in water using thin layer chromatography and spectrofluorometry*. In: Egan, H., Castegnaro, M., Bogovski, P., Kunte, H. & Walker, E.A., eds, *Environmental Carcinogens - Selected Methods of Analysis*, Vol. 3, *Analysis of Polycyclic Aromatic Hydrocarbons in Environmental Samples* (*IARC Scientific Publications No. 29*), Lyon, International Agency for Research on Cancer, pp. 129-139

Brockhaus, A. & Tomingas, R. (1976) Emission of polycyclic hydrocarbons during burning processes in small heating installations and their concentration in the atmosphere (Ger.). *Staub-Reinhalt. Luft*, *36*, 96-101

Brune, H.F.K. (1977) *Experimental results with percutaneous applications of automobile exhaust condensates in mice*. In: Mohr, U., Schmähl, D. & Tomatis, L., eds, *Air Pollution and Cancer in Man* (*IARC Scientific Publications No. 16*), Lyon, International Agency for Research on Cancer, pp. 41-47

Brune, H., Habs, M. & Schmähl, D. (1978) The tumor-producing effect of automobile exhaust condensate and fractions thereof. II. Animal studies. *J. environ. Pathol. Toxicol.*, *1*, 737-746

Brunnemann, K.D., Hoffmann, D., Wynder, E.L. & Gori, G.B. (1976) *Determination of tar, nicotine, and carbon monoxide in cigarette smoke. A comparison of international smoking conditions of chemical studies on tobacco smoke*. In: Wynder, E.L., Hoffmann, D. & Gori, G.B., eds, *Smoking and Health. I. Modifying the Risk for the Smoker* (*DHEW Publ. No. (NIH) 76-1221*), Washington DC, US Government Printing Office, pp. 441-449

Busbee, D.L., Rankin, P.W., Payne, D.M. & Jasheway , D.W. (1982) Binding of benzo[a]pyrene and intracellular transport of a bound electrophilic benzo[a]pyrene metabolite by lipoproteins. *Carcinogenesis*, *3*, 1107-1112

Butler, J.D. & Crossley, P. (1981) Reactivity of polycyclic aromatic hydrocarbons adsorbed on soot particles. *Atmos. Environ.*, *15*, 91-94

Caldwell, J. (1982) Conjugation reactions in foreign-compound metabolism: Definition, consequences and species variations. *Drug Metab. Rev.*, *13*, 745-777

Campbell, J.M. & Lindsey, A.J. (1957) Polycyclic hydrocarbons in cigar smoke. *Br. J. Cancer*, *11*, 192-195

Cantuti, V., Cartoni, G.P., Liberti, A. & Torri, A.G. (1965) Improved evaluation of polynuclear hydrocarbons in atmospheric dust by gas chromatography. *J. Chromatogr.*, *17*, 60-65

Cautreels, W. & Van Cauwenberghe, K. (1977) Fast quantitative analysis of organic compounds in airborne particulate matter by gas chromatography with selective mass spectrometric detection. *J. Chromatogr.*, *131*, 253-264

Cautreels, W. & Van Cauwenberghe, K. (1978) Experiments on the distribution of organic pollutants between airborne particulate matter and the corresponding gas phase. *Atmos. Environ.*, *12*, 1133-1141

Chipman, J.K., Hirom, P.C., Frost, G.S. & Millburn, P. (1982) *Benzo[a]pyrene metabolism and enterohepatic circulation in the rat*. In: Snyder, R., Parke, D.V., Kocsis, J.J., Jollow, D.J., Gibson, C.G. & Witmer, C.M., eds, *Biological Reactive Intermediates*, II. *Chemical Mechanisms and Biological Effects*, Part A, New York, Plenum Press, pp. 761-768

Ciusa, W., Nebbia, G., Buccelli, A. & Volpones, E. (1965) Research on polycyclic aromatic hydrocarbons present in olive oil (Ital.). *Riv. ital. Sost. grasse*, *42*, 175-179

Cohen, G.M., Haws, S.M., Moore, B.P. & Bridges, J.W. (1976) Benzo(*a*)pyren-3-yl hydrogen sulphate, a major ethyl acetate-extractable metabolite of benzo(*a*)pyrene in human, hamster and rat lung cultures. *Biochem. Pharmacol.*, *25*, 2561-2570

Commission of the European Communities (1979) *Concerted Action. Analysis of Organic Micropollutants in Water* (*COST 64b bis*), 3rd ed., Vol. II, Luxembourg

Commission of the European Communities (1980) Council Directive of 15 July 1980 relating to the quality of water intended for human consumption. *Off. J. Eur. Communities*, *L229*, 11-29

Committee on Biological Effects of Atmospheric Pollutants (1972) *Particulate Polycyclic Organic Matter*, Washington DC, National Academy of Sciences

Conney, A.H. (1967) Pharmacological implications of microsomal enzyme induction. *Pharmacol. Rev.*, *19*, 317-366

Conney, A.H. (1982) Induction of microsomal enzymes by foreign chemicals and carcinogenesis by polycyclic aromatic hydrocarbons: G.A.H. Clowes Memorial Lecture. *Cancer Res.*, *42*, 4875-4917

Cook, J.W., Carruthers, W. & Woodhouse, D.L. (1958) Carcinogenicity of mineral oil fractions. *Br. med. Bull.*, *14*, 132-135

Cooke, M., Dennis, A.J. & Fisher, G.L., eds (1982) *Polynuclear Aromatic Hydrocarbons - Physical and Biological Chemistry, 6th Int. Symposium*, Columbus, OH, Battelle Press

Cooper, C.S., Grover, P.L. & Sims, P. (1983) *The metabolism and activation of benzo[*a*]pyrene.* In: Bridges, J.W. & Chaseand, L.F., eds, *Progress in Drug Metabolism*, Vol. 7, New York, John Wiley & Sons, pp. 295-395

Cooper, J.A. (1980) Environmental impact of residential wood combustion emissions and its implications. *J. Air Pollut. Control Assoc.*, *30*, 855-861

Creasia, D.A., Poggenburg, J.K., Jr & Nettesheim, P. (1976) Elution of benzo[*a*]pyrene from carbon particles in the respiratory tract of mice. *J. environ. Health*, *1*, 967-975

Cuffe, S.T., Gerstle, R.W., Orning, A.A. & Schwartz, C.H. (1964) Air pollutant emissions from coal-fired power plants. Report No. 1. *J. Air Pollut. Control Assoc.*, *14*, 353-362

Daisey, J.M. & Leyko, M.A. (1979) Thin-layer gas chromatographic method for the determination of polycyclic aromatic and aliphatic hydrocarbons in airborne particulate matter. *Anal. Chem.*, *51*, 24-26

Davies, I.W., Harrison, R.M., Perry, R., Ratnayaka, D & Wellings, R.A. (1976) Municipal incinerator as source of polynuclear aromatic hydrocarbons in environment. *Environ. Sci. Technol.*, *10*, 451-453

Davis, B.R., Whitehead, J.K., Gill, M.E., Lee, P.N., Butterworth, A.D. & Roe, F.J.C. (1975) Response of rat lung to tobacco smoke condensate or fractions derived from it administered repeatedly by intratracheal instillation. *Br. J. Cancer*, *31*, 453-461

DePierre, J.W. & Ernster, L. (1978) The metabolism of polycyclic hydrocarbons and its relationship to cancer. *Biochim. biophys. Acta*, *473*, 149-186

Diamond, L., Kruszewski, F., Aden, D.P., Knowles, B.B. & Baird, W.M. (1980) Metabolic activation of benzo[*a*]pyrene by a human hepatoma cell line. *Carcinogenesis*, *1*, 871-875

Diehl, E.K., du Breuil, F. & Glenn, R.A. (1967) Polynyclear hydrocarbon emission from coal-fired installations. *J. Eng. Power*, *89*, 276-282

Dipple, A. (1976) *Polynuclear aromatic carcinogens*. In: Searle, C.E., ed., *Chemical Carcinogens* (*ACS Monograph No. 173*), Washington DC, American Chemical Society, pp. 245-314

Doerjer, G., Diessner, H., Bücheler, J. & Kleihues, P. (1978) Reaction of 7,12-dimethylbenz(*a*)anthracene with DNA of fetal and maternal rat tissues *in vivo*. *Int. J. Cancer*, *22*, 288-291

Dube, M.F. & Green, C.R. (1982) Methods of collection of smoke for analytical purposes. *Recent Adv. Tobacco Sci.*, *8*, 42-102

Egan, H., Castegnaro, M., Bogovski, P., Kunte, H. & Walker, E.A., eds (1979) *Environmental Carcinogens. Selected Methods of Analysis*, Vol. 3, *Analyses of Polycyclic Aromatic Hydrocarbons in Environmental Samples* (*IARC Scientific Publications No. 29*), Lyon, International Agency for Research on Cancer

Eiceman, G.A., Clement, R.F. & Karasek, F.W. (1979) Analysis of fly ash from municipal incinerators for trace organic compounds. *Anal. Chem.*, *51*, 2343-2350

Ellington, J.J., Schlotzhauer, P.F. & Schepartz, A.I. (1978) Quantitation of hexane-extractable lipids in serial samples of flue-cured tobaccos. *J. Food agric. Chem.*, *26*, 270-273

Elliot, L.P. & Rowe, D.R. (1975) Air quality during public gatherings. *J. Air Pollut. Control Assoc.*, *25*, 635-363

Elmenhorst, H. & Stadler, L. (1967) Problem of mechanical pipe smoke (Ger.). *Beitr. Tabakforsch.*, *4*, 21-28

Emerole, G.O. (1980) Carcinogenic polycyclic aromatic hydrocarbons in some Nigerian food. *Bull. environ. Contam. Toxicol.*, *24*, 641-646

Engst, R. & Fritz, W. (1975) Contamination of food with carcinogenic hydrocarbons of environmental origin. *Rocz. Panst. Zakl. Hig.*, *26*, 113-118

Fábián, B. (1968) Carcinogenic substances in edible fats and oils. IV. Studies on margarine, vegetable shortenings and butter (Ger.). *Arch. Hyg. (München), 152,* 231-237

Fábián, B. (1969) Carcinogenic substances in edible fats and oils. VI. New studies on margarine and chocolate (Ger.). *Arch. Hyg. (München)*, *153*, 21-24

Falk, H.L., Markul, I. & Kotin, P. (1956) Aromatic hydrocarbons. IV. Their fate following emission into the atmosphere and experimental exposure to washed air and synthetic smog. *Arch. ind. Health*, *13*, 13-17

Fox, C.H., Selkirk, J.K., Price, F.M., Croy, R.G., Sanford, K.K. & Cottler-Fox, M. (1975) Metabolism of benzo(*a*)pyrene by human epithelial cells *in vitro*. *Cancer Res.*, *35*, 3551-3557

Fretheim, K. (1976) Carcinogenic polycyclic aromatic hydrocarbons in Norwegian smoked meat. *J. agric. Food Chem.*, *24*, 976-979

Freudenthal, R.I. & Jones, P.W., eds (1976) *Carcinogenesis - A Comprehensive Survey*, Vol. 1, *Polynuclear Aromatic Hydrocarbons: Chemistry, Metabolism, and Carcinogenesis*, New York, Raven Press

Fritz, W. (1969) Solubility of polyaromatics after boiling of coffee substitutes and pure coffee (Ger.). *Dtsch. Lebensm.-Rundsch.*, *65*, 83-85

Fritz, W. (1972) A contribution to the contamination of food with carcinogenic hydrocarbons during processing and cooking (Ger.). *Arch. Geschwulstforsch.*, *40*, 81-90

Fritz, W. & Engst, R. (1971) Contamination of the environment with carcinogenic hydrocarbons (Ger.). *Z. ges. Hyg.*, *17*, 271-275

Funcke, W., König, J., Balfanz, E., Romanowski, T. & Grossmann, I. (1982) Analysis of polycyclic aromatic hydrocarbon content of airborne particulate matter from the Ruhr area and one rural town (Ger.). *Staub-Reinhalt. Luft*, *42*, 192-197

Galuskinova, V. (1964) 3,4-Benzpyrene determination in the smoking atmosphere of social meeting rooms and restaurants. Contribution to the problem of the noxiousness of so-called passive smoking. *Neoplasm*, *11*, 465-468

Gelboin, H.V. & Ts'O, P.O.P., eds (1978) *Polycyclic Hydrocarbons and Cancer*, Vol. 1, *Environment, Chemistry and Metabolism*, Vol. 2, *Molecular and Cell Biology*, New York, Academic Press

Gerstle, R.W., Cuffe, S.T., Orning, A.A. & Schwartz, C.H. (1965) Air pollution emissions from coal-fired power plants. Report No. 2. *J. Air Pollut. Control Assoc.*, *15*, 59-64

Giger, W. & Blumer, M. (1974) Polycyclic aromatic hydrocarbons in the environment: Isolation and characterization by chromatography, visible, ultraviolet, and mass spectrometry. *Anal. Chem.*, *46*, 1663-1671

Goldsmith, J.R. & Friberg, L.T. (1977) *Effects of air pollution on human health*. In: Stern, A.C., ed., *Air Pollution*, 3rd ed., Vol. II, New York, Academic Press, pp. 457-610

Gräf, W. (1970) Levels of 3,4-benzopyrene in human organs of different ages. Second communication. *Arch. Hyg.*, *4*, 331-335

Gräf, W., Eff, H. & Schormair, S. (1975) Levels of carcinogenic, polycyclic aromatic hydrocarbons in human and animal tissues. Third communication. *Zbl. Bakt. Hyg., Abt. Orig. B.*, *161*, 85-103

Gray, J.I. & Morton, I.D. (1981) Some toxic compounds produced in food by cooking and processing. *J. human Nutr.*, *35*, 5-23

Griest, W.H. & Guerin, M.R. (1979) *Identification and Quantification of Polynuclear Organic Matter (POM) on Particulates from a Coal-Fired Power Plant* (*APRI report No. EA-1095*), Washington DC, National Technical Information Service [*Chem. Abstr.*, *92*, 1522322]

Grimmer, G. (1968) Carcinogenic hydrocarbons in the human environment (Ger.). *Dtsch. Apotheker-Z.*, *108*, 529-533

Grimmer, G.E. (1983a) *Environmental Carcinogens: Polycyclic Aromatic Hydrocarbons*, Boca Raton, FL, CRC Press

Grimmer, G. (1983b) *Profile analysis of polycyclic aromatic hydrocarbons in air.* In: Bjørseth, A., ed., *Handbook of PAH*, New York, Marcel Dekker, pp. 149-181

Grimmer, G. & Böhnke, H. (1972) Determination of polycyclic aromatic hydrocarbons in automobile exhaust and air pollutants by capillary-gas chromatography (Ger.). *Z. anal. Chem.*, *261*, 310-314

Grimmer, G. & Böhnke, H. (1975) Profile analysis of polycyclic aromatic hydrocarbons and metal content in sediment layers of a lake. *Cancer Lett.*, *1*, 75-84

Grimmer, G. & Böhnke, H. (1977) Investigation on drilling cores of sediments of Lake Constance. I. Profiles of polycyclic aromatic hydrocarbons (Ger.). *Z. Naturforsch.*, *32c*, 703-711

Grimmer, G. & Böhnke, H. (1978) The tumor-producing effect of automobile exhaust condensate and fractions thereof. I. Chemical studies. *J. environ. Pathol. Toxicol.*, *1*, 661-667

Grimmer, G. & Hildebrandt, A. (1965) Hydrocarbons in the human environment. II. The content of polycyclic hydrocarbons in bread-grains from various localities (Ger.). *Z. Krebsforsch.*, *67*, 272-277

Grimmer, G. & Hildebrandt, A. (1966) Content of polycyclic hydrocarbons in coffee and tea (Ger.). *Dtsch. Lebensm.-Rundsch.*, *62*, 19-21

Grimmer, G. & Hildebrandt, A. (1967a) Hydrocarbons in the human environment. V. The content of polycyclic hydrocarbons in meat and smoked goods (Ger.). *Z. Krebsforsch.*, *69*, 223-229

Grimmer, G. & Hildebrandt, A. (1967b) Content of polycyclic hydrocarbons in crude vegetable oils. *Chem. Ind.*, 25 November, 2000-2002

Grimmer, G. & Naujack, K.-W. (1979) Gas chromatographic profile analysis of polycyclic aromatic hydrocarbons in water (Ger.). *Wasser*, *53*, 1-8

Grimmer, G., Jacob, J. & Hildebrandt, A. (1972) Hydrocarbons in the human environment. 9. Content of polycyclic hydrocarbons in Iceland soil (Ger.). *Z. Krebsforsch.*, *78*, 65-72

Grimmer, G., Hildebrandt, A. & Böhnke, H. (1973a) Investigations on the carcinogenic burden by air pollution in man. III. Sampling and analytics of polycyclic aromatic hydrocarbons in automobile exhaust gas. 1. Optimization of the collecting arrangement. *Zbl. Bakt. Hyg. 1 Abt. Orig. B*, *158*, 22-34

Grimmer, G., Hildebrandt, A. & Böhnke, H. (1973b) Investigations on the carcinogenic burden by air pollution in man. III. Sampling and analytics of polycyclic aromatic hydrocarbons in automobile exhaust gas. 2. Enrichment of the PNA and separation of the mixture of all PHA. *Zbl. Bakt. Hyg. 1 Abt. Orig. B*, *158*, 35-49

Grimmer, G., Böhnke, H. & Harke, H.-P. (1977a) Passive smoking: Measuring of concentrations of polycyclic aromatic hydrocarbons in rooms after machine smoking of cigarettes (Ger.). *Int. Arch. occup. environ. Health*, *40*, 83-92

Grimmer, G., Böhnke, H. & Harke, H.-P. (1977b) Passive smoking: Intake of polycyclic aromatic hydrocarbons by breathing of cigarette smoke containing air (Ger.). *Int. Arch. occup. environ. Health*, *40*, 93-99

Grimmer, G., Böhnke, H. & Glaser, A. (1977c) Investigation on the carcinogenic burden by air pollution in man. XV. Polycyclic aromatic hydrocarbons in automobile exhaust gas - An inventory. *Zbl Bakt. Hyg., I. Abt. Orig. B*, *164*, 218-234

Grimmer, G., Böhnke, H. & Borwitzky, H. (1978) Gas-chromatographic profile analysis of polycyclic aromatic hydrocarbons in sewage sludge (Ger.). *Fresenius Z. anal. Chem.*, *289*, 91-95

Grimmer, G., Naujack, K.-W. & Schneider, D. (1980a) *Changes in PAH-profiles in different areas of a city during the year.* In: Bjørseth, A. & Dennis, A.J., eds, *Polynuclear Aromatic Hydrocarbons: Chemistry and Biological Effects, 4th Int. Symposium*, Columbus, OH, Battelle Press, pp. 107-125

Grimmer, G., Hilge, G. & Niemitz, W. (1980b) Comparison of the profile of polycyclic aromatic hydrocarbons of sewage sludge samples from 25 filter plants (Ger.). *Wasser*, *54*, 255-272

Grimmer, G., Dettbarn, G. & Schneider, D. (1981a) Capillary-gas chromatographic profile analysis of polycyclic aromatic hydrocarbons in drinking and industrial water (Ger.). *Z. Wasser Abwasser Forsch.*, *14*, 100-106

Grimmer, G., Naujack, K.-W. & Schneider, D. (1981b) Comparison of the profiles of polycyclic aromatic hydrocarbons in different areas of a city by glass-capillary-gas-chromatography in the nanogram-range. *Int. J. environ. anal. Chem.*, *10*, 265-276

Grimmer, G., Schneider, D. & Dettbarn, G. (1981c) The load of different rivers in the Federal Republic of Germany by polycyclic aromatic hydrocarbons (Ger.). *Wasser*, *56*, 131-144

Grimmer, G., Buck, M. & Ixfeld, H. (1981d) *Immissionsmessungen von Polycyclischen Aromatischen Kohlenwasserstoffen (PAH)* [Measures of emissions of polycyclic aromatic hydrocarbons], Minister für Arbeit, Gesundheit und Soziales des Landes Nordrheim-Westfalen, Schottedruck, Krefeld

Grimmer, G., Naujack, K.-W., Dettbarn, G., Brune, H., Deutsch-Wenzel, R. & Misfeld, J. (1982a) *Analysis of balance of carcinogenic impact from emission condensates of automobile exhaust, coal heating, and used engine oil by mouse-skin-painting as a carcinogen-specific detector*. In: Bjørseth, A. & Dennis, A.J., eds, *Polynuclear Aromatic Hydrocarbons, Physical and Biological Chemistry, 6th Int. Syposium*, Colombus, OH, Battelle Press, pp. 335-345

Grimmer, G., Detttbarn, G., Brune, H., Deutsch-Wenzel, R. & Misfeld, J. (1982b) Quantification of the carcinogenic effect of polycyclic aromatic hydrocarbons in used engine oil by topical application onto the skin of mice. *Int. Arch. occup. environ. Health*, *50*, 95-100

Grimmer, G., Naujack, K.-W., Dettbarn, G., Brune, H., Deutsch-Wenzel, R. & Misfeld, J. (1982c) Carcinogenic action of lubricating oil from motor vehicles (Ger.). *Erdöl Kohle. Erdgas. Petrochem. ver. Brennstoff. Chem., 35,* 466-472

Grimmer, G., Naujack, K.-W. & Schneider, D. (1982d) Profile analysis of polycyclic aromatic hydrocarbons by glass capillary gas chromatography in atmospheric suspended particulate matter in the nanogram range collecting 10 m^3 of air. *Fresenius Z. anal. Chem.*, *311*, 475-484

Grimmer, G., Jacob, J., Naujack, K.-W. & Dettbarn, G. (1983a) Determination of polycyclic aromatic compounds emitted from brown-coal-fired residential stoves by gas chromatography/mass spectrometry. *Anal. Chem.*, *55*, 892-900

Grimmer, G., Naujack, K.-W., Dettbarn, G., Brune, H., Deutsch-Wenzel, R. & Misfeld, J. (1983b) *Analysis of the balance of carcinogenic impact from emission condensate of automobile exhaust by implantation into the lung of rats as a carcinogen specific detector*. In: Cooke, M., ed., *Polynuclear Aromatic Hydrocarbons: 7th Int. Symposium*, Columbus, OH, Battelle Press (in press)

Grob, K. & Voellmin, J.A. (1970) GC-MS analysis of 'semi-volatiles' of cigarette smoke. *J. chromatogr. Sci.*, *8*, 218-220

Grover, P.L., Hewer, A. & Sims, P. (1973) K-Region epoxides of polycyclic hydrocarbons: Formation and further metabolism of benz[*a*]anthracene-5,6 oxide by human lung preparations. *FEBS Lett.*, *34*, 63-68

Grover, P.L., MacNicoll, A.D., Sims, P., Easty, G.C. & Neville, A.M. (1980) Polycyclic hydrocarbon activation and metabolism in epithelial cell aggregates prepared from human mammary tissue. *Int. J. Cancer*, *26*, 467-479

Guggenberger, J., Krammer, G. & Lindenmüller, W. (1981) Contribution to the determination of the emission of polycyclic aromatic hydrocarbons from large capacity furnaces (Ger.). *Staub-Reinhalt. Luft*, *41*, 339-344

Gurtoo, H.L., Vaught, J.B., Marinello, A.J., Paigen, B., Gessner, T. & Bolanowska, W. (1980) High-pressure liquid chromatographic analysis of benzo(*a*)pyrene metabolism by human lymphocytes from donors of different aryl hydrocarbon hydroxylase inducibility and antipyrene half-lives. *Cancer Res.*, *40*, 1305-1310

Gurtoo, H.L., Williams, C.J., Gottlieb, K., Mulhern, A., Caballes, L., Vaught, J.B., Marinello, A.J. & Bansal, S.K. (1983) Population distribution of placental benzo(*a*)pyrene metabolism in smokers. *Int. J. Cancer*, *31*, 29-37

Hagenmaier, H., Feierabend, R. & Jäger, W. (1977) Determination of polycyclic aromatic hydrocarbons in water with high-performance liquid chromatography (Ger.). *Z. Wasser Abwasser-Forsch.*, *10*, 99-104

Hangebrauck, R.P., von Lehmden, D.J. & Meeker, J.E. (1964) Emissions of polynuclear hydrocarbons and other pollutants from heat-generation and incineration processes. *J. Air Pollut. Control Assoc.*, *14*, 267-278

Harris, C.C., Autrup, H., Connor, R., Barrett, L.A., McDowell, E.M. & Trump, B.F. (1976) Interindividual variation in binding of benzo[*a*]pyrene to DNA in cultured human bronchi. *Science*, *194*, 1067-1069

Harris, C.C., Autrup, H., Stoner, G., Yang, S.K., Leutz, J.C., Gelboin, H.V., Selkirk, J.K., Connor, R.J., Barrett, L.A., Jones, R.T., McDowell, E. & Trump, B.F. (1977) Metabolism of benzo[*a*]pyrene and 7,12-dimethylbenz[*a*]anthracene in cultured human bronchus and pancreatic duct. *Cancer Res.*, *37*, 3349-3355

Harris, C.C., Hsu, I.C., Stoner, G.B., Trump, B.F. & Selkirk, J.K. (1978) Human pulmonary alveolar macrophages metabolise benzo[*a*]pyrene to proximate and ultimate mutagens. *Nature*, *272*, 633-634

Harris, C.C., Autrup, H., Stoner, G.D., Trump, B.F., Hillman, E., Schafer, P.W. & Jeffrey, A.M. (1979) Metabolism of benzo(*a*)pyrene, *N*-nitrosodimethylamine and *N*-nitrosopyrrolidine and identification of the major carcinogen-DNA adducts formed in cultured human esophagus. *Cancer Res.*, *39*, 4401-4406

Häsänen, E., Pohjola, V., Pyysalo, H. & Wickström, K. (1983) Polycyclic aromatic hydrocarbons in the Finnish wood-heated sauna. *Kemia-Kemi*, *1*, 27-29

Hecht, S.S., Bondinell, W.E. & Hoffmann, D. (1974) Chrysene and methylchrysenes: Presence in tobacco smoke and carcinogenicity. *J. natl Cancer Inst.*, *53*, 1121-1133

Hecht, S.S., Grabowski, W. & Groth, K. (1979) Analysis of faeces for benzo[*a*]pyrene after consumption of charcoal-broiled beef by rats and humans. *Food Cosmet. Toxicol.*, *17*, 223-227

Heidelberger, C. (1976) *Studies on the mechanisms of carcinogenesis by polycyclic aromatic hydrocarbons and their derivatives*. In: Freudenthal, R.I. & Jones, P.W., eds, *Carcinogenesis*, Vol. 1, *Polynuclear Aromatic Hydrocarbons: Chemistry, Metabolism and Carcinogenesis*, New York, Raven Press, pp. 1-8

Heidelberger, C. & Weiss, S.M. (1951) The distribution of radioactivity in mice following administration of 3,4-benzpyrene-5-C^{14} and 1,2,5,6-dibenzanthracene-9,10-C^{14}. *Cancer Res.*, *11*, 885-891

Helberg, D. (1964) Assay of paraffin and microcrystalline wax from cancerogenic polycyclic aromatic hydrocarbons (Ger.). *Dtsch. Lebensm.-Rundsch.*, *60*, 345-347

Henry, M.C. & Kaufman, D.G. (1973) Clearance of benzo[*a*]pyrene from hamster lungs after administration on coated particles. *J. natl Cancer Inst.*, *51*, 1961-1964

Hoffmann, D. & Wynder, E.L. (1960) On the isolation and identification of polycyclic aromatic hydrocarbons. *Cancer*, *13*, 1062-1073

Hoffmann, D. & Wynder, E.L. (1971) A study of tobacco carcinogenesis. XI. Tumor initiators, tumor accelerators, and tumor promoting activity of condensate fractions. *Cancer*, *27*, 848-864

Hoffmann, D. & Wynder, E.L. (1972) Smoke of cigarettes and little cigars: An analytical comparison. *Science*, *178*, 1197-1199

Hoffmann, D. & Wynder, E.L. (1976) *Environmental respiratory carcinogenesis*. In: Searle, C.E., ed., *Chemical Carcinogens* (*ACS Monograph No. 173*), Washington DC, American Chemical Society, pp. 324-265

Hoffmann, D. & Wynder, E.L. (1977) *Organic particulate pollutants - Chemical analysis and bioassay for carcinogenicity*. In: Stern, A.H., ed., *Air Pollution*, 3rd ed., Vol. II, New York, Academic Press, pp. 361-455

Hoffmann, D., Rathkamp, G. & Wynder, E.L. (1963) Comparison of the yields of several selected components in the smoke from different tobacco products. *J. natl Cancer Inst.*, *31*, 627-637

Hoffmann, D., Rathkamp, G. & Woziwodzki, H. (1968) Chemical studies on tobacco smoke. VI. The determination of carbazoles in cigarette smoke. *Beitr. Tabakforsch.*, *4*, 253-263

Hoffmann, D., Rathkamp, G., Nesnow, S. & Wynder, E.L. (1972) Fluoranthenes: Quantitative determination in cigarette smoke, formation by pyrolysis, and tumor-initiating activity. *J. natl Cancer Inst.*, *49*, 1165-1175

Hoffmann, D., Sanghvi, L.D. & Wynder, E.L. (1974) Comparative chemical analysis of Indian bidi and American cigarette smoke. *Int. J. Cancer*, *14*, 49-53

Hoffmann, D., Brunnemann, K.D., Gori, G.B. & Wynder, E.L. (1975) *On the carcinogenicity of marijuana smoke*. In: Runeckles, V.C., ed., *Recent Advances in Phytochemistry*, Vol. 9, New York, Plenum Press, pp. 63-81

Holder, G., Yagi, H., Dansette, P., Jerina, D.M., Levin, W., Lu, A.Y.H. & Conney, A.H. (1974) Effects of inducers of epoxide hydrase on the metabolism of benzo[*a*]pyrene by liver microsomes and a reconstituted system: Analysis by high pressure liquid chromatography. *Proc. natl Acad. Sci. USA*, *71*, 4356-4360

Hueper, W.C., Kotin, P., Tabor, E.C., Payne, W.W., Falk, H. & Sawicki, E. (1962) Carcinogenic bioassays on air pollutants. *Arch. Pathol.*, *74*, 89-116

Huggins, C. & Yang, N.C. (1962) Induction and extinction of mammary cancer. *Science*, *137*, 257-262

Hughes, M.M., Natusah, D.F.S., Taylor, D.R. & Zeller, M.V. (1980) *Chemical transformations of particulate polycyclic organic matter*. In: Bjørseth, A. & Dennis, A.J., eds, *Polynuclear Aromatic Hydrocarbons, Chemistry and Biological Effects, 4th Int. Symposium*, Columbus, OH, Battelle Press, pp. 1-8

IARC (1973) *IARC Monographs on the Evaluation of Carcinogenic Risk of Chemicals to Man*, Vol. 3, *Certain Polycyclic Aromatic Hydrocarbons and Heterocyclic Compounds*, Lyon

IARC (1979) *IARC Monographs on the Evaluation of the Carcinogenic Risk of Chemicals to Humans*, Supplement 1, *Chemicals and Industrial Processes Associated with Cancer in Humans. IARC Monographs, Volumes 1 to 20*, Lyon

IARC (1980) *IARC Monographs on the Evaluation of the Carcinogenic Risk of Chemicals to Humans*, Supplement 2, *Long-term and Short-term Screening Assays for Carcinogens: A Critical Appraisal*, Lyon

Ilnitsky, A.P., Mischenko, V.S. & Shabad, L.M. (1977) New data on volcanoes as natural sources of carcinogenic substances. *Cancer Lett.*, *3*, 227-230

International Committee for Cigar Smoke Study (1974) Machine smoking of cigar. *Coresta Inf. Bull.*, *1*, 31-34

Jacob, J., Grimmer, G. & Schmoldt, A. (1981) The influence of polycyclic aromatic hydrocarbons as inducers of monooxygenases on the metabolite profile of benz[*a*]anthracene in the rat liver microsomes. *Cancer Lett.*, *14*, 175-185

Jellinek, J.S. (1970) *Formulation and Function of Cosmetics*, London, John Wiley & Sons, pp. 108-197

Jerina, D.M. & Daly, J.W. (1974) Arene oxides: A new aspect of drug metabolism. *Science*, *185*, 573-582

Jerina, D.M., Lehr, R.E., Yagi, H., Hernandez, O., Dansette, P.M., Wislocki, P.G., Wood, A.W., Chang, R.L., Levin, W. & Conney, A.H. (1976) *Mutagenicity of benzo[*a*]pyrene derivatives and the description of a quantum mechanical model which predicts the ease of carbonium ion formation from diol epoxides*. In: de Serres, F.J., Fouts, J.R., Bend, J.R. & Philpot, R.M., eds, In vitro *Metabolic Activation in Mutagenesis Testing*, Amsterdam, Elsevier/North Holland, pp. 159-178

Jerina, D.M., Sayer, J.M., Thakker, D.R., Yagi, H., Levin, W., Wood, A.W. & Conney, A.H. (1980) *Carcinogenicity of polycyclic aromatic hydrocarbons: the bay-region theory*. In: Pullman, B., Ts'O, P.O.P. & Gelboin, H., eds, *Carcinogenesis: Fundamental Mechanisms and Environmental Effects*, Hingham, MA, D. Reidel Publishing Co., pp. 1-12

Jones, P.W. & Freudenthal, R.I., eds (1978) *Carcinogenesis. A Comprehensive Survey*, Vol. 3, *Polynuclear Aromatic Hydrocarbons*, New York, Raven Press

Jones, P.W. & Leber, P., eds (1979) *Polynuclear Aromatic Hydrocarbons*, Ann Arbor, MI, Ann Arbor Science

Jones, P.W., Giammar, R.D., Strup, P.E. & Stanford, T.B. (1976) Efficient collection of polycyclic organic compounds from combustion effluents. *Environ. Sci. Technol.*, *10*, 806-810

Josefsson, B. (1982) *Recent concepts in sampling methodology*. In: Bjørseth, A. & Angeletti, G., eds, *Analysis of Organic Micropollutants in Water*, Dordrecht, The Netherlands, D. Reidel, pp. 7-15

Just, J., Borkowska, M. & Maziarka, S. (1972) Air pollution by tobacco smoke in Warsaw coffee houses (Pol.). *Rocz. Panéstw. Zakł. Hyg.*, *23*, 129-135

Kelman, B.J. & Springer, D.L. (1982) Movements of benzo[*a*]pyrene across the hemochorial placenta of the guinea pig. *Proc. Soc. exp. biol. Med.*, *169*, 58-62

King, C.M., Wang, C.Y. & Warmer, P.O. (1980) Evidence for the presence of nitroaromatics in airborne particulates (Abstract no. 334). *Proc. Am. Assoc. Cancer Res.*, *21*, 83

Kiryu, S. & Kuratsune, M. (1966) Polycyclic aromatic hydrocarbons in cigarette tar produced by human smoking. *Gann*, *57*, 317-322

König, J., Funcke, W., Balfanz, E., Grosch, B. & Pott, F. (1980) Testing a high volume air sampler for quantitative collection of polycyclic aromatic hydrocarbons. *Atmos. Environ.*, *14*, 609-613

Korfmacher, W.A., Wehry, E.L., Mamantov, G. & Natusch, D.F.S. (1980) Resistance to photochemical decomposition of polycyclic aromatic hydrocarbons vapor-adsorbed on coal fly ash. *Environ. Sci. Technol.*, *14*, 1094-1099

Kotin, P. & Falk, H.L. (1959) The role and action of environmental agents in the pathogenesis of lung cancer. I. Air pollutants. *Cancer*, *12*, 147-163

Kotin, P. & Falk, H.L. (1960) The role and action of environmental agents in the pathogenesis of lung cancer. II. Cigarette smoke. *Cancer*, *13*, 250-262

Kotin, P., Falk, H.L. & Thomas, M. (1954) Aromatic hydrocarbons. II. Presence in the particulate phase of gasoline-engine exhausts and the carcinogenicity of exhaust extracts. *A.M.A. Arch. ind. Hyg.*, *9*, 164-177

Kotin, P., Falk, H.L. & Thomas, M. (1955) Aromatic hydrocarbons. III. Presence in the particulate phase of diesel-engine exhausts and the carcinogenicity of exhaust extracts. *A.M.A. Arch. ind. Health*, *11*, 113-120

Kotin, P., Falk, H.L. & Busser, R. (1959) Distribution, retention and elimination of C^{14}-3,4-benzpyrene after administration to mice and rats. *J. natl Cancer Inst.*, *23*, 541-555

Kraft, E.R., Hoch, S.G., Quisno, R.A. & Newcomb, E.A. (1972) The importance of the vehicle in formulating sunscreen and tanning preparations. *J. Soc. Cosmet. Chem.*, *23*, 383-391

Kubitscheck, H.E. & Williams, D.M. (1980) Mutagenicity of fly ash from fluidized-bed combuster during start-up and steady operating conditions. *Mutat. Res.*, *77*, 287-291

Kuratsune, M. & Hueper, W.C. (1960) Polycyclic aromatic hydrocarbons in roasted coffee. *J. natl Cancer Inst.*, *24*, 463-469

Kuroki, T., Nemoto, N. & Kitano, Y. (1980) Metabolism of benzo[*a*]pyrene in human epidermal keratinocytes in culture. *Carcinogenesis*, *1*, 559-565

Kveseth, K., Sortland, B. & Bokn, T. (1982) Polycyclic aromatic hydrocarbons in sewage, mussels and tap water. *Chemosphere*, *11*, 623-639

La Du, B.N., Mandel, H.G. & Way, E.L. (1971) *Fundamentals of Drug Metabolism and Drug Disposition*, Baltimore, MD, Williams & Wilkins

Laflamme, R.E. & Hites, R.A. (1979) Tetra- and pentacyclic, naturally-occurring, aromatic hydrocarbons in recent sediments. *Geochim. cosmochim. Acta*, *43*, 1687-1691

Lao, R.C. & Thomas, R.S. (1980) *The volatility of PAH and possible losses in ambient sampling.* In: Bjørseth, A. & Dennis, A.J., eds, *Polynuclear Aromatic Hydrocarbons: Chemistry and Biological Effects, 4th Int. Symposium*, Columbus, OH, Battelle Press, pp. 829-839

Lao, R.C., Thomas, R.S., Oja, H. & Dubois, L. (1973) Application of a gas chromatograph-mass spectrometer-data processor combination to the analysis of the polycyclic aromatic hydrocarbon content of airborne pollutants. *Anal. Chem.*, *45*, 908-915

Lawther, P.J. & Waller, R.E. (1976) *Coal fires, industrial emissions and motor vehicles as sources of environmental carcinogens.* In: Rosenfeld, C. & Davis, W., eds, *Environmental Pollution and Carcinogenic Risks* (*IARC Scientific Publications No. 13*), Lyon, International Agency for Research on Cancer, pp. 27-40

Lazar, P., Libermann, C., Chouroulinkov, I. & Guerin, M. (1963) Skin tests on mice to determine carcinogenic activities: Methods onset (Fr.). *Bull. Cancer*, *50*, 567-577

Lee, F.S.-C., Pierson, W.R. & Ezike, J. (1980a) *The problem of PAH degradation during filter collection of airborne particulates - An evaluation of several commonly used filter media.* In: Bjørseth, A. & Dennis, A.J., eds, *Polynuclear Aromatic Hydrocarbons: Chemistry and Biological Effects, 4th Int. Symposium*, Columbus, OH, Battelle Press, pp. 543-563

Lee, F.S.-C., Harvey, T.M., Prater, T.J., Paputa, M.C. & Schuetzle, D. (1980b) *Chemical analysis of diesel particulate matter and an evaluation of artifact formation.* In: *Sampling and Analysis of Toxic Organics in the Atmosphere* (*ASTM STP721*), Philadelphia, PA, American Society for Testing and Materials, pp. 92-110

Lee, M.L., Bartle, K.D. & Novotny, M.V. (1975) Profiles of the polynuclear aromatic fraction from engine oils obtained by capillary-column gas-liquid chromatography and nitrogen-selective detection. *Anal. Chem.*, *47*, 540-548

Lee, M.L., Novotny, M.V. & Bartle, K.D. (1976) Gas chromatography/mass spectrometric and nuclear magnetic resonance spectrometric studies of carcinogenic polynuclear aromatic hydrocarbons in tobacco and marijuana smoke condensates. *Anal. Chem.*, *48*, 405-416

Lee, M.L., Novotny, M.V. & Bartle, K.D. (1981) *Analytical Chemistry of Polycyclic Aromatic Compounds*, New York, Academic Press

Lee, S.D. & Grant, L., eds (1981) Health and ecological assessment of polynuclear aromatic hydrocarbons. *J. environ. Pathol. Toxicol.*, *5*, 1-364

Levin, W., Buening, M.K., Wood, A.W., Chang, R.L., Kedzierski, B., Thakker, D.R., Boyd, D.R., Gadaginamath, G.S., Armstrong, R.N., Yagi, H., Karle, J.M., Slaga, T.J., Jerina, D.M. & Conney, A.H. (1980) An enantiomeric interaction in the metabolism and tumorigenicity of (+)- and (-)-benzo[*a*]pyrene 7,8-oxide. *J. biol. Chem.*, *255*, 9067-9074

Levin, W., Wood, A., Chang, R.L., Ryan, D., Thomas, P.E., Yagi, H., Thakker, D.R., Vyas, K., Boyd, C., Chu, S.-Y., Conney, A.H. & Jerina, D.M. (1982) Oxidative metabolism of polycyclic aromatic hydrocarbons to ultimate carcinogens. *Drug Metab. Rev.*, *13*, 555-580

Liberti, A., Cartoni, G.P. & Cantuti, V. (1964) Gas chromatographic determination of polynuclear hydrocarbons in dust. *J. Chromatogr.*, *15*, 141-148

Lijinsky, W. & Ross, A.E. (1967) Production of carcinogenic polynuclear hydrocarbons in the cooking of food. *Food Cosmet. Toxicol.*, *5*, 343-347

Lijinsky, W. & Shubik, P. (1964) Benzo(a)pyrene and other polynuclear hydrocarbons in charcoal-broiled meat. *Science*, *145*, 53-55

Lijinsky, W. & Shubik, P. (1965) Polynuclear hydrocarbon carcinogens in cooked meat and smoked food. *Ind. Med. Surg.*, *34*, 152-154

Lijinsky, W., Domsky, I., Mason, G., Ramahi, H.Y. & Safavi, T. (1963) The chromatographic determination of trace amounts of polynuclear hydrocarbons in petrolatum, mineral oil, and coal tar. *Anal. Chem.*, *35*, 952-956

Lintas, C., De Matthaeis, M.C. & Merli, F. (1979) Determination of benzo[*a*]pyrene in smoked, cooked and toasted food products. *Food Cosmet. Toxicol.*, *17*, 325-328

Locati, G., Fantuzzi, A., Consonni, G., Li Gotti, I. & Bonomi, G. (1979) Identification of polycyclic aromatic hydrocarbons in carbon black with reference to carcinogenic risk in tire production. *Am. ind. Hyg. Assoc. J.*, *40*, 644-652

Lunde, G. (1976) Long-range aerial transmission of organic micropollutants. *Ambio*, *5*, 207-208

Lunde, G. & Bjørseth, A. (1977) Polycyclic aromatic hydrocarbons in long-range transported aerosols. *Nature*, *268*, 518-519

MacLeod, M.D., Cohen, G.M. & Selkirk, J.K. (1979) Metabolism and macromolecular binding of the carcinogen benzo(*a*)pyrene and its relatively inert isomer benzo(*e*)pyrene by hamster embryo cells. *Cancer Res.*, *39*, 3463-3470

MacLeod, M.C., Levin, W., Conney, A.H., Lehr, R.E., Mansfield, B.K., Jerina, D.M. & Selkirk, J.K. (1980) Metabolism of benzo(*e*)pyrene by rat liver microsomal enzymes. *Carcinogenesis*, *1*, 165-173

MacNicoll, A.D., Easty, G.C., Neville, A.M., Grover, P.L. & Sims, P. (1980) Metabolism and activation of carcinogenic polycyclic hydrocarbons by human mammary cells. *Biochem. biophys. Res. Commun.*, *95*, 1599-1606

Malanoski, A.J., Greenfield, E.L., Barnes, C.J., Worthington, J.M. & Joe, F.L., Jr (1968) Survey of polycyclic aromatic hydrocarbons in smoked foods. *J. Assoc. off. anal. Chem.*, *51*, 114-121

Mallet, L. & Héros, M. (1962) Pollution of plant soils by polybenzenic hydrocarbons like benzo[*a*]pyrene (Fr.). *C.R. Acad. Sci. (Paris)*, *254*, 958-960

Marnett, L.J. & Reed, G.A. (1979) Peroxidatic oxidation of benzo[*a*]pyrene and prostaglandin biosynthesis. *Biochemistry*, *18*, 2923-2929

Marnett, L.J., Reed, G. & Johnson, J.T. (1977) Prostaglandin synthetase dependent benzo[*a*]pyrene oxidation: Products of the oxidation and inhibition of their formation by antioxidants. *Biochem. biophys. Res. Commun.*, *79*, 569-576

Marshall, M.V., McLemore, T.L., Martin, R.R., Jenkins, W.T., Snodgrass, D.R., Corson, M.A., Arnott, M.S., Wray, N.P. & Griffin, A.C. (1979) Patterns of benzo[*a*]pyrene metabolism in normal human pulmonary alveolar macrophages. *Cancer Lett.*, *8*, 103-109

Mass, M.J., Rodgers, N.T. & Kaufman, D.G. (1981) Benzo[*a*]pyrene metabolism in organ cultures of human endometrium. *Chem.-biol. Interactions*, *33*, 195-205

Masuda, Y. & Kuratsune, M. (1971) Polycyclic aromatic hydrocarbons in smoked fish, 'katsuobushi'. *Gann*, *62*, 27-30

Masuda, Y. & Kuratsune, M. (1972) Comparison of the yield of polycyclic aromatic hydrocarbons in smoke from Japanese tobacco. *Jpn. J. Hyg.*, *27*, 339-341

Masuda, Y., Mori, K., Hirohata, T. & Kuratsune, M. (1966a) Carcinogenesis in the esophagus. III. Polycyclic aromatic hydrocarbons and phenols in whisky. *Gann*, *57*, 549-557

Masuda, Y., Mori, K. & Kuratsune, M. (1966b) Polycyclic aromatic hydrocarbons in common Japanese foods. I. Broiled fish, roasted barley, shoyu, and caramel. *Gann*, *57*, 133-142

McCarthy, J.P. (1976) Lanolin derivatives in sunscreen preparations. *Household pers. Prod. Ind.*, *13*, 44246

McKay, J.F. & Latham, D.R. (1973) Polyaromatic hydrocarbons in high-boiling petroleum distillates. Isolation by gel permeation chromatography and identification by fluorescence spectrometry. *Anal. Chem.*, *45*, 1050-1055

McMahon, C.K. & Tsoukalas, S.N. (1978) *Polynuclear aromatic hydrocarbons in forest fire smoke*. In: Jones, P.W. & Freudenthal, R.I., eds, *Carcinogenesis*, Vol. 3, *Polynuclear Aromatic Hydrocarbons*, New York, Raven Press, pp. 61-73

Mehta, R. & Cohen, G.M. (1979) Major differences in the extent of conjugation with glucuronic acid and sulphate in human peripheral lung. *Biochem. Pharmacol.*, *28*, 2479-2484

Mehta, R., Meredith-Brown, M. & Cohen, G.M. (1979) Metabolism and covalent binding of benzo[*a*]pyrene in human peripheral lung. *Chem.-biol. Interactions*, *28*, 345-358

Mermelstein, R., Kiriazides, D.K., Butler, M., McCoy, E.C. & Rosenkranz, H.S. (1981) The extraordinary mutagenicity of nitropyrenes in bacteria. *Mutat. Res.*, *89*, 187-196

Miller, E.C. & Miller, J.A. (1982) *Reactive metabolites as key intermediates in pharmacologic and toxicologic responses: Examples from chemical carcinogenesis*. In: Snyder, R., Parke, D.V., Kocsis, J.J., Jollow, D.J., Gibson, C.G. & Witmer, C.M., eds, *Biological Reactive Intermediates*, II. *Chemical Mechanisms and Biological Effects*, Part A, New York, Plenum, pp. 1-21

Miller, J.A. (1970) Carcinogenesis by chemical: An overview - G.H.A. Clowes Memorial Lecture. *Cancer Res.*, *30*, 559-576

Mitchell, C.E. (1982) Distribution and retention of benzo(a)pyrene in rats after inhalation. *Toxicol. Lett.*, *11*, 35-42

Mitchell, C.E. & Tu, K.W. (1979) Distribution, retention and elimination of pyrene in rats after inhalation. *J. Toxicol. environ. Health*, *5*, 1171-1179

Monarca, S., Fagioli, F. & Morozzi, G. (1981) Evaluation of the potential carcinogenicity of paraffins for medical and cosmetic uses - Determination of polycyclic aromatic hydrocarbons. *Sci. total Environ.*, *17*, 83-93

Monarca, S., Sforzolini, G.S. & Fagioli, F. (1982) Presence of benzo[*a*]pyrene and other polycyclic aromatic hydrocarbons in suntan oils. *Food chem. Toxicol.*, *20*, 183-187

Müller, G., Grimmer, G. & Böhnke, H. (1977) Sedimentary record of heavy metals and polycyclic aromatic hydrocarbons in Lake Constance. *Naturwissenschaften*, *64*, 427-431

Müller, K. & Meyer, J.P. (1974) *Einfluss von Ottokraftstoffen auf die Emission von polynuklearen aromatischen Kohlenwasserstoffen in Automobilabgasen im Europa-Test.* [Effect of gasoline components on emission of polynuclear aromatic hydrocarbons in car exhaust in the Europa test] (*Forschungsbericht 4568*), Hamburg, Deutsche Gesellschaft für Mineralölwissenschaft und Kohlechemie e.V.

Müller, K.H., Neurath, G. & Horstmann, H. (1964) Influence of air permeability of cigarette paper on the yield of smoke and on its composition (Ger.). *Beitr. Tabakforsch.*, *2*, 271-281

Müller, R., Moldenhauer, W. & Schlemmer, P. (1967) Experiences with the quantitative determination of polycyclic hydrocarbons in tobacco smoke (Ger.). *Ber. Inst. Tabakforsch. Dresden*, *14*, 159-173

Nagel, D.L., Stenbäck, F., Clayson, D.B. & Wallcave, L. (1976) Intratracheal instillation studies with 7H-dibenzo[*c,g*]carbazole in Syrian hamster. *J. natl Cancer Inst.*, *57*, 119-123

Namkung, M.J. & Juchau, M.R. (1980) On the capacity of human placental enzymes to catalyse the formation of diols from benzo[*a*]pyrene. *Toxicol. appl. Pharmacol.*, *55*, 253-259

National Research Council (1981) *Health Effects of Exposure to Diesel Exhaust*, Washington DC, National Academy Press

National Research Council (1982) *Diet, Nutrition, and Cancer*, Washington DC, National Academy Press, pp. 14-25 - 14-28

Nebert, D.W. (1980) Pharmacogenetics: An approach to understanding chemical and biologic aspects of cancer. *J. natl Cancer Inst.*, *64*, 1279-1290

Neff, J.M. (1979) *Polycyclic Aromatic Hydrocarbons in the Aquatic Environment. Sources, Fates and Biological Effects*, London, Applied Science Publ., p. 51

Neff, J.M., Cox, B.A., Dixit, D. & Anderson, J.W. (1976) Accumulation and release of petroleum-derived aromatic hydrocarbons by four species of marine animals. *Marine Biol.*, *38*, 279-289

Nielsen, T. (1981) *Nordic PAH-Project. A Study of the Reactivity of Polycyclic Aromatic Hydrocarbons*, Oslo, Central Institute for Industrial Research

Obana, H., Hori, S., Kashimoto, T. & Kunita, N. (1981) Polycyclic aromatic hydrocarbons in human fat and liver. *Bull. environ. Contam. Toxicol.*, *27*, 23-27

Okano, P., Miller, H.N., Robinson, R.C. & Gelboin, H.V. (1979) Comparison of benzo(*a*)pyrene and (-)-*trans*-7,8-dihydroxy-7,8-dihydrobenzo(*a*)pyrene metabolism in human blood monocytes and lymphocytes. *Cancer Res.*, *39*, 3184-3193

Olufsen, B. (1980) *Polynuclear aromatic hydrocarbons in Norwegian drinking water resources.* In: Bjørseth, A. & Dennis, A.J., eds, *Polynuclear Aromatic Hydrocarbons: Chemistry and Biological Effects*, *4th Int. Symposium,* Columbus, OH, Battelle Press, PP. 333-343

Pal, K., Grover, P.L. & Sims, P. (1975) The metabolism of carcinogenic polycyclic hydrocarbons by tissues of the respiratory tract. *Biochem. Soc. Trans.*, *3*, 174-175

Pancirov, R.J. & Brown, R.A. (1977) Polynuclear aromatic hydrocarbons in marine tissues. *Environ. Sci. Technol.*, *11*, 989-992

Pelkonen, O. (1979) *Prenatal and neonatal development of drug and carcinogen metabolism.* In: Estabrook, R.W. & Lindenbaur, E., eds, *The Induction of Drug Metabolism*, Stuttgart-New York, F.K. Schattauer Verlag, pp. 507-516

Pelkonen, O. & Nebert, D.W. (1982) Metabolism of polycyclic aromatic hydrocarbons: Etiologic role in carcinogenesis. *Pharmacol. Rev.*, *34*, 189-222

Pelkonen, O. & Saarni, H. (1980) Unusual patterns of benzo[*a*]pyrene metabolites and DNA-benzo[*a*]pyrene adducts produced by human placental microsomes *in vitro. Chem.-biol. Interactions*, *30*, 287-296

Pelkonen, O., Sotaniemi, E. & Mokka, R. (1977) The in vitro oxidative metabolism of benzo(*a*)pyrene in human liver measured by different assays. *Chem.-biol. Interactions*, *16*, 13-21

Perry, J. (1973) Fasten your seat belts: No smoking. *Br. Columbia med. J.*, *15*, 304-305

Pierce, R.C. & Katz, M. (1975) Dependency of polynuclear aromatic hydrocarbon content on size distribution of atmospheric aerosols. *Environ. Sci. Technol.*, *9*, 347-353

Pitts, J.N., Jr (1979) Photochemical and biological implications of the atmospheric reactions of amines and benzo(*a*)pyrene. *Phil. Trans. R. Soc. London Ser. A*, *290*, 551-576

Pitts, J.N., Jr, Lokensgard, D.M., Harger, W., Fisher, T.S., Mejia, V., Schuler, J.J., Scorziell, G.M. & Katzenstein, Y.A. (1982) Mutagens in diesel exhaust particulate. Identification and direct activities of 6-nitrobenzo[*a*]pyrene, 9-nitroanthracene, 1-nitropyrene and 5*H*-phenanthro[4,5-*bcd*]pyran-5-one. *Mutat. Res.*, *103*, 241-249

Poucher, W.A. (1974) *Perfumes, Cosmetics and Soaps*, 7th ed., London, Chapman & Hall, pp. 292-293

Prough, R.A., Patrizi, V.W., Chita, R.T., Masters, B.S.S. & Jakobsson, S.W. (1979) Characteristics of benzo(*a*)pyrene metabolism by kidney, liver, and lung microsomal fractions from rodents and humans. *Cancer Res.*, *39*, 1199-1206

Pupp, C., Lao, R.C., Murray, J.J. & Pottie, R.F. (1974) Equilibrium vapour concentrations of some polycyclic aromatic hydrocarbons and the collection efficiencies of these air pollutants, As_4O_6 and SeO_2. *Atmos. Environ.*, *8*, 915-925

Purchase, I.F.H., Longstaff, E., Ashby, J., Styles, J.A., Anderson, D., Lefevre, P.A. & Westwood, F.R. (1978) An evaluation of six short-term tests for detecting organic chemical carcinogens. *Br. J. Cancer*, *37*, 873-959

Pyriki, C. (1963) Polycyclic and aliphatic hydrocarbons in tobacco smoke (Ger.). *Nahrung*, *7*, 439-448

Ratajczak, E.A., Ahland, E., Grimmer, G. & Dettbarn, G. (1982a) PAH- and S-PAC emissions from oil-fired vaporizing pot burners (Ger.). *Erdöl und Kohle-Erdgas-Petrochem. Brennstoff-Chem., 35,* 530

Ratajczak, E.A., Ahland, E., Grimmer, G. & Dettbarn, G. (1982b) *Untersuchung zur PAH-Messung an Heizöl-Verdampferöfen* (Determination of PAHs in the flue gas of oil-fired vaporizing pot burners) (*BMI-DGMK-Projekt 111-03*), Hamburg, Deutsche Gesellschaft für Mineralölwissenschaft und Kohlechemie

Rathkamp, G., Tso, T.C. & Hoffmann, D. (1973) Chemical analysis on tobacco smoke. XX. Smoke analysis of cigarettes made from Bright tobaccos differing in variety and stalk position. *Beitr. Tabakforsch.*, *7*, 179-189

Rees, E.D., Mandelstam, P., Lowry, J.Q. & Lipscomb, H. (1971) A study of the mechanism of intestinal absorption of benzo(*a*)pyrene. *Biochim. biophys. Acta*, *225*, 96-107

Rigdon, R.H. & Neal, J. (1966) Gastric carcinomas and pulmonary adenomas in mice fed benzo(a)pyrene. *Texas Rep. Biol. Med.*, *24*, 195-207

Romanowski, T., Funcke, W., König, J. & Balfanz, E. (1982) Isolation of polycyclic aromatic hydrocarbons in air particulate matter by liquid chromatography. *Anal. Chem.*, *54*, 1285-1287

Rondia, D. (1965) On the volatility of PAH (Fr.). *Int. J. Air Water Poll.*, *9*, 113-121

Sabadie, N., Richter-Reichhelm, H.B., Saracci, R., Mohr, U. & Bartsch, H. (1981) Inter-individual differences in oxidative benzo(*a*)pyrene metabolism by normal and tumorous surgical lung specimens from 105 lung cancer patients. *Int. J. Cancer*, *27*, 417-425

Santodonato, J., Basu, D. & Howard, P.H. (1980) *Multimedia human exposure and carcinogenic risk assessment for environmental PAH.* In: Bjørseth, A. & Dennis, A.J., eds, *Polynuclear Aromatic Hydrocarbons: Chemistry and Biological Effects, 4th Int. Symposium*, Columbus, OH, Battelle Press, pp. 435-454

Sato, R. & Omura, T. (1978) *Cytochrome P-450*, New York, Academic Press

Sawicki, E. (1967) Airborne carcinogens and allied compounds. *Arch. environ. Health*, *14*, 46-53

Sawicki, E. (1976) *Analysis of atmospheric carcinogens and their cofactors*. In: Rosenfeld, C. & Davis, W., eds, *Environmental Pollution and Carcinogenic Risks* (*IARC Scientific Publications No. 13*), Lyon, International Agency for Research on Cancer, pp. 297-354

Sawicki, E., Hauser, T.R., Elbert, W.C., Fox, F.T. & Meeker, J.E. (1962) Polynuclear aromatic hydrocarbon composition of the atmosphere in some large American cities. *Am. ind. Hyg. Assoc. J.*, *23*, 137-144

Saxena, J., Basu, D.K. & Schwartz, D.J. (1980) *Method development and monitoring of polynuclear aromatic hydrocarbons in selected U.S. waters*. In: Albaiges, J., ed., *Analytical Techniques in Environmental Chemistry, Proceedings of the International Congress, Barcelona*, Oxford, Pergamon Press, pp. 119-126

Schlede, E., Kuntzman, R., Haber, S. & Conney, A.H. (1970a) Effect of enzyme induction on the metabolism and tissue distribution of benzo[α]pyrene. *Cancer Res.*, *30*, 2893-2897

Schlede, E., Kuntzman, R. & Conney, A.H. (1970b) Stimulatory effect of benzo[α]pyrene and phenobarbital pretreatment on the biliary excretion of benzo[α]pyrene metabolites in the rat. *Cancer Res.*, *30*, 2898-2904

Schmidt, D.J., Petti, R.W., Rowley, B.B. & Pelton, D.J. (1977) *Benzo(a)pyrene. Final Report Tasks 1,2* (*Contract No. 68-03-2504*), Research Triangle Park, NC, US Environmental Protection Agency

Schmoldt, A., Jacob, J. & Grimmer, G. (1981) Dose-dependent induction of rat liver microsomal aryl hydrocarbon monooxygenase by benzo[*k*]fluoranthene. *Cancer Lett.*, *13*, 249-257

Seelkopf, C., Ricken, W. & Dhom, G. (1963) Carcinogenic properties of tar from cigarettes (Ger.). *Z. Krebsforsch.*, *65*, 241-249

Selkirk, J.K. (1977) Benzo[*a*]pyrene carcinogenesis: A biochemical selection mechanism. *J. Toxicol. environ. Health*, *2*, 1245-1258

Selkirk, J.K., Croy, R.G., Whitlock, J.P., Jr & Gelboin, H.V. (1975) *In vitro* metabolism of benzo(*a*)pyrene by human liver microsomes and lymphocytes. *Cancer Res.*, *35*, 3651-3655

Selkirk, J.K., Nikbakht, A. & Stoner, G.D. (1983) Comparative metabolism and macromolecular binding of benzo[*a*]pyrene in explant cultures of human bladder, skin, bronchus and esophagus from eight individuals. *Cancer Lett.*, *18*, 11-19

Severson, R.F., Arrendale, R.F., Chaplin, J.F. & Williamson, R.E. (1979) Use of pale-yellow tobacco to reduce smoke polynuclear aromatic hydrocarbons. *J. agric. Food Chem.*, *27*, 896-900

Shabad, L.M. (1967) Studies in the USSR on the distribution, circulation and fate of carcinogenic hydrocarbons in the human environment and the role of their deposition in tissues in carcinogenesis: A review. *Cancer Res.*, *27*, 1132-1137

Shabad, L.M. (1968) On the distribution and the fate of the carcinogenic hydrocarbon benz(*a*)pyrene (3,4 benzpyrene) in the soil. *Z. Krebsforsch.*, *70*, 204-210

Shabad, L.M. & Smirnov, G.A. (1972) Aircraft engines as a source of carcinogenic pollution of the environment [Benzo(*a*)pyrene-studies]. *Atmos. Environ.*, *6*, 153-164

Shabad, L.M., Khesina, A.J., Linnik, A.B. & Serkovskaya, G.S. (1970) Possible carcinogenic hazards of several tars and of locacorten-tar ointment (spectro-fluorescent investigations and experiments in animals). *Int. J. Cancer*, *6*, 314-318

Shabad, L.M., Cohan, Y.L., Ilnitsky, A.P., Khesina, A.Y., Shcherbak, N.P. & Smirnov, G.A. (1971) The carcinogenic hydrocarbon benzo[*a*]pyrene in the soil. *J. natl Cancer Inst.*, *47*, 1179-1191

Shendrikova, I.A. & Aleksandrov, V.A. (1974) Comparative characteristics of penetration of polycyclic hydrocarbons through the placenta to the fetus in rats. *Bull. exp. Biol. Med.*, *77*, 77-79

Shendrikova, I.A., Ivanov-Golitsyn, M.N., Anisimov, V.N. & Likhachev, A.Ya. (1973) Dynamics of the transplacental penetration of 7,12-dimethylbenz(*a*)anthracene in mice (Russ.). *Vopr. Onkol.*, *19*, 75-79

Shendrikova, I.A., Ivanov-Golitsyn, M.N. & Likhachev, A.Ya. (1974) The transplacental penetration of benzo(*a*)pyrene in mice (Russ.). *Vopr. Onkol.*, *20*, 53-56

Siegfried, R. (1975) 3,4-Benzopyrene in oils and fats (Ger.). *Naturwissenschaften*, *62*, 576

Sims, P. (1982) *The metabolic activation of some polycyclic hydrocarbons: The role of dihydrodiols and diol-epoxides*. In: Synder, R., Parke, D.V., Kocsis, J.J., Jollow, D.J., Gibson, C.G. & Witmer, C.M., eds, *Biological Reactive Intermediates - II. Chemical Mechanisms and Biological Effects*, Part A, New York, Plenum Press, pp. 487-500

Sims, P. & Grover, P.L. (1974) Epoxides in polycyclic aromatic hydrocarbon metabolism and carcinogenesis. *Adv. Cancer Res.*, *20*, 165-274

Sims, P. & Grover, P.L. (1981) *Involvement of dihydrodiols and diol epoxides in the metabolic activation of polycyclic hydrocarbons other than benzo[*a*]pyrene*. In: Gelboin, H.V. & Ts'O, P.O.P., eds, *Polycylic Hydrocarbons and Cancer*, Vol. 3, New York, Academic Press, pp. 117-181

Sims, P., Grover, P.L., Swaisland, A., Pal, K. & Hewer, A. (1974) Metabolic activation of benzo(*a*)pyrene proceeds by a diol-epoxide. *Nature*, *252*, 326-328

Šipal, Z., Ahlenius, T., Bergstrand, A., Rodriguez, L. & Jakobsson, S.W. (1979) Oxidative biotransformation of benzo(*a*)pyrene by human lung microsomal fractions prepared from surgical specimens. *Xenobiotica*, *9*, 633-345

Snook, M.E., Severson, R.F., Arrendale, R.F., Higman, H.C. & Chortyk, O.T. (1977) The identification of high molecular weight polynuclear aromatic hydrocarbons in a biologically active fraction of cigarette smoke condensate. *Beitr. Tabakforsch.*, *9*, 79-101

Snook, M.E., Severson, R.F., Arrendale, R.F., Higman, C.H. & Chortyk, O.T. (1978) Multi-alkylated polynuclear aromatic hydrocarbons of tobacco smoke: Separation and identification. *Beitr. Tabakforsch.*, *9*, 222-247

Sonnichsen, T.W., McElroy, M.W. & Bjørseth, A. (1980) *Use of PAH tracers during sampling of coal fired boilers*. In: Bjørseth, A. & Dennis, A.J., eds, *Polynuclear Aromatic Hydrocarbons, Chemistry and Biological Effects, 4th Int. Symposium*, Columbus, OH, Battelle Press, pp. 617-632

Starkey, R. & Warpinski, J. (1974) Size distribution of particulate benzo[*a*]pyrene. *J. environ. Health*, *36*, 503-505

Steinig, J. (1976) 3,4-Benzopyrene contents in smoked fish depending on smoking procedure (Ger.). *Z. Lebensm. Unters.-Forsch.*, *162*, 235-242

Stich, H.F., Acton, A.B. & Dunn, B.P. (1976) *Carcinogens in estuaries, their monitoring and possible hazard to man*. In: Rosenfeld, C. & Davis, W., eds, *Environmental Pollution and Carcinogenic Risks* (*IARC Scientific Publications No. 13*), Lyon, International Agency for Research on Cancer, pp. 83-93

Stoner, G.D., Harris, C.C., Autrup, H., Trump, B.F., Kingsbury, E.W. & Myers, G.A. (1978) Explant culture of human peripheral lung. I. Metabolism of benzo[*a*]pyrene. *Lab. Invest.*, *38*, 685-692

Strand, J.W. & Andren, A.W. (1980) *Polyaromatic hydrocarbons in aerosols over Lake Michigan, fluxes to the lake.* In: Bjørseth, A. & Dennis, A.J., eds, *Polynuclear Aromatic Hydrocarbons: Chemistry and Biological Effects, 4th Int. Symposium*, Columbus, OH, Battelle Press, pp. 127-137

Styles, J.A. (1977) A method of detecting carcinogenic organic chemicals using mammalian cells in culture. *Br. J. Cancer*, *36*, 558-563

Swarin, S.J. & Williams, R.L. (1980) *Liquid chromatographic determination of benzo[*a*]pyrene in diesel exhaust particulate: Verification of the collection and analytical methods.*In: Bjørseth, A. & Dennis, A.J., eds, *Polynuclear Aromatic Hydrocarbons: Chemistry and Biological Effects, 4th Int. Symposium*, Columbus, OH, Battelle Press, pp. 771-790

Talcott, R. & Wei, E. (1977) Airborne mutagens bioassayed in *Salmonella typhimurium*. *J. natl Cancer Inst.*, *58*, 449-451

Taylor, G.T., Redington, T.E., Bailey, M.J., Buddingh, F. & Nau, C.A. (1980) Solvent extracts of carbon black - Determination of total extractables and analysis for benzo(α)pyrene. *Am. ind. Hyg. Assoc. J.*, *41*, 819-825

Tebbens, B.D., Mukai, M. & Thomas, J.F. (1971) Fate of arenes incorporated with airborne soot: Effect of irradiation. *Am. ind. Hyg. Assoc. J.*, *32*, 365-372

Testa, B. & Jenner, P. (1976) *Drugs and the Pharmaceutical Sciences*, Vol. 4, *Drug Metabolism: Chemical and Biochemical Aspects*, New York, Marcel Dekker

Thakker, D.R., Yagi, H., Akagi, H., Korveda, M., Lu, A.Y.H., Levin, W., Wood, A.W., Conney, A.H. & Jerina, D.M. (1977) Metabolism of benzo[*a*]pyrene. VI. Stereoselective metabolism of benzo[*a*]pyrene and benzo[*a*]pyrene 7,8-dihydrodiol to diol epoxides. *Chem.-biol. Interactions*, *16*, 281-300

Thakker, D.R., Levin, W., Buening, M., Yagi, H., Lehr, R.E., Wood, A.W., Conney, A.H. & Jerina, D.M. (1981) Species-specific enhancement by 7,8-benzoflavone of hepatic microsomal metabolism of benzo[*e*]pyrene 9,10-dihydrodiol to bay-region diol epoxides. *Cancer Res.*, *41*, 1389-1396

Thomas, J.F., Mukai, M. & Tebbens, B.D. (1968) Fate of airborne benzo[*a*]pyrene. *Environ. Sci. Technol.*, *2*, 33-39

Tokiwa, H., Morita, K., Takeyoshi, H., Takahashi, K. & Ohnishi, Y. (1977) Detection of mutagenic activity in particulate air pollutants. *Mutat. Res.*, *48*, 237-248

Tomatis, L. (1973) *Transplacental carcinogenesis.* In: Raven, R.W., ed., *Modern Trends in Oncology*, Part 1, *Research Progress*, London, Butterworths, pp. 99-126

Tomatis, L. (1974) *Role of prenatal events in determining cancer risks.* In: Boyland, E. & Goulding, R., eds, *Modern Trends in Toxicology*, Vol. 2, London, Butterworths, pp. 163-178

Tomingas, R., Pott, F. & Dehnen, W. (1976) Polycyclic aromatic hydrocarbons in human bronchial carcinoma. *Cancer Lett.*, *1*, 189-196

Tóth, L. (1971) Polycyclic hydrocarbons in smoked ham and belly fat (Ger.). *Fleischwirtschaft*, *7*, 1069-1070

US Department of Health & Human Services (1982) *The Health Consequences of Smoking. Cancer. A Report of the Surgeon General* (*DHHS (PHS) 82-50179*), Washington DC, US Government Printing Office

US Environmental Protection Agency (1971) National primary and secondary ambient air quality standards. *Fed. Regist.*, *36*, 8186

US Environmental Protection Agency (1975) *Scientific and Technical Assessment Report on Particulate Polycyclic Organic Matter (PPOM)* (*EPA-600/6-75-001*), Washington DC

Vainio, H., Uotila, P., Hartiala, J. & Pelkonen, O. (1976) The fate of intratracheally instilled benzo(*a*)pyrene in the isolated perfused rat lung of both control and 20-methylcholanthrene pretreated rats. *Res. Commun. chem. Pathol. Pharmacol., 13,* 259-271

Van Duuren, B.L. (1958a) Identification of smoke polynuclear aromatic hydrocarbons in cigarette-smoke condensate. *J. natl Cancer Inst.*, *21*, 1-16

Van Duuren, B.L. (1958b) The polynuclear aromatic hydrocarbons in cigarette smoke condensate. II. *J. natl Cancer Inst.*, *21*, 623-630

Van Duuren, B.L., Bilbao, J.A. & Joseph, C.A. (1960) The carcinogenic nitrogen heterocyclics in cigarette-smoke condensate. *J. natl Cancer Inst.*, *25*, 53-61

Vaught, J.B., Gurtoo, H.L., Paigen, B., Minowada, J. & Sartori, P. (1978) Comparison of benzo[*a*]pyrene metabolism by human peripheral blood lymphocytes and monocytes. *Cancer Lett.*, *5*, 261-268

Vermorken, A.J.M., Goos, C.M.A.A., Roelofs, H.M.J., Henderson, P.Th. & Bloemendal, H. (1979) Metabolism of benzo[*a*]pyrene in isolated human scalp hair follicles. *Toxicology*, *14*, 109-116

Wakeham, S.G., Schaffner, C. & Giger, W. (1980) Polycyclic aromatic hydrocarbons in recent lake sediments - II. Compounds derived from biogenic precursors during diagenesis. *Geochim. cosmiochim. Acta*, *44*, 415-429

Wattenberg, L.W. (1978) Inhibitors of chemical carcinogenesis. *Adv. Cancer Res.*, *26*, 197-226

Weinstein, I.B., Jeffrey, A.M., Leffler, S., Pulkrabek, P., Yamasaki, H. & Grunberger, D. (1978) *Interactions between polycyclic aromatic hydrocarbons and cellular macromolecules.* In: Ts'O, P.O.P. & Gelboin, H.V., eds, *Polycyclic Hydrocarbons and Cancer*, Vol. 2, New York, Academic Press, pp. 4-36

Weldre, J.A., Rachu, M.A., Ilnitzky, A.P., Lochow, L.G. & Schereweschew, N.J. (1977) On the investigation of carcinogenic hydrocarbons, especially benzo[*a*]pyrene in water in the ESSR (Estonian SSR) (Russ.). *Water Res.*, *3*, 147-152

Whitehead, J.K. (1977) *An Appraisal of Various Approaches to the Fractionation of Tobacco Smoke Condensate* (*Research Paper 13*), London, Tobacco Research Council

WHO (1964) Prevention of Cancer. Report of a WHO Expert Committee (*WHO Tech. Rep. Ser. No. 276*), Geneva

WHO (1971) *International Standards for Drinking Water*, 3rd ed., Geneva, p. 37

de Wiest, F., Rondia, D., Gol-Winkler, R. & Gielen, J. (1982) Mutagenic activity of non-volatile organic matter associated with suspended matter in urban air. *Mutat. Res.*, *104*, 201-207

Williams, R.L. & Swarin, S.J. (1979) *Benzo(a)pyrene Emissions from Gasoline and Diesel Automobiles* (*GMR-2881R*), Warren, MI, General Motors Research Laboratories

Woidich, W., Pfannhauser, W., Blaicher, G. & Tiefenbacher, K. (1976) Analysis of polycyclic aromatic hydrocarbons in drinking and utilitary water (Ger.). *Lebensmittelchem. gerichtl. Chem.*, *30*, 141-146

Wood, A.W., Levin, W., Thakker, D.R., Yagi, H., Chang, R.L., Ryan, D.E., Thomas, P.E., Dansette, P.M., Whittaker, N., Turujman, S., Lehr, R.E., Kumar, S., Jerina, D.M. & Conney, A.H. (1979) Biological activity of benzo[*e*]pyrene. *J. biol. Chem.*, 4408-4415

Wszolek, P.C. & Wachs, T. (1982) Occurrence of polycyclic aromatic hydrocarbons in municipal sewage sludge ashes. *Arch. environ. Contam. Toxicol.*, *11*, 69-72

Wynder, E.L. & Hoffmann, D. (1959) The carcinogenicity of benzofluoranthenes. *Cancer*, *12*, 1194-1199

Wynder, E.L. & Hoffmann, D. (1961) Present status of laboratory studies on tobacco carcinogenesis. *Acta pathol. microbiol. scand.*, *52*, 119-132

Wynder, E.L. & Hoffmann, D. (1962) A study of air pollution carcinogenesis. III. Carcinogenic activity of gasoline engine exhaust condensate. *Cancer*, *15*, 103-108

Wynder, E.L. & Hoffmann, D. (1963) Experimental contribution to tobacco smoke carcinogenesis (Ger.). *Dtsch. med. Wochenschr.*, *88*, 623-628

Wynder, E.L. & Wright, E. (1957) A study of tobacco carcinogenesis. I. The primary fractions. *Cancer*, *10*, 255-271

Yang, S.K., Roller, P.P. & Gelboin, H.V. (1978) *Benzo[*a*]pyrene metabolism: Mechanism in the formation of epoxides, phenols, dihydrodiols and the 7,8-diol-9,10-epoxides*. In: Jones, P.W. & Freudenthal, R.I., eds, *Carcinogenesis*, Vol. 3, *Polynuclear Aromatic Hydrocarbons*, New York, Raven Press, pp. 285-301

Zdrazil, J. & Pícha, F. (1966) The occurrence of the carcinogenic compounds 3,4-benzpyrene and arsenic in the soil. *Neoplasma*, *13*, 49-55

Zelenski, S.G., Hunt, G.T. & Pangaro, N. (1980) *Comparison of SIM GC/MS and HPLC for the detection of polynuclear aromatic hydrocarbons in fly ash collected from stationary combustion sources*. In: Bjørseth, A. & Dennis, A.J., eds, *Polynuclear Aromatic Hydrocarbons: Chemistry and Biological Effects, 4th Int. Symposium*, Columbus, OH, Battelle Press, pp. 589-597

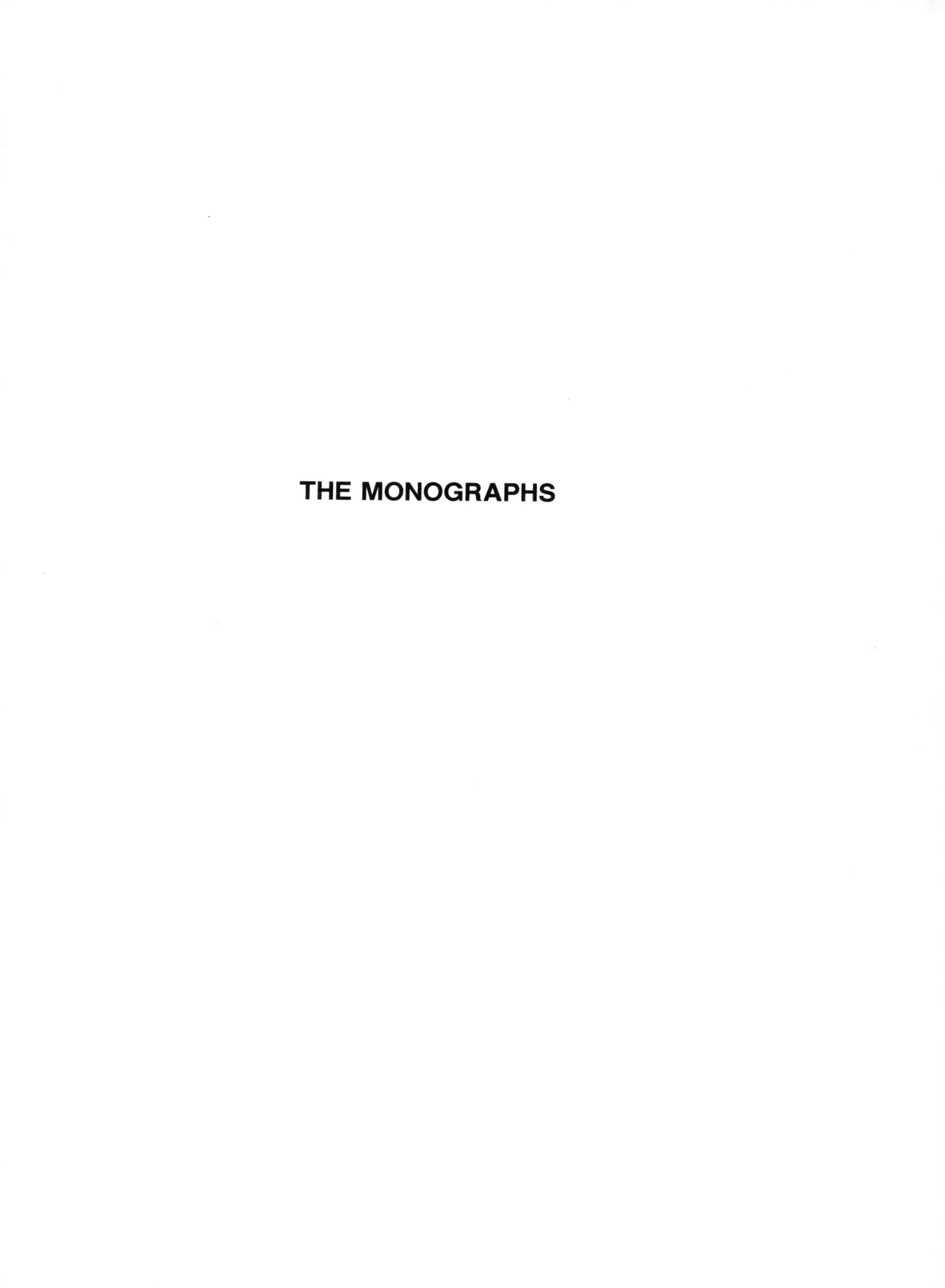

THE MONOGRAPHS

ANTHANTHRENE

1. Chemical and Physical Data

1.1 Synonyms and trade names

Chem. Abstr. Services Reg. No.: 191-26-4

Chem. Abstr. Name: Dibenzo(def,mno)chrysene

IUPAC Systematic Name: Dibenzo[*def,mno*]chrysene

Synonyms: Anthanthren; dibenzo(cd,jk)pyrene

1.2 Structural and molecular formulae and molecular weight

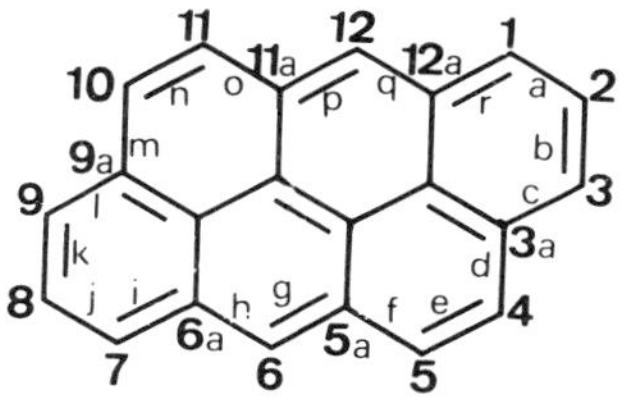

$C_{22}H_{12}$

Mol. wt: 276.3

1.3 Chemical and physical properties of the pure substance

From Clar (1964), unless otherwise specified

(*a*) *Description:* Golden-yellow plates (recrystallized from xylene)

(*b*) *Melting-point*: 264°C (Karcher *et al.*, 1983)

(*c*) *Spectroscopy data*: λ_{max} 231, 254, 257, 282, 293, 305, 320, 360, 379, 382, 399, 405, 420, 429 nm (in cyclohexane) (Karcher *et al.*, 1983). Mass and nuclear magnetic resonance spectra have been tabulated (NIH/EPA Chemical Information System, 1982; Karcher *et al.*, 1983). Infrared spectra have been reported (Buckingham, 1982).

(*d*) *Solubility*: Soluble in 1,4-dioxane (Hoffmann & Wynder, 1966), benzene, toluene (Lijinsky & Garcia, 1972) and olive oil (Lacassagne *et al.*, 1958)

(*e*) *Stability*: No data were available.

(*f*) *Reactivity*: Can be halogenated and nitrated; reacts with sulphuryl chloride to give 6,12-dichloroanthanthrene; reacts with NO and NO_2 to form nitro derivatives (Butler & Crossley, 1981)

2. Production, Use, Occurrence and Analysis

2.1 Production and use

There is no commercial production or known use of this compound. A reference material of certified purity is available (Community Bureau of Reference, 1982).

2.2 Occurrence and analysis

Data on occurrence and methods of analysis are summarized in the 'General Remarks on the Substances Considered', p. 35.

Anthanthrene occurs ubiquitously as a product of incomplete combustion. It also occurs in fossil fuels. It is found in relatively high concentrations in coal-tar (Lang & Eigen, 1967). It has been identified in mainstream cigarette smoke (0.2 μg/100 g of burnt material) (Masuda & Kuratsune, 1972), (22 ng/cigarette) (Grimmer *et al.*, 1977a); sidestream cigarette smoke (39 ng/cigarette) (Grimmer *et al.*, 1977a); cigarette-smoke polluted rooms (3 ng/m^3) (Grimmer *et al.*, 1977b); the air of restaurants (0.5-1.9 ng/m^3) (Just *et al.*, 1972); mainstream smoke of marijuana cigarettes (0.5 μg/100 cigarettes) (Lee *et al.*, 1976); urban air (0.26-6 ng/m^3) (Hoffmann & Wynder, 1976); gasoline engine exhaust tar (44 μg/kg) (Hoffmann & Wynder, 1962); the exhaust of various burnt coals (0-0.9 mg/kg) (Brockhaus & Tomingas, 1976); charcoal-broiled steaks (2 μg/kg) (Lijinsky & Shubik, 1964); high-protein foods, vegetable oils and fats (Grimmer & Böhnke, 1975); surface water (0.2-10.9 ng/l) (Grimmer *et al.*, 1981a); waste water (0.04-0.6 μg/l) (Grimmer *et al.*, 1981a); freeze-dried sewage sludge (0-2100 μg/kg) (Grimmer *et al.*, 1980); dried sediment from lakes (10-212 μg/kg) (Grimmer & Böhnke, 1977); various lubricating oils (0.002-0.03 mg/kg) (Grimmer *et al.*, 1981b) and used motor oils (0.03-14.66 mg/kg) (Grimmer *et al.*, 1981c).

3. Biological Data Relevant to the Evaluation of Carcinogenic Risk to Humans

3.1 Carcinogenicity studies in animals[1]

(a) Skin application

Mouse: No skin tumour was reported in 30 mice [strain, sex, age and body weight unspecified] given skin applications of 0.3% anthanthrene in benzene [dose and purity unspecified] twice weekly for life. The last mouse died on day 712; one mouse had a lung adenoma. Some other derivatives of cholanthrene or benzo[*a*]pyrene tested under the same experimental conditions induced skin tumours at an incidence of up to 30% (10/30) (Badger *et al.*, 1940).

Five groups of 20 female Ha/ICR/Mil (Swiss albino) mice, seven to eight weeks of age, received skin applications of 0.05 or 0.1% anthanthrene (purified by column chromatography and crystallization) [doses unspecified] in 1,4-dioxane or 1,4-dioxane alone (vehicle controls), or of 0.05 or 0.1% benzo[*a*]pyrene in 1,4-dioxane (positive controls) thrice weekly for 12 months. Survivors were killed at the end of the fifteenth month, when all positive controls and 8/20 vehicle controls had died [mortality rate of anthanthrene-treated group unspecified]. The numbers of skin tumour-bearing mice were 0/20, 0/20, 0/20, 17/20 and 19/20 in the low- and high-dose anthanthene-treated, vehicle control and low- and high-dose benzo[*a*]pyrene-treated groups, respectively (Hoffmann & Wynder, 1966).

A group of 30 female random-bred Swiss mice, eight to ten weeks of age, were given skin applications of 43 μg anthanthrene purified by crystallization and chromatography in toluene (16-20 μl of a 0.2% solution of anthanthrene in toluene) (total dose, 6.5 mg/animal) twice weekly for 75 weeks. A group of 30 mice of the same strain, sex and age received toluene alone (vehicle controls) twice weekly for 72 weeks. All animals were observed until death. Mortality rates were 9/30 in the anthanthrene-treated group and 6/30 in the control group after 60 weeks, 13/30 and 15/30 after 80 weeks, and 27/30 and 30/30 after 100 weeks. In the treated group one animal had a skin carcinoma (latent period, 58 weeks) and of the controls two had a skin tumour, one a carcinoma and one a papilloma (latent period, 58 weeks). Hydrogenated derivatives of anthanthrene tested under the same experimental conditions gave rise to skin tumours (papillomas and carcinomas) at incidences of up to 67% (Lijinsky & Garcia, 1972).

Five groups of 40 female random-bred Swiss mice, seven weeks of age, were given skin applications of 109.3 μg anthanthrene (purity, 98.65%; impurities were 1,2,3,7,8,9-hexahydroanthanthrene, 4,5,6,10,11,12-hexahydroanthanthrene, tetrahydroanthanthrene and 4,5-dihydroanthranthrene) suspended in 16.7 μl acetone, or a corresponding molar amount of dibenzo[*a,h*]pyrene, benzo[*a*]pyrene or benz[*a*]anthracene in 16.7 μl acetone, or 16.7 μl acetone alone (vehicle controls) twice weekly for 30 weeks. After 70 weeks all sur-

[1] The Working Group was aware of a study in progress by skin application in mice (IARC, 1983).

vivors were killed. Mean survival times ranged from 38 ± 5 weeks in the benzo[*a*]pyrene-treated group to 61 ± 15 weeks in the anthanthrene-treated group and 65 ± 11 weeks in the vehicle control group. The numbers of skin tumour-bearing mice were 18/38, 35/39, 30/38, 1/39 and 0/29 in the anthanthrene-treated, dibenzo[*a,h*]pyrene-treated, benzo[*a*]pyrene-treated, benz[*a*]anthracene-treated and vehicle-control groups, respectively. Anthanthrene induced a total of seven papillomas, two keratoacanthomas, 14 carcinomas and one sebaceous-gland adenoma; dibenzo[*a,h*]pyrene induced 17 papillomas, 13 keratoacanthomas, 45 carcinomas, two fibrosarcomas and two sebaceous-gland adenomas; benzo[*a*]pyrene induced seven papillomas, seven keratoacanthomas, 36 carcinomas and one malignant schwannoma; benz[*a*]anthracene induced one papilloma. Tumours at other sites (lung adenomas and malignant lymphomas) occurred at similar incidence rates in control and test groups; these tumours could not be associated with treatment (Cavalieri *et al.*, 1977).

In order to test the initiating activity of anthanthrene, two groups of 30 Ha/ICR/Mil (Swiss albino) mice [sex, age and body weight unspecified] received skin applications of 0.025 ml 0.1% chromatographically purified anthanthrene in 1,4-dioxane [25 μg; total dose, 0.25 mg/animal] or 0.025 ml 0.1% benzo[*a*]pyrene in 1,4-dioxane (positive controls) once every three days for 28 days. On day 28, the skin was painted with 2.5% (2.3 mg) croton oil in acetone [volume, number, frequency and duration of treatments not given]. A group of 30 mice was treated with croton oil solution alone (promoter controls), and another group of 20 mice received skin applications of 1,4-dioxane alone (vehicle controls). All survivors were killed after six months, at which time mortality rates were 5/30, 2/30, 4/30 and 2/20, and the number of skin papilloma-bearing mice 2/30, 24/30, 2/30 and 0/20 in the anthanthrene-treated, positive-control, promoter control and vehicle control groups, respectively (Hoffmann & Wynder, 1966). [The Working Group noted the short duration of the study.]

In order to examine the initiating activity of anthanthrene, two groups of 13 female ICR/Ha Swiss mice, eight weeks of age [body weight unspecified], were given four consecutive skin applications of 0.25 mg 'rigorously purified' anthanthrene in 0.1 ml benzene [total dose, 1 mg]. In one group, this treatment was followed two weeks later by skin applications of 25 μg croton resin in 0.1 ml acetone thrice weekly for life. These were four control groups: one group of 20 untreated females, one group of 20 females treated with acetone only, and two groups of 20 females treated with croton resin in acetone. The duration of the experiments varied from 59-66 weeks [mortality rates unspecified]. No skin tumour was found in animals treated with anthanthrene only, in untreated controls or in controls treated with acetone alone. Skin papillomas developed in 2/13 animals treated with anthanthrene followed by croton resin. However, in one of the control groups treated with croton resin alone, 5/20 animals had a skin papilloma and 1/20 a skin carcinoma; in the other control group, 1/20 animals had a skin papilloma. 6-Methylanthanthrene tested under the same experimental conditions showed clear tumour initiating activity (6/16 mice had 14 skin papillomas and 4/16 mice a skin carcinoma) (Van Duuren *et al.*, 1968).

In order to examine the initiating activity of anthanthrene, a group of 30 female CD-1 mice, eight weeks of age, received a single skin application of 2.5 μmol [690 μg] chromatographically purified anthanthrene in benzene, followed one week later by applications of 5 μmol 12-*O*-tetradecanoylphorbol-13-acetate (TPA) twice weekly for 34 weeks. A control group of 30 female mice of the same strain was treated with 10 μmol TPA only twice weekly for 34 weeks. At week 35, 28/30 animals in the anthanthrene-treated group and all controls were still alive. At that time, 5/28 (18%) anthanthrene-treated mice had developed a skin papilloma, whereas skin tumours had occurred in 3% of TPA controls (Scribner, 1973). [The Working Group noted that the dose of TPA was probably 5 μg (Scribner & Süss, 1978).]

(*b*) *Subcutaneous and/or intramuscular administration*

Mouse: No local sarcoma was found in seven male and seven female young adult XVII mice [body weights unspecified] given s.c. injections of 0.6 mg anthanthrene [purity unspecified] in 0.2 ml olive oil [total dose, 1.8 mg/animal] once a month for three months [mortality rate unspecified]. Of 14 male and 16 female mice of the same strain treated with benzo[*a*]pyrene under the same experimental conditions (positive controls), 13 males (93%) and eight females (50%) developed local sarcomas (Lacassagne *et al.*, 1958). [The Working Group noted the small number of animals used and the short duration of treatment.]

(*c*) *Other experimental systems*

Intrapulmonary injection: Two groups of 35 female Osborne-Mendel *rats*, three months of age, body weight 245 g, were injected directly into the lung after thoracotomy (Stanton *et al.*, 1972) with 0.05 ml of a mixture of beeswax + tricaprylin containing 0.16 or 0.83 mg anthanthrene (purity, 99.4%) [0.65 or 3.4 mg/kg bw]. Three similar groups of 35 rats received 0.1, 0.3 or 1.0 mg benzo[*a*]pyrene (positive controls), and further groups of rats served as untreated or vehicle controls. The animals were observed until spontaneous death: mean survival times were 102, 88, 111, 77, 54, 118 and 104 weeks for the low- and high-dose anthanthrene-treated, the low-, mid- and high-dose benzo[*a*]pyrene-treated, untreated and vehicle-control groups, respectively. The numbers of anthanthrene-treated rats bearing lung tumours were 1/35 and 19/35 in the low- and high-dose groups, respectively. All tumours were squamous-cell carcinomas. The incidences of lung carcinomas in the positive-control groups were 4/35, 21/35 and 33/35 in the low-, mid- and high-dose groups, respectively. In addition, six low-dose rats and two mid-dose rats from the positive-control groups had developed a pulmonary fibrosarcoma. No lung tumour was seen in any of the vehicle or untreated controls (Deutsch-Wenzel *et al.*, 1983).

3.2 Other relevant biological data[1]

(*a*) *Experimental systems*

No data were available to the Working Group on toxic effects, effects on reproduction and prenatal toxicity, or absorption, distribution, excretion and metabolism.

Mutagenicity and other short-term tests

Results from short-term tests are given in Table 1.

Table 1. Results from short-term tests: Anthanthrene[a]

Test	Organism/assay[b]	Exogenous metabolic system[b]	Reported result	Comments	References
PROKARYOTES					
Mutation	*Salmonella typhimurium* (his⁻/his⁺)	Aro-R-PMS	Positive	At 20 μg/plate in strain TA100	Andrews *et al.* (1978);
				At 1 nmol/plate in strain TA98	Hermann (1981)
	Salmonella typhimurium (8AG^S/8AG^R)	Aro-R-PMS	Positive	At 40 nmol/ml in strain TM677	Kaden *et al.* (1979)

[a] This table comprises selected assays and references and is not intended to be a complete review of the literature

[b] For an explanation of the abbreviations, see the Appendix, p. 452.

[1] See also 'General Remarks on the Substances Considered', p. 53.

(*b*) *Humans*

No data were available to the Working Group.

3.3 Case reports and epidemiological studies of carcinogenicity to humans

No data were available to the Working Group.

4. Summary of Data Reported and Evaluation[1]

4.1 Experimental data

Anthanthrene was tested for carcinogenicity by skin application in mice in four studies. In one study, relatively high doses of anthanthrene in acetone produced skin tumours; in the other studies, no increased incidence of tumours was observed. It was also tested in the mouse-skin initiation-promotion assay in three studies. In two of the studies for initiating activity, anthanthrene gave negative results; in one, the results were inconclusive.

An experiment involving subcutaneous injection of anthanthrene to mice was inadequate. In a study using direct injection into the pulmonary tissue of rats, anthanthrene produced pulmonary squamous-cell carcinomas in a dose-related fashion.

No data on the teratogenicity of this compound were available.

Anthanthrene was mutagenic to *Salmonella typhimurium* in the presence of an exogenous metabolic system.

There is *inadequate evidence* that anthanthrene is active in short-term tests.

[1] For definitions of the italicized terms, see Preamble, p. 19.

4.2 Human data[2]

Anthanthrene is present as a minor component of the total content of polynuclear aromatic compounds of the environment. Human exposure to anthanthrene occurs primarily through the smoking of tobacco, inhalation of polluted air and by ingestion of food and water contaminated with combustion products (for details, see 'General Remarks on the Substances Considered', p. 35).

4.3 Evaluation

There is *limited evidence* that anthanthrene is carcinogenic to experimental animals.

5. References

Andrews, A.W., Thibault, L.H. & Lijinsky, W. (1978) The relationship between carcinogenicity and mutagenicity of some polynuclear hydrocarbons. *Mutat. Res.*, *51*, 311-318

Badger, G.M., Cook, J.W., Hewett, C.L., Kennaway, E.L., Kennaway, N.M., Martin, R.H. & Robinson, A.M. (1940) The production of cancer by pure hydrocarbons. V. *Proc. R. Soc. London*, *129*, 439-467

Brockhaus, A. & Tomingas, R. (1976) Emission of polycyclic hydrocarbons during burning processes in small heating installations and their concentration in the atmosphere (Ger.). *Staub-Reinhalt. Luft*, *36*, 96-101

Buckingham, J., ed. (1982) *Dictionary of Organic Compounds*, 5th ed., Vol. 2, New York, Chapman & Hall, p. 1596

[2] Studies on occupational exposure to polynuclear aromatic compounds will be considered in future *IARC Monographs*.

Butler, J.D. & Crossley, P. (1981) Reactivity of polycyclic aromatic hydrocarbons adsorbed on soot particles. *Atmos. Environ.*, *15*, 91-94

Cavalieri, E., Mailander, P. & Pelfrene, A. (1977) Carcinogenic activity of anthanthrene on mouse skin. *Z. Krebsforsch.*, *89*, 113-118

Clar, E., ed. (1964) *Polycyclic Hydrocarbons*, Vol. 2, New York, Academic Press, pp. 206-208

Community Bureau of Reference (1982) *Polycyclic Aromatic Hydrocarbon Reference Materials of Certified Purity* (*Information Handout, No. 22*), Brussels, Commission of the European Communities

Deutsch-Wenzel, R.P., Brune, H., Grimmer, G., Dettbarn, G. & Misfeld, J. (1983) Experimental studies in rat lungs on the carcinogenicity and dose-response relationships of eight frequently occurring environmental polycyclic aromatic hydrocarbons. *J. natl Cancer Inst.*, *71*, 539-544

Grimmer, G. & Böhnke, H. (1975) Polycyclic aromatic hydrocarbon profile analysis of high-protein foods, oils, and fats by gas chromatography. *J. Assoc. off. anal. Chem.*, *58*, 725-733

Grimmer, G. & Böhnke, H. (1977) Investigation on drilling cores of sediments of Lake Constance. I. Profiles of the polycyclic aromatic hydrocarbons. *Z. Naturforsch.*, *32c*, 703-711

Grimmer, G., Böhnke, H. & Harke, H.-P. (1977a) Passive smoking: Measuring the concentrations of polycyclic aromatic hydrocarbons in rooms after machine smoking of cigarettes (Ger.). *Int. Arch. occup. environ. Health*, *40*, 83-92

Grimmer, G., Böhnke, H. & Harke, H.-P. (1977b) Passive smoking: Intake of polycyclic aromatic hydrocarbons by breathing of cigarette smoke containing air (Ger.). *Int. Arch. occup. environ. Health*, *40*, 93-99

Grimmer, G., Hilge, G. & Niemitz, W. (1980) Comparison of polycyclic aromatic hydrocarbons in sewage sludge samples from 25 sewage treatment works (Ger.). *Wasser*, *54*, 255-272

Grimmer, G., Schneider, D. & Dettbarn, G. (1981a) The load of different rivers in the Federal Republic of Germany by PAH (PAH - profiles of surface water) (Ger.). *Wasser*, *56*, 131-144

Grimmer, G., Jacob, J. & Naujack, K.-W. (1981b) Profile of the polycyclic aromatic hydrocarbons from lubricating oils. Inventory by GCGC/MS - PAH in environmental materials, Part 1. *Fresenius Z. anal. Chem.*, *306*, 347-355

Grimmer, G., Jacob, J., Naujack, K.-W. & Dettbarn, G. (1981c) Profile of the polycyclic aromatic hydrocarbons from used engine oil - Inventory by GCGC/MS - PAH in environmental materials, Part 2. *Fresenius Z. anal. Chem.*, *309*, 13-17

Hermann, M. (1981) Synergistic effects of individual polycyclic aromatic hydrocarbons on the mutagenicity of their mixtures. *Mutat. Res.*, *90*, 399-409

Hoffmann, D. & Wynder, E.L. (1962) A study of air pollution carcinogenesis. II. The isolation and identification of polycyclic aromatic hydrocarbons from gasoline engine exhaust condensate. *Cancer*, *15*, 93-102

Hoffmann, D. & Wynder, E.L. (1966) Contribution to the carcinogenic action of dibenzopyrenes (Ger.). *Z. Krebsforsch.*, *68*, 137-149

Hoffmann, D. & Wynder, E.L. (1976) *Environmental respiratory carcinogenesis*. In: Searle, C.E., ed., *Chemical Carcinogens* (*ACS Monograph 173*), Washington DC, American Chemical Society, p. 341

IARC (1983) *Information Bulletin on the Survey of Chemicals Being Tested for Carcinogenicity*, No. 10, Lyon, p. 17

Just, J., Borkowska, M. & Maziarka, S. (1972) Air pollution by tobacco smoke in Warsaw coffee houses (Pol.). *Rocz. Panstw. Zakł. Hyg.*, *23*, 129-135

Kaden, D.A., Hites, R.A. & Thilly, W.G. (1979) Mutagenicity of soot and associated polycyclic aromatic hydrocarbons to *Salmonella typhimurium*. *Cancer Res.*, *39*, 4152-4159

Karcher, W., Fordham, R., Dubois, J. & Gloude, P. (1983) *Spectral Atlas of Polycyclic Aromatic Compounds*, Dordrecht, The Netherlands, D. Reidel (in press)

Lacassagne, A., Buu-Hoï, N.P. & Zajdela, F. (1958) Relationship between molecular structure and carcinogenic activity in three series of hexacyclic aromatic hydrocarbons (Fr.). *C.R. Acad. Sci. Paris*, *246*, 1477-1480

Lang, K.F. & Eigen, I. (1967) *Organic compounds in coal tar* (Ger.). In: Heilbronner, E., Hofmann, U., Schäfer, K. & Wittig, G., eds, *Fortschrifte der chemischer Forschung* (Advances in chemical research), Vol. 8, Berlin, Springer, pp. 91-170

Lee, M.L., Novotny, M. & Bartle, K.D. (1976) Gas chromatography/mass spectrometric and nuclear magnetic resonance spectrometric studies of carcinogenic polynuclear aromatic hydrocarbons in tobacco and marijuana smoke condensates. *Anal. Chem.*, *48*, 405-416

Lijinsky, W. & Garcia, H. (1972) Skin carcinogenesis tests of hydrogenated derivatives of anthanthrene and other polynuclear hydrocarbons. *Z. Krebsforsch.*, *77*, 226-230

Lijinsky, W. & Shubik, P. (1964) Benzo(*a*)pyrene and other polynuclear hydrocarbons in charcoal-broiled meat. *Science*, *145*, 53-55

Masuda, Y. & Kuratsune, M. (1972) Comparison of the yield of polycyclic aromatic hydrocarbons in smoke from Japanese tobacco. *Jpn. J. Hyg.*, *27*, 339-341

NIH/EPA Chemical Information System (1982) *Mass Spectral Search System*, Washington, DC, CIS Project, Information Services Corporation

Scribner, J.D. (1973) Brief communication: Tumor initiation by apparently noncarcinogenic polycyclic aromatic hydrocarbons. *J. natl Cancer Inst.*, *50*, 1717-1719

Scribner, J.D. & Süss, R. (1978) Tumour initiation and promotion. *Int. Rev. exp. Pathol.*, *18*, 137-198

Stanton, M.F., Miller, E., Wrench, C. & Blackwell, R. (1972) Experimental induction of epidermoid carcinoma in the lungs of rats by cigarette smoke condensate. *J. natl Cancer Inst.*, *49*, 867-877

Van Duuren, B.L., Sivak, A., Langseth, L., Goldschmidt, B.M. & Segal, A. (1968) Initiators and promotors in tobacco carcinogenesis. *Natl Cancer Inst. Monogr.*, *28*, 173-180

ANTHRACENE

1. Chemical and Physical Data

1.1 Synonyms and trade names

Chem. Abstr. Services Reg. No.: 120-12-7

Chem. Abstr. Name: Anthracene

IUPAC Systematic Name: Anthracene

Synonyms: Anthracin; green oil; paranaphthalene

Trade Name: Tetra Olive N2G

1.2 Structural and molecular formulae and molecular weight

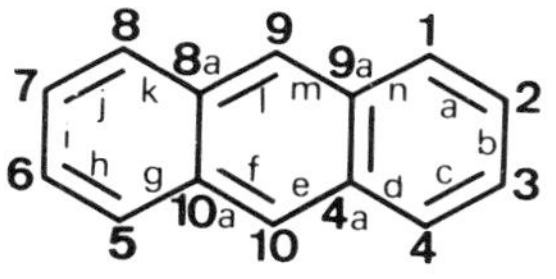

$C_{14}H_{10}$ Mol. wt: 178.2

1.3 Chemical and physical properties of the pure substance

From Windholz (1976), unless otherwise specified

(*a*) *Description*: Tablets or monoclinic prisms (recrystallized from ethanol); when pure, colourless with violet fluorescence; when impure (due to tetracene and naphthacene), yellow with green fluorescence

(*b*) *Boiling-point*: 342°C

(*c*) *Melting-point*: 218°C

(*d*) *Spectroscopy data*: λ_{max} 251.5, 308, 323, 338, 354.5, 374.5 nm (in methanol/ethanol) (Clar, 1964). Mass, nuclear magnetic resonance and infrared spectra have been tabulated (NIH/EPA Chemical Information System, 1982).

(*e*) *Solubility*: Virtually insoluble in water (44-75 μg/l) (Davis *et al.*, 1942; May *et al.*, 1978); 1 g dissolves in 62 ml benzene, 31 ml carbon disulphide, 86 ml carbon tetrachloride, 85 ml chloroform, 200 ml diethyl ether, 67 ml absolute ethanol, 70 ml methanol and 125 ml toluene

(*f*) *Volatility*: Vapour pressure, 1 mm at 145°C (sublimes) (National Library of Medicine, 1982)

(*g*) *Stability*: Darkens in sunlight. When solutions of crude anthracene in coal tar naphtha are exposed to ultraviolet radiation, the anthracene is precipitated as dianthracene, which is reconverted to anthracene by sublimation (Windholz, 1976). Unstable in cyclohexane under fluorescent light; undergoes photo-oxidation under sunlight in solution (Kuratsune & Hirohata, 1962). Undergoes photodecomposition (Korfmacher *et al.*, 1980)

(*h*) *Reactivity*: Undergoes numerous addition and substitution reactions, as for instance with chlorine or bromine. Oxidized readily to anthraquinone (Clar, 1964)

1.4 Technical products and impurities

Anthracene was reported to be available until 1982 from one US producer, as refined anthracene, with the following specifications: 90-95% purity by weight; carbazole, 3% max; sublimation residue, 0.5% max; pyridine, 0.2% max; ash, 0.1% max; and iron, 0.03% max (Allied Chemicals, undated). Typical properties of this refined anthracene were: melting-point, 216°C; boiling-point, 340°C; specific gravity, 1.25; vapour density, 6.15; and vapour pressure, $<3 \times 10^{-4}$ mm Hg (20°C).

'Anthracene oil' is a high-boiling (270-360°C) (Hawley, 1981) fraction of coal-tar, consisting of anthracene, phenanthrene and other solid hydrocarbons, as well as acridine (Gosselin *et al.*, 1981).

2. Production, Use, Occurrence and Analysis

2.1 Production and use

A review on anthracene has been published (Clar, 1964).

(*a*) *Production*

Anthracene was first isolated from coal-tar by Dumas in 1833 (Windholz, 1976). It was first synthesized in 1866 by heating benzyl chloride with water (Clar, 1964).

Anthracene has been produced commercially by recovery from the coal-tar distillation fraction known as 'anthracene oil' or 'green oil'. Yields of 6-10% of crude anthracene containing 15-30% pure anthracene are obtained by crystallization from anthracene oil. Further concentration to 40-50% purity is achieved by washing the crystals with solvent naphtha and pressing out, while hot, or centrifuging. The major contaminant of this crude anthracene is carbazole (see p.), which is converted to potassium carbazole by heating with potassium hydroxide. Potassium carbazole remains behind in a subsequent vacuum distillation. Further purification of anthracene is achieved by crystallization from solvents such as a mixture of pyridine bases (Clar, 1964).

Commercial production of anthracene was first reported in the USA in 1918 (US Tariff Commission, 1919). In 1977, one US company reported production in the range of 907-9070 thousand kg, and another company reported production of 0-450 kg. An additional US company reported production of anthracene in 1977 but did not report the amount produced (US Environmental Protection Agency, 1982). It is believed that the last known US anthracene producer stopped production in 1982.

In 1981, US imports of anthracene (all from the Federal Republic of Germany) totalled 21 thousand kg, down sharply from the 510 thousand kg imported in 1979 (US Bureau of the Census, 1980, 1982). Separate data on US exports were not available.

(*b*) *Use*

Anthracene has been used primarily as an intermediate in dye production. It has also been used in smoke screens, scintillation counter crystals and organic semiconductor research (Hawley, 1981).

According to the Society of Dyers and Colourists (1971), two dyes can be prepared directly from anthracene. However, only one of these, Vat Green 7, has been produced commercially in the USA in recent years; one company reported commercial production in 1981 (US International Trade Commission, 1982). In 1977, one US company imported an undisclosed amount of Vat Green 14, the other anthracene-based dye (US Environmental Protection Agency, 1982).

In the past, large quantities of anthracene were used in the commercial production of anthraquinone for use in the synthesis of anthraquinone dyes; however, it is believed that anthracene is no longer used for this purpose in the USA. The preferred method for the manufacture of anthraquinone in the USA for many years has been the reaction of phthalic anhydride with benzene (Chung, 1978).

Anthracene oil, which contains a significant amount of anthracene, has reportedly been used as a diluting agent for wood preservatives used in the furniture industry (see also IARC, 1981). Anthracene oil has also been used as a pesticide; however, all registrations for uses connected with food production or storage were cancelled in the USA in 1969, and newer, more effective chemicals have replaced anthracene oil for this purpose (Meister, 1982).

There is no evidence that national occupational exposure limits have been set for anthracene.

2.2 Occurrence and analysis

Data on occurrence and methods of analysis are presented in the 'General Remarks on the Substances Considered', p. 35.

Anthracene occurs ubiquitously as a product of incomplete combustion; it also occurs in fossil fuels. Anthracene has been identified in mainstream smoke of cigarettes (2.3 μg/100 cigarettes) (Lee *et al.*, 1976), (8.4-35 μg/1000 cigarettes) (Kiryu & Kuratsune, 1966), (30.9 μg/100 g of burnt material) (Masuda & Kuratsune, 1972); smoke of cigars (11.9 μg/100 g) and pipes (110 μg/100 g) (Campbell & Lindsay, 1957); and mainstream smoke of marijuana cigarettes (3.3 μg/100 cigarettes) (Lee *et al.*, 1976); exhaust emissions from gasoline engines (534-642 μg/l fuel burned) (Grimmer *et al.*, 1977); samples of charcoal-broiled steaks (4.5 μg/kg) (Lijinsky & Shubik, 1964); edible oils (0.2-402 μg/kg) (Grimmer & Hildebrandt, 1967); surface water (identified) (Grob *et al.*, 1975); tap water (1.1-59.7 ng/l) (Melchiorri *et al.*, 1973); waste water (1.6-7.0 mg/l) (Wedgwood & Cooper, 1956); and dried sediment of lakes (30-650 μg/kg) (Grimmer & Böhnke, 1975).

3. Biological Data Relevant to the Evaluation of Carcinogenic Risk to Humans

3.1 Carcinogenicity studies in animals

(*a*) *Oral administration*

Rat: A group of 28 BDI or BDIII rats, 14 weeks old [sex unspecified], received a diet containing initially 5 mg and later 15 mg 'highly purified' anthracene in oil on six days/week for 78 weeks. The total dose was 4.5 g/rat. The animals were observed until natural death; mean survival time was 700 days. Malignant tumours developed in two rats: in one a sarcoma of the liver after 18 months and in the other an adenocarcinoma of the uterus after 25 months. [No concurrent control was used.] (Schmähl, 1955).

(*b*) *Skin application*

Mouse: No skin tumour was found in 100 mice [strain, sex, age and body weight unspecified] in which the skin was painted with a 40% suspension of anthracene [purity, dose and number of applications unspecified] in lanolin ; 55 mice had died after six months of treatment (Kennaway, 1924a,b). Another skin-painting study with anthracene in mice was also negative (Maisin *et al.*, 1926, 1927).

No skin tumour was reported in 41 albino mice of a pure strain [strain, sex, age and body weight unspecified] given skin applications of a solution of anthracene in benzene or sesame oil [purity, dose and number of applications unspecified]. Positive results were obtained when 1,2,5,6-dibenzanthracene [dibenz[*a*,*h*]anthracene] was tested in the same study (Pollia, 1939).

No skin tumour was reported in five female Swiss mice [age and body weight unspecified] receiving skin applications of a 10% solution of anthracene in acetone [purity and dose unspecified] on the back thrice weekly for life. The mice died 10-20 months after the start of treatment. Groups of 20-30 mice treated with solutions of 0.01, 0.005 or 0.001% benzo[*a*]pyrene in acetone under the same experimental conditions developed both skin papillomas and skin carcinomas; incidences ranged from 95% and 95% in the high-dose group to 43% and 3% in the low-dose group (Wynder & Hoffmann, 1959).

In order to examine the tumour-initiating activity of anthracene, 20 'S' mice [sex, age and body weight unspecified] were given 20 skin applications of 0.3 ml 0.5% anthracene [purity unspecified] in acetone (two applications with an interval of 30 min, thrice weekly; total dose, 30 mg/animal) followed by 18 weekly skin applications consisting of: one 0.3-ml application of 0.17% croton oil, two 0.3-ml applications of 0.085% croton oil and a further 15 0.3 ml applications of 0.17% croton oil in acetone, beginning 25 days after the first anthracene application. One concurrent control group of 20 'S' mice received the same treatment of croton oil in acetone only. All survivors of the anthracene-treated group were killed at the end of the croton oil treatment; mortality rates at that time were 3/20 in the anthracene-treated group and 1/20 in the control group. In the anthracene-treated group, three mice had skin papillomas (a total of four papillomas); four skin papilloma were found in four mice in the control group (Salaman & Roe, 1956).

In order to examine the initiating activity of anthracene, a group of 30 female CD-1 mice, eight weeks of age, [body weight unspecified] received a single skin application of 10 μmol [1782 μg] chromatographically purified anthracene in benzene, followed one week later by skin application of 5 μmol 12-*O*-tetradecanoylphorbol-13-acetate (TPA) twice weekly for 34 weeks. A control group of 30 female CD-1 mice was treated twice weekly with 10 μmol TPA only for 34 weeks. At week 35, 28/30 animals in the anthracene-treated group and 30/30 controls were still alive. At that time, 4/28 (14%) mice treated with anthracene and TPA applications had each developed a skin papilloma, whereas skin tumour had occurred in 3% of TPA controls [$p = 0.05$] (Scribner, 1973). [The Working Group noted that the dose of TPA was probably 5 μg (Scribner & Süss, 1978).]

No skin tumour was reported in 44 mice [strain, sex, age and body weight unspecified] receiving skin applications of a 5% solution of anthracene [purity and dose unspecified] in petroleum jelly-olive oil on the ears thrice weekly for life. After 11 months, only one mouse was still alive. In another group of 44 mice similarly treated with anthracene and, in addition, with ultraviolet radiation (wave length > 320 nm) for 40 or 60 min two hours after skin application, no skin tumour was found. Of this group, only five mice were still alive after seven months, and the last mouse died during the ninth month. Another group of 100 mice were treated similarly, with anthracene but received 90 min ultraviolet radiation (wave length >320 nm); again, no skin tumour developed, and mortality was high (93/100 died after seven months; the last mouse died during the ninth month) (Miescher, 1942).

In a study in white mice involving skin application of a solution of 10% anthracene [purity and dose unspecified] in petroleum jelly-olive oil followed by either ultraviolet radiation (wave length 405-320 nm) alone or with exposure to visible light, a high incidence of skin tumours was found five to eight weeks after the start of the treatment. Many of the skin tumours were carcinomas, several of which had metastasized. In three control groups treated with anthracene, ultraviolet radiation or ultraviolet radiation plus visible light, respectively, no skin tumour occurred (Heller, 1950). [The Working Group noted the reported unusually short latency of skin tumours and the inadequacy of the reporting of the histopathology.]

In order to examine the enhancing activity of anthracene on photocarcinogenesis, three groups of 24 male and female outbred Skh:hairless-1 mice, six weeks old, [sex distribution and body weight unspecified] were given skin applications of 40 μl methanol containing 0 (vehicle controls) or 4 μg anthracene [purity unspecified; total dose, 0.76 mg/animal], or 4 μg 8-methoxypsoralen (positive controls) once daily, on five days/week, for 38 weeks, followed by two hours of 300 J/m^2 ultraviolet radiation (wave length >290 nm) after each application. Survivors were killed after 38 weeks; mortality rates were 4/24, 5/24 and 8/24 in the vehicle-control, anthracene-treated and positive-control groups, respectively. The times to 50% skin tumour prevalence (point of time when 50% of the survivors had a skin tumour) for the vehicle-treated and anthracene groups were 27.2 and 28.2 weeks, respectively. The difference was not statistically significant, but the time of 20.0 weeks to 50% skin tumour prevalence for the positive controls differed significantly from those of the two other groups ($p < 0.01$; Wilcoxon rank sum test). Of the skin tumours examined histologically from each group, the majority were squamous-cell carcinomas (Forbes *et al.*, 1976).

(*c*) *Subcutaneous and/or intramuscular administration*

Rat: No subcutaneous sarcoma was reported in 10 rats [strain, sex, age and body weight unspecified] given weekly s.c. injections of 2 ml of a 0.05% suspension of anthracene in water for life [purity unspecified] (maximum total dose, 103 mg/animal). Mortality rates were: 0/10 after six months, 7/10 after 12 months and 8/10 after 18 months. Treatment of several groups of 10 or 18 rats with 1,2,5,6-dibenzanthracene (dibenz[*a*,*h*]anthracene) under the same experimental conditions produced subcutaneous sarcomas in at least 6/10 and 9/18 rats (Boyland & Burrows, 1935). [The Working Group noted the small number of animals used.]

No tumour was reported in five Wistar rats [sex unspecified], six to eight weeks of age, given six or seven s.c. injections of 0.5 ml of a solution of anthracene [purity unspecified] in sesame oil (5 mg anthracene/injection) once a week. At 10 months, four of the five remaining rats were killed. Of five rats treated four to eight times with 1,2,5,6-dibenzanthracene (dibenz[*a*,*h*]anthracene), two developed a large subcutaneous tumour [type unspecified] (Pollia, 1941). [The Working Group noted the small number of animals used and the short duration of the study.]

A group of 10 BDI and BDIII rats [sex unspecified], 14 weeks of age, received s.c. injections of 1 ml of 2% highly purified anthracene in oil (20 mg anthracene/injection; total dose, 660 mg/animal) once a week for 33 weeks. The animals were observed until natural death. Tumours (fibromas partly with sarcomatous areas) developed at the site of injection in 5/9 rats. The first tumour (myxofibroma/sarcoma) was found after 17 months; the mean latency of the tumours was 26 months [no concurrent control was used]. The tumours were not attributed to the solvent, because another group of 10 rats treated similarly with a solution of naphthalene in oil [type unspecified] in the same study did not develop tumours at the site of injection (Schmähl, 1955). [The Working Group noted the lack of vehicle controls and the very high dose of anthracene used.]

(*d*) *Intraperitoneal administration*

Rat: A group of 10 BDI or BDIII rats [sex unspecified], 14 weeks of age, received i.p. injections of 1 ml of 2% highly purified anthracene (20 mg anthracene/injection; total dose, 660 mg/animal) in oil [type unspecified] once a week for 33 weeks. The animals were observed until natural death. The mean survival time was slightly over two years. Only one rat had a tumour, which was a spindle-cell sarcoma in the abdominal cavity found after two years [no concurrent control was used] (Schmähl, 1955).

(*e*) *Other experimental systems*

Pulmonary injection: No lung tumour was found in 60 female Osborne-Mendel *rats*, three to six months of age, given one direct injection into the lungs of 0.05 ml of a mixture of beeswax and tricaprylin (1:1) containing 0.5 mg anthracene [purity unspecified]. Nearly half the rats [number unspecified] were killed at the end of one year. Treatment of several groups of rats with 3-methylcholanthrene (doses ranging from 5-200 μg) under the same experimental conditions produced pulmonary epidermoid carcinomas before the end of the first year (incidences ranging from 5/89 in the low-dose group to 62/126 in the high-dose group) (Stanton *et al.*, 1972).

Implantation: No glioma was reported in nine *rabbits* of various breeds, ages and weights [strain, sex, age and body weight unspecified] which received an implant of a 4-20-mg pellet of pure anthracene [purity unspecified] into the cerebrum, cerebellum or eye. Animals died or were killed between 20-54 months after implantation. Negative results were also obtained with a group of 20 rabbits that received implants of cholanthrene (Russell, 1947). [The Working Group noted the small number of animals used.]

3.2 Other relevant biological data[1]

(*a*) *Experimental systems*

Toxic effects

The LD_{50} for the mouse (i.p.) is greater than 430 mg/kg bw (Salamone, 1981). The ID_{50} (skin irritant activity) for the mouse is 6.6×10^{-4} mmol/ear (Brune *et al.*, 1978).

Of a group of ten mice, nine survived when anthracene was administered i.p. for seven days in daily doses of 500 mg/kg bw (Gerarde, 1960). When the skin of hairless mice was pretreated with anthracene dissolved in methanol (0.1 g/l; 40 μl/20 cm^2 skin), ultraviolet irradiation (>290 nm) provoked a more severe inflammatory response than after methanol alone (Forbes *et al.*, 1976).

---

[1] See also 'General Remarks on the Substances Considered', p. 53.

Goblet-cell hyperplasia and cases of so-called 'transitional hyperplasia' were observed when anthracene was implanted (in a beeswax pellet) into isogenically transplanted rat tracheas (Topping *et al.*, 1978). The growth rate of mouse ascites sarcoma cells in culture was slightly (12%) inhibited when anthracene was added at a concentration of 1 mmol/l dissolved in dimethyl sulphoxide (Pilotti *et al.*, 1975).

Effects on reproduction and prenatal toxicity

When anthracene, 8 mg/mouse (strains BALB/C; C3H/A; C57BL × CBA F1 hybrids), was given either s.c. daily or as a single oral dose during the last week of gestation, greater survival and hyperplastic changes were seen than in untreated controls in explants of embryonic kidney in organ culture. The changes seen were qualitatively similar but less marked than those produced by treatment with 1-8 mg 7,12-dimethylbenz[*a*]anthracene (Shabad *et al.*, 1972). Benzo[*a*]pyrene hydroxylase activity of the rat placenta can be induced by anthracene (Welch *et al.*, 1969). No other data on reproduction were available.

Metabolism and activation

The 1,2-dihydrodiol has been identified as the major metabolite of anthracene following incubation of this compound with rat-liver preparations (Akhtar *et al.*, 1979). The 1,2-dihydrodiol, 9,10-anthraquinone, 9,10-dihydrodiol and 2,9,10-trihydroxyanthracene have been identified as metabolites in rat urine (Sims, 1964), together with conjugates consistent with the formation of the 1,2-oxide.

No information was available to the Working Group on the biological activity of anthracene metabolites.

Mutagenicity and other short-term tests

Results from short-term tests are given in Table 1.

Table 1. Results from short-term tests: Anthracene[a]

Test	Organism/assay[b]	Exogenous metabolic system[b]	Reported result	Comments	References
PROKARYOTES					
DNA damage	*Escherichia coli* (polA$^+$/polA$^-$)	UI-R-PMS	Negative	Tested at up to 250 μg/ml [The Working Group noted that UI-PMS was used]	Rosenkranz & Poirier (1979)
	Bacillus subtilis (rec$^+$/rec$^-$)	Aro-R-PMS	Negative	Tested at 62 μg/well	McCarroll *et al.* (1981)
Mutation	*Salmonella typhimurium* (his$^-$/his$^+$)	Aro-R-PMS, 3MC-R-PMS	Negative	Tested at up to 1000 μg/plate in strains TA1535, TA1537, TA1538, TA98 and TA100	McCann *et al.* (1975); Simmon (1979*a*); LaVoie *et al.* (1979); Salamone *et al.* (1979); Ho *et al.* (1981)
		Various	Negative	International collaborative programme	de Serres & Ashby (1981)

Test	Organism/assay[b]	Exogenous metabolic system[b]	Reported result	Comments	References
	Salmonella typhimurium (8AG^S/8AG^R)	Aro-R-PMS, PB-R-PMS	Negative	Tested at up to 225 nmol/ml	Kaden *et al.* (1979)
FUNGI					
Mutation	*Saccharomyces cerevisiae* D3 (mitotic recombination)	Aro-R-PMS	Negative	Tested at up to 5% w/v (sic)	Simmon (1979b)
	Saccharomyces cerevisiae (various end-points)	Various	Negative	International collaborative programme	de Serres & Ashby (1981)
MAMMALIAN CELLS *IN VITRO*					
DNA damage	Primary rat hepatocytes (unscheduled DNA synthesis)	--	Negative	Tested at up to 1 μmol/ml	Williams (1977); Probst *et al.* (1981); Tong *et al.* (1981a)
	HeLa cells (unscheduled DNA synthesis)	PB-R-PMS, 3MC-R-PMS	Negative	Tested at up to 100 μg/ml	Martin *et al.* (1978); Martin & McDermid (1981)
Mutation	Chinese hamster V79 cells (6TG^S/6TG^R)	R-PMS	Negative	Tested at up to 125 μg/ml [PMS induction unspecified]	Knaap *et al.* (1981)
	Mouse lymphoma L5178Y cells (TFT^S/TFT^R)	Aro-R-PMS	Negative	Tested at up to 100 nmol/ml	Amacher & Turner (1980); Amacher *et al.* (1980)
	Human lymphoblastoid TK6 cells (TFT^S/TFT^R)	Aro-R-PMS	Negative	Tested at up to 200 nmol/ml	Barfknecht *et al.* (1981)
Chromosome effects	Chinese hamster D6 cells (sister chromatid exchange; breaks)	None	Negative	Tested at up to 1 μmol/ml	Abe & Sasaki (1977)
	Rat liver epithelial ARL 18 cells (sister chromatid exchange)	--	Negative	Tested at up to 1 μmol/ml	Tong *et al.* (1981b)
Cell transformation	Mouse BALB/3T3 cells (morphological)	--	Negative	Tested at 10 μg/ml	DiPaolo *et al.* (1972)
	Guinea-pig foetal cells (morphological)	--	Negative	Tested at 0.5 μg/ml	Evans & DiPaolo (1975)
	Syrian hamster embryo cells (morphological)	--	Negative	Tested at up to 50 μg/ml	Pienta *et al.* (1977)
MAMMALIAN CELLS *IN VIVO*					
Chromosome effects	Chinese hamster bone-marrow cells (sister chromatid exchange; aberrations)	--	Negative	Treated i.p. with 2 × 450 mg/kg bw	Roszinsky-Kocher *et al.* (1979)
	Mouse bone-marrow cells (micronuclei)	--	Negative	Treated i.p. with 344 mg/kg bw	Salamone *et al.* (1981)
Cell transformation	Chinese hamster embryo cells (morphological)	--	Negative	Pregnant females treated i.p. with up to 30 mg/kg bw	DiPaolo *et al.* (1973)

[a] This table comprises selected assays and references and is not intended to be a complete review of the literature
[b] For an explanation of the abbreviations, see the Appendix, p. 452.

(*b*) *Humans*

Toxic effects

In studies on the treatment of psoriasis, anthracene solubilized in an alcohol *N*-methyl-2-pyrrolidone vehicle induced photosensitive reactions when administered topically in low concentrations (≃ 0.25%) to humans in combination with ultraviolet radiation (320-400 nm) (Urbanek, 1980; Walter, 1980).

No data were available to the Working Group on effects on reproduction and prenatal toxicity, absorption, distribution, excretion and metabolism or mutagenicity and chromosomal effects.

3.3 Case reports and epidemiological studies of carcinogenicity to humans

Three cases of epithelioma of the hand, cheek and wrist, respectively, were reported in men handling crude anthracene in an alizarin factory (Kennaway, 1924a,b).

4. Summary of Data Reported and Evaluation[1]

4.1 Experimental data

Anthracene was tested for carcinogenicity in mice by skin application in several studies, and in the mouse-skin initiation-promotion assay in two studies. The results were not indicative of a carcinogenic effect or of initiating activity.

It was tested in rats by oral, subcutaneous, intraperitoneal and intrapulmonary administration, and in rabbits by implantation into the brain or eyes. The studies involving oral or intrapulmonary administration produced no evidence of carcinogenicity. The studies in rats by subcutaneous or intraperitoneal administration and in rabbits by implantation into the brain or eyes were inadequate for evaluation.

When anthracene was administered by skin application to mice together with exposure to ultraviolet radiation, contradictory results were obtained.

No data on the teratogenicity of this compound were available.

Anthracene was negative in an assay for differential survival using DNA repair-proficient/-deficient strains of *Bacillus subtilis*. It did not induce mutations in bacteria or yeast nor unscheduled DNA synthesis or mutations in cultured mammalian cells. No

[1] For definitions of the italicized terms, see Preamble, p. 19.

cytogenetic effect in mammalian cells was observed *in vitro* or *in vivo*, and assays for morphological transformation were negative.

There is no evidence that anthracene is active in short-term tests.

4.2 Human data[1]

Anthracene is present as a major component of the total content of polynuclear aromatic compounds in the environment and has been produced in commercial quantities. Human exposure to anthracene occurs primarily through the smoking of tobacco, inhalation of polluted air and ingestion of food or water contaminated by combustion effluents (for details, see 'General Remarks on the Substances Considered', p. 35).

No relevant case report or epidemiological study on exposure to anthracene alone was available to the Working Group.

4.3 Evaluation

The available data provide *no evidence* that anthracene is carcinogenic to experimental animals.

5. References

Abe, S. & Sasaki, M. (1977) Studies on chromosomal aberrations and sister chromatid exchanges induced by chemicals. *Proc. Jpn Acad.*, *53*, 46-49

Akhtar, M.N., Hamilton, J.G., Boyd, D.R., Braunstein, A., Seifried, H.E. & Jerina, D.M. (1979) Anthracene 1,2-oxide: Synthesis and role in the metabolism of anthracene in mammals. *J. chem. Soc. Perkin 1*, 1422-1446

Allied Chemicals (undated) *Tar Products. Technical Data Report*, Morristown, NJ, Semet Solvay Division

Amacher, D.E. & Turner, G.N. (1980) Promutagen activation by rodent-liver postmitochondrial fractions in the L5178Y/TK cell mutation assay. *Mutat. Res.*, *74*, 485-501

Amacher, D.E., Paillet, S.C., Turner, G.N., Ray, V.A. & Salsburg, D.S. (1980) Point mutations at the thymidine kinase locus in L5178Y mouse lymphoma cells. II. Test validation and interpretation. *Mutat. Res.*, *72*, 447-474

[1] Studies on occupational exposure to polynuclear aromatic compounds will be considered in future *IARC Monographs*.

Barfknecht, T.R., Andon, B.M., Thilly, W.G. & Hites, R.A. (1981) *Soot and mutation in bacteria and human cells*. In: Cooke, M. & Dennis, A.J., eds, *Chemical Analysis and Biological Fate: Polynuclear Aromatic Hydrocarbons, 5th Int. Symposium*, Columbus, OH, Battelle Press, pp. 231-242

Boyland, E. & Burrows, H. (1935) The experimental production of sarcoma in rats and mice by a colloidal aqueous solution of 1:2:5:6-dibenzanthracene. *J. Pathol. Bacteriol.*, *41*, 231-238

Brune, K., Kalin, H., Schmidt, R. & Hecker, E. (1978) Inflammatory, tumor initiating and promoting activities of polycyclic aromatic hydrocarbons and diterpene esters in mouse skin as compared with their prostaglandin releasing potency *in vitro*. *Cancer Lett.*, *4*, 333-342

Campbell, J.M. & Lindsey, A.J. (1957) Polycyclic hydrocarbons in cigar smoke. *Br. J. Cancer*, *11*, 192-195

Chung, R.H. (1978) *Anthraquinone*. In: Kirk, R.E. & Othmer, D.F., eds, *Encyclopedia of Chemical Technology*, 3rd ed., Vol. 2, New York, John Wiley & Sons, pp. 700-707

Clar, E. (1964) *Polycyclic Hydrocarbons*, Vol. 1, London, Academic Press, pp. 288-307

Davis, W.W., Krahl, M.E. & Clowes, G.H.A. (1942) Solubility of carcinogenic and related hydrocarbons in water. *J. Am. chem. Soc.*, *64*, 108-110

DiPaolo, J.A., Takano, K. & Popescu, N.C. (1972) Quantitation of chemically induced neoplastic transformation of BALB/3T3 cloned cell lines. *Cancer Res.*, *32*, 2686-2695

DiPaolo, J.A., Nelson, R.L., Donovan, P.J. & Evans, C.H. (1973) Host-mediated *in vivo-in vitro* assay for chemical carcinogenesis. *Arch. Pathol.*, *95*, 380-385

Evans, C.H. & DiPaolo, J.A. (1975) Neoplastic transformation of guinea pig fetal cells in culture induced by chemical carcinogens. *Cancer Res.*, *35*, 1035-1044

Forbes, P.D., Davies, R.E. & Urbach, F. (1976) Phototoxicity and photocarcinogenesis: Comparative effects of anthracene and 8-methoxypsoralene in the skin of mice. *Food Cosmet. Toxicol.*, *14*, 303-306

Gerarde, H.W. (1960) *Toxicology and biochemistry of aromatic hydrocarbons*. In: Browning, E., ed., *Elsevier Monographs on Toxic Agents*, Amsterdam, Elsevier, pp. 240-321

Gosselin, R.E., Hodge, H.C., Smith, R.P. & Gleason, M.N. (1981) *Clinical Toxicology of Commercial Products: Acute Poisoning*, 4th ed., Baltimore, MD, Williams & Wilkins, p. 108

Grimmer, G. & Böhnke, H. (1975) Profile analysis of polycyclic aromatic hydrocarbons and metal content in sediment layers of a lake. *Cancer Lett.*, *1*, 75-84

Grimmer, G. & Hildebrandt, A. (1967) Content of polycyclic hydrocarbons in crude vegetable oils. *Chem. Ind.*, 25 November, 2000-2002

Grimmer, G., Böhnke, H. & Glaser, A. (1977) Investigation on the carcinogenic burden by air pollution in man. XV. Polycyclic aromatic hydrocarbons in automobile exhaust gas - An inventory. *Zbl. Bakt. Hyg., 1 Abt., Orig. B164*, 218-234

Grob, K., Grob, K., Jr & Grob, G. (1975) Organic substances in potable water and its precursor. III. The closed-loop stripping procedure compared with rapid liquid extraction. *J. Chromatogr.*, *106*, 299-315

Hawley, G.G., ed. (1981) *The Condensed Chemical Dictionary*, 10th ed., New York, Van Nostrand Reinhold, p. 75

Heller, W. (1950) Experimental study on tumours produced by light. 2. Tumours produced by light through photosensibilisation (Ger.). *Strahlentherapie*, *81*, 529-548

Ho, C.-H., Clark, B.R., Guerin, M.R., Barkenbus, B.D., Roa, T.K. & Epler, J.L. (1981) Analytical and biological analyses of test materials from the synthetic fuel technologies. IV. Studies of chemical structure-mutagenic activity relationships of aromatic nitrogen compounds relevant to synfuels. *Mutat. Res.*, *85*, 335-345

IARC (1981) *IARC Monographs on the Evaluation of the Carcinogenic Risk of Chemicals to Humans*, Vol. 25, *Wood, Leather and Some Associated Industries*, Lyon, p. 314

Kaden, D.A., Hites, R.A. & Thilly, W.G. (1979) Mutagenicity of soot and associated polycyclic aromatic hydrocarbons to *Salmonella typhimurium*. *Cancer Res.*, *39*, 4152-4159

Kennaway, E.L. (1924a) On cancer-producing tars and tar-fractions. *J. ind. Hyg.*, *5*, 462-488

Kennaway, E.L. (1924b) On cancer-producing factor in tar. *Br. med. J.*, *i*, 564-567

Kiryu, S. & Kuratsune, M. (1966) Polycyclic aromatic hydrocarbons in the cigarette tar produced by human smoking. *Gann*, *57*, 317-322

Knaap, A.G.A.C., Goze, C. & Simons, J.W.I.M. (1981) *Mutagenic activity of seven coded samples in V79 Chinese hamster cells*. In: de Serres, F.J. & Ashby, J., eds, *Evaluation of Short-term Tests for Carcinogens. Report of the International Collaborative Program. Progress in Mutation Research*, Vol. 1, Amsterdam, Elsevier/North Holland, pp. 608-613

Korfmacher, W.A., Wehry, E.L., Mamantov, G. & Natusch, D.F.S. (1980) Resistance to photochemical decomposition of polycyclic aromatic hydrocarbons vapor-adsorbed on coal fly ash. *Environ. Sci. Technol.*, *14*, 1094-1099

Kuratsune, M. & Hirohata, T. (1962) Decomposition of polycyclic aromatic hydrocarbons under laboratory illuminations. *Natl Cancer Inst. Monogr.*, *9*, 117-125

LaVoie, E., Bedenko, V., Hirota, N., Hecht, S.S. & Hoffmann, D. (1979) *A comparison of the mutagenicity, tumor-initiating activity and complete carcinogenicity of polynuclear aromatic hydrocarbons*. In: Jones, P.W. & Leber, P., eds, *Polynuclear Aromatic Hydrocarbons*, Ann Arbor, MI, Ann Arbor Science Publishers, pp. 705-721

Lee, M.L., Novotny, M. & Bartle, K.D. (1976) Gas chromatography/mass spectrometry and nuclear magnetic resonance spectrometric studies of carcinogenic polynuclear aromatic hydrocarbons in tobacco and marijuana smoke condensates. *Anal. Chem.*, *48*, 405-416

Lijinsky, W. & Shubik, P. (1964) Benzo(*a*)pyrene and other polynuclear hydrocarbons in charcoal-broiled meat. *Science*, *145*, 53-55

Maisin, J., Rome, M. & Jacqmin, L. (1926) Method for obtaining non-carcinogenic tars (Fr.). *C.R. Soc. Biol.*, *94*, 767-769

Maisin, J., Desmedt, P. & Jacqmin, L. (1927) Carcinogenic action of carbazole (Fr.). *C.R. Soc. Biol.*, *96*, 1056-1058

Martin, C.N. & McDermid, A.C. (1981) *Testing of 42 coded compounds for their ability to induce unscheduled DNA repair synthesis in HeLa cells.* In: de Serres, F.J. & Ashby, J., eds, *Evaluation of Short-term Tests for Carcinogens. Report of the International Collaborative Program. Progress in Mutation Research*, Vol. 1, Amsterdam, Elsevier/North Holland, pp. 533-537

Martin, C.N., McDermid, A.C. & Garner, R.C. (1978) Testing of known carcinogens and noncarcinogens for their ability to induce unscheduled DNA synthesis in HeLa cells. *Cancer Res.*, *38*, 2621-2627

Masuda, Y. & Kuratsune, M. (1972) Comparison of the yield of polycyclic aromatic hydrocarbons in smoke from Japanese tobacco. *Jpn. J. Hyg.*, *27*, 339-341

May, W.E., Wasik, S.P. & Freeman, D.H. (1978) Determination of the aqueous solubility of polynuclear aromatic hydrocarbons by a coupled column liquid chromatographic technique. *Anal. Chem.*, *50*, 175-179

McCann, J., Choi, E., Yamasaki, E. & Ames, B.N. (1975) Detection of carcinogens as mutagens in the *Salmonella*/microsome test: Assay of 300 chemicals. *Proc. natl Acad. Sci. USA*, *72*, 5135-5139

McCarroll, N.E., Keech, B.H. & Piper, C.E. (1981) A micro-suspension adaptation of the *Bacillus subtilis* 'rec' assay. *Environ. Mutagenesis*, *3*, 607-616

Meister, R.T., ed. (1982) *1982 Farm Chemical Handbook*, Willoughby, OH, Meister Publishing Co., pp. C18, C279

Melchiorri, C., Chiacchiarini, L., Grella, A. & D'Arca, S.U. (1973) Identification and determination of polycyclic aromatic hydrocarbons in some tap waters of Rome (Ital.). *Nuovi Ann. Ig. Microbiol.*, *24*, 279-301

Miescher, G. (1942) Experimental studies on the formation of tumours by photosensitization (Ger.). *Schweiz. med. Wochenschr.*, *72*, 1082-1084

National Library of Medicine (1982) *Toxicology Data Bank*, Bethesda, MD, National Library of Medicine Specialized Information Services, Toxicology Information Program

NIH/EPA Chemical Information System (1982) *Mass Spectral Search System and Infrared Spectral Search System*, Washington DC, CIS Project, Information Services Corporation

Pienta, R.J., Poiley, J.A. & Lebherz III, W.B. (1977) Morphological transformation of early passage golden Syrian hamster embryo cells derived from cryopreserved primary cultures as a reliable *in vitro* bioassay for identifying diverse carcinogens. *Int. J. Cancer*, *19*, 642-655

Pilotti, A., Ancker, K., Arrhenius, E. & Enzell, C. (1975) Effects of tobacco and tobacco smoke constituents on cell multiplication *in vitro*. *Toxicology*, *5*, 49-62

Pollia, J.A. (1939) Investigations on the possible carcinogenic effect of anthracene and chrysene and some of their compounds. I. The effect of painting on the skin of mice. *J. ind. Hyg. Toxicol.*, *21*, 219-220

Pollia, J.A. (1941) Investigations on the possible carcinogenic effect of anthracene and chrysene and some of their compounds. II. The effect of subcutaneous injection in rats. *J. ind. Hyg. Toxicol.*, *23*, 449-451

Probst, G.S., McMahon, R.E., Hill, L.E., Thompson, C.Z., Epp, J.K. & Neal, S.B. (1981) Chemically-induced unscheduled DNA synthesis in primary rat hepatocyte cultures: A comparison with bacterial mutagenicity using 218 compounds. *Environ. Mutagenesis*, *3*, 11-32

Rosenkranz, H.S. & Poirier, L.A. (1979) Evaluation of the mutagenicity and DNA-modifying activity of carcinogens and noncarcinogens in microbial systems. *J. natl Cancer Inst.*, *62*, 873-892

Roszinsky-Köcher, G., Basler, A. & Röhrborn, G. (1979) Mutagenicity of polycyclic hydrocarbons. V. Induction of sister-chromatid exchanges *in vivo*. *Mutat. Res.*, *66*, 65-67

Russell, H. (1947) An unsuccessful attempt to induce gliomata in rabbits with cholanthrene. *J. Pathol. Bacteriol.*, *59*, 481-483

Salaman, M.H. & Roe, F.J.C. (1956) Further tests for tumour-initiating activity: *N,N*-Di-(2-chloroethyl)-*p*-aminophenyl-butyric acid (CB1348) as an initiator of skin tumour formation in the mouse. *Br. J. Cancer*, *10*, 363-378

Salamone, M.F. (1981) Toxicity of 41 carcinogens and noncarcinogenic analogs. *Prog. Mutat. Res.*, *1*, 682-685

Salamone, M.F., Heddle, J.A. & Katz, M. (1979) The mutagenic activity of thirty polycyclic aromatic hydrocarbons (PAH) and oxides in urban airborne particulates. *Environ. Int.*, *2*, 37-43

Salamone, M.F., Heddle, J.A. & Katz, M. (1981) *Mutagenic activity of 41 compounds in the* in vivo *micronucleus assay*. In: de Serres, F.J. & Ashby, J., eds, *Evaluation of Short-term Tests for Carcinogens. Report of the International Collaborative Program. Progress in Mutation Research*, Vol. 1, Amsterdam, Elsevier/North Holland, pp. 686-697

Schmähl, D. (1955) Examination of the carcinogenic action of naphthalene and anthracene in rats (Ger.). *Z. Krebsforsch.*, *60*, 697-710

Scribner, J.D. (1973) Brief communication: Tumor initiation by apparently noncarcinogenic polycyclic aromatic hydrocarbons. *J. natl Cancer Inst.*, *50*, 1717-1719

Scribner, J.D. & Süss, R. (1978) Tumour initation and promotion. *Int. Rev. exp. Pathol.*, *18*, 137-198

de Serres, F.J. & Ashby, J., eds (1981) *Evaluation of Short-term Tests for Carcinogens. Report of the International Collaborative Program. Progress in Mutation Research*, Vol. 1, Amsterdam, Elsevier/North Holland

Shabad, L.M., Sorokina, J.D., Golub, N.I. & Bogovski, S.P. (1972) Transplacental effect of some chemical compounds on organ cultures of embryonic kidney tissue. *Cancer Res.*, *32*, 617-627

Simmon, V.F. (1979a) *In vitro* mutagenicity assays of chemical carcinogens and related compounds with *Salmonella typhimurium*. *J. natl Cancer Inst.*, *62*, 893-899

Simmon, V.F. (1979b) *In vitro* assays for recombinogenic activity of chemical carcinogens and related compounds with *Saccharomyces cerevisiae D3*. *J. natl Cancer Inst.*, *62*, 901-909

Sims, P. (1964) Metabolism of polycyclic compounds. 25. The metabolism of anthracene and some related compounds in rats. *Biochem. J.*, *92*, 621-631

Stanton, M.F., Miller, E., Wrench, C. & Blackwell, R. (1972) Experimental induction of epidermoid carcinoma in the lungs of rats by cigarette smoke condensate. *J. natl Cancer Inst.*, *49*, 867-877

The Society of Dyers and Colourists (1971) *Colour Index*, 3rd ed., Vol. 4, Bradford, UK, Lund Humphries, pp. 4521, 4522

Tong, C., Laspia, M.F., Telang, S. & Williams, G.M. (1981a) The use of adult rat liver cultures in the detection of the genotoxicity of various polycyclic aromatic hydrocarbons. *Environ. Mutagenesis*, *3*, 477-487

Tong, C., Brat, S.V. & Williams, G.M. (1981b) Sister-chromatid exchange induction by polycyclic aromatic hydrocarbons in an intact cell system of adult rat-liver epithelial cells. *Mutat. Res.*, *91*, 467-473

Topping, D.C., Pal, B.C., Martin, D.H., Nelson, F.R. & Nettesheim, P. (1978) Pathologic changes induced in respiratory tract mucosa by polycyclic hydrocarbons of differing carcinogenic activity. *Am. J. Pathol.*, *93*, 311-324

Urbanek, R.W. (1980) Side effects of anthracene. *J. Am. Acad. Dermatol.*, *2*, 240

US Bureau of the Census (1980) *US Imports for Consumption and General Imports, TSUSA Commodity by Country of Origin*, FT246/Annual 1979, Washington DC, US Government Printing Office, p. 1-231

US Bureau of the Census (1982) *US Imports for Consumption and General Imports, TSUSA Commodity by Country of Origin*, IM146/December 1981, Washington DC, US Government Printing Office, p. 1383

US Environmental Protection Agency (1982) *Chemicals in Commerce Information System (CICIS)*, Washington DC, Office of Pesticides and Toxic Substances, Chemical Information Division

US International Trade Commission (1982) *Synthetic Organic Chemicals, US Production and Sales, 1981* (*USITC Publ. 1292*), Washington DC, US Government Printing Office, p. 86

US Tariff Commission (1919) *Report on Dyes and Related Coal-Tar Chemicals, 1918*, Washington DC, US Government Printing Office, p. 18

Walter, J.F. (1980) Side effects of anthracene (Reply). *J. Am. Acad. Dermatol.*, *2*, 240-243

Wedgwood, P. & Cooper, R.L. (1956) Detection and determination of traces of polynuclear hydrocarbons in industrial effluents and sewage. IV. The quantitative examination of effluents. *Analyst*, *81*, 42-44

Welch, R.M., Harrison, Y.E., Gommi, B.W., Poppers, P.J., Finster, M. & Conney, A.H. (1969) Stimulatory effect of cigarette smoking on the hydroxylation of 3,4-benzpyrene and the *N*-demethylation of 3-methyl-4-monomethylaminoazobenzene by enzymes in human placenta. *Clin. Pharmacol.*, *10*, 100-109

Williams, G.M. (1977) Detection of chemical carcinogens by unscheduled DNA synthesis in rat liver primary cell cultures. *Cancer Res.*, *37*, 1845-1851

Windholz, M., ed. (1976) *The Merck Index*, 9th ed., Ranway, NJ, Merck & Co., p. 93

Wynder, E.L. & Hoffmann, D. (1959) A study of tobacco carcinogenesis VII. The role of higher polycyclic hydrocarbons. *Cancer*, *12*, 1079-1086

BENZ[*a*]ACRIDINE

1. Chemical and Physical Data

1.1 Synonyms and trade names

Chem. Abstr. Services Reg. No.: 225-11-6

Chem. Abstr. Name: Benz(a)acridine

IUPAC Systematic Name: Benz[*a*]acridine

Synonyms: 7-Azabenz(a)anthracene; 1,2-benzacridine

1.2 Structural and molecular formulae and molecular weight

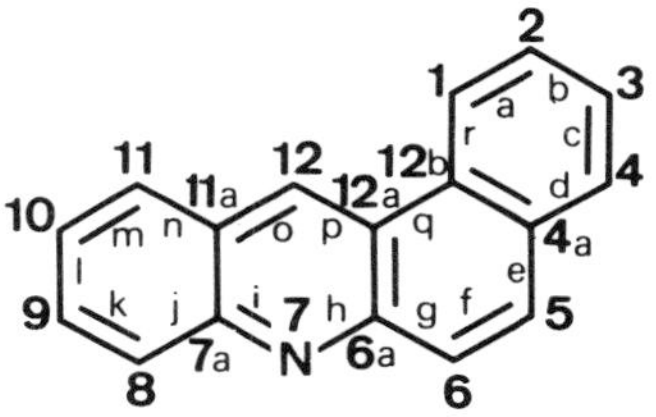

1 2 3 4 4a 5 6 6a 7 N 7a 8 9 10 11 11a 12 12a 12b
a b c d e f g h i j k l m n o p q r

$C_{17}H_{11}N$ Mol. wt: 229.3

1.3 Chemical and physical properties of the pure substance

(*a*) *Melting-point*: 131°C (Karcher *et al.*, 1983)

(*b*) *Spectroscopy data*: λ_{max} 224, 234, 242, 274, 284, 316, 330, 339, 344, 354, 362, 372, 381 nm (in cyclohexane). Nuclear magnetic resonance spectra have been tabulated (Karcher *et al.*, 1983).

(*c*) *Solubility*: Very slightly soluble in water, soluble in benzene and acetone (Lacassagne *et al.*, 1956)

(*d*) *Stability*: Very stable (Lacassagne *et al.*, 1956)

(*e*) *Reactivity*: pKa, 3.95 (Pagès-Flon *et al.*, 1953). The heterocyclic nitrogen atom reacts with halogen and alkyl sulphates to form quaternary acridinium salts (Lacassagne *et al.* 1956). See also Rawlins (1973) and Selby (1973).

2. Production, Use, Occurrence and Analysis

2.1 Production and use

There is no commercial production or known use of this compound.

2.2 Occurrence and analysis

Data on occurrence and methods of analysis are summarized in the 'General Remarks on the Substances Considered', p. 35.

Benz[*a*]acridine has been identified in coal-tar (Kruber, 1941; Lang & Eigen, 1967); coke-plant emission (Burchill *et al.*, 1978); emissions from coal conversion processes (Bridbord & French, 1978); polluted air from coal heating (7.7-26 μg/m^3) and industrial effluent (18 μg/m^3) (Sawicki *et al.*, 1965).

3. Biological Data Relevant to the Evaluation of Carcinogenic Risk to Humans

3.1 Carcinogenicity studies in animals

Skin application

Mouse: A solution of 0.3% benz[*a*]acridine [purity unspecified] in acetone was applied twice weekly for life to the dorsal skin of 12 XVII strain mice; six animals that lived longer than 90 days (165-379 days) had no lesion at the application site (Lacassagne *et al.*, 1956). [The Working Group noted the small number of animals used.]

3.2 Other relevant biological data[1]

(*a*) *Experimental systems*

No data were available to the Working Group on toxic effects or on effects on reproduction and prenatal toxicity.

Metabolism and activation

The 5,6-dihydrodiol has been detected as a metabolite of benz[*a*]acridine in studies in which this compound was incubated with rat-liver and -lung microsomes (Jacob *et al.*, 1982).

Mutagenicity and other short-term tests

Benz[*a*]acridine was tested in *Salmonella typhimurium* strain TA98 (*his*$^-$/*his*$^+$) at concentrations of up to 0.5 mg/plate in the presence of an exogenous metabolic system (postmitochondrial supernatant from Aroclor-induced rat liver), with inconclusive results (Ho *et al.*, 1981).

(*b*) *Humans*

No data were available to the Working Group.

3.3 Case reports and epidemiological studies of carcinogenicity to humans

No data were available to the Working Group.

4. Summary of Data Reported and Evaluation[2]

4.1 Experimental data

Benz[*a*]acridine was inadequately tested for carcinogenicity in one experiment by skin application to mice.

No data on the teratogenicity of this compound were available.

The single report of the mutagenicity of benz[*a*]acridine in *Salmonella typhimurium* was inconclusive.

There is *inadequate evidence* that benz[*a*]acridine is active in short-term tests.

---

[1] See also 'General Remarks on the Substances considered', p. 53.

[2] For definitions of the italicized terms, see Preamble, p. 19.

4.2 Human data[1]

Benz[*a*]acridine is present as a minor component of the total content of polynuclear aromatic compounds in the environment. Human exposure to benz[*a*]acridine occurs primarily through inhalation of polluted air and by ingestion of food and water contaminated with combustion products (for details, see 'General Remarks on the Substances Considered', p. 35).

4.3 Evaluation

The available data are inadequate to permit the evaluation of the carcinogenicity of benz[*a*]acridine in experimental animals.

5. References

Bridbord, K. & French, J.G. (1978) *Carcinogenic and mutagenic risks associated with fossil fuels*. In: Jones, P.W. & Freudenthal, R.I, eds, *Carcinogenesis*, Vol. 3, *Polynuclear Aromatic Hydrocarbons*, New York, Raven Press, pp. 451-463

Burchill, P., Herod, A.A. & James, R.G. (1978) *A comparison of some chromatographic methods for estimation of polynuclear aromatic hydrocarbons in pollutants*. In: Jones, P.W. & Freudenthal, R.I., eds, *Carcinogenesis*, Vol. 3, *Polynuclear Aromatic Hydrocarbons*, New York, Raven Press, pp. 35-45

Ho, C.H., Clark, B.R., Guerin, M.R., Barkenbus, B.D., Tao, T.K. & Epler, J.L. (1981) Analytical and biological analyses of test materials from the synthetic fuel technologies. IV. Studies of chemical structure - mutagenic activity relationships of aromatic nitrogen compounds relevant to synfuels. *Mutat. Res.*, *85*, 335-345

Jacob, J., Schmoldt, A., Kohbrok, W., Raab, G. & Grimmer, G. (1982) On the metabolic activation of benz[*a*]acridine and benz[*c*]acridine by rat liver and lung microsomes. *Cancer Lett.*, *16*, 297-306

Karcher, W., Fordham, R., Dubois, J. & Gloude, P. (1983) *Spectral Atlas of Polycyclic Aromatic Compounds*, Dordrecht, The Netherlands, D. Reidel (in press)

Kruber, O. (1941) Determination of chrysene-fraction in coal-tar (Ger.). *Chem. Ber.*, *74*, 1688-1692

---

[1] Studies on occupational exposure to polynuclear aromatic compounds will be considered in future *IARC Monographs*.

Lacassagne, A., Buu-Hoï, N.P., Daudel, R. & Zajdela, F. (1956) *The relation between carcinogenic activity and the physical and chemical properties of angular benzoacridines.* In: Greenstein, J.P. & Haddow, A., eds, *Advances in Cancer Research*, Vol. 4, New York, Academic Press, pp. 315-369

Lang, K.F. & Eigen, I. (1967) *Organic compounds in coal tar* (Ger.). In: Heilbronner, E., Hofmann, U., Schäfer, K. & Wittig, G., eds, *Fortschrifte der chemische Forschung* (Advances in chemical research), Vol. 8, Berlin, Springer, pp. 91-170

Pagès-Flon, M., Buu Hoï, N.P. & Daudel, R. (1953) Relationship between pK and carcinogenic action for two series of benzacridines (Fr.). *C.R. Acad. Sci.*, *236*, 2182-2184

Raulins, N.R. (1973) *Acridines*. In: Acheson, R.M., ed., *Acridines*, 2nd ed., New York, Interscience, pp. 63-67

Sawicki, E., Meeker, J.E. & Morgan, M.J. (1965) The quantitative composition of air pollution source effluents in terms of aza heterocyclic compounds and polynuclear aromatic hydrocarbons. *Int. J. Air Water Pollut.*, *9*, 291-298

Selby, I.A. (1973) *Acridinium salts and reduced acridines*. In: Acheson, R.M., ed., *Acridines*, 2nd ed., New York, Interscience, pp. 434-445

BENZ[*c*]ACRIDINE

This compound was considered by a previous working group, in December 1972 (IARC, 1973). Data that have become available since that time have been incorporated in the present monograph and taken into consideration in the evaluation.

1. Chemical and Physical Data

1.1 Synonyms and trade names

Chem. Abstr. Services Reg. No.: 225-51-4

Chem. Abstr. Name: Benz(c)acridine

IUPAC Systematic Name: Benz[*c*]acridine

Synonyms: 12-Azabenz(a)anthracene; B(c)AC; 3,4-benzacridine; 3,4-benzoacridine; α-chrysidine; α-naphthacridine

1.2 Structural and molecular formulae and molecular weight

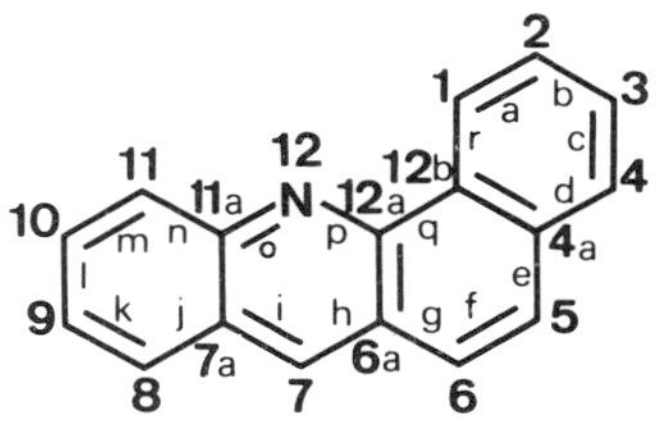

$C_{17}H_{11}N$ Mol. wt: 229.3

1.3 Chemical and physical properties of the pure substance

From National Library of Medicine (1982), unless otherwise specified

(*a*) *Description*: Yellow needles

(*b*) *Melting-point*: 108°C (Karcher *et al.*, 1983)

(*c*) *Spectroscopy data*: λ_{max} 204, 219, 222, 232, 266, 273, 284, 315, 330, 338, 346, 355, 364, 374, 383 (in cyclohexane); nuclear magnetic resonance spectra have been tabulated (Karcher *et al*., 1983).

(*d*) *Solubility*: Very slightly soluble in water; soluble in benzene and acetone (Lacassagne *et al*., 1956)

(*e*) *Stability*: Very stable (Lacassagne *et al*., 1956)

(*f*) *Reactivity*: pKa, 3.24 (Pagès-Flon *et al*., 1953); the heterocyclic nitrogen atom reacts with halogen and alkyl sulphates to form quaternary acridinium salts (Lacassagne *et al*., 1956). See also Raulins (1973) and Selby (1973).

2. Production, Use, Occurrence and Analysis

2.1 Production and use

There is no commercial production or known use of this compound.

2.2 Occurrence and analysis

Data on occurrence and methods of analysis are summarized in the 'General Remarks on the Substances Considered', p. 35.

Benz[*c*]acridine has been identified in urban atmospheres (0.1-1.5 ng/m^3) (Stanley *et al*., 1968); polluted air from coal heating (15-18 μg/m^3) and industrial effluents (60 μg/m^3) (Sawicki *et al*., 1965b); gasoline engine exhaust (200 ng/m^3) (Sawicki *et al*., 1965a); and coal-tar (Lang & Eigen, 1967).

3. Biological Data Relevant to the Evaluation of Carcinogenic Risk to Humans

3.1 Carcinogenicity studies in animals

(*a*) *Skin application*

Mouse: A solution of 0.3% benz[*c*]acridine [purity unspecified] in acetone was applied twice weekly for life to the dorsal skin of 12 mice of the XVII strain. Seven animals that lived longer than 90 days (230-394 days) had no lesion at the application site (Lacassagne *et al*., 1956). [The Working Group noted the small number of animals used.]

One drop of a solution of 0.3% benz[*c*]acridine [purity unspecified] in benzene was applied thrice weekly for life on the neck skin of 64 male and female mice (Swiss and Haffkine mice and the hybrids). Five skin epitheliomas [squamous-cell carcinomas] were found in 19 mice surviving more than 400 days. In a second experiment, 24 mice of the same origin were treated identically and received, in addition, weekly applications of a 0.5% solution of croton oil in acetone. Eighteen mice survived 180 days; three survived until the day the first tumour was found [day not specified]; two epitheliomas [squamous-cell carcinomas] appeared in this latter group. In the third experiment, 24 mice of the same origin received two applications of one drop of 0.3% benz[*c*]acridine solution in benzene; one month later, the mice were treated weekly with 0.5% croton oil in acetone. Sixteen mice survived 150 days; in four that survived 400 days, no tumour was produced. No tumour was found in 12 mice given skin applications of a solution of 0.5% croton oil in benzene alone, four of which survived 180 days and two for 400 days. In several hundred mice [number not specified] maintained as controls, no skin cancer was found (Hakim, 1968).

(*b*) *Other experimental systems*

Bladder implantation: Paraffin wax pellets containing benz[*c*]acridine [concentration and purity not specified] were implanted into the bladder of 58 *rats* [strain, age, sex and body weight not specified]. A few rats had died after one year; all animals were killed after l6 months. At that time, 21 papillomas and eight 'cancers' were observed. In 64 control rats given paraffin wax pellets, two papillomas of the bladder were observed (Hakim, 1968).

3.2 Other relevant biological data[1]

(*a*) *Experimental systems*

No data were available to the Working Group on toxic effects or effects on reproduction or prenatal toxicity.

Metabolism and activation

The 5,6-dihydrodiol of benz[*c*]acridine has been detected as a metabolite together with other, unidentified dihydrodiols and monohydroxyderivatives in studies in which benz[*c*]acridine was incubated with rat-liver and -lung microsomal preparations (Jacob *et al.*, 1982).

The 5,6-oxide was weakly mutagenic to bacterial cells, but the 5,6-dihydrodiol was inactive in the presence of an exogenous metabolic system (Wood *et al.*, 1983).

---

[1] See also 'General Remarks on the Substances Considered', p. 53.

Mutagenicity and other short-term tests

Results of short-term tests are given in Table 1.

Table 1. Results from short-term tests: Benz[*c*]acridine

Test	Organism/assay[a]	Exogenous metabolic system[a]	Reported result	Comments	References
PROKARYOTES					
Mutation	*Salmonella typhimurium* (his⁻/his⁺)	PCB-R-PMS	Positive	At 25 μg/plate in strain TA100	Okano *et al.* (1979)
		3MC-GP-PMS, Aro-GP-PMS	Positive	At 25 μg/plate in strain TA100	Baker *et al.* (1980)

[a] For an explanation of the abbreviations see the Appendix, p. 452.

(*b*) *Humans*

No data were available to the Working Group.

3.3 Case reports and epidemiological studies of carcinogenicity to humans

No data were available to the Working Group.

4. Summary of Data Reported and Evaluation[1]

4.1 Experimental data

Benz[*c*]acridine was tested for carcinogenicity in one experiment in mice by skin application and produced a low incidence of skin tumours. It was also tested in the mouse-skin initiation-promotion assay and was inactive as an initiator. It was also tested in rats by bladder implantation in wax pellets; an increased incidence of bladder tumours was observed.

No data on the teratogenicity of this compound were available.

[1] For definitions of the italicized terms, see Preamble, p. 19.

Benz[*c*]acridine was mutagenic to *Salmonella typhimurium* in the presence of an exogenous metabolic system.

There is *inadequate evidence* that benz[*c*]acridine is active in short-term tests.

4.2 Human data[1]

Benz[*c*]acridine is present as a minor component of the total content of polynuclear aromatic compounds in the environment. Human exposure to benz[*c*]acridine occurs primarily through the inhalation of polluted air and by ingestion of food and water contaminated with combustion products (for details, see 'General Remarks on the Substances Considered', p. 35).

4.3 Evaluation

There is *limited evidence* that benz[*c*]acridine is carcinogenic to experimental animals.

5. References

Baker, R.S.U., Bonin, A.M., Stupans, I. & Holder, G.M. (1980) Comparison of rat and guinea pig as sources of the S9 fraction in the *Salmonella*/mammalian microsome mutagenicity test. *Mutat. Res.*, *71*, 43-52

Hakim, S.A.E. (1968) Sanguinarine - a carcinogenic contaminant in Indian edible oils. *Indian J. Cancer*, *5*, 183-197

IARC (1973) *IARC Monographs on the Evaluation of Carcinogenic Risk of Chemicals to Man*, Vol. 3, *Certain polycyclic aromatic hydrocarbons and heterocyclic compounds*, Lyon, pp. 241-246

Jacob, J., Schmoldt, A., Kohbrok, W., Raab, G. & Grimmer, G. (1982) On the metabolic activation of benz[*a*]acridine and benz[*c*]acridine by rat liver and lung microsomes. *Cancer Lett.*, *16*, 297-306

[1] Studies on occupational exposure to polynuclear aromatic compounds will be considered in future *IARC Monographs*.

Karcher, W., Fordham, R., Dubois, J. & Gloude, P. (1983) *Spectral Atlas of Polycyclic Aromatic Compounds*, Dordrecht, The Netherlands, D. Reidel (in press)

Lacassagne, A., Buu-Hoï, N.P., Daudel, R. & Zajdela, F. (1956) *The relation between carcinogenic activity and the physical and chemical properties of angular benzoacridines.* In: Greenstein, J.P. & Haddow, A., eds, *Advances in Cancer Research*, Vol. 4, New York, Academic Press, pp. 315-369

Lang, K.F. & Eigen, I. (1967) *Organic compounds in coal tar* (Ger.). In: Heilbronner, E., Hofmann, U., Schäfer, K. & Wittig, G., eds, *Fortschrifte der chemischer Forschung* (Advances in chemical research), Vol. 8, Berlin, Springer, pp. 91-170

National Library of Medicine (1982) *Toxicology Data Bank*, Bethesda, MD, National Library of Medicine Specialized Information Services, Toxicology Information Program

Okano, T., Horie, T., Koike, T. & Motohashi, N. (1979) Relationship of carcinogenicity, mutagenicity, and K-region reactivity in benz[*c*]acridines. *Gann*, *70*, 749-754

Pagès-Flon, M., Bun Hoï, N.P. & Daudel, R. (1953) Relationship between pK and carcinogenic action for two series of benzacridines (Fr.). *C.R. Acad. Sci.*, *236*, 2182-2184

Raulins, N.R. (1973) *Acridines*. In: Acheson, R.M., ed., *Acridines*, 2nd ed., New York, Interscience, pp. 63-67

Sawicki, E., Meeker, J.E. & Morgan, M. (1965a) Polynuclear aza compounds in automotive exhaust. *Arch. environ. Health*, *11*, 773-775

Sawicki, E., Meeker, J.E. & Morgan, M.J. (1965b) The quantitative composition of air pollution source effluents in terms of aza heterocyclic compounds and polynuclear aromatic hydrocarbons. *Int. J. Air Water Pollut.*, *9*, 291-298

Selby, I.A. (1973) *Acridinium salts and reduced acridines*. In: Acheson, R.M., ed., *Acridines*, 2nd ed., New York, Interscience, pp. 434-445

Stanley, T.W., Morgan, M.J. & Grisby, E.M. (1968) Application of a rapid thin-layer chromatographic procedure to the determination of benz[*a*]pyrene, benz[*c*]acridine, and 7H-benz[*de*]anthracen-7-one in airborne particulates from many American cities. *Environ. Sci. Technol.*, *2*, 699-702

Wood, A.W., Chang, R.L., Levin, W., Ryan, D.E., Thomas, P.E., Lehr, R.E., Kumar, S., Schaefer-Ridder, M., Engelhardt, U., Yagi, H., Jerina, D.M. & Conney, A.H. (1983) Mutagenicity of diol epoxides and tetrahydroepoxides of benz[*a*]acridine and benz[*c*]acridine in bacteria and in mammalian cells. *Cancer Res.*, *43*, 1656-1662 (and 'Errata', p. 3978)

BENZ[*a*]ANTHRACENE

This compound was considered by a previous working group, in December 1972 (IARC, 1973).

1. Chemical and Physical Data

1.1 Synonyms and trade names

Chem. Abstr. Services Reg. No.: 56-55-3

Chem. Abstr. Name: Benz(a)anthracene

IUPAC Systematic Name: Benz[*a*]anthracene

Synonyms: BA; benzanthracene; 1,2-benz(a)anthracene; benzanthrene; 1,2-benzanthracene; benzoanthracene; 1,2-benzoanthracene; benzo(a)anthracene; 2,3-benzophenanthrene; benz(b)phenanthrene; 2,3-benzphenanthrene; naphthanthracene; tetraphene

1.2 Structural and molecular formulae and molecular weight

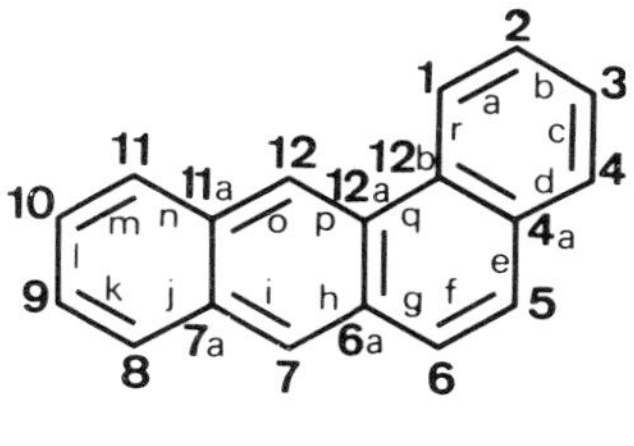

$C_{18}H_{12}$

Mol. wt: 228.3

1.3 Chemical and physical properties of the pure substance

From National Library of Medicine (1982), unless otherwise specified

(*a*) *Description*: Colourless plates with greenish-yellow fluorescence (recrystallized from glacial acetic acid or ethanol)

(*b*) *Boiling-point*: 435°C

(*c*) *Melting-point*: 167°C

(*d*) *Spectroscopy data*: λ_{max} 222, 227, 254 and 267 nm (in ethanol); 280, 290, 316, 329, 344, 359 and 385 nm (in benzene) (Clar, 1964). Infrared spectra have been reported (Fuson & Josien, 1956; API Research Project 44, 1960). Mass spectra have been tabulated (NIH/EPA Chemical Information System, 1982).

(*e*) *Solubility*: Virtually insoluble in water (9-14 μg/l) (Davis *et al.*, 1942; May *et al.*, 1978); slightly soluble in acetic acid; soluble in acetone and diethyl ether; very soluble in benzene

(*f*) *Stability*: Does not undergo photo-oxidation in organic solvents under fluorescent light or indoor sunlight (Kuratsune & Hirohata, 1962)

(*g*) *Reactivity*: Reacts as a diene in the Diels-Alder reaction to add maleic anhydride in the 7,12-position (Jones *et al.*, 1948). Can be hydrogenated catalytically to octadecahydrobenz[*a*]anthracene (Jarman, 1971). Oxidized by sodium dichromate in glacial acetic acid to give the 7,12-quinone (Clar, 1964). Oxidized by osmium tetroxide at the 5,6-position, yielding the *cis*-dihydrodiol (Cook & Schoental, 1948). Reacts with NO and NO_2 to form nitro derivatives (Butler & Crossley, 1981)

2. Production, Use, Occurrence and Analysis

2.1 Production and use

There is no commercial production or known use of this compound.

2.2 Occurrence and analysis

Data on occurrence and methods of analysis are summarized in the 'General Remarks on the Substances Considered', p. 35.

Benz[*a*]anthracene occurs ubiquitously in products of incomplete combustion; it is also found in fossil fuels. It has been identified in mainstream cigarette smoke (2.6 μg/100 cigarettes) (Lee *et al.*, 1976) (3.9-9.4 μg/1000 cigarettes) (Kiryu & Kuratsune, 1966); cigar smoke (2.5-3.9 μg/cigar) (Hoffmann & Wynder, 1972); mainstream smoke of marijuana cigarettes (3.3 μg/100 cigarettes (Lee *et al.*, 1976); urban atmospheres (0.1-21.6 ng/m^3) (Hoffmann & Wynder, 1976); exhaust emissions from gasoline engines (50-83 μg/l fuel) (Grimmer *et al.*, 1977); emissions from burnt coals (0.04-9.4 mg/kg) (Brockhaus & Tomingas, 1976); charcoal-broiled steaks (4.5 μg/kg) (Lijinsky & Shubik, 1964); different broiled and smoked foods (up to 189 μg/kg) (Masuda & Kuratsune, 1971), as well as fresh foods (up to 230 μg/kg) (Hettche, 1971); edible oils (0.1-98.3 μg/kg) (Grimmer & Hildebrandt, 1967); surface water, river water (0.4-30.6 ng/l) (Grimmer *et al.*, 1981); tap water (0.4-10.7 ng/l) (Melchiorri *et al.*, 1973); rainfall (3.2-12.3 ng/l) (Woidich *et al.*, 1976); subterranean water (0-1.3 ng/l) (Woidich *et al.*, 1976); waste water (0.5-4.9 μg/l) (Grimmer *et al.*, 1981); sludge (230-1760 μg/kg) (Borneff & Kunte, 1967); freeze-dried sewage sludge (0.62-19 mg/kg) (Grimmer *et al.*, 1980); and crude oils (1-6.7 mg/kg) (Grimmer *et al.*, 1983).

3. Biological Data Relevant to the Evaluation of Carcinogenic Risk to Humans

3.1 Carcinogenicity studies in animals

There is *sufficient evidence* that benz[*a*]anthracene is carcinogenic to experimental animals (see 'General Remarks on the Substances Considered', p. 33).

3.2 Other relevant biological data[1]

(*a*) *Experimental systems*

Toxic effects

The growth rate of mouse ascites sarcoma cells in culture was inhibited (27%) when benz[*a*]anthracene was added at a level of 1 μmol/ml dissolved in dimethyl sulphoxide (Pilotti *et al.*, 1975).

Benz[*a*]anthracene, but not anthracene, produced ultrastructural changes in columnar respiratory epithelium of rat tracheas maintained in organ culture (Dirksen & Crocker, 1968).

Goblet-cell hyperplasia and cases of so-called 'transitional hyperplasia' were observed when benz[*a*]anthracene was implanted (in a beeswax pellet) into isogenically-transplanted rat tracheas (Topping *et al.*, 1978).

Effects on reproduction and prenatal toxicity

S.c. injections of 5 mg/rat benz[*a*]anthracene daily to two rats from the first day of pregnancy resulted in fetal death and resorption (Wolfe & Bryan, 1939). I.p. administration of

[1] See also 'General Remarks on the Substances Considered', p. 53.

40 mg/kg benz[*a*]anthracene to Wistar rats can reduce the potency of the synthetic oestrogen triphenylethylene, presumably as a result of enzyme induction (Buu-Hoï & Hieu, 1972). It has been shown to induce benzo[*a*]pyrene hydroxylase activity in rat placenta (Welch *et al.*, 1969).

Metabolism and activation

Benz[*a*]anthracene is metabolized to all five of its dihydrodiols as well as to a number of phenolic metabolites and conjugates (Jacob *et al.*, 1981; Sims & Grover, 1981; Thakker *et al.*, 1982).

The metabolites of benz[*a*]anthracene have been evaluated extensively for their ability to induce mutations, cell transformation and tumours, and to bind covalently to nucleic acids (reviewed in Sims & Grover, 1974; Sims & Grover, 1981; Conney, 1982). The 3,4-diol-1,2-epoxide metabolite is mutagenic; its metabolic precursor, the 3,4-dihydrodiol, is also mutagenic in the presence of a rat-liver preparation. The 3,4-dihydrodiol and a 3,4-diol-1,2-epoxide also have high tumorigenic activity.

Analyses of DNA from the skin of mice or cultured cells treated with benz[*a*]anthracene indicate that a 3,4-diol-1,2-epoxide and a 8,9-diol-10,11-epoxide form nucleic acid adducts covalently (see reviews cited above).

Mutagenicity and other short-term tests

Results from short-term tests are given in Table 1.

Table 1. Results from short-term tests: Benz[*a*]anthracene

Test	Organism/assay[a]	Exogenous metabolic system[a]	Reported result	Comments	References
PROKARYOTES					
DNA damage	*Escherichia coli* (polA$^+$/polA$^-$)	UI-R-PMS	Negative	Tested at up to 250 μg/ml [The Working Group noted that UI-PMS was used]	Rosenkranz & Poirier (1979)
Mutation	*Salmonella typhimurium* (his$^-$/his$^+$)	Aro-R-PMS	Positive	At 20 μg/plate in strain TA100	McCann *et al.* (1975); Coombs *et al.* (1976); Simmon (1979a); Salamone *et al.* (1979)
		3MC-R-PMS	Positive	At 0.07 μmol/plate in strain TA100	Bartsch *et al.* (1980)
	Salmonella typhimurium (8AG^S/8AG^R)	Aro-R-PMS	Positive	At 65 nmol/ml in strain TM677	Kaden *et al.* (1979)
FUNGI					
Mutation	*Saccharomyces cerevisiae* D3 (mitotic recombination)	Aro-R-PMS	Negative	Tested at up to 5% w/v (*sic*)	Simmon (1979b)
INSECTS					
Mutation	*Drosophila melanogaster* (lethals, visibles, bobbed mutants)	--	Positive	Administered by microinjection	Fahmy & Fahmy (1973)

Test	Organism/assay[a]	Exogenous metabolic system[a]	Reported result	Comments	References
MAMMALIAN CELLS *IN VITRO*					
DNA damage	Primary rat hepatocytes (unscheduled DNA synthesis)	--	Positive	At 100 nmol/ml	Probst *et al.* (1981); Tong *et al.* (1981)
	HeLa cells (unscheduled DNA synthesis)	3MC-R-PMS	Positive	At up to 100 pmol/ml	Martin *et al.* (1978)
Mutation	Chinese hamster V79 cells (OUA^S/OUA^R)	SHE feeder layer	Positive	At 44 nmol/ml	Slaga *et al.* (1978)
	Chinese hamster V79 cells ($6TG^S/6TG^R$)	3MC-R-PMS	Positive	At 46 nmol/ml	Krahn & Heidelberger (1977)
	Mouse lymphoma L5178Y cells (TFT^S/TFT^R)	Various	Positive	At 40 nmol/ml [The Working Group noted only a small increase in mutation frequency]	Amacher *et al.* (1980); Amacher & Turner (1980)
	Rat liver epithelial ARL 18 cells ($6TG^S/6TG^R$)	--	Negative	Tested at up to 100 nmol/ml	Tong *et al.* (1981)
Chromosome effects	Chinese hamster ovary cells (sister chromatid exchange)	None	Positive	At 2 μg/ml	Pal (1981)
Cell transformation	Syrian hamster embryo cells (morphological)	--	Positive	At 0.1 μg/ml At 10 μg/ml	Pienta *et al.* (1977); DiPaolo *et al.* (1969, 1971)
	Mouse prostate C3HG23 cells (morphological)	--	Positive	At 1 μg/ml [Test protocol modified from that of an earlier study in which negative results were obtained (Marquardt *et al.*, 1972)]	Marquardt & Heidelberger (1972)
	Mouse C3H/10T1/2 cells (morphological)	--	Negative	Tested at up to 100 nmol/ml	Nesnow & Heidelberger (1976)
MAMMALIAN CELLS *IN VIVO*					
Chromosome effects	Chinese hamster bone-marrow cells (sister chromatid exchange)	--	Positive	At 2 × 450 mg/kg bw, i.p.	Roszinsky-Kocher *et al.* (1979)
	(aberrations)	--	Negative	Treated i.p. with 2 × 450 mg/kg bw	
	Long-Evans rat bone-marrow cells (aberrations)	--	Negative	Treated i.v. with 50 mg/kg bw	Sugiyama *et al.* (1973)
	Chinese hamster bone-marrow cells (micronuclei; aberrations)	--	Positive	At 2 x 900 mg/kg bw, i.p.	Péter *et al.* (1979)
	NMRI mice (metaphase II oocytes) (aberrations)	--	Positive	At 450 mg/kg bw orally	Péter *et al.* (1979)

[a] For an explanation of the abbreviations, see the Appendix, p. 452.

(*b*) *Humans*

No data were available to the Working Group.

3.3 Case reports and epidemiological studies of carcinogenicity to humans

No data were available to the Working Group.

4. Summary of Data Reported and Evaluation[1]

4.1 Experimental data

Benz[*a*]anthracene has been shown to be carcinogenic to experimental animals (see 'General Remarks on the Substances Considered', p. 33 and IARC, 1973).

The available data on reproductive toxicity and teratogenicity were inadequate for evaluation.

Benz[*a*]anthracene was mutagenic to *Salmonella typhimurium* in the presence of an exogenous metabolic system and was mutagenic to *Drosophila melanogaster*; it was also mutagenic to mammalian cells *in vitro* in the presence of an exogenous metabolic system. This compound was positive in one study of sister chromatid exchange. It induced unscheduled DNA synthesis in cultured mammalian cells and morphological transformation. In one in-vivo study, it induced sister chromatid exchange in hamsters; reports from in-vivo studies on the induction of chromosomal aberrations were conflicting.

There is *sufficient evidence* that benz[*a*]anthracene is active in short-term tests.

4.2 Human data[2]

Benz[*a*]anthracene is present as a major component of the total content of polynuclear aromatic compounds in the environment. Human exposure to benz[*a*]anthracene occurs primarily through smoking of tobacco, inhalation of polluted air and by ingestion of food and water contaminated by combustion effluents (for details, see 'General Remarks on the Substances Considered', p. 35).

---

[1] For definitions of the italicized terms, see Preamble, p. 19.

[2] Studies on occupational exposure to polynuclear aromatic compounds will be considered in future *IARC Monographs*.

4.3 Evaluation[1]

There is *sufficient evidence* that benz[*a*]anthracene is carcinogenic to experimental animals.

5. References

Amacher, D.E. & Turner, G.N. (1980) Promutagen activation by rodent-liver postmitochondrial fractions in the L5178Y/TK cell mutation assay. *Mutat. Res.*, *74*, 485-501

Amacher, D.E., Paillet, S.C., Turner, G.N., Ray, V.A. & Salsburg, D.S.(1980) Point mutations at the thymidine kinase locus in L5178Y mouse lymphoma cells. II. Test validation and interpretation. *Mutat. Res.*, *72*, 447-474

API Research Project 44 (1960) *Selected Infrared Spectral Data*, Vol. VI, No. 2236, Washington DC, American Petroleum Institute

Bartsch, H., Malaveille, C., Camus, A.-M., Martel-Planche, G., Brun, G., Hautefeuille, A., Sabadie, N., Barbin, A., Kuroki, T., Drevon, C., Piccoli, C. & Montesano, R. (1980) Validation and comparative studies on 180 chemicals with *S. typhimurium* strains and V79 Chinese hamster cells in the presence of various metabolizing systems. *Mutat. Res.*, *76*, 1-50

Borneff, J. & Kunte, H. (1967) Carcinogenic substances in water and soil. XIX. Action of sewage purification on polycyclic hydrocarbons. *Arch. Hyg. Bakt.*, *151*, 202-210

Brockhaus, A. & Tomingas, R. (1976) Emission of polycyclic hydrocarbons during burning processes in small heating installations and their concentration in the atmosphere (Ger.). *Staub-Reinhalt. Luft*, *36*, 96-101

Butler, J.D. & Crossley, P. (1981) Reactivity of polycyclic aromatic hydrocarbons adsorbed on soot particles. *Atmos. Environ.*, *15*, 91-94

Buu-Hoï, N.P.& Hieu, H.T. (1972) Inhibition of the estrogenic activity of triphenylethylene by 6-aminochrysene and other inducers of drug-metabolizing enzymes. *Horm. Metab. Res.*, *4*, 119-121

[1] In the absence of adequate data on humans, it is reasonable, for practical purposes, to regard chemicals for which there is *sufficient evidence* of carcinogenicity in animals as if they presented a carcinogenic risk to humans (see also Preamble, p. 20).

Clar, E., ed. (1964) *Polycyclic Hydrocarbons*, Vol. 1, New York, Academic Press, pp. 307-322

Conney, A.H. (1982) Induction of microsomal enzymes by foreign chemicals and carcinogenesis by polycyclic aromatic hydrocarbons: G.H.A. Clowes Memorial Lecture. *Cancer Res.*, *42*, 4875-4917

Cook, J.W. & Schoental, R. (1948) Oxidation of carcinogenic hydrocarbons by osmium tetroxide. *J. chem. Soc.*, 170-173

Coombs, M.M., Dixon, C. & Kissonerghis, A.-M. (1976) Evaluation of the mutagenicity of compounds of known carcinogenicity, belonging to the benz[*a*]anthracene, chrysene, and cyclopenta[*a*]phenanthrene series, using Ames's test. *Cancer Res.*, *36*, 4525-4529

Davis, W.W., Krahl, M.E. & Clowes, G.H.A. (1942) Solubility of carcinogenic and related hydrocarbons in water. *J. Am. chem. Soc.*, *64*, 108-110

DiPaolo, J.A., Donovan, J.P. & Nelson, R.L. (1969) Quantitative studies of *in vitro* transformation by chemical carcinogens. *J. natl Cancer Inst.*, *42*, 867-874

DiPaolo, J.A., Donovan, J.P. & Nelson, R.L. (1971) Transformation of hamster cells *in vitro* by polycyclic hydrocarbons without cytotoxicity. *Proc. natl Acad. Sci. USA*, *68*, 2958-2961

Dirksen, E.R. & Crocker, T.T. (1968) Ultrastructural alterations produced by polycyclic aromatic hydrocarbons on rat tracheal epithelium in organ culture. *Cancer Res.*, *28*, 906-923

Fahmy, O.G. & Fahmy, M.J. (1973) Oxidative activation of benz[*a*]anthracene and methylated derivatives in mutagenesis and carcinogenesis. *Cancer Res.*, *33*, 2354-2361

Fuson, N. & Josien, M.-L. (1956) Infrared spectra of polynuclear aromatic compounds. I. 1,2-Benzanthracene, the monomethyl-1,2-benzanthracenes and some dimethyl-1,2benzanthracenes. *J. Am. chem. Soc.*, *78*, 3049-3060

Grimmer, G. & Hildebrandt, A. (1967) Content of polycyclic hydrocarbons in crude vegetable oils. *Chem. Ind.*, 25 November, 2000-2002

Grimmer, G., Böhnke, H. & Glaser, A. (1977) Investigation on the carcinogenic burden by air pollution in man. XV. Polycyclic aromatic hydrocarbons in automobile exhaust gas - An inventory. *Zbl. Bakt. Hyg., 1 Orig. B164*, 218-234

Grimmer, G., Hilge, G. & Niemitz, W. (1980) Comparison of the profile of polycyclic aromatic hydrocarbons in sewage sludge samples from 25 sewage treatment works (Ger.). *Wasser*, *54*, 255-272

Grimmer, G., Schneider, D. & Dettbarn, G. (1981) The load of different rivers in the Federal Republic of Germany by PAH (PAH-profiles of surface water) (Ger.). *Wasser*, *56*, 131-144

Grimmer, G., Jacob, J. & Naujack, K.-W. (1983) Profile of the polycyclic aromatic compounds from crude oils. Inventory by GCGC/MS - PAH in environmental materials, Part 3. *Fresenius Z. anal. Chem.*, *314*, 29-36

Hettche, H.O. (1971) Plant waxes as collectors of polycyclic aromatics in the air of residential areas. *Staub (Engl. Transl.)*, *31*, 34-41

Hoffmann, D. & Wynder, E.L. (1972) Smoke of cigarettes and little cigars: An analytical comparison. *Science*, *178*, 1197-1199

Hoffmann, D. & Wynder, E. (1976) *Environmental respiratory carcinogenesis*. In: Searle, C.E., ed., *Chemical Carcinogens* (*ACS Monograph 173*), Washington DC, American Chemical Society, p. 341

IARC (1973) *IARC Monographs on the Evaluation of Carcinogenic Risk of Chemicals to Man*, Vol. 3, *Certain Polycyclic Aromatic Hydrocarbons and Heterocyclic Compounds*, Lyon, pp. 46-68

Jacob, J., Grimmer, G. & Schmoldt, A. (1981) The influence of polycyclic aromatic hydrocarbons as inducers of monooxygenases on the metabolite profile of benz[*a*]anthracene in rat liver microsomes. *Cancer Lett.*, *14*, 175-185

Jarman, M. (1971) Total reduction of benz[*a*]anthracene: The preparation of octadecahydrobenz[*a*]anthracene. *Chem. Ind.*, *8*, 228

Jones, R.N., Gogek, C.J. & Sharpe, R.W. (1948) The reaction of maleic anhydride with polynuclear aromatic hydrocarbons. *Can. J. Res.*, *26*, 719-727

Kaden, D.A., Hites, R.A. & Thilly, W.G. (1979) Mutagenicity of soot and associated polycyclic aromatic hydrocarbons to *Salmonella typhimurium*. *Cancer Res.*, *39*, 4152-4159

Kiryu, S. & Kuratsune, M. (1966) Polycyclic aromatic hydrocarbons in the cigarette tar produced by human smoking. *Gann*, *57*, 317-322

Krahn, D.F. & Heidelberger, C. (1977) Liver homogenate-mediated mutagenesis in Chinese hamster V79 cells by polycyclic aromatic hydrocarbons and aflatoxins. *Mutat. Res.*, *46*, 27-44

Kuratsune, M. & Hirohata, T. (1962) Decomposition of polycyclic aromatic hydrocarbons under laboratory illuminations. *Natl Cancer Inst. Monogr.*, *9*, 117-125

Lee, M.L., Novotny, M. & Bartle, K.D. (1976) Gas chromatography/mass spectrometry and nuclear magnetic resonance spectrometric studies of carcinogenic polynuclear aromatic hydrocarbons in tobacco and marijuana smoke condensates. *Anal. Chem.*, *48*, 405-416

Lijinsky, W. & Shubik, P. (1964) Benzo(*a*)pyrene and other polynuclear hydrocarbons in charcoal-broiled meat. *Science*, *145*, 53-55

Marquardt, H. & Heidelberger, C. (1972) Influence of 'feeder cells' and inducers and inhibitors of microsomal mixed-function oxidases on hydrocarbon-induced malignant transformation of cells derived from C3H mouse prostate. *Cancer Res.*, *32*, 721-725

Marquardt, H., Kuroki, T., Huberman, E., Selkirk, J.K., Heidelberger, C., Grover, P.L. & Sims, P. (1972) Malignant transformation of cells derived from mouse prostate by epoxides and other derivatives of polycyclic hydrocarbons. *Cancer Res.*, *32*, 716-720

Martin, C.N., McDermid, A.C. & Garner, R.C. (1978) Testing of known carcinogens and noncarcinogens for their ability to induce unscheduled DNA synthesis in HeLa cells. *Cancer Res.*, *38*, 2621-2627

Masuda, Y. & Kuratsune, M. (1971) Polycyclic aromatic hydrocarbons in smoked fish, 'katsuobushi'. *Gann*, *62*, 27-30

May, W.E., Wasik, S.P. & Freeman, D.H. (1978) Determination of the solubility behavior of some polynuclear aromatic hydrocarbons in water. *Anal. Chem.*, *50*, 997-1000

McCann, J., Choi, E., Yamasaki, E. & Ames, B.N. (1975) Detection of carcinogens as mutagens in the *Salmonella*/microsome test: Assay of 300 chemicals. *Proc. natl Acad. Sci. USA*, *72*, 5135-5139

Melchiorri, C., Chiacchiarini, L., Grella, A. & D'Arca, S.U. (1973) Identification and determination of polycyclic aromatic hydrocarbons in some tap waters of Roma city (Ital.). *Nuovi Ann. Ig. Microbiol.*, *24*, 279-301

National Library of Medicine (1982) *Toxicology Data Bank*, Bethesda, MD, National Library of Medicine Specialized Information Services, Toxicology Information Program

Nesnow, S. & Heidelberger, C. (1976) The effect of modifiers of microsomal enzymes on chemical oncogenesis in cultures of C3H mouse cell lines. *Cancer Res.*, *36*, 1801-1808

NIH/EPA Chemical Information System (1982) *Mass Spectral Search System*, Washington DC, CIS Project, Information Services Corporation

Pal, K. (1981) The induction of sister-chromatid exchanges in Chinese hamster ovary cells by K-region epoxides and some dihydrodiols derived from benz[*a*]anthracene, dibenz[*a*,*c*]anthracene and dibenz[*a*,*h*]anthracene. *Mutat. Res.*, *84*, 389-398

Péter, S., Palme, G.E. & Röhrborn, G. (1979) Mutagenicity of polycyclic hydrocarbons. III. Monitoring genetic hazards of benz(*a*)anthracene. *Acta morphol. acad. sci. hung.*, *27*, 199-204

Pienta, R.J., Poiley, J.A. & Lebherz III, W.B. (1977) Morphological transformation of early passage golden Syrian hamster embryo cells derived from cryopreserved primary cultures as a reliable *in vitro* bioassay for identifying diverse carcinogens. *Int. J. Cancer*, *19*, 642-655

Pilotti, A., Ancker, K., Arrhenius, E. & Enzell, C. (1975) Effects of tobacco and tobacco smoke constituents on cell multiplication *in vitro*. *Toxicology*, *5*, 49-62

Probst, G.S., McMahon, R.E., Hill, L.E., Thompson, C.Z., Epp, J.K. & Neal, S.B. (1981) Chemically-induced unscheduled DNA synthesis in primary rat hepatocyte cultures: A comparison with bacterial mutagenicity using 218 compounds. *Environ. Mutagenesis*, *3*, 11-32

Rosenkranz, H.S. & Poirier, L.A. (1979) Evaluation of the mutagenicity and DNA-modifying activity of carcinogens and noncarcinogens in microbial systems. *J. natl Cancer Inst.*, *62*, 873-892

Roszinsky-Köcher, G., Basler, A. & Röhrborn, G. (1979) Mutagenicity of polycyclic hydrocarbons. V. Induction of sister-chromatid exchanges *in vivo*. *Mutat. Res.*, *66*, 65-67

Salamone, M.F., Heddle, J.A. & Katz, M. (1979) The mutagenic activity of thirty polycyclic aromatic hydrocarbons (PAH) and oxides in urban airborne particulates. *Environ. Int.*, *2*, 37-43

Simmon, V.F. (1979a) *In vitro* mutagenicity assays of chemical carcinogens and related compounds with *Salmonella typhimurium*. *J. natl Cancer Inst.*, *62*, 893-899

Simmon, V.F. (1979b) *In vitro* assays for recombinogenic activity of chemical carcinogens and related compounds with *Saccharomyces cerevisiae D3*. *J. natl Cancer Inst.*, *62*, 901-909

Sims, P. & Grover, P. (1974) Epoxides in polycyclic aromatic hydrocarbon metabolism and carcinogenesis. *Adv. Cancer Res.*, *20*, 165-274

Sims, P. & Grover, P. (1981) *Involvement of dihydrodiols and diol expoxides in the metabolic activation of polycyclic hydrocarbons other than benzo[a]pyrene*. In: Gelboin, H.V. & Ts'O, P.O.P., eds, *Polycyclic Hydrocarbons and Cancer*, Vol. 3, New York, Academic Press, pp. 117-181

Slaga, T.J., Huberman, E., Selkirk, J.K., Harvey, R.G. & Bracken, W.M. (1978) Carcinogenicity and mutagenicity of benz(*a*)anthracene diols and diol-epoxides. *Cancer Res.*, *38*, 1699-1704

Sugiyama, T. (1973) Chromosomal aberrations and carcinogenesis by various benz[*a*]anthracene derivatives. *Gann*, *64*, 637-639

Thakker, D.R., Levin, W., Yagi, H., Tada, M., Ryan, D.E., Thomas, P.E., Conney, A.H. & Jerina, D.M. (1982) Stereoselective metabolism of the (+)- and (-)-enantiomers of *trans*-3,4-dihydroxy-3,4-dihydrobenz[*a*]anthracene by rat liver microsomes and by a purified and reconstituted cytochrome P-450 system. *J. biol. Chem.*, *257*, 5103-5110

Tong, C., Laspia, M.F., Telang, S. & Williams, G.M. (1981) The use of adult rat liver cultures in the detection of the genotoxicity of various polycyclic aromatic hydrocarbons. *Environ. Mutagenesis*, *3*, 477-487

Topping, D.C., Pal, B.C., Martin, D.H., Nelson, F.R. & Nettesheim, P. (1978) Pathologic changes induced in respiratory tract mucosa by polycyclic hydrocarbons of differing carcinogenic activity. *Am. J. Pathol.*, *93*, 311-324

Welch, R.M., Harrison, Y.E., Gommi, B.W., Poppers, P.J., Finster, M. & Conney, A.H. (1969) Stimulatory effect of cigarette smoking on the hydroxylation of 3,4-benzpyrene and the *N*-demethylation of 3-methyl-4-monomethylaminoazobenzene by enzymes in human placenta. *Clin. Pharmacol. Ther.*, *10*, 100-109

Woidich, H., Pfannhauser, W., Blaicher, G. & Tiefenbacher, K. (1976) Analysis of polycyclic aromatic hydrocarbons in drinking and industrial water (Ger.). *Lebensmittelchem. gerichtl. Chem.*, *30*, 141-160

Wolfe, J.M. & Bryan, W.R. (1939) Effects induced in pregnant rats by injection of chemically pure carcinogenic agents. *Am. J. Cancer*, *36*, 359-368

BENZO[*b*]FLUORANTHENE

This compound was considered by a previous working group, in December 1972 (IARC, 1973).

1. Chemical and Physical Data

1.1 Synonyms and trade names

Chem. Abstr. Services Reg. No : 205-99-2

Chem. Abstr. Name: Benz(e)acephenanthrylene

IUPAC Systematic Name: Benz[*e*]acephenanthrylene

Synonyms: 3,4-Benz(e)acephenanthrylene; 2,3-benzfluoranthene; 3,4-benzfluoranthene; 2,3-benzofluoranthene; 3,4-benzofluoranthene; benzo(e)fluoranthene; B(b)F

1.2 Structural and molecular formulae and molecular weight

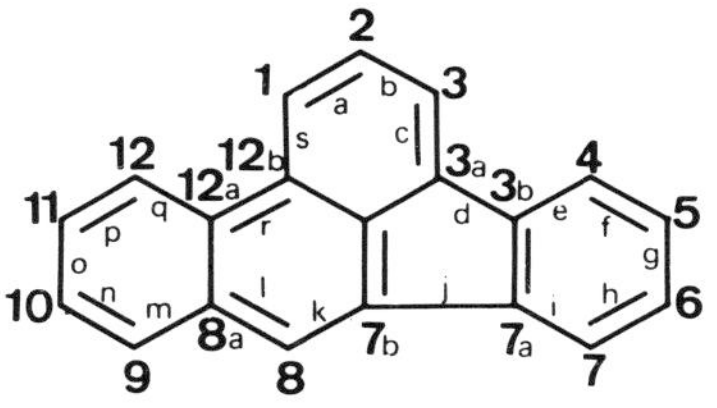

$C_{20}H_{12}$

Mol. wt: 252.3

1.3 Chemical and physical properties of the pure substance

From National Library of Medicine (1982), unless otherwise specified

(*a*) *Description*: Needles (recrystallized from benzene), colourless needles (recrystallized from toluene or glacial acetic acid)

(*b*) *Melting-point*: 168.3°C (Karcher *et al.*, 1983)

(*c*) *Spectroscopy data*: λ_{max} 255, 275, 288, 300, 340, 349, 367 nm (in cyclohexane) (Karcher *et al.*, 1983). The infrared absorption spectrum of the solid compound and of solutions in carbon tetrachloride and carbon disulphide have been published (API Research Project 44, 1961). Mass and nuclear magnetic resonance spectra have been tabulated (NIH/EPA Chemical Information System, 1982; Karcher *et al.*, 1983)

(*d*) *Solubility*: Virtually insoluble in water; slightly soluble in benzene and acetone

(*e*) *Stability*: No data were available

(*f*) *Reactivity*: No data were available

2. Production, Use, Occurrence and Analysis

2.1 Production and use

There is no commercial production or known use of this compound. A reference material of certified high purity is available (Karcher *et al.*, 1980; Community Bureau of Reference, 1982).

2.2 Occurrence and analysis

Data on occurrence and methods of analysis are summarized in the 'General Remarks on the Substances Considered', p. 35.

Benzo[*b*]fluoranthene occurs ubiquitously in products of incomplete combustion; it also occurs in fossil fuels. It has been identified in mainstream cigarette smoke (0.1 mg/kg cigarette smoke condensate) (Wynder & Hoffmann, 1959), (2.9 μg/100 g of burnt material) (Masuda & Kuratsune, 1972), (1.2-6.5 μg/1000 cigarettes) (Kiryu & Kuratsune, 1966); urban air (2.3-7.4 ng/m^3) (Hoffmann & Wynder, 1976); gasoline engine exhaust (19-48 μg/l fuel) (Grimmer *et al.*, 1977); emissions from the burning of various types of coal (0.05-17.2 mg/kg with benzo[*j*]-,-[*k*]fluoranthenes) (Brockhaus & Tomingas, 1976); emissions from oil-fired heating (0.00003-0.405 mg/kg) (Behn & Grimmer, 1981); broiled and smoked food (up to 15.1 μg/kg) (Toth, 1971) and oils and margarine (up to 14.5 μg/kg) (Fábián, 1968); surface water (0.6-1.1 ng/l) (Woidich *et al.,* 1976); tap water (0.4-5.4 ng/l) (Borneff & Kunte, 1969; Melchiorri *et al.,* 1973); rain water (4.4-14.6 ng/l) (Woidich *et al.*, 1976); subterranean water (0.6-9.0 ng/l) (Borneff & Kunte, 1969); waste water (0.04-23.7 μg/l) and sludge (510-2160 μg/kg) (Borneff & Kunte, 1967); gasolines (0.19-1.34 mg/kg) (Müller & Meyer, 1974); crude oil (7.4 mg/kg) (Grimmer *et al.*, 1983).

3. Biological Data Relevant to the Evaluation of Carcinogenic Risk to Humans

3.1 Carcinogenicity studies in animals

There is *sufficient evidence* that benzo[*b*]fluoranthene is carcinogenic to experimental animals (see 'General Remarks on the Substances Considered', p. 33).

3.2 Other relevant biological data[1]

(*a*) *Experimental system*

No data were available to the Working Group on toxic effects or on effects on reproduction and prenatal toxicity.

Metabolism and activation

The 1,2- and 11,12-dihydrodiols as well as the 4-(or 7-), 5-(or 6-)monohydroxy derivatives of benzo[*b*]fluoranthene have been detected as metabolites (Amin *et al.*, 1982).

Mutagenicity and other short-term tests

Results from short-term tests are given in Table 1.

Table 1. Results from short-term tests: Benzo[*b*]fluoranthene

Test	Organism/assay[a]	Exogenous metabolic system[a]	Reported result	Comments	References
PROKARYOTES					
Mutation	*Salmonella typhimurium* (his⁻/his⁺)	Aro-R-PMS	Negative	Tested at up to 100 μg/ml in strain TA100	Mossanda *et al.* (1979)
		Aro-R-PMS	Positive	At 100 μg/plate in strain TA100 At 7 nmol/plate in strain TA98	LaVoie *et al.* (1979); Hermann (1981)
MAMMALIAN CELLS *IN VIVO*					
Chromosome effects	Chinese hamster bone-marrow cells: sister chromatid exchange	--	Positive	At 2 x 450 mg/kg bw, i.p.	Roszinsky-Kocher *et al.* (1979)
	aberrations	--	Negative	Treated i.p. with 2 x 450 mg/kg bw	

[a] For an explanation of the abbreviations, see the Appendix, p. 452.

[1] See also 'General Remarks on the Substances Considered', p. 53.

(*b*) *Humans*

No data were available to the Working Group.

3.3 Case reports and epidemiological studies of carcinogenicity to humans

No data were available to the Working Group.

4. Summary of Data Reported and Evaluation[1]

4.1 Experimental data

Benzo[*b*]fluoranthene has been shown to be carcinogenic to experimental animals (see 'General Remarks on the Substances Considered', p. 33 and IARC, 1973).

No data on the teratogenicity of this compound were available.

Benzo[*b*]fluoranthene was mutagenic to *Salmonella typhimurium* in the presence of an exogenous metabolic system. In the one available study it was reported to induce sister chromatid exchange but not chromosomal aberrations in bone-marrow cells of hamsters treated *in vivo*.

There is *inadequate evidence* that benzo[*b*]fluoranthene is active in short-term tests.

4.2 Human data[2]

Benzo[*b*]fluoranthene is present as a major component of the total content of polynuclear aromatic compounds in the environment. Human exposure to benzo[*b*]fluoranthene occurs

---

[1] For definitions of the italicized terms, see Preamble, p. 19.

[2] Studies on occupational exposure to polynuclear aromatic compounds will be considered in future *IARC Monographs*.

primarily through the smoking of tobacco, inhalation of polluted air and by ingestion of food and water contaminated by combustion effluents (for details, see 'General Remarks on the Substances Considered', p. 35).

4.3 Evaluation[1]

There is *sufficient evidence* that benzo[*b*]fluoranthene is carcinogenic to experimental animals.

5. References

Amin, S., LaVoie, E.J. & Hecht, S.S. (1982) Identification of metabolites of benzo[*b*]fluoranthene. *Carcinogenesis*, *3*, 171-174

API Research Project 44 (1961) *Selected Infrared Spectral Data*, Vol. VI, No. 2312, Washington DC, American Petroleum Institute

Behn, U. & Grimmer, G. (1981) *Bestimmung polycyclischer aromatischer Kohlenwasserstoffe (PAH) in Abgas einer Zentralheizungsanlage auf Heizöl EL-Basis* [Determination of polycyclic aromatic hydrocarbons (PAH) in the flue gas of an oil-fired heating unit] (*BMI-DGMK Projekt 111-02*), Hamburg, Deutsche Gesellschaft für Mineralölwissenschaft und Kohlechemie e.V.

Borneff, J. & Kunte, H. (1967) Carcinogenic substances in water and soil. XIX. Action of sewage purification on polycylic hydrocarbons. *Arch. Hyg. Bakt.*, *151*, 202-210

Borneff, J. & Kunte, H. (1969) Carcinogenic substances in water and soil. 25. A routine method for the determination of polynuclear aromatic hydrocarbons in water (Ger.). *Arch. Hyg. Bakt.*, *153*, 220-229

---

[1] In the absence of adequate data on humans, it is reasonable, for practical purposes, to regard chemicals for which there is *sufficient evidence* of carcinogenicity in animals as if they presented a carcinogenic risk to humans (see also Preamble, p. 20).

Brockhaus, A. & Tomingas, R. (1976) Emission of polycyclic hydrocarbons during burning processes in small heating installations and their concentration in the atmosphere (Ger.). *Staub-Reinhalt. Luft*, *36*, 96-101

Community Bureau of Reference (1982) *Polycyclic Aromatic Hydrocarbon Reference Materials of Certified Purity* (*Information Handout, No. 22*), Brussels, Commission of the European Communities

Fábián, B. (1968) Carcinogenic substances in cooking fat and oil. IV. Studies on margarine, vegetable fat and butter (Ger.). *Arch. Hyg. Bakt.*, *152*, 231-237

Grimmer, G., Böhnke, H. & Glaser, A. (1977) Investigation on the carcinogenic burden by air pollution in man. XV. Polycyclic aromatic hydrocarbons in automobile exhaust gas - An inventory. *Zbl. Bakt. Hyg., 1. Abt. Orig. B164*, 218-234

Grimmer, G., Jacob, J. & Naujack, K.-W. (1983) Profile of the polycyclic aromatic compounds from crude oils. Inventory by GCGC/MS - PAH in environmental materials, Part 3. *Fresenius Z. anal. Chem.*, *314*, 29-36

Hermann, M. (1981) Synergistic effects of individual polycyclic aromatic hydrocarbons on the mutagenicity of their mixtures. *Mutat. Res.*, *90*, 399-409

Hoffmann, D. & Wynder, E.. (1976) *Environmental respiratory carcinogenesis*. In: Searle, C.E., ed., *Chemical Carcinogens* (*ACS Monograph 173*), Washington DC, American Chemical Society, p. 341

IARC (1973) *IARC Monographs on the Evaluation of Carcinogenic Risk of Chemicals to Man*, Vol. 3, *Certain Polycyclic Aromatic Hydrocarbons and Heterocyclic Compounds*, Lyon, pp. 69-81

Karcher, W., Jacob, J. & Haemers, L. (1980) *The Certification of Eight Polycyclic Aromatic Hydrocarbon Materials (PAH) (BCR Reference Materials Nos. 46, 47, 48, 49, 50, 51, 52 and 53)* (*EUR 6967 EN*), Luxembourg, Commission of the European Communities

Karcher, W., Fordham, R., Dubois, J. & Gloude, P. (1983) *Spectral Atlas of Polycyclic Aromatic Compounds*, Dordrecht, The Netherlands, D. Reidel (in press)

Kiryu, S. & Kuratsune, M. (1966) Polycyclic aromatic hydrocarbons in the cigarette tar produced by human smoking. *Gann*, *57*, 317-322

LaVoie, E., Bedenko, V., Hirota, N., Hecht, S.S. & Hoffmann, D. (1979) *A comparison of the mutagenicity, tumor-initiating activity and complete carcinogenicity of polynuclear aromatic hydrocarbons*. In: Jones, P.W. & Leber, P., eds, *Polynuclear Aromatic Hydrocarbons*, Ann Arbor, MI, Ann Arbor Science Publishers, pp. 705-721

Masuda, Y. & Kuratsune, M. (1972) Comparison of the yield of polycyclic aromatic hydrocarbons in smoke from Japanese tobacco. *Jpn. J. Hyg.*, *27*, 339-341

Melchiorri, C., Chiacchiarini, L., Grella, A. & D'Arca, S.U. (1973) Research and determination of polycyclic aromatic hydrocarbons in some drinking waters of the city of Roma (Ital.). *Nuovi Ann. Ig. Microbiol.*, *24*, 279-301

Mossanda, K., Poncelet, F., Fouassin, A. & Mercier, M. (1979) Detection of mutagenic polycyclic aromatic hydrocarbons in African smoked fish. *Food Cosmet. Toxicol.*, *17*, 141-143

Müller, K. & Meyer, J.P. (1974) *Einfluss von Ottokraftstoffen auf die Emission von polynuklearen aromatischen Kohlenwasserstoffen in Automobilabgasen im Europa-test* (Effect of gasoline components on emission of polynuclear aromatic hydrocarbons in car exhaust in the Europa test) (*Forschungsbericht 4568*), Hamburg, Deutsche Gesellschaft für Mineralölwissenschaft und Kohlechemie e.V.

National Library of Medicine (1982) *Toxicology Data Bank*, Bethesda, MD, National Library of Medicine Specialized Information Services, Toxicology Information Program

NIH/EPA Chemical Information System (1982) *Mass Spectral Search System*, Washington DC, CIS Project, Information Services Corporation

Roszinsky-Köcher, G., Basler, A. & Röhrborn, G. (1979) Mutagenicity of polycyclic hydrocarbons. V. Induction of sister-chromatid exchanges *in vivo*. *Mutat. Res.*, *66*, 65-67

Toth, L. (1971) Polycyclic hydrocarbons in smoked ham and bacon (Ger.). *Fleischwirtschaft*, *51*, 1069-1070

Woidich, H., Pfannhauser, W., Blaicher, G. & Tiefenbacher, K. (1976) Analysis of polycyclic aromatic hydrocarbons in drinking and industrial water (Ger.). *Lebensmittelchem. gerichtl. Chem.*, *30*, 141-160

Wynder, E.L. & Hoffmann, D. (1959) The carcinogenicity of benzofluoranthenes. *Cancer*, *12*, 1194-1199

BENZO[*j*]FLUORANTHENE

This compound was considered by a previous working group, in December 1972 (IARC, 1973). Data that have become available since that time have been incorporated into the present monograph and taken into consideration in the evaluation.

1. Chemical and Physical Data

1.1 Synonyms and trade names

Chem. Abstr. Services Reg. No.: 205-82-3

Chem. Abstr. Name: Benzo(j)fluoranthene

IUPAC Systematic Name: Benzo[*j*]fluoranthene

Synonyms: 7,8-Benzofluoranthene; 10,11-benzofluoranthene; benzo(l)fluoranthene; benzo-12,13-fluoranthene; B(j)F; dibenzo(a,jk)fluorene

1.2 Structural and molecular formulae and molecular weight

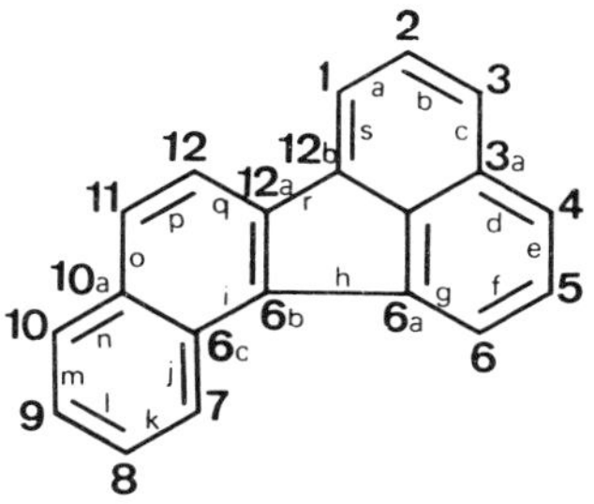

$C_{20}H_{12}$

Mol. wt: 252.3

1.3 Chemical and physical properties of the pure substance

(*a*) *Description*: Yellow plates (recrystallized from ethanol); needles (recrystallized from acetic acid) (National Library of Medicine, 1982)

(*b*) *Melting-point*: 165.4°C (Karcher *et al.*, 1983)

(*c*) *Spectroscopy data*: λ_{max} 239, 280, 291, 306, 317, 331, 362, 373, 382 nm (in cyclohexane) (Karcher *et al.*, 1983). Mass and nuclear magnetic resonance spectra have been tabulated (NIH/EPA Chemical Information System, 1982; Karcher *et al.*, 1983).

(*d*) *Solubility*: Virtually insoluble in water; slightly soluble in acetic acid and ethanol (National Library of Medicine, 1982)

(*e*) *Stability*: No data were available.

(*f*) *Reactivity*: No data were available.

2. Production, Use, Occurrence and Analysis

2.1 Production and use

There is no commercial production or known use of this compound. A reference material of certified high purity is available (Karcher *et al.*, 1980; Community Bureau of Reference, 1982).

2.2 Occurrence and analysis

Data on occurrence and methods of analysis are summarized in the 'General Remarks on the Substances Considered', p. 35.

Benzo[*j*]fluoranthene occurs ubiquitously as a product of incomplete combustion; it also occurs in fossil fuels. It has been identified in mainstream smoke of cigarettes (0.15-0.2 mg/kg cigarette smoke condensate) (Wynder & Hoffmann, 1959), (2.1 μg/100 cigarettes) (Lee *et al.*, 1976); mainstream smoke of marijuana cigarettes (3.0 μg/100 cigarettes) (Lee *et al.*, 1976); air (0.8-4.4 ng/m^3) (Hoffmann & Wynder, 1976); exhaust from gasoline engines (11-27 μg/l fuel) (Grimmer *et al.*, 1977); with benzo[*b*]-,-[*k*]fluoranthenes in emissions from the burning of various types of coal (0.05-17.2 mg/kg) (Brockhaus & Tomingas, 1976) and in emissions from oil-fired heating (0.00003-0.405 mg/kg) (Behn & Grimmer, 1981); used motor oils (9.95 mg/kg) (Grimmer *et al.*, 1981); crude oils (7.4 mg/kg) (Grimmer *et al.*, 1983); coal-tar (0.45-0.63 g/kg) (Lijinsky *et al.*, 1963); smoked and broiled fish (at levels as high as 23 μg/kg) (Masuda & Kuratsune, 1971); various oils and margarines (at levels as high as 10.5 μg/kg) (Fábián, 1968); surface water (0.6-1.2 ng/l), rainwater (2.6-11.1 ng/l) and subterranean water (0.6-1.3 ng/l) (Woidich *et al.*, 1976); tap water (0.07 ng/l) (Olufsen, 1980); waste water (0.02-29.6 μg/l) and sludge (230-2060 μg/kg) (Borneff & Kunte, 1967).

3. Biological Data Relevant to the Evaluation of Carcinogenic Risk to Humans

3.1 Carcinogenicity studies in animals[1]

(a) Skin application

Mouse: Five groups of 20 female Swiss mice [age and body weight unspecified] were painted on the skin with 0.1 or 0.5% 'highly purified' benzo[*j*]fluoranthene in acetone or 0.01, 0.05 or 0.5% benzo[*a*]pyrene in acetone (positive controls) thrice weekly for life [no vehicle control was used]. By the end of seven months, 11, 20, 1, 20 and 20 animals had died in the low- and high-dose benzo[*j*]fluoranthene-treated and low-, mid- and high-dose benzo[*a*]pyrene-treated groups, respectively. At nine months, all animals, except for eight low-dose benzo[*a*]pyrene-treated controls, had died. The incidences of skin papillomas were 70, 95, 85, 70 and 75%, and those of skin carcinomas 100, 95, 85, 95 and 75% in the low- and high-dose benzo[*j*]fluoranthene-treated and in the low-, mid- and high-dose benzo[*a*]pyrene-treated groups, respectively (Wynder & Hoffmann, 1959).

Seven groups of 40 female NMRI mice, 10 weeks of age, received skin applications of 0.02 ml acetone [vehicle controls]; 3.4, 5.6 or 9.2 μg benzo[*j*]fluoranthene in 0.02 ml acetone (purity, >96%); or 1.7, 2.8 or 4.6 μg benzo[*a*]pyrene in 0.02 ml acetone (positive controls) twice weekly for life. Information on mortality was presented in the form of curves. After one year, mortality was about 8% in controls, 5-20% in benzo[*j*]fluoranthene-treated groups and 12-20% in positive controls; after two years, mortality was about 70% in controls, 75-90% in the benzo[*j*]fluoranthene-treated groups and 80-100% in the positive controls (all animals of the high-dose benzo[*a*]pyrene-treated group had died before week 85). A total of four mice treated with benzo[*j*]fluoranthene developed local tumours: 1/38 low-dose animals had a sarcoma at the site of application (2.6%), 1/35 mid-dose animals had a local tumour (2.9%) [site and type unspecified], and 1/38 high-dose animals had a papilloma and 1/38 a carcinoma of the dorsal skin (5.3%). No local tumour was found in acetone-treated controls. The incidence of local tumours in the various positive-control groups ranged from 8/34 to 24/35. There was a high rate of 'spontaneous' systemic tumours (about 70%). After age-standardization of incidences, the differences between the groups were not statistically significant (Habs *et al.*, 1980).

In order to examine the initiating activity of benzo[*j*]fluoranthene, five groups of 20 female Cr1:CD-1(ICR)BR outbred albino mice, 50-55 days of age [body weights unspecified], were given skin applications of 0.1 ml acetone (vehicle controls), 3 μg benzo[*a*]pyrene (positive controls), or 3, 10 or 100 μg benzo[*j*]fluoranthene (purity, >99%) in 0.1 ml acetone [total doses, 30, 100 or 1000 μg/animal] once every two days for 20 days. Promotion was begun ten days after initiation was completed, all mice receiving skin applications of 2.5 μg 12-*O*-tetradecanoylphorbol-13-acetate (TPA) in 0.1 ml acetone thrice weekly for 20 weeks. At

[1] The Working Group was aware of a study by s.c. injection in mice which has recently been completed (IARC, 1983).

the end of the treatment period, all survivors were killed [mortality rates unspecified]. The percentages of animals bearing skin tumours (predominantly squamous-cell papillomas and a few keratoacanthomas) were 0, 85, 30, 55 and 95; and the numbers of skin tumours/animal were 0, 4.9, 0.6, 1.9 and 7.2 in the vehicle control, positive-control and the low-, mid- and high-dose benzo[*j*]fluoranthene-treated groups, respectively. No other tumour was found, apart from one malignant lymphoma in a mouse in the high-dose group (LaVoie *et al.*, 1982).

(*b*) *Other experimental systems*

Intrapulmonary injection: Three groups of 35 female Osborne-Mendel *rats*, three months old, body weight 245 g, were injected directly into the pulmonary tissue after thoracotomy (Stanton *et al.*, 1972) [with 0.05 ml] of a mixture of beeswax + tricaprylin (1:1) containing 0.2, 1.0 or 5.0 mg benzo[*j*]fluoranthene (purity, 99.9%) [0.8, 4 or 20 mg/kg bw]. Three similar groups of 35 rats received 0.1, 0.3 or 1.0 mg benzo[*a*]pyrene (positive controls), and further groups of 35 rats served as untreated or vehicle controls. The rats were observed until spontaneous death: mean survival times were 110, 117, 89, 111, 77, 54, 118 and 104 weeks for the low-, mid- and high-dose benzo[*j*]fluoranthene-treated, the low-, mid- and high-dose positive-control, the untreated and the vehicle-control groups, respectively. Pulmonary squamous-cell carcinomas were found in each of the benzo[*j*]fluoranthene-treated groups, with incidences of 1/35, 3/35 and 18/35 in the low-, mid- and high-dose groups, respectively. The incidences of pulmonary carcinomas in the positive control groups were 4/35, 21/35 and 33/35 in the low-, mid- and high-dose groups, respectively. In addition, six low-dose rats and two mid-dose rats of the positive-control groups developed a pulmonary fibrosarcoma. No lung tumour was observed in the vehicle-control or untreated groups (Deutsch-Wenzel *et al.*, 1983).

3.2 Other relevant biological data[1]

(*a*) *Experimental systems*

No data were available to the Working Group on toxic effects or effects on reproduction and prenatal toxicity.

Metabolism and activation

The 9,10-dihydrodiol has been detected as a metabolite of benzo[*j*]fluoranthene in rat-liver preparations (Hecht *et al.*, 1980). This metabolite has been found to be mutagenic to bacterial cells in the presence of an exogenous metabolic system (LaVoie *et al.*, 1980) and to have tumour-initiating activity in mouse skin (LaVoie *et al.*, 1982).

Mutagenicity and other short-term tests

Benzo[*j*]fluoranthene was reported to induce mutations in *Salmonella typhimurium* strain TA100 (his^-/his^+) at a concentration of 10 μg/plate in the presence of an exogenous metabolic system (a postmitochondrial supernatant from Aroclor-induced rat liver) (LaVoie *et al.*, 1980).

(*b*) *Humans*

No data were available to the working Group.

---

[1] See also 'General Remarks on the Substances Considered', p. 53.

3.3 Case reports and epidemiological studies of carcinogenicity to humans

No data were available to the Working Group.

4. Summary of Data Reported and Evaluation[1]

4.1 Experimental data

Benzo[*j*]fluoranthene was tested for carcinogenicity in female mice by skin application and produced benign and malignant skin tumours. It was also tested in a mouse-skin initiation-promotion assay and was active as an initiator. In one study in rats involving direct injection of benzo[*j*]fluoranthene into the pulmonary tissue, it produced squamous-cell carcinomas in a dose-related manner.

No study on the teratogenicity of this compound was available.

In the one available study, benzo[*j*]fluoranthene was mutagenic to *Salmonella typhimurium* in the presence of an exogenous metabolic system.

There is *inadequate evidence* that benzo[*j*]fluoranthene is active in short-term tests.

4.2 Human data[2]

Benzo[*j*]fluoranthene is present as a component of the total content of polynuclear aromatic compounds of the environment. Human exposure to benzo[*j*]fluoranthene occurs primarily through the smoking of tobacco, inhalation of polluted air and by ingestion of food and water contaminated by combustion effluents (for details, see 'General Remarks on the Substances Considered', p. 35).

4.3 Evaluation[3]

There is *sufficient evidence* that benzo[*j*]fluoranthene is carcinogenic to experimental animals.

---

[1] For definitions of the italicized terms, see Preamble, p. 19.

[2] Studies on occupational exposure to polynuclear aromatic compounds will be considered in future *IARC Monographs*.

[3] In the absence of adequate data on humans, it is reasonable, for practical purposes, to regard chemicals for which there is *sufficient evidence* of carcinogenicity in animals as if they presented a carcinogenic risk to humans (see also Preamble, p. 20).

5. References

Behn, U. & Grimmer, G. (1981) *Bestimmung polycyclischer aromatischer Kohlenwasserstoffe (PAH) in Abgas einer Zentralheizungsanlage auf Heizöl EL-Basis* [Determination of polycyclic aromatic hydrocarbons (PAH) in the flue gas of an oil-fired heating unit], (*BMI-DGMK Projekt 111-02*), Hamburg, Deutsche Gesellschaft für Mineralölwissenschaft und Kohlechemie e.V.

Borneff, J. & Kunte, H. (1967) Carcinogenic substances in water and soil. XIX. Action of sewage purification on polycylic hydrocarbons (Ger.). *Arch. Hyg. Bakt.*, *151*, 202-210

Brockhaus, A. & Tomingas, R. (1976) Emission of polycyclic hydrocarbons during burning processes in small heating installations and their concentration in the atmosphere (Ger.). *Staub-Reinhalt. Luft*, *36*, 96-101

Community Bureau of Reference (1980) *Polycyclic Aromatic Hydrocarbon Reference Materials of Certified Purity* (*Information Handout, No. 22*), Brussels, Commission of the European Communities

Deutsch-Wenzel, R.P., Brune, H., Grimmer, G., Dettbarn, G. & Misfeld, J. (1983) Experimental studies in rat lungs on the carcinogenicity and dose-response relationships of eight frequently occurring environmental polycyclic aromatic hydrocarbons. *J. natl Cancer Inst.*, *71*, 539-544

Fábián, B. (1968) Carcinogenic substances in cooking fat and oil. IV. Studies on margarine, vegetable fat and butter (Ger.). *Arch. Hyg. Bakt.*, *152*, 231-237

Grimmer, G., Böhnke, H. & Glaser, A. (1977) Investigation on the carcinogenic burden by air pollution in man. XV. Polycyclic aromatic hydrocarbons in automobile exhaust gas - An inventory. *Zbl. Bakt. Hyg.*, *1 Abt., Orig. B164*, 218-234

Grimmer, G., Jacob, J., Naujack, K.-W. & Dettbarn, G. (1981) Profile of the polycyclic aromatic compounds from used engine oil - Inventory by GCGC/MS - PAH in environmental materials, Part 2. *Fresenius Z. anal. Chem.*, *309*, 13-19

Grimmer, G., Jacob, J. & Naujack, K.-W. (1983) Profile of the polycyclic aromatic compounds from crude oils. Inventory by GCGC/MS - PAH in environmental materials, Part 3. *Fresenius Z. anal. Chem.*, *314*, 29-36

Habs, M., Schmähl, D., Misfeld, J. (1980) Local carcinogenicity of some environmentally relevant polycyclic hydrocarbons after lifelong topical application to mouse skin. *Arch. Geschwulstforsch.*, *50*, 266-274

Hecht, S.S., LaVoie, E., Amin, S., Bedenko, V. & Hoffmann, D. (1980) *On the metabolic activation of the benzofluoranthenes*. In: Bjørseth, A. & Dennis, A.J., eds, *Polynuclear Aromatic Hydrocarbons. Chemistry and Biological Effects, 4th Int. Symposium*, Columbus, OH, Battelle Press, pp. 417-433

Hoffmann, D. & Wynder, E.L. (1976) *Environmental respiratory carcinogenesis*. In: Searle, C.E., ed., *Chemical Carcinogens* (*ACS Monograph 173*), Washington DC, American Chemical Society, p. 341

IARC (1973) *IARC Monographs on the Evaluation of Carcinogenic Risk of Chemicals to Man*, Vol. 3, *Certain Polycyclic Aromatic Hydrocarbons and Heterocyclic Compounds*, Lyon, pp. 82-90

IARC (1983) *Information Bulletin on the Survey of Chemicals Being Tested for Carcinogenicity*, No. 10, Lyon, p. 5

Karcher, W., Jacob, J. & Haemers, L. (1980) *The Certification of Eight Polycyclic Aromatic Hydrocarbon Materials (PAH) (BCR Reference Materials Nos 46, 47, 48, 49, 50, 51, 52 and 53*) (*EUR 6967EN*), Luxembourg, Commission of the European Communities

Karcher, W., Fordham, R., Dubois, J. & Gloude, P. (1983) *Spectral Atlas of Polycyclic Aromatic Compounds*, Dordrecht, The Netherlands, D. Reidel (in press)

LaVoie, E.J., Hecht, S.S., Amin, S., Bedenko, V. & Hoffmann, D. (1980) Identification of mutagenic dihydrodiols as metabolites of benzo(*j*)fluoranthene and benzo(*k*)fluoranthene. *Cancer Res.*, *40*, 4528-4532

LaVoie, E.J., Amin, S., Hecht, S.S., Furuya, K. & Hoffmann, D. (1982) Tumour initiating activity of dihydrodiols of benzo[*b*]fluoranthene, benzo[*j*]fluoranthene and benzo[*k*]fluoranthene. *Carcinogenesis*, *3*, 49-52

Lee, M.L., Novotny, M. & Bartle, K.D. (1976) Gas chromatography/mass spectrometric and nuclear magnetic resonance spectrometric studies of carcinogenic polynuclear aromatic hydrocarbons in tobacco and marijuana smoke condensates. *Anal. Chem.*, *48*, 405-416

Lijinsky, W., Domsky, I., Mason, G., Ramahi, H.Y. & Safavi, T. (1963) The chromatographic determination of trace amounts of polynuclear hydrocarbons in petrolatum, mineral oil, and coal tar. *Anal. Chem.*, *35*, 952-956

Masuda, Y. & Kuratsune, M. (1971) Polycyclic aromatic hydrocarbons in smoked fish, 'Katsuobushi'. *Gann*, *62*, 27-30

National Library of Medicine (1982) *Toxicology Data Bank*, Bethesda, MD, National Library of Medicine Specialized Information Services, Toxicology Information Program

NIH/EPA Chemical Information System (1982) *Mass Spectral Search System*, Washington DC, CIS Project, Information Services Corporation

Olufsen, B. (1980) *Polynuclear aromatic hydrocarbons in Norwegian drinking water resources*. In: Bjørseth, A. & Dennis, A.J., eds, *Polynuclear Aromatic Hydrocarbons: Chemistry and Biological Effects*, *4th Int. Symposium*, Columbus, OH, Battelle Press, pp. 333-343

Stanton, M.F., Miller, E., Wrench, C. & Blackwell, R. (1972) Experimental induction of epidermoid carcinoma in the lungs of rats by cigarette smoke condensate. *J. natl Cancer Inst.*, *49*, 867-877

Woidich, H., Pfannhauser, W., Blaicher, G. & Tiefenbacher, K. (1976) Analysis of polycyclic aromatic hydrocarbons in drinking and industrial water (Ger.). *Lebensmittelchem. gerichtl. Chem.*, *30*, 141-160

Wynder, E.L. & Hoffmann, D. (1959) The carcinogenicity of benzofluoranthenes. *Cancer*, *12*, 1194-1199

BENZO[k]FLUORANTHENE

1. Chemical and Physical Data

1.1 Synonyms and trade names

Chem. Abstr. Services Reg. No.: 207-08-9

Chem. Abstr. Name: Benzo(k)fluoranthene

IUPAC Systematic Name: Benzo[k]fluoranthene

Synonyms: 8,9-Benzfluoranthene; 8,9-benzofluoranthene; 11,12-benzofluoranthene; 2,3,1′,8′-binaphthylene; dibenzo(b,jk)fluorene

1.2 Structural and molecular formulae and molecular weight

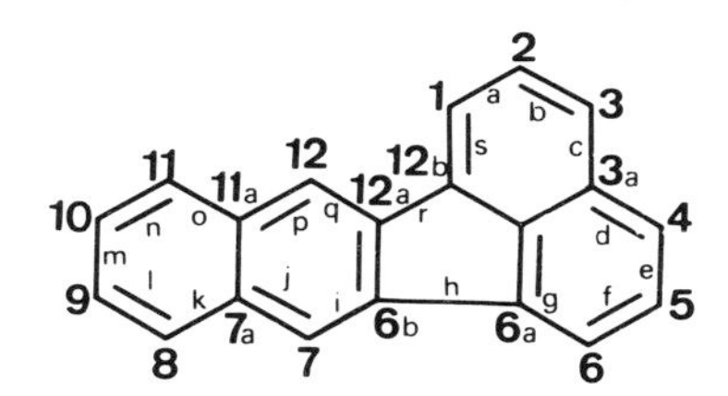

$C_{20}H_{12}$

Mol. wt: 252.3

1.3 Chemical and physical properties of the pure substance

(*a*) *Description*: Pale-yellow needles (recrystallized from benzene) (Weast, 1975)

(*b*) *Boiling-point*: 480°C (Weast, 1975)

(*c*) *Melting-point*: 215.7°C (Karcher *et al.*, 1983); 217°C (Weast, 1975)

(*d*) *Spectroscopy data*: λ_{max} 245, 266, 282, 295, 306, 358, 370, 378, 400 nm (in cyclohexane) (Karcher *et al.*, 1983). Mass and nuclear magnetic resonance spectra have been tabulated (NIH/EPA Chemical Information System, 1982; Karcher *et al.*, 1983).

(*e*) *Solubility*: Virtually insoluble in water; soluble in acetic acid, benzene and ethanol (Weast, 1975)

(*f*) *Stability*: Does not undergo photo-oxidation in organic solvents under fluorescent light or indoor sunlight (Kuratsune & Hirohata, 1962)

(*g*) *Reactivity*: Can be oxidized to a quinone and methylated at the 8-position (Clar, 1964)

2. Production, Use, Occurrence and Analysis

2.1 Production and use

There is no commercial production or known use of this compound. A reference material of certified high purity is available (Karcher *et al.*, 1980; Community Bureau of Reference, 1982).

2.2 Occurrence and analysis

Data on occurrence and methods of analysis are summarized in the 'General Remarks on the Substances Considered', p. 35.

Benzo[*k*]fluoranthene occurs ubiquitously as a product of incomplete combustion. It also occurs in fossil fuels. It has been identified in mainstream cigarette smoke (0.15-0.2 mg/kg cigarette smoke condensate) (Wynder & Hoffmann, 1959), (1.2 μg/100 cigarettes) (Lee *et al.*, 1976), (0.8 μg/100 g of burnt material) (Masuda & Kuratsune, 1972), (0.7-1.4 μg/1000 cigarettes) (Kiryu & Kuratsune, 1966); mainstream smoke of marijuana cigarettes (1.1 μg/100 cigarettes) (Lee *et al.*, 1976); air (1.3 ng/m^3) (Greenberg *et al.*, 1980; Grimmer *et al.*, 1980); exhaust emissions from gasoline engines (7-17 μg/l of fuel) (Grimmer *et al.*, 1977); with benzo[*b*],-[*j*]fluoranthene in emissions from the burning of various types of coals (0.05-17.2 mg/kg) (Brockhaus & Tomingas, 1976) and from oil-fired stoves (0.00003-0.405 mg/kg) (Behn & Grimmer, 1981); lubricating oils (0.04 mg/kg) (Grimmer *et al.*, 1981a) and used motor oils (1.5-36.8 mg/kg) (Grimmer *et al.*, 1981b); coal-tar (Lang & Eigen, 1967); crude oils (15.9 mg/kg) (Grimmer *et al.*, 1983); surface water (0.2-0.8 ng/l) (Woidich *et al.*, 1976); tap water (1-3.4 ng/l) (Borneff & Kunte, 1969), (0.07 ng/l) (Olufsen, 1980); rain water (1.6-10.1 ng/l) (Woidich *et al.*, 1976); subterranean water (1-3.5 ng/l) (Borneff & Kunte, 1969); effluent discharge (0.01-8 μg/l) and sludge (150-1270 μg/kg) (Borneff & Kunte, 1967).

3. Biological Data Relevant to the Evaluation of Carcinogenic Risk to Humans

3.1 Carcinogenicity studies in animals

(*a*) *Skin application*

Mouse: Five groups of 20 female Swiss mice [age and body weight unspecified] were painted on the skin with 0.1 or 0.5% highly purified benzo[*k*]fluoranthene in acetone or 0.01, 0.05 or 0.5% benzo[*a*]pyrene in acetone (positive controls) thrice weekly for life [no vehicle control was used]. At the end of the thirteenth month, all survivors - 8/20 and 3/20 mice treated with 0.1 and 0.5% benzo[*k*]fluoranthene, respectively - were killed; all mice treated with benzo[*a*]pyrene had died before the end of a year. Skin papillomas developed in 2/20 mice treated with 0.5% benzo[*k*]fluoranthene; no skin tumour occurred in the low-dose group. Skin tumour incidences in the positive-control groups ranged from 70-85% (papillomas) and 75-95% (carcinomas) (Wynder & Hoffmann, 1959).

Seven groups of 40 female NMRI mice, 10 weeks old, were given skin applications of 0.02 ml acetone (vehicle controls), 3.4, 5.6 or 9.2 μg benzo[*k*]fluoranthene (purity, >96%) in acetone, or 1.7, 2.8 or 4.6 μg benzo[*a*]pyrene in 0.02 ml acetone (positive controls) twice weekly for life. Information on mortality was presented in the form of curves. After one year, mortality rates were about 8%, 2-10% and 12-20% and after two years, mortality rates were about 70%, 75-85% and 80-100% in the vehicle-control, benzo[*k*]fluoranthene-treated and positive-control groups, respectively. (All animals in the high-dose benzo[*a*]pyrene-treated group died before week 85). No local tumour was found in vehicle controls or in benzo[*k*]fluoranthene-treated mice, except for one tumour [type unspecified] in a mouse of the low-dose group. The incidence of local tumours in the positive-control groups ranged from 8/34 to 24/35. There was a high rate of 'spontaneous' systemic tumours (about 70%) in all groups. After age standardization of incidences, the differences between the groups were not statistically significant (Habs *et al*., 1980).

In order to examine the initiating activity of benzo[*k*]fluoranthene, five groups of 20 female Cr1:CD-1(ICR)BR outbred albino mice, 50-55 days of age, were given skin applications of 0.1 ml acetone (vehicle controls), 3 μg benzo[*a*]pyrene (positive controls), or 3, 10 or 100 μg benzo[*k*]fluoranthene (purity, >99%) in 0.1 ml acetone [total doses, 30, 100 or 1000 μg/animal] once every two days for 20 days. Promotion was begun ten days after initiation was completed, all mice receiving skin applications of 2.5 μg 12-*O*-tetradecanoylphorbol-13-acetate (TPA) in 0.1 ml acetone thrice weekly for 20 weeks. At the end of the treatment period, all survivors were killed [mortality rates unspecified]. The percentages of animals bearing skin tumours (predominantly squamous-cell papillomas and some keratoacanthomas) were 0, 85, 5, 25 and 75, and the numbers of skin tumours/animal 0, 4.9, 0.1, 0.4 and 2.8, in the vehicle control, positive-control, and low-, mid- and high-dose benzo[*k*]fluoranthene-treated groups, respectively. No other tumour was found, except for one endometrial carcinoma of the uterus in the high-dose group (LaVoie *et al*., 1982).

(*b*) *Subcutaneous and/or intramuscular administration*

Mouse: A group of 16 male and 14 female XVII nc/Z mice [age and body weight unspecified] were given s.c. injections of 0.6 mg benzo[*k*]fluoranthene [purity unspecified] in olive oil once a month for three months [total dose, 1.8 mg/animal]. Eight males and five females developed a sarcoma at the site of injection. When the first sarcoma appeared, all males and 11 females were still alive. The average latency of the sarcomas was 203 days in males and 210 days in females. In studies carried out during the three preceding years in which positive controls were treated with benzo[*a*]pyrene under similar experimental conditions, subcutaneous sarcomas were found in all of 154 male mice and in 112/162 female mice. The average latency of the sarcomas was 110 days in males and 146 days in females [no vehicle-treated control was used] (Lacassagne *et al*., 1963).

(*c*) *Other experimental systems*

Intrapulmonary injection: Three groups of 27-35 female Osborne-Mendel *rats*, three months of age, body weight 245 g, were injected directly into the pulmonary tissue after thoracotomy (Stanton *et al*., 1972) [with 0.05 ml] of a mixture of beeswax + tricaprylin (1:1) containing 0.16, 0.83 or 4.15 mg benzo[*k*]fluoranthene (purity, 99.5%) [0.65, 3.4 or 17 mg/kg bw]. Three similar groups of 35 rats received 0.1, 0.3 or 1.0 mg benzo[*a*]pyrene (positive controls), and further groups of 35 rats served as untreated or vehicle controls. The rats were observed until spontaneous death: mean survival times were 114, 95, 98, 111, 77, 54, 118 and 104 weeks for the low-, mid- and high-dose benzo[*k*]fluoranthene-treated, the low-, mid- and high-dose benzo[*a*]pyrene-treated, untreated and vehicle-control groups, respectively. The numbers of benzo[*k*]fluoranthene-treated rats bearing lung tumours were 0/35, 3/31 and 12/27 in the low-, mid- and high-dose groups, respectively. All tumours were squamous-cell carcinomas. The incidences of pulmonary carcinomas in the positive-control groups were 4/35, 21/35 and 33/35 in the low-, mid- and high-dose groups, respectively. In addition, six low-dose rats and two mid-dose rats of the positive-control groups developed a pulmonary fibrosarcoma. No lung tumour was seen in any of the vehicle or untreated controls (Deutsch-Wenzel *et al*., 1983).

3.2 Other relevant biological data[1]

(*a*) *Experimental systems*

No data were available to the Working Group on toxic effects or on effects on reproduction and prenatal toxicity.

Metabolism and activation

The 8,9-dihydrodiol has been detected as a metabolite of benzo[*k*]fluoranthene in rat-liver preparations. This metabolite was reported to be a bacterial mutagen in the presence of an exogenous metabolic system (LaVoie *et al*., 1980).

Mutagenicity and other short-term tests

Benzo[*k*]fluoranthene was reported to induce mutations in *Salmonella typhimurium* strain TA100 (*his*$^-$/*his*$^+$) at a concentration of 10 μg/plate (LaVoie *et al*., 1980) and in strain TA98 at a concentration of 5 μg/plate (Hermann *et al*., 1980) in the presence of an exogenous metabolic system (a postmitochondrial supernatant from Aroclor-induced rat liver).

[1] See also 'General Remarks on the Substances Considered', p. 53.

(*b*) *Humans*

No data were available to the Working Group.

3.3 Case reports and epidemiological studies of carcinogenicity to humans

No data were available to the Working Group.

4. Summary of Data Reported and Evaluation[1]

4.1 Experimental data

Benzo[*k*]fluoranthene was tested for carcinogenicity in females of two strains of mice by skin application and produced a few skin tumours. It was also tested in a mouse-skin initiation-promotion assay and was active as an initiator. In one experiment involving subcutaneous injection of benzo[*k*]fluoranthene to mice, it produced sarcomas at the site of injection. Benzo[*k*]fluoranthene produced squamous-cell carcinomas of the lung in rats in a dose-related manner following its direct injection into pulmonary tissue.

No data on the teratogenicity of this compound were available.

Benzo[*k*]fluoranthene was mutagenic to *Salmonella typhimurium* in the presence of an exogenous metabolic system.

There is *inadequate evidence* that benzo[*k*]fluoranthene is active in short-term tests.

4.2 Human data[2]

Benzo[*k*]fluoranthene is present as a component of the total content of polynuclear aromatic compounds in the environment. Human exposure to benzo[*k*]fluoranthene occurs primarily through the smoking of tobacco, inhalation of polluted air and by ingestion of food and water contaminated by combustion effluents (for details, see 'General Remarks on the Substances Considered', p. 35).

---

[1] For definitions of the italicized terms, see Preamble, p. 19.

[2] Studies on occupational exposure to polynuclear aromatic compounds will be considered in future *IARC Monographs*.

4.3 Evaluation[1]

There is *sufficient evidence* that benzo[*k*]fluoranthene is carcinogenic to experimental animals.

5. References

Behn, U. & Grimmer, G. (1981) *Bestimmung polycyclischer aromatischer Kohlenwasserstoffe (PAH) in Abgas einer Zentralheizungsanlage auf Heizöl EL-Basis* [Determination of polycyclic aromatic hydrocarbons (PAH) in the flue gas of an oil-fired heating unit], (*BMI-DGMK Projekt 111-02*), Hamburg, Deutsche Gesellschaft für Mineralölwissenschaft und Kohlechemie e.V.

Borneff, J. & Kunte, H. (1967) Carcinogenic substances in water and soil. XIX. Action of sewage purification on polycylic hydrocarbons (Ger.). *Arch. Hyg. Bakt.*, *151*, 202-210

Borneff, J. & Kunte, H. (1969) Carcinogenic substances in water and soil. 25. A routine method for the determination of polynuclear aromatic hydrocarbons in water (Ger.). *Arch. Hyg. Bakt.*, *153*, 220-229

Brockhaus, A. & Tomingas, R. (1976) Emission of polycyclic hydrocarbons during burning processes in small heating installations and their concentration in the atmosphere (Ger.). *Staub-Reinhalt. Luft*, *36*, 96-101

Clar, E., ed. (1964) *Polycyclic Hydrocarbons*, Vol. 2, New York, Academic Press, p. 315

Community Bureau of Reference (1982) *Polycyclic Aromatic Hydrocarbon Reference Materials of Certified Purity* (*Information Handout, No. 22*), Brussels, Commission of the European Communities

Deutsch-Wenzel, R.P., Brune, H., Grimmer, G., Dettbarn, G. & Misfeld, J. (1983) Experimental studies in rat lungs on the carcinogenicity and dose-response relationships of eight frequently occurring environmental polycyclic aromatic hydrocarbons. *J. natl Cancer Inst.*, *71*, 539-544

---

[1] In the absence of adequate data on humans, it is reasonable, for practical purposes, to regard chemicals for which there is *sufficient evidence* of carcinogenicity in animals as if they presented a carcinogenic risk to humans (see also Preamble, p. 20).

Greenberg, A., Yokoyama, R., Giorgio, P. & Cannova, F. (1980) *Analysis of polynuclear aromatic hydrocarbons on the airborne particulates of urban New Jersey*. In: Bjørseth, A. & Dennis, A.J., eds, *Polynuclear Aromatic Hydrocarbons: Chemistry and Biological Effects, 4th Int. Symposium*, Columbus, OH, Battelle Press, pp. 193-198

Grimmer, G., Böhnke, H. & Glaser, A. (1977) Investigation on the carcinogenic burden by air pollution in man. XV. Polycyclic aromatic hydrocarbons in automobile exhaust gas - An inventory. *Zbl. Bakt. Hyg., 1 Abt., Orig. B164*, 218-234

Grimmer, G., Naujack, K.-W. & Schneider, D. (1980) *Changes in PAH-profiles in different areas of a city during the year*. In: Bjørseth, A. & Dennis, A.J., eds, *Polynuclear Aromatic Hydrocarbons: Chemistry and Biological Effects, 4th Int. Symposium*, Columbus, OH, Battelle Press, pp. 107-125

Grimmer, G., Jacob, J. & Naujack, K.-W. (1981a) Profile of the polycyclic aromatic compounds from lubricating oils. Inventory by GCGC/MS - PAH in environmental materials, Part 1. *Fresenius Z. anal. Chem., 306*, 347-355

Grimmer, G., Jacob, J. & Naujack, K.-W. (1981b) Profile of the polycyclic aromatic compounds from used engine oil. Inventory by GCGC/MS - PAH in environmental materials, Part 2. *Fresenius Z. anal. Chem., 309*, 13-19

Grimmer, G., Jacob, J. & Naujack, K.-W. (1983) Profile of the polycyclic aromatic compounds from crude oils. Inventory by GCGC/MS - PAH in environmental materials, Part 3. *Fresenius Z. anal. Chem., 314*, 29-36

Habs, M., Schmähl, D. & Misfeld, J. (1980) Local carcinogenicity of some environmentally relevant polycyclic aromatic hydrocarbons after lifelong topical application to mouse skin. *Arch. Geschwulstforsch., 50*, 266-274

Hermann, M., Durand, J.P., Charpentier, J.M., Chaudé, O., Hofnung, M., Pétroff, N., Vandecasteele, J.-P. & Weill, N. (1980) *Correlations of mutagenic activity with polynuclear aromatic hydrocarbon content of various mineral oils*. In: Bjørseth, A. & Dennis, A.J., eds, *Polynuclear Aromatic Hydrocarbons: Chemistry and Biological Effects, 4th Int. Symposium*, Columbus, OH, Battelle Press, pp. 899-916

Karcher, W., Jacob, J. & Haemers, L. (1980) *The Certification of Eight Polycyclic Aromatic Hydrocarbon Materials (PAH) (BCR Reference Materials Nos. 46, 47, 48, 49, 50, 51, 52 and 53)* (*EUR 6967 EN*), Luxembourg, Commission of the European Communities

Karcher, W., Fordham, R., Dubois, J. & Gloude, P. (1983) *Spectral Atlas of Polycyclic Aromatic Compounds*, Dordrecht, The Netherlands, D. Reidel (in press)

Kiryu, S. & Kuratsune, M. (1966) Polycyclic aromatic hydrocarbons in the cigarette tar produced by human smoking. *Gann, 57*, 317-322

Kuratsune, M. & Hirohata, T. (1962) Decomposition of polycyclic aromatic hydrocarbons under laboratory illuminations. *Natl Cancer Inst. Monogr., 9*, 117-125

Lacassagne, A., Buu-Hoï, N.P., Zajdela, F., Lavit-Lamy, D. & Chalvet, O. (1963) Carcinogenic activity of polycyclic aromatic hydrocarbons with a fluoranthene group (Fr.). *Unio int. contra cancrum Acta, 19*, 490-496

Lang, K.F. & Eigen, I. (1967) *Organic compounds in coal tar* (Ger.). In: Heilbronner, E., Hofmann, U., Schäfer, K. & Wittig, G., eds, *Fortschrifte der chemischer Forschung* (Advances in chemical research), Vol. 8, Berlin, Springer, pp. 91-170

LaVoie, E.J., Hecht, S.S., Amin, S., Bedenko, V. & Hoffmann, D. (1980) Identification of mutagenic dihydrodiols as metabolites of benzo[*j*]fluoranthene and benzo[*k*]fluoranthene. *Cancer Res.*, *40*, 4528-4532

LaVoie, E.J., Amin, S., Hecht, S.S., Furuya, L. & Hoffmann, D. (1982) Tumour initiating activity of dihydrodiols of benzo(*b*)fluoranthene, benzo(*j*)fluoranthene, and benzo(*k*)fluoranthene. *Carcinogenesis*, *3*, 49-52

Lee, M.L., Novotny, M. & Bartle, R.K. (1976) Gas chromatography/mass spectrometric and nuclear magnetic resonance spectrometric studies of carcinogenic polynuclear aromatic hydrocarbons in tobacco and marijuana smoke condensates. *Anal. Chem.*, *48*, 405-416

Masuda, Y. & Kuratsune, M. (1972) Comparison of the yield of polycyclic aromatic hydrocarbons in smoke from Japanese tobacco. *Jpn. J. Hyg.*, *27*, 339-341

NIH/EPA Chemical Information System (1982) *Mass Spectral Search System*, Washington DC, CIS Project, Information Services Corporation

Olufsen, B. (1980) *Polynuclear aromatic hydrocarbons in Norwegian drinking water resources.* In: Bjørseth, A. & Dennis, A.J., eds, *Polynuclear Aromatic Hydrocarbons: Chemistry and Biological Effects*, *4th Int. Symposium*, Columbus, OH, Battelle Press, pp. 333-343

Stanton, M.F., Miller, E., Wrench, C. & Blackwell, R. (1972) Experimental induction of epidermoid carcinoma in the lungs of rats by cigarette smoke condensate. *J. natl Cancer Inst.*, *49*, 867-877

Weast, R.C., ed. (1975) *CRC Handbook of Chemistry and Physics*, 56th ed., Cleveland, OH, Chemical Rubber Co., p. C-174

Woidich, H., Pfannhauser, W., Blaicher, G. & Tiefenbacher, K. (1976) Analysis of polycyclic aromatic hydrocarbons in drinking and industrial water (Ger.). *Lebensmittelchem. gerichtl. Chem.*, *30*, 141-160

Wynder, E.L. & Hoffmann, D. (1959) The carcinogenicity of benzofluoranthenes. *Cancer*, *12*, 1194-1199

BENZO[*ghi*]FLUORANTHENE

1. Chemical and Physical Data

1.1 Synonyms and trade names

Chem. Abstr. Services Reg. No.: 203-12-3

Chem. Abstr. Name: Benzo(ghi)fluoranthene

IUPAC Systematic Name: Benzo[*ghi*]fluoranthene

Synonyms: Benzo(mno)fluoranthene; 7,10-benzofluoranthene

1.2 Structural and molecular formulae and molecular weight

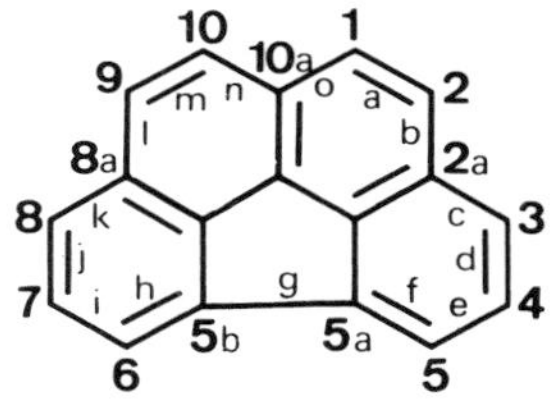

$C_{18}H_{10}$ Mol. wt: 226.3

1.3 Chemical and physical properties of the pure substance

(*a*) *Description:* Yellow needles with greenish-yellow fluorescence [recrystallized from petroleum ether (a mixture of low-boiling hydrocarbons)]; blue fluorescence in solution (Buckingham, 1982)

(*b*) *Melting-point*: 149°C (Karcher *et al.*, 1983)

(*c*) *Spectroscopy data*: λ_{max} 215, 231, 244, 249, 259, 278, 289, 322, 329, 347 nm (in cyclohexane) (Karcher *et al.*, 1983). Mass and nuclear magnetic resonance spectra have been tabulated (NIH/EPA Chemical Information System, 1982; Karcher *et al.*, 1983).

(*d*) *Stability*: No data were available.

(*e*) *Reactivity*: No data were available.

2. Production, Use, Occurrence and Analysis

2.1 Production and use

There is no commercial production or known use of this compound.

2.2 Occurrence and analysis

Data on occurrence and methods of analysis are summarized in the 'General Remarks on the Substances Considered', p. 35.

Benzo[*ghi*]fluoranthene occurs ubiquitously as a product of incomplete combustion; it also occurs in fossil fuels. It has been identified in mainstream cigarette smoke (0.1 μg/100 cigarettes) (Van Duuren, 1958), (0.1 mg/kg cigarette smoke condensate) (Wynder & Hoffmann, 1959); mainstream smoke of marijuana cigarettes (Lee *et al.*, 1976); polluted air (Grimmer *et al.*, 1980); gasoline engine exhaust (112-244 μg/l fuel) (Grimmer *et al.*, 1977); fuel combustion (Prado *et al.*, 1981); emissions from the burning of various types of coals (0.01-8.9 mg/kg burned fuel) (Brockhaus & Tomingas, 1976) and from oil-fired stoves (0.15-0.47 mg/kg) (Behn & Grimmer, 1981); surface water (1.0-11.2 ng/l) (Grimmer *et al.*, 1981); effluent discharge (0.042-0.663 μg/l) (Grimmer *et al.*, 1981); and dried sediment (2-364 μg/kg) (Grimmer & Böhnke, 1975, 1977).

3. Biological Data Relevant to the Evaluation of Carcinogenic Risk to Humans

3.1 Carcinogenicity studies in animals

Skin application

Mouse: Five groups of 20 female Swiss mice [age and body weight unspecified] were painted on the skin with 0.1 or 0.5% highly purified benzo[*ghi*]fluoranthene in acetone or 0.01, 0.05 or 0.5% benzo[*a*]pyrene in acetone (positive controls) thrice weekly for life [no vehicle-treated control was used]. At the end of the thirteenth month, all survivors - 12/20 and 15/20 low-dose and high-dose benzo[*ghi*]fluoranthene-treated mice, respectively - were killed; all mice treated with benzo[*a*]pyrene had died before the end of a year. No skin tumour was found in animals treated with the test compound; skin tumour incidences in the positive control groups ranged from 70-85% (papillomas) and 75-95% (carcinomas) (Wynder & Hoffmann, 1959).

3.2 Other relevant biological data[1]

(*a*) *Experimental systems*

No data were available to the Working Group on toxic effects, on effects on reproduction and prenatal toxicity or on metabolism and activation.

Mutagenicity and other short-term tests

Benzo[*ghi*]fluoranthene was reported to induce mutations in *Salmonella typhimurium* strain TA100 (*his*$^-$/*his*$^+$), at a concentration of 20 μg/plate in the presence of an exogenous metabolic system (postmitochondrial supernatant from Aroclor-induced rat liver) (LaVoie *et al.*, 1979).

(*b*) *Humans*

No data were available to the Working Group.

3.3 Case reports and epidemiological studies of carcinogenicity to humans

No data were available to the Working Group.

4. Summary of Data Reported and Evaluation[2]

4.1 Experimental data

Benzo[*ghi*]fluoranthene was tested for carcinogenicity in one study in female mice by skin painting; no skin tumour was observed.

No data on the teratogenicity of this compound were available.

---

[1] See also 'General Remarks on the Substances Considered', p. 53.

[2] For definitions of the italicized terms, see Preamble p. 19.

In the one available study, benzo[*ghi*]fluoranthene was mutagenic to *Salmonella typhimurium* in the presence of an exogenous metabolic system.

There is *inadequate evidence* that benzo[*ghi*]fluoranthene is active in short-term tests.

4.2 Human data[1]

Benzo[*ghi*]fluoranthene is present as a component of the total content of polynuclear aromatic compounds in the environment. Human exposure to benzo[*ghi*]fluoranthene occurs primarily through the smoking of tobacco, inhalation of polluted air and by ingestion of food and water contaminated by combustion effluents (for details, see 'General Remarks on the Substances Considered', p. 35).

4.3 Evaluation

The available data are inadequate to permit an evaluation of the carcinogenicity of benzo[*ghi*]fluoranthene to experimental animals.

5. References

Behn, U. & Grimmer, G. (1981) *Bestimmung polycyclischer aromatischer Kohlenwasserstoffe (PAH) in Abgas einer Zentralheizungsanlage auf Heizöl EL-Basis* [Determination of polycyclic aromatic hydrocarbons (PAH) in the flue gas of an oil-fired heating unit], (*BMI-DGMK-Projekt 111-02*), Hamburg, Deutsche Gesellschaft für Mineralölwissenschaft und Kohlechemie e.V.

Brockhaus, A. & Tomingas, R. (1976) Emission of polycyclic hydrocarbons during burning processes in small heating installations and their concentration in the atmosphere (Ger.). *Staub-Reinhalt. Luft*, *36*, 96-101

[1] Studies on occupational exposure to polynuclear aromatic compounds will be considered in future *IARC Monographs*.

Buckingham, J., ed. (1982) *Dictionary of Organic Compounds*, 5th ed., Vol. 1, New York, Chapman & Hall, p. 556

Grimmer, G. & Böhnke, H. (1975) Profile analysis of polycyclic aromatic hydrocarbons and metal content in sediment layers of a lake. *Cancer Lett.*, *1*, 75-84

Grimmer, G. & Böhnke, H. (1977) Investigation on drilling cores of sediments of Lake Constance. I. Profiles of the polycyclic aromatic hydrocarbons (Ger.). *Z. Naturforsch.*, *32c*, 703-711

Grimmer, G., Böhnke, H. & Glaser, A. (1977) Investigation on the carcinogenic burden by air pollution in man. XV. Polycyclic aromatic hydrocarbons in automobile exhaust gas - An inventory. *Zbl. Bakt. Hyg., 1 Abt., Orig. B164*, 218-234

Grimmer, G., Naujack, K.-W. & Schneider, D. (1980) *Changes in PAH-profiles in different areas of a city during the year.* In: Bjørseth, A. & Dennis, A.J., eds, *Polynuclear Aromatic Hydrocarbons: Chemistry and Biological Effects*, *4th Int. Symposium*, Columbus, OH, Battelle Press, pp. 107-125

Grimmer, G., Schneider, D. & Dettbarn, G. (1981) The load of different rivers in the Federal Republic of Germany by PAH (PAH-profiles of surface water) (Ger.). *Wasser*, *56*, 131-144

Karcher, W., Fordham, R., Dubois, J. & Gloude, P. (1983) *Spectral Atlas of Polycyclic Aromatic Compounds*, Dordrecht, The Netherlands, D. Reidel (in press)

LaVoie, E., Bedenko, V. Hirota, N., Hecht, S.S. & Hoffmann, D. (1979) *A comparison of the mutagenicity, tumor-initiating activity and complete carcinogenicity of polynuclear aromatic hydrocarbons.* In: Jones, P.W. & Leber, P., eds, *Polynuclear Aromatic Hydrocarbons*, Ann Arbor, MI, Ann Arbor Science Publishers, pp. 705-721

Lee, M.L., Novotny, M. & Bartle, K.D. (1976) Gas chromatography/mass spectrometry and nuclear magnetic resonance spectrometric studies of carcinogenic polynuclear aromatic hydrocarbons in tobacco and marijuana smoke condensates. *Anal. Chem.*, *48*, 405-416

NIH/EPA Chemical Information System (1982) *Mass Spectral Search System*, Washington DC, CIS Project, Information Services Corporation

Prado, G., Westmoreland, P.R., Andon, B.M., Leary, J.A., Biemann, K., Thilly, W.G., Longwell, J.P. & Howard, J.B. (1981) *Formation of polycyclic aromatic hydrocarbons in premixed flames. Chemical analysis and mutagenicity.* In: Cooke, M. & Dennis, A.J., eds, *Chemical Analysis and Biological Fate of Polynuclear Aromatic Hydrocarbons*, *5th Int. Symposium*, Columbus, OH, Battelle Press, pp. 189-198

Van Duuren, B.L. (1958) The polynuclear aromatic hydrocarbons in cigarette smoke condensate. II. *J. natl Cancer Inst.*, *21*, 623-630

Wynder, E.L. & Hoffmann, D. (1959) The carcinogenicity of benzofluoranthenes. *Cancer*, *12*, 1194-1199

BENZO[a]FLUORENE

1. Chemical and Physical Data

1.1 Synonyms and trade names

Chem. Abstr. Services Reg. No.: 238-84-6

Chem. Abstr. Name: 11H-Benzo(a)fluorene

IUPAC Systematic Name: 11*H*-Benzo[*a*]fluorene

Synonyms: 1,2-Benzofluorene; chrysofluorene; α-naphthofluorene

1.2 Structural and molecular formulae and molecular weight

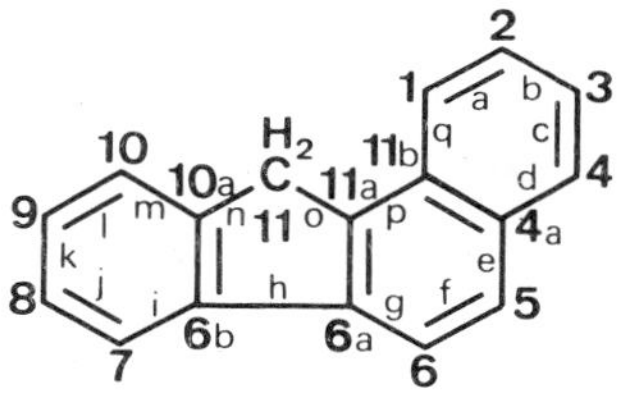

$C_{17}H_{12}$

Mol. wt: 216:3

1.3 Chemical and physical properties of the pure substance

From Weast (1975), unless otherwise specified

(*a*) *Description*: Plates (recrystallized from acetone or acetic acid)

(*b*) *Boiling-point*: 398-400°C (Rappoport, 1967); 413°C

(*c*) *Melting-point*: 189-190°C

(*d*) *Spectroscopy data*: λ_{max} 263, 296, 316 nm (in ethanol). Mass spectra have been tabulated (NIH/EPA Chemical Information System, 1982).

(*e*) *Solubility*: Slightly soluble in ethanol; soluble in hot benzene, chloroform and diethyl ether

(*f*) *Stability*: No data were available.

(*g*) *Reactivity*: No data were available.

2. Production, Use, Occurrence and Analysis

2.1 Production and use

There is no commercial production or known use of this compound.

2.2 Occurrence and analysis

Data on occurrence and methods of analysis are summarized in the 'General Remarks on the Substances Considered', p. 35.

Benzo[*a*]fluorene occurs ubiquitously in products of incomplete combustion; it also occurs in fossil fuels. It has been identified in mainstream cigarette smoke (4.1 μg/100 cigarettes) (Ayres & Thornton, 1965), (4.9 μg/100 cigarettes) (Lee *et al*., 1976), (184 ng/cigarette) (Grimmer *et al*., 1977b); sidestream cigarette smoke (751 ng/cigarette) (Grimmer *et al*., 1977a); cigarette smoke-polluted rooms (39 ng/m^3) (Grimmer *et al*., 1977c); smoke of marijuana cigarettes (4.2 μg/100 cigarettes) (Lee *et al*., 1976); gasoline engine exhaust (82-136 μg/l fuel) (Grimmer *et al*., 1977a); lubricating oils (2.67 mg/kg) (Grimmer *et al*., 1981) and crude oils (10.8 mg/kg) (Grimmer *et al*., 1983); tap water (<0.05 ng/l) (Olufsen, 1980); freeze-dried sewage sludge (280-9000 μg/kg) (Grimmer *et al*., 1980) and dried sediment from lakes (2-1400 μg/kg) (Grimmer & Böhnke, 1975, 1977).

3. Biological Data Relevant to the Evaluation of Carcinogenic Risk to Humans

3.1 Carcinogenicity studies in animals

(*a*) *Skin application*

No skin tumour was reported in 20 stock mice painted with 0.3% benzo[*a*]fluorene in benzene twice weekly for life. Mortality rates were 10/20 after six months and 14/20 after 12 months; the last mouse died on day 600. Four mice developed a lung adenoma and one a sebaceous adenoma. A group of 10 mice treated with dibenzo[*a*,*g*]carbazole under the same experimental conditions all died within the first year; five developed skin tumours (Badger *et al*., 1942).

A group of 20 female Swiss albino (Ha/ICR) mice received skin applications of 100 μl of a 0.1% solution of benzo[*a*]fluorene (purity >99.5%, LaVoie *et al.*, 1981a) in acetone ten times on alternate days (total dose, 1 mg). Starting ten days after the last dose, thrice-weekly applications of 2.5 μg 12-*O*-tetradecanoylphorbol-13-acetate (TPA) in 100 μl acetone were given for 20 weeks. A group of 20 controls received TPA alone [mortality unspecified]. At week 20, 2/20 benzo[*a*]fluorene-treated mice had skin tumours (0.15 tumours/mouse) compared to 1/20 in controls (0.05 tumours/mouse) ($p > 0.05$) (LaVoie *et al.*, 1981b).

(*b*) *Subcutaneous and/or intramuscular administration*

No sarcoma was reported in 10 stock mice [sex, age and body weight unspecified] given s.c. injections of 5 mg benzo[*a*]fluorene [purity unspecified] in 0.2 ml sesame oil at intervals of a few weeks for life. Mortality rates were 2/10 after six months and 7/10 after 12 months; the last mouse died on day 690. One mouse developed a lung adenoma (Badger *et al.*, 1942).

3.2 Other relevant biological data[1]

(*a*) *Experimental systems*

No data were available to the Working Group on toxic effects, on effects on reproduction and prenatal toxicity or on metabolism and activation.

Mutagenicity and other short-term tests

Benzo[*a*]fluorene did not induce mutations in *Salmonella typhimurium* strains TA1535, TA1537, TA1538, TA98 and TA100 (*his*$^-$/*his*$^+$) at concentrations of up to 1000 μg/plate in the presence of an exogenous metabolic system (postmitochondrial supernatant from Aroclor-induced rat liver) (Salamone *et al.*, 1979). In another study in which benzo[*a*]fluorene was tested at concentrations of up to 200 μg/plate in strains TA98 and TA100 in the presence of the same exogenous metabolic system, the results were inconclusive (LaVoie *et al.*, 1981a).

(*b*) *Humans*

No data were available to the Working Group.

3.3 Case reports and epidemiological studies of carcinogenicity in humans

No data were available to the Working Group.

---

[1] See also 'General Remarks on the Substances Considered', p. 53.

4. Summary of Data Reported and Evaluation[1]

4.1 Experimental data

Benzo[*a*]fluorene was tested for carcinogenicity in mice in one study by skin application and in a mouse-skin initiation-promotion assay. Negative results were obtained in both studies. In a study involving subcutaneous administration of benzo[*a*]fluorene to mice, no injection-site tumour was observed.

No data on the teratogenicity of this chemical were available.

The available data were inadequate to evaluate the mutagenicity of benzo[*a*]fluorene to *Salmonella typhimurium*.

There is *inadequate evidence* that benzo[*a*]fluorene is active in short-term tests.

4.2 Human data[2]

Benzo[*a*]fluorene is present as a minor component of the total content of polynuclear aromatic compounds in the environment. Human exposure to benzo[*a*]fluorene occurs primarily through the smoking of tobacco, inhalation of polluted air and by ingestion of food and water contaminated with combustion products (for details, see 'General Remarks on the Substances Considered', p. 35).

4.3 Evaluation

The available data were inadequate to permit an evaluation of the carcinogenicity of benzo[*a*]fluorene to experimental animals.

---

[1] For definitions of the italicized terms, see Preamble, p. 19.

[2] Studies on occupational exposure to polynuclear aromatic compounds will be considered in future *IARC Monographs*.

5. References

Ayres, C.I. & Thornton, R.E. (1965) Determination of benzo(*a*)pyrene and related compounds in cigarette smoke. *Beitr. Tabakforsch.*, *3*, 285-290

Badger, G.M., Cook, J.W., Hewett, C.L., Kennaway, E.L., Kennaway, N.M. & Martin, R.H. (1942) The production of cancer by pure hydrocarbons. VI. *Proc. R. Soc. London*, *131*, 170-182

Grimmer, G. & Böhnke, H. (1975) Profile analysis of polycyclic aromatic hydrocarbons and metal content in sediment layers of a lake. *Cancer Lett.*, *1*, 75-84

Grimmer, G. & Böhnke, H. (1977) Investigation on drilling cores of sediments of Lake Constance. I. Profiles of the polycyclic aromatic hydrocarbons. *Z. Naturforsch.*, *32c*, 703-711

Grimmer, G., Böhnke, H. & Glaser, A. (1977a) Investigation on the carcinogenic burden by air pollution in man. XV. Polycyclic aromatic hydrocarbons in automobile exhaust gas - An inventory. *Zbl. Bakt. Hyg.*, *1 Abt.*, *Orig. B164*, 218-234

Grimmer, G., Böhnke, H. & Harke, H.-P. (1977b) Passive smoking: Measuring of concentrations of polycyclic aromatic hydrocarbons in rooms after machine smoking of cigarettes (Ger.). *Int. Arch. occup. environ. Health*, *40*, 83-92

Grimmer, G., Böhnke, H. & Harke, H.-P. (1977c) Passive smoking: Intake of polycyclic aromatic hydrocarbons by breathing of cigarette smoke containing air (Ger.). *Int. Arch. occup. environ. Health*, *40*, 93-99

Grimmer, G., Hilge, G. & Niemtz, W. (1980) Comparison of the profile of polycyclic aromatic hydrocarbons in sewage sludge samples from 25 sewage treatment works (Ger.). *Wasser*, *54*, 255-272

Grimmer, G., Jacob, J., Naujack, K.-W. & Dettbarn, G. (1981) Profile of the polycyclic aromatic hydrocarbons from lubricating oils - Inventory by GCGC/MS - PAH in environmental materials, Part 1. *Fresenius Z. anal. Chem.*, *306*, 347-355

Grimmer, G., Jacob, J. & Naujack, K.-W. (1983) Profile of the polycyclic aromatic compounds from crude oils. Inventory by GCGC/MS - PAH in environmental materials, Part 3. *Fresenius Z. anal. Chem.*, *314*, 29-36

LaVoie, E.J., Tulley, L., Bedenko, V. & Hoffmann, D. (1981a) Mutagenicity of methylated fluorenes and benzofluorenes. *Mutat. Res.*, *91*, 167-176

LaVoie, E.J., Tulley-Freiler, L., Bedenko, V., Girach, Z. & Hoffmann, D. (1981b) *Comparative studies on the tumor initiating activity and metabolism of methylfluorenes and methylbenzofluorenes*. In: Cooke, M. & Dennis, A.J., eds, *Chemical Analysis and Biological Fate: Polynuclear Aromatic Hydrocarbons*, *5th Int. Symposium*, Columbus, OH, Battelle Press, pp. 417-427

Lee, M.L., Novotny, M. & Bartle, K.D. (1976) Gas chromatography/mass spectrometric and nuclear magnetic resonance spectrometric studies of carcinogenic polynuclear aromatic hydrocarbons in tobacco and marijuana smoke condensates. *Anal. Chem.*, *48*, 405-416

NIH/EPA Chemical Information System (1982) *Mass Spectral Search System*, Washington DC, CIS Project, Information Services Corporation

Olufsen, B. (1980) *Polynuclear aromatic hydrocarbons in Norwegian drinking water resources.* In: Bjørseth, A. & Dennis, A.J., eds, *Polynuclear Aromatic Hydrocarbons: Chemistry and Biological Effects, 4th Int. Symposium*, Columbus, OH, Battelle Press, pp. 333-343

Rappoport, Z., ed. (1967) *CRC Handbook of Tables for Organic Compound Identification*, 3rd ed., Boca Raton, FL, Chemical Rubber Co., p. 50

Salamone, M.F., Heddle, J.A. & Katz, M. (1979) The mutagenic activity of thirty polycyclic aromatic hydrocarbons (PAH) and oxides in urban airborne particulates. *Environ. Int.*, *2*, 37-43

Weast, R.C., ed. (1975) *CRC Handbook of Chemistry and Physics*, 56th ed., Cleveland, OH, Chemical Rubber Co., p. C-174

BENZO[*b*]FLUORENE

1. Chemical and Physical Data

1.1 Synonyms and trade names

Chem. Abstr. Services Reg. No.: 243-17-4

Chem. Abstr. Name: 11H-Benzo(b)fluorene

IUPAC Systematic Name: 10*H*-Benzo[*b*]fluorene

Synonym: 2,3-Benzofluorene

1.2 Structural and molecular formulae and molecular weight

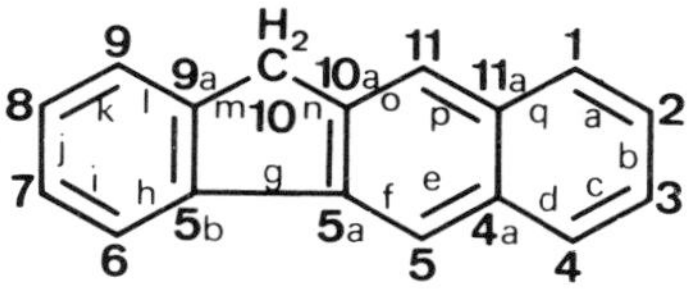

$C_{17}H_{12}$

Mol. wt: 216.3

1.3 Chemical and physical properties of the pure substance

From Buckingham (1982), unless otherwise specified

(*a*) *Description:* Crystals [crystallized from petroleum ether (a mixture of low-boiling hydrocarbons) or acetic acid]

(*b*) *Boiling-point*: 401-402°C

(*c*) *Melting-point*: 208-209°C

(*d*) *Spectroscopy data:* λ_{max} 217, 257, 263, 275, 286, 291, 304, 318, 324, 331, 340 nm (in ethanol)(Friedel & Orchin, 1951). Mass spectra have been tabulated (NIH/EPA Chemical Information System, 1982).

(*e*) *Solubility*: Soluble in acetone (LaVoie *et al.*, 1981a)

(*f*) *Stability*: No data were available.

(*g*) *Reactivity*: No data were available.

2. Production, Use, Occurrence and Analysis

2.1 Production and use

There is no commercial production or known use of this compound.

2.2 Occurrence and analysis

Data on occurrence and methods of analysis are summarized in the 'General Remarks on the Substances Considered', p. 35.

It should be noted that it is at present difficult to distinguish benzo[*b*]fluorene analytically from benzo[*c*]fluorene by gas chromatography.

Benzo[*b*]fluorene has been identified in mainstream cigarette smoke (0.5 mg/kg cigarette smoke condensate) (Hoffmann & Wynder, 1960); mainstream smoke of marijuana cigarettes (Lee *et al.*, 1976); gasoline engine exhaust (65-112 μg/l fuel) (Grimmer *et al.*, 1977); tap water (< 0.05 ng/l) (Olufsen, 1980); and freeze-dried sewage sludge (140-8600 μg/kg) (Grimmer *et al.*, 1980).

3. Biological Data Relevant to the Evaluation of Carcinogenic Risk to Humans

3.1 Carcinogenicity studies in animals

Skin application

Mouse: A group of 20 female Swiss albino (Ha/ICR) mice received skin applications of 100 μl of a 0.1% solution of benzo[*b*]fluorene (purity >99.5%, LaVoie *et al.*, 1981b) in acetone ten times on alternate days (total dose, 1 mg). Starting ten days after the last dose, thrice weekly applications of 2.5 μg 12-*O*-tetradecanoylphorbol-13-acetate (TPA) in 100 μl acetone were given for 20 weeks. A group of 20 controls received TPA alone [mortality unspecified]. At week 20, 4/20 benzo[*b*]fluorene-treated mice had skin tumours (0.35 tumours/mouse) compared with 1/20 in controls (0.05 tumours/mouse) ($p > 0.05$) (LaVoie *et al.*, 1981b).

3.2 Other relevant biological data[1]

(a) Experimental systems

No data were available to the Working Group on toxic effects, on effects on reproduction and prenatal toxicity or on metabolism and activation.

Mutagenicity and other short-term tests

Results from short-term tests are given in Table 1.

Table 1. Results from short-term tests: Benzo[*b*]fluorene

Test	Organism/assay[a]	Exogenous metabolic system[a]	Reported result	Comments	References
PROKARYOTES					
Mutation	*Salmonella typhimurium* (his^-/his^+)	Aro-R-PMS	Inconclusive	Tested at up to 200 μg/plate in strains TA98 and TA100	LaVoie *et al.* (1981a)
		Aro-R-PMS	Negative	Tested at up to 2000 nmol/plate in strain TA98	Hermann (1981)
		Aro-R-PMS	Negative	Tested at up to 1000 μg/plate in strains TA1535, TA1537, TA1538, TA98 and TA100	Salamone *et al.* (1979)
	Salmonella typhimurium ($8AG^S/8AG^R$)	Aro-R-PMS	Positive	At 25 nmol/ml in strain TM677 [purity not specified]	Kaden *et al.* (1979)

[a] For an explanation of the abbreviations, see the Appendix, p. 452.

(b) Humans

No data were available to the Working Group.

[1] See also 'General Remarks on the Substances Considered', p. 53.

3.3 Case reports and epidemiological studies of carcinogenicity to humans

No data were available to the Working Group.

4. Summary of Data Reported and Evaluation[1]

4.1 Experimental data

Benzo[*b*]fluorene was tested in a mouse-skin initiation-promotion assay; it did not show initiating activity.

No data on the teratogenicity of this compound were available.

There are conflicting reports regarding the mutagenicity of benzo[*b*]fluorene to *Salmonella typhimurium*.

There is *inadequate evidence* that benzo[*b*]fluorene is active in short-term tests.

4.2 Human data[2]

Benzo[*b*]fluorene is present as a minor component of the total content of polynuclear aromatic compounds in the environment. Human exposure to benzo[*b*]fluorene occurs primarily through the smoking of tobacco, inhalation of polluted air and by ingestion of food and water contaminated with combustion products (for details, see 'General Remarks on the Substances Considered', p. 35).

4.3 Evaluation

The available data were inadequate to permit an evaluation of the carcinogenicity of benzo[*b*]fluorene to experimental animals.

---

[1] For definitions of the italicized terms, see Preamble, p. 19.

[2] Studies on occupational exposure to polynuclear aromatic compounds will be considered in future *IARC Monographs*.

5. References

Buckingham, J., ed. (1982) *Dictionary of Organic Compounds*, 5th ed., Vol. 1, New York, Chapman & Hall, p. 557

Friedel, R.A. & Orchin, M., eds (1951) *Ultraviolet Spectra of Aromatic Compounds*, New York, John Wiley & Sons, p. 427

Grimmer, G., Böhnke, H. & Glaser, A. (1977) Investigation on the carcinogenic burden by air pollution in man. XV. Polycyclic aromatic hydrocarbons in automobile exhaust gas - An inventory. *Zbl. Bakt. Hyg., 1 Abt., Orig. B164*, 218-234

Grimmer, G., Hilge, G. & Niemitz, W. (1980) Comparison of the profile of polycyclic aromatic hydrocarbons in sewage sludge samples from 25 sewage treatment works (Ger.). *Wasser*, *54*, 255-272

Hermann, M. (1981) Synergistic effects of individual polycyclic aromatic hydrocarbons on the mutagenicity of their mixtures. *Mutat. Res.*, *90*, 399-409

Hoffmann, D. & Wynder, E.L. (1960) On the isolation and identification of polycyclic aromatic hydrocarbons. *Cancer*, *13*, 1062-1073

Kaden, D.A., Hites, R.A. & Thilly, W.G. (1979) Mutagenicity of soot and associated polycyclic aromatic hydrocarbons to *Salmonella typhimurium*. *Cancer Res.*, *39*, 4152-4159

LaVoie, E.J., Tulley, L., Bedenko, V. & Hoffmann, D. (1981a) Mutagenicity of methylated fluorenes and benzofluorenes. *Mutat. Res:.*, *91*, 167-176

LaVoie, E.J., Tulley-Freiler, L., Bedenko, V., Girach, Z. & Hoffmann, D. (1981b) *Comparative studies on the tumor initiating activity and metabolism of methylfluorenes and methylbenzofluorenes.* In: Cooke, M. & Dennis, A.J., eds, *Chemical Analysis and Biological Fate: Polynuclear Aromatic Hydrocarbons*, *5th Int. Symposium*, Columbus, OH, Battelle Press, pp. 417-427

Lee, M.L., Novotny, M. & Bartle, K.D. (1976) Gas chromatography/mass spectrometric and nuclear magnetic resonance spectrometric studies of carcinogenic polynuclear aromatic hydrocarbons in tobacco and marijuana smoke condensates. *Anal. Chem.*, *48*, 405-416

NIH/EPA Chemical Information System (1982) *Mass Spectral Search System*, Washington DC, CIS Project, Information Services Corporation

Olufsen, B. (1980) *Polynuclear aromatic hydrocarbons in Norwegian drinking water resources.* In: Bjørseth, A. & Dennis, A.J., eds, *Polynuclear Aromatic Hydrocarbons: Chemistry and Biological Effects*, *4th Int. Symposium*, Columbus, OH, Battelle Press, pp. 333-343

Salamone, M.F., Heddle, J.A. & Katz, M. (1979) The mutagenic activity of thirty polycyclic aromatic hydrocarbons (PAH) and oxides in urban airborne particulates. *Environ. Int.*, *2*, 37-43

BENZO[*c*]FLUORENE

1. Chemical and Physical Data

1.1 Synonyms and trade names

Chem. Abstr. Services Reg. No.: 205-12-9

Chem. Abstr. Name: 7H-Benzo(c)fluorene

IUPAC Systematic Name: 7*H*-Benzo[*c*]fluorene

Synonym: 3,4-Benzofluorene

1.2 Structural and molecular formulae and molecular weight

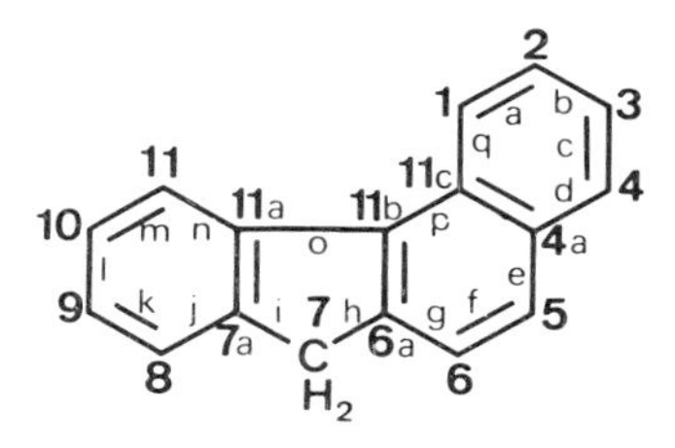

$C_{17}H_{12}$ Mol. wt: 216.3

1.3 Chemical and physical properties of the pure substance

From Buckingham (1982), unless otherwise specified

(*a*) *Description*: Plates (recrystallized from ethanol)

(*b*) *Melting-point*: 124-125°C

(*c*) *Spectroscopy data*: λ_{max} 231, 251, 298, 311, 320, 328, 335 nm (in ethanol) (Friedel & Orchin, 1951). Mass spectra have been tabulated (NIH/EPA Chemical Information System, 1982).

(*d*) *Solubility*: Soluble in benzene (Bachmann *et al.*, 1937) and acetone (LaVoie *et al.*, 1981a)

(*e*) *Stability*: No data were available.

(*f*) *Reactivity*: No data were available.

2. Production, Use, Occurrence and Analysis

2.1 Production and use

There is no commercial production or known use of this compound.

2.2 Occurrence and analysis

Data on occurrence and methods of analysis are summarized in the 'General Remarks on the Substances Considered', p. 35.

Benzo[*c*]fluorene occurs ubiquitously in products of incomplete combustion; it also occurs in fossil fuels.

It should be noted that this compound is difficult to distinguish analytically from benzo[*b*]fluorene by gas chromatography.

Benzo[*c*]fluorene has been identified in mainstream cigarette smoke (Hoffmann & Wynder, 1971; Snook *et al.*, 1977).

3. Biological Data Relevant to the Evaluation of Carcinogenic Risk to Humans

3.1 Carcinogenicity studies in animals

Skin application

Mouse: No skin tumour was reported in a group of 20 mice painted on the skin with 0.3% benzo[*c*]fluorene in benzene twice weekly for life. Mortality rates were 7/20 after six months and 14/20 after 12 months; the last mouse died on day 537. All mice of a group of 20 treated with dibenzo[*a,g*]fluoranthrene under the same experimental conditions died within the first year; four had developed skin tumours (Bachmann *et al.*, 1937).

A group of 20 female Swiss albino (Ha/ICR) mice received skin applications of 100 μl of a 0.1% solution of benzo[*c*]fluorene (purity >99.5%; LaVoie *et al.*, 1981b) in acetone ten times on alternate days (total dose, 1 mg). Starting ten days after the last dose, thrice-weekly applications of 2.5 μg 12-*O*-tetradecanoylphorbol-13-acetate (TPA) in 100 μl acetone were given for 20 weeks. A group of 20 controls received TPA alone [mortality unspecified]. At week 20, 5/20 benzo[*c*]fluorene-treated mice had skin tumours (0.25 tumours/mouse) compared with 1/20 in controls (0.05 tumours/mouse) ($p > 0.05$) (LaVoie *et al.*, 1981a).

3.2 Other relevant biological data[1]

(*a*) *Experimental systems*

No data were available to the Working Group on toxic effects, on effects on reproduction and prenatal toxicity or on absorption, distribution, excretion and metabolism.

Mutagenicity and other short-term tests

The mutagenicity of benzo[*c*]fluorene was studied in *Salmonella typhimurium* strains TA100 and TA98 (*his*$^-$/*his*$^+$) at concentrations of up to 200 μg/plate in the presence of an exogenous metabolic system (post-mitochondrial supernatant from Aroclor-induced rat liver), with inconclusive results (LaVoie *et al.*, 1981b).

(*b*) *Humans*

No data were available to the Working Group.

3.3 Case reports and epidemiological studies of carcinogenicity to humans

No data were available to the Working Group.

4. Summary of Data Reported and Evaluation[2]

4.1 Experimental data

Benzo[*c*]fluorene was tested for carcinogenicity in mice in one experiment by skin application and in one mouse-skin initiation-promotion assay. There was no indication of carcinogenic or initiating activity.

---

[1] See also 'General Remarks on the Substances Considered', p. 53.

[2] For definitions of the italicized terms, see Preamble, p. 19.

No data on the teratogenicity of this compound were available.

The single report of mutagenicity of benzo[*c*]fluorene in *Salmonella typhimurium* was inconclusive.

There is *inadequate evidence* that benzo[*c*]fluorene is active in short-term tests.

4.2 Human data[1]

Benzo[*c*]fluorene is present as a minor component of the total content of polynuclear aromatic compounds in the environment. Human exposure to benzo[*c*]fluorene occurs primarily through the smoking of tobacco, inhalation of polluted air and by ingestion of food and water contaminated with combustion products (for details, see 'General Remarks on the Substances Considered', p. 35).

4.3 Evaluation

The available data were inadequate to permit an evaluation of the carcinogenicity of benzo[*c*]fluorene to experimental animals.

5. References

Bachmann, W.E., Cook, J.W., Dansi, A., de Worms, C.G.M., Haslewood, G.A.D., Hewett, C.L. & Robinson, A.M. (1937) The production of cancer by pure hydrocarbons - IV. *Proc. R. Soc. London, Ser. B*, *123*, 343-368

Buckingham, J., ed. (1982) *Dictionary of Organic Compounds*, 5th ed., Vol. 1, New York, Chapman & Hall, p. 557

Friedel, R.A. & Orchin, M., eds (1951) *Ultraviolet Spectra of Aromatic Compounds*, New York, John Wiley & Sons, p. 430

Hoffmann, D. & Wynder, E.L. (1971) A study of tobacco carcinogenesis, IX. Tumor initiators, tumor accelerators, and tumor promoting activity of condensate fractions. *Cancer*, *27*, 848-864

[1] Studies on occupational exposure to polynuclear aromatic compounds will be considered in future *IARC Monographs*.

LaVoie, E.J., Tulley-Freiler, L., Bedenko, V., Girach, Z. & Hoffmann, D. (1981a) *Comparative studies on the tumor initiating activity and metabolism of methylfluorenes and methylbenzofluorenes*. In: Cooke, M. & Dennis, A.J., eds, *Chemical Analysis and Biological Fate: Polynuclear Aromatic Hydrocarbons, 5th Int. Symposium*, Columbus, OH, Battelle Press, pp. 417-427

LaVoie, E.J., Tulley, L., Bedenko, V. & Hoffmann, D. (1981b) Mutagenicity of methylated fluorenes and benzofluorenes. *Mutat. Res.*, *91*, 167-176

NIH/EPA Chemical Information System (1982) *Mass Spectral Search System*, Washington DC, CIS Project, Information Services Corporation

Snook, M.E., Severson, R.F., Arrendale, R.F., Higman, H.C. & Chortyk, O.T. (1977) The identification of high molecular weight polynuclear aromatic hydrocarbons in biologically active fraction of cigarette smoke condensate. *Beitr. Tabakforsch.*, *9*, 79-101

BENZO[*ghi*]PERYLENE

1. Chemical and Physical Data

1.1 Synonyms and trade names

Chem. Abstr. Services Reg. No.: 191-24-2

Chem. Abstr. Name: Benzo(ghi)perylene

IUPAC Systematic Name: Benzo[*ghi*]perylene

Synonyms: 1,12-Benzoperylene; 1,12-benzperylene

1.2 Structural and molecular formulae and molecular weight

1 2 2a 3 4 4a 5 6 7 7a 7b 8 9 10 10a 11 12 12a
a b c d e f g h i j k l m n o p q r

$C_{22}H_{12}$ Mol. wt: 276.3

1.3 Chemical and physical properties of the pure substance

(*a*) *Description*: Large, pale yellow-green plates (recrystallized from xylene) (Clar, 1964)

(*b*) *Melting-point*: 278.3°C (Karcher *et al.*, 1983)

(*c*) *Spectroscopy data*: λ_{max} 222, 252, 275, 288, 299, 311, 324, 328, 344, 362, 383, 392 and 396 (in cyclohexane) (Karcher *et al.*, 1983). Mass and nuclear magnetic resonance spectra have been tabulated (NIH/EPA Chemical Information System, 1982; Karcher *et al.*, 1983).

(*d*) *Solubility*: Soluble in 1,4-dioxane (Hoffmann & Wynder, 1966), dichloromethane (Müller, 1968), benzene (Van Duuren *et al*., 1970) and acetone (Van Duuren *et al*., 1973)

(*e*) *Stability*: Stable to photo-oxidation (Clar, 1964)

(*f*) *Reactivity*: Reaction with bromomaleic anhydride forms a dibromobenzo[*ghi*]perylene byproduct as well as coronene-dicarboxylic anhydride derivatives (Clar, 1964); reacts with NO and NO_2 to form nitro derivatives (Butler & Crossley, 1981)

2. Production, Use, Occurrence and Analysis

2.1 Production and use

There is no commercial production or known use of this compound. A reference material of certified high purity is available (Karcher *et al*., 1980; Community Bureau of Reference, 1982).

2.2 Occurrence and analysis

Data on occurrence and methods of analysis are summarized in the 'General Remarks on the Substances Considered', p. 35.

Benzo[*ghi*]perylene occurs ubiquitously in products of incomplete combustion; it also occurs in considerable amounts in coal-tar and is an important component of gasoline engine exhaust.

It has been identified in mainstream cigarette smoke (0.5 μg/500 cigarettes) (Cooper & Lindsey, 1955), (0.15-0.2 mg/kg cigarette smoke condensate) (Hoffmann & Wynder, 1960), (0.3 μg/100 cigarettes) (Lee *et al*., 1976), (39 ng/cigarette) (Grimmer *et al*., 1977b); sidestream cigarette smoke (98 ng/cigarette) (Grimmer *et al*., 1977b); smoke-filled rooms (17 ng/m^3) (Grimmer *et al*., 1977c); the air of restaurants (2-11.4 ng/m^3)(Just *et al*., 1972); mainstream smoke from marijuana cigarettes (0.7 μg/100 cigarettes) (Lee *et al*., 1976); exhaust emissions from gasoline engines (115-333 μg/l fuel) (Grimmer *et al*., 1977a); emissions from the burning of various types of coal (0.02-2.7 mg/kg) (Brockhaus & Tomingas, 1976); crude oils (0.6-5.0 mg/kg oil) (Grimmer *et al*., 1983); lubricating oils (0.07 mg/kg) (Grimmer *et al*., 1981b) and used motor oils (333.9 mg/kg) (Grimmer *et al*., 1981c); gasoline (0.32-2.66 mg/kg) (Müller & Meyer, 1974); charcoal-broiled steaks (4.5 μg/kg) (Lijinsky & Shubik, 1964); edible oils (0-18 μg/kg) (Grimmer & Hildebrandt, 1967); surface water (0.3-28.5 ng/l) (Woidich *et al*., 1976; Grimmer *et al*., 1981a); tap water (0.8-7.1 ng/l) (Borneff & Kunte, 1969); rain water (2.3-10.8 ng/l) (Woidich *et al*., 1976); subterranean water (0.7-6.4 ng/l) (Borneff & Kunte, 1969); waste water (0.4-2.8

μg/l) (Grimmer *et al.*, 1981a); sludge (200-1220 μg/kg) (Borneff & Kunte, 1967); freeze-dried sewage sludge samples (400-8700 μg/kg) (Grimmer *et al.*, 1980); and dried sediments from lakes (1-1930 μg/kg) (Grimmer & Böhnke, 1975, 1977).

3. Biological Data Relevant to the Evaluation of Carcinogenic Risk to Humans

3.1 Carcinogenicity studies in animals[1]

(*a*) *Skin application*

Mouse: No skin tumour was reported in female Swiss mice [age, number and body weight unspecified] given skin applications of a saturated (0.38%) solution of benzo[*ghi*]perylene [purity, dose and number of applications unspecified] in benzene for life (Lijinsky & Saffiotti, 1965). [The Working Group noted the incomplete reporting of this experiment.]

Five groups of 20 female Ha/ICR/Mil Swiss albino mice, seven to eight weeks of age, were painted on the skin with 0.05 or 0.1% chromatographically purified benzo[*ghi*]perylene in dioxane [1,4-dioxane] [dose unspecified], dioxane alone (vehicle controls), or 0.05 or 0.1% benzo[*a*]pyrene in dioxane (positive controls) thrice weekly for 12 months. Survivors were killed at the end of the fifteenth month, at which time all positive controls and 8/20 vehicle controls had died [mortality rates of the benzo[*ghi*]perylene-treated groups unspecified]. The numbers of skin tumour-bearing mice were 1/20 (one papilloma found in the tenth month), 0/20, 0/20, 17/20 and 19/20 in the low- and high-dose benzo[*ghi*]perylene-treated, vehicle control, and low- and high-dose benzo[*a*]pyrene-treated groups, respectively (Hoffmann & Wynder, 1966).

In a comparative study, the initiating activity of benzo[*ghi*]perylene was compared with that of benzo[*a*]pyrene using two groups of 30 female Ha/ICR/Mil Swiss albino mice, seven to eight weeks of age, which received skin applications of 0.025 ml 0.1% chromatographically purified benzo[*ghi*]perylene [25 μg/application; total dose, 0.25 mg/animal] in dioxane [1,4-dioxane] or 0.025 ml 0.1% benzo[*a*]pyrene in dioxane (positive controls) once every three days for 28 days. On day 28, skin painting with 2.5% (2.3 mg) croton oil in acetone was started [frequency and duration of treatments unspecified]. A group of 30 mice was treated similarly with a solution of 2.5% croton oil alone (promoter controls), and another group of 20 mice received skin applications of dioxane alone (vehicle controls). All survivors were killed after six months. The mortality rates at six months were 3/30, 2/30, 4/30 and 2/20, and the numbers of skin papilloma-bearing mice were 2/30, 24/30, 2/30 and 0/20 in the benzo[*ghi*]perylene-treated, positive-control, promoter control and vehicle-control groups, respectively (Hoffmann & Wynder, 1966).

The treatments and results of a study undertaken in nine groups of female NMRI mice, three to four months old, initial body weight, 25-30 g, are given in Table 1 (Müller, 1968).

---

[1] The Working Group was aware of a carcinogenicity study in progress by skin application in mice (IARC, 1983).

Table 1. Carcinogenicity of benzo[*ghi*]perylene in female NMRI mice (Müller, 1968)

Skin applications given twice weekly for 25 weeks	Initial no. of mice	No. of deaths by study termination[a]	No. of mice with tumours	No. of mice with papillomas and carcinomas of the skin[b]	No. of mice with benign tumours at other sites[c]	No. of mice with malignant tumours at other sites[c]
0.2 ml 0.01% benzo[*ghi*]perylene in dichloromethane (total dose, 1 mg)	50	32	3	0	0	3
0.4 ml 0.5% benzo[*ghi*]perylene in dichloromethane (total dose, 100 mg)	50	36	6	0	2	4
0.8 ml 0.5% benzo[*ghi*]perylene in dichloromethane (total dose, 200 mg)	50	33	4	0	3	1
1 mg benzo[*ghi*]perylene in dichloromethane followed by 0.2 ml 0.5% croton oil [frequency and duration of treatment unspecified]	50	38	5	0	4	1
2 mg benzo[*ghi*]perylene in dichloromethane followed by 0.2 ml 0.5% croton oil [frequency and duration of treatment unspecified]	50	29	4	1[d]	2	1
0.2 ml 0.01% benzo[*a*]pyrene in dichloromethane (total dose, 1 mg; positive controls)	50	50	50	50	6	3
0.2 ml 0.1% benzo[*a*]pyrene in dichloromethane (total dose, 10 mg; positive controls)	50	50	50	50	6	3
0.2 ml dichloromethane (vehicle controls)	50	33	7	0	3	4
0.2 ml 0.5% croton oil (promoter controls)	50	32	3	1[d]	2	0

[a] Survivors were killed on day 675 after the first application
[b] Numbers of papillomas and carcinomas unspecified
[c] Site and type unspecified
[d] Papillomas

In order to examine the initiating activity of benzo[*ghi*]perylene, a group of 20 female ICR/Ha Swiss mice, six to eight weeks of age, received a single skin application of 0.8 mg rigorously purified benzo[*ghi*]perylene [about 40 mg/kg bw], dissolved in 0.2 ml spectroscopic-grade benzene. Another group of ICR/Ha Swiss mice, of the same age and sex, received the same treatment, followed two weeks later by skin applications of 2.5 μg phorbol myristyl acetate (PMA) in 0.1 ml spectroscopic-grade acetone thrice weekly for the duration of the experiment. Three control groups were available: one group of 100 untreated females; one group of 20 females treated with PMA in acetone (promoter controls); and one group of 20 females treated with acetone only (vehicle controls). The test groups had median survival times of 360-400 days; those of the control groups were 406-420 days. In the group treated with benzo[*ghi*]perylene followed by PMA-treatment, 3/20 mice had a papilloma and one of these also had a squamous-cell carcinoma; the first papilloma was found on day 183. In the group treated with PMA alone, 1/20 mice had two skin papillomas, the first appearing on day 338. No skin tumour was found in the other groups. Dibenz[*a*,*c*]anthracene and 7,12-dimethylbenz[*a*]anthracene were tested in the same study under similar experimental conditions and showed notable tumour-initiating activity (Van Duuren *et al.*, 1970).

In order to examine benzo[*ghi*]perylene for cocarcinogenic activity, three groups of 50 female ICR/Ha Swiss mice, seven weeks old, received skin applications of 5 μg benzo[*a*]pyrene in 0.1 ml acetone alone, simultaneously with 21 μg of the highly pure test compound [total dose, 3.3 mg/animal] dissolved in 0.1 ml acetone, or simultaneously with 80 μg anthralin in 0.1 ml acetone, thrice weekly for 52 weeks. Three control groups received benzo[*ghi*]perylene without benzo[*a*]pyrene, acetone alone or no treatment. The numbers of survivors after 52 weeks were 42, 37 and 37 in the group treated with benzo[*a*]pyrene alone, the group given benzo[*a*]pyrene and benzo[*ghi*]perylene and the group treated with benzo[*a*]pyrene and anthralin, respectively. The incidences of skin papillomas (and total numbers of papillomas) were 13/50 (14), 20/50 (33) and 26/50 (58) in the three groups, respectively; and the incidences of squamous-cell skin carcinomas were 10/50, 17/50 and 15/50 in the three groups, respectively. No skin tumour was observed in any of the control groups [average survival times unspecified] (Van Duuren *et al.*, 1973). [The Working Group noted that details of the experimental conditions of the controls were not reported.]

(*b*) *Subcutaneous and/or intramuscular administration*

Mouse: Three groups of 50 female NMRI mice, three to four months of age, received s.c. injections of 0 (controls), 0.83 mg or 16.7 mg benzo[*ghi*]perylene suspended in 0.15 ml 10% aqueous gelatine once a fortnight for six months [total doses, 0, 10 and 200 mg/animal]. Survivors were killed on day 675 after the first injection; at that time, 32/50 mice in each of the three groups had died. No tumour was found at the site of injection in any of the animals. The numbers of animals with tumours at other sites were 4/50, 5/50 and 4/40 in the control, low- and high-dose groups, respectively (Müller, 1968).

Seven groups of 20 NMRI mice, six to seven months old [sex unspecified], were given s.c. injections of 0.15 ml 10% aqueous gelatine containing 0 (control), 0.1, 1 or 10 mg suspended benzo[*ghi*]perylene [total doses, 0, 1, 10 or 100 mg/animal] or 0.1, 1 or 10 mg suspended benzo[*a*]pyrene (positive controls) once a fortnight for 20 weeks. The animals were observed until spontaneous death. Survival was not adversely affected by treatment with benzo[*ghi*]perylene (the last animal died 22 months after the start of the study), whereas mice treated with benzo[*a*]pyrene died much earlier than the controls, the last animals dying in months 17, 7 and 6 in the low-, mid- and high-dose groups, respectively. No skin or subcutaneous tumour was found in mice treated with benzo[*ghi*]perylene or gelatine; all mice given s.c. injections of benzo[*a*]pyrene developed sarcomas at the injection site. Only a few tumours were found in organs other than the skin or subcutaneous tissue, and the incidences in the benzo[*ghi*]perylene-treated groups were not different from those in the gelatine controls (Müller, 1968).

(*c*) *Other experimental systems*

Intrapulmonary injection: Three groups of 34 or 35 female Osborne-Mendel *rats*, three months of age, body weight 245 g, were injected directly into the lung after thoracotomy (Stanton *et al.*, 1972) [with 0.05 ml] of a mixture of beeswax + tricaprylin (1:1) containing 0.16, 0.83 or 4.15 mg benzo[*ghi*]perylene (purity, 98.5%) [0.65, 3.4 or 17 mg/kg bw]. Three similar groups of 35 rats received 0.1, 0.3 or 1.0 mg benzo[*a*]pyrene (positive controls), and further groups of 35 rats served as untreated or vehicle controls. The rats were observed until spontaneous death. Mean survival times were 109, 114, 106, 111, 77, 54, 118 and 104 weeks for the low-, mid- and high-dose benzo[*ghi*]perylene-treated, the low-, mid- and high-dose benzo[*a*]pyrene-treated, untreated and vehicle-control groups, respectively. The numbers of benzo[*ghi*]perylene-treated rats bearing lung tumours were 0/35, 1/35 and 4/34 in the low-, mid- and high-dose groups, respectively. All tumours were squamous-cell carcinomas. The incidences of lung carcinomas in the positive control groups were 4/35, 21/35 and 33/35 in the low-, mid- and high-dose groups, respectively. In addition, six low-dose rats and two mid-dose rats of the positive-control groups developed a pulmonary fibrosarcoma. No lung tumour was seen in any of the vehicle or untreated controls (Deutsch-Wenzel *et al.*, 1983). [The Working Group noted that, according to the authors, the weak effect observed after intrapulmonary administration of the test compound might have been caused by impurities in the benzo[*ghi*]perylene used.]

3.2 Other relevant biological data[1]

(*a*) *Experimental systems*

No data were available to the Working Group on toxic effects, on effects on reproduction and prenatal toxicity or on distribution, absorption, excretion and metabolism.

Mutagenicity and other short-term tests

Results from short-term tests are given in Table 2.

Table 2. Results from short-term tests: Benzo[*ghi*]perylene

Test	Organism/assay[a]	Exogenous metabolic system[a]	Reported result	Comments	References
PROKARYOTES					
Mutation	*Salmonella typhimurium* (his⁻/his⁺)	Aro-R-PMS	Positive	At 20 μg/plate in strain TA100	Andrews *et al.* (1978); Mossanda *et al.* (1979);
			Positive	In the range 0.1-1000 μg/plate in strain TA98	Salamone *et al.* (1979)
	Salmonella typhimurium ($8AG^S/8AG^R$)	Aro-R-PMS	Positive	At 72 nmol/ml in strain TM677	Kaden *et al.* (1979)
MAMMALIAN CELLS *IN VIVO*					
Cell transformation	Hamster embryo cells (morphological; growth in soft agar)	--	Negative	Pregnant females treated i.p. with up to 30 mg/kg bw	Quarles *et al.* (1979)

[a] For an explanation of the abbreviations, see the Appendix, p. 452.

(*b*) *Humans*

No data were available to the Working Group.

3.3 Case reports and epidemiological studies of carcinogenicity to humans

No data were available to the Working Group.

[1] See also 'General Remarks on the Substances Considered', p. 53.

4. Summary of Data Reported and Evaluation[1]

4.1 Experimental data

Benzo[*ghi*]perylene was tested for carcinogenicity in two studies by skin application to female mice, and no carcinogenic effect was observed. It was tested in three studies in the mouse-skin initiation-promotion assay, also with negative results. In two studies in mice by subcutaneous injection, no tumour was observed at the injection site. The results of a test using intrapulmonary injection in rats were inadequate for evaluation, although some pulmonary tumours occurred. When benzo[*ghi*]perylene was administered simultaneously with benzo[*a*]pyrene to the skin of mice, an increased number of skin tumours was observed over that with benzo[*a*]pyrene alone.

No data on teratogenicity of this compound were available.

Benzo[*ghi*]perylene was mutagenic to *Salmonella typhimurium* in the presence of an exogenous metabolic system. It was negative in one study of morphological transformation in mammalian cells.

There is *inadequate evidence* that benzo[*ghi*]perylene is active in short-term tests.

4.2 Human data[2]

Benzo[*ghi*]perylene is present as a major component of the total content of polynuclear aromatic compounds in the environment. Human exposure to benzo[*ghi*]perylene occurs primarily through the smoking of tobacco, inhalation of polluted air and by ingestion of food and water contaminated by combustion effluents (for details, see 'General Remarks on the Substances Considered', p. 35).

4.3 Evaluation

The available data were inadequate to permit an evaluation of the carcinogenicity of benzo[*ghi*]perylene to experimental animals.

---

[1] For definitions of the italicized terms, see Preamble, p. 19.

[2] Studies on occupational exposure to polynuclear aromatic compounds will be considered in future *IARC Monographs*.

5. References

Andrews, A.W., Thibault, L.H. & Lijinsky, W. (1978) The relationship between carcinogenicity and mutagenicity of some polynuclear hydrocarbons. *Mutat. Res.*, *51*, 311-318

Borneff, J. & Kunte, H. (1967) Carcinogenic substances in water and soil. XIX. Action of sewage purification on polycylic hydrocarbons. *Arch. Hyg. Bakt.*, *151*, 202-210

Borneff, J. & Kunte, H. (1969) Carcinogenic substances in water and soil. 25. A routine method for the determination of aromatic hydrocarbons in water (Ger.). *Arch. Hyg. Bakt.*, *153*, 220-229

Brockhaus, A. & Tomingas, R. (1976) Emission of polycyclic hydrocarbons during burning processes in small heating installations and their concentration in the atmosphere (Ger.). *Staub-Reinhalt. Luft*, *36*, 96-101

Butler, J.D. & Crossley, P. (1981) Reactivity of polycyclic aromatic hydrocarbons adsorbed on soot particles. *Atmos. Environ.*, *15*, 91-94

Clar, E., ed. (1964) *Polycyclic Hydrocarbons*, Vol. 2, New York, Academic Press, pp. 60-63

Community Bureau of Reference (1982) *Polycyclic Aromatic Hydrocarbon Reference Materials of Certified Purity* (*Information Handout, No. 22*), Brussels, Commission of the European Communities

Cooper, R.L. & Lindsey, A.J. (1955) 3:4-Benzpyrene and other polycyclic hydrocarbons in cigarette smoke. *Br. J. Cancer*, *9*, 304-309

Deutsch-Wenzel, R.P., Brune, H., Grimmer, G., Dettbarn, G. & Misfeld, J. (1983) Experimental studies in rat lungs on the carcinogenicity and dose-response relationships of eight frequently occurring environmental polycyclic aromatic hydrocarbons. *J. natl Cancer Inst.*, *71*, 539-544

Grimmer, G. & Böhnke, H. (1975) Profile analysis of polycyclic aromatic hydrocarbons and metal content in sediment layers of a lake. *Cancer Lett.*, *1*, 75-84

Grimmer, G. & Böhnke, H. (1977) Investigation on drilling cores of sediments of Lake Constance. I. Profiles of the polycyclic aromatic hydrocarbons (Ger.). *Z. Naturforsch.*, *32c*, 703-711

Grimmer, G. & Hildebrandt, A. (1967) Content of polycyclic hydrocarbons in crude vegetable oils. *Chem. Ind.*, 25 November, 2000-2002

Grimmer, G., Böhnke, H. & Glaser, A. (1977a) Investigation on the carcinogenic burden by air pollution in man. XV. Polycyclic aromatic hydrocarbons in automobile exhaust gas - An inventory. *Zbl. Bakt. Hyg., 1 Abt., Orig. B164*, 281-234

Grimmer, G., Böhnke, H. & Harke, H.-P. (1977b) Passive smoking: Intake of polycyclic aromatic hydrocarbons by breathing of air containing cigarette smoke (Ger.). *Int. Arch. occup. environ. Health*, *40*, 93-99

Grimmer, G., Böhnke, H. & Harke, H.-P. (1977c) Passive smoking: Measuring concentrations of polycyclic aromatic hydrocarbons in rooms after machine smoking of cigarettes (Ger.). *Int. Arch. occup. environ. Health*, *40*, 83-92

Grimmer, G., Hilge, G. & Niemitz, W. (1980) Comparison of the profile of polycyclic aromatic hydrocarbons in sewage sludge samples from 25 sewage treatment works (Ger.). *Wasser*, *54*, 255-272

Grimmer, G., Schneider, D. & Dettbarn, G. (1981a) The load of different rivers in the Federal Republic of Germany by PAH (PAH - profiles of surface waters) (Ger.). *Wasser*, *56*, 131-144

Grimmer, G., Jacob, J. & Naujack, K.-W. (1981b) Profile of the polycyclic aromatic compounds in lubricating oils. Inventory by GCGC/MS - PAH in environmental materials, Part 1. *Fresenius Z. anal. Chem.*, *306*, 347-355

Grimmer, G., Jacob, J., Naujack, K.-W. & Dettbarn, G. (1981c) Profile of the polycyclic aromatic compounds from used engine oil - Inventory by GCGC/MS - PAH in environmental materials, Part 2. *Fresenius Z. anal. Chem.*, *309*, 13-19

Grimmer, G., Jacob, J. & Naujack, K.-W. (1983) Profile of the polycyclic aromatic compounds from crude oils. Inventory by GCGC/MS - PAH in environmental materials, Part 3. *Fresenius Z. anal. Chem.*, *314*, 29-36

Hoffmann, D. & Wynder, E.L. (1960) On the isolation and identification of polycyclic aromatic hydrocarbons. *Cancer*, *13*, 1062-1073

Hoffmann, D. & Wynder, E.L. (1966) Contribution to the carcinogenic action of dibenzopyrenes (Ger.). *Z. Krebsforsch.*, *68*, 137-149

IARC (1983) *Information Bulletin on the Survey of Chemicals Being Tested for Carcinogenicity*, No. 10, Lyon, p. 15

Just, J., Borkowska, M. & Maziarka, S. (1972) Air pollution by tobacco smoke in Warsaw coffee houses (Pol.). *Rocz. Panstw. Zakl/. Hyg.*, *23*, 129-135

Kaden, D.A., Hites, R.A. & Thilly, W.G. (1979) Mutagenicity of soot and associated polycyclic aromatic hydrocarbons to *Salmonella typhimurium*. *Cancer Res.*, *39*, 4152-4159

Karcher, W., Jacob, J. & Haemers, L. (1980) *The Certification of Eight Polycyclic Aromatic Hydrocarbon Materials (PAH) (BCR Reference Materials Nos. 46, 47, 48, 49, 50, 51, 52 and 53)* (*EUR 6967EN*), Luxembourg, Commission of the European Communities

Karcher, W., Fordham, R., Dubois, J. & Gloude, P. (1983) *Spectral Atlas of Polycyclic Aromatic Compounds*, Dordrecht, The Netherlands, D. Reidel (in press)

Lee, M.L., Novotny, M. & Bartle, K.D. (1976) Gas chromatography/mass spectrometry and nuclear magnetic resonance spectrometric studies of carcinogenic polynuclear aromatic hydrocarbons in tobacco and marijuana smoke condensates. *Anal. Chem.*, *48*, 405-416

Lijinsky, W. & Saffiotti, U. (1965) Relationships between structure and skin tumorigenic activity among hydrogenated derivatives of several polycyclic aromatic hydrocarbons. *Ann. ital. Dermatol. clin. sper.*, *19*, 34-41

Lijinsky, W. & Shubik, P. (1964) Benzo(*a*)pyrene and other polynuclear hydrocarbons in charcoal-broiled meat. *Science*, *145*, 53-55

Mossanda, K., Poncelet, F., Fouassin, A. & Mercier, M. (1979) Detection of mutagenic polycyclic aromatic hydrocarbons in African smoked fish. *Food Cosmet. Toxicol.*, *17*, 141-143

Müller, E. (1968) Carcinogenic substances in water and soils. XX. Studies on the carcinogenic properties of 1,12-benzoperylene (Ger.). *Arch. Hyg.*, *152*, 23-36

Müller, K. & Meyer, J.P. (1974) *Einfluss von Ottokraftstoffen auf die Emission von polynuklearen aromatischen Kohlenwasserstoffen in Automobilabgasen im Europa-Test* (Effect of gasoline components on emission of polynuclear aromatic hydrocarbons in car exhaust in the Europa test) (*Forschungsbericht 4568*), Hamburg, Deutsche Gesellschaft für Mineralölwissenschaft und Kohlechemie e.V.

NIH/EPA Chemical Information System (1982) *Mass Spectral Search System*, Washington DC, CIS Project, Information Sciences Corporation

Quarles, J.M., Sega, M.W., Schenley, C.K. & Lijinsky, W. (1979) Transformation of hamster fetal cells by nitrosated pesticides in a transplacental assay. *Cancer Res.*. *39*, 4525-4533

Salamone, M.F., Heddle, J.A. & Katz, M. (1979) The mutagenic activity of thirty polycyclic aromatic hydrocarbons (PAH) and oxides in urban airborne particulates. *Environ. Int.*, *2*, 37-43

Stanton, M.F., Miller, E., Wrench, C. & Blackwell, R. (1972) Experimental induction of epidermoid carcinoma in the lungs of rats by cigarette smoke condensate. *J. natl Cancer Inst.*, *49*, 867-877

Van Duuren, B.L., Sivak, A., Goldschmidt, B.M., Katz, C. & Melchionne, S. (1970) Initiating activity of aromatic hydrocarbons in two-stage carcinogenesis. *J. natl Cancer Inst.*, *44*, 1167-1173

Van Duuren, B.L., Katz, C. & Goldschmidt, B.M. (1973) Brief communication: Cocarcinogenic agents in tobacco carcinogenesis. *J. natl Cancer Inst.*, *51*, 703-705

Woidich, H., Pfannhauser, W., Blaicher, G. & Tiefenbacher, K. (1976) Analysis of polycyclic aromatic hydrocarbons in drinking and industrial water (Ger.). *Lebensmittelchem. gerichtl. Chem.*, *30*, 141-145

BENZO[*c*]PHENANTHRENE

1. Chemical and Physical Data

1.1 Synonyms and trade names

Chem. Abstr. Services Reg. No.: 195-19-7

Chem. Abstr. Name: Benzo(c)phenanthrene

IUPAC Systematic Name: Benzo[*c*]phenanthrene

Synonyms: 3,4-Benzophenanthrene; 3,4-benzphenanthrene; tetrahelicene

1.2 Structural and molecular formulae and molecular weight

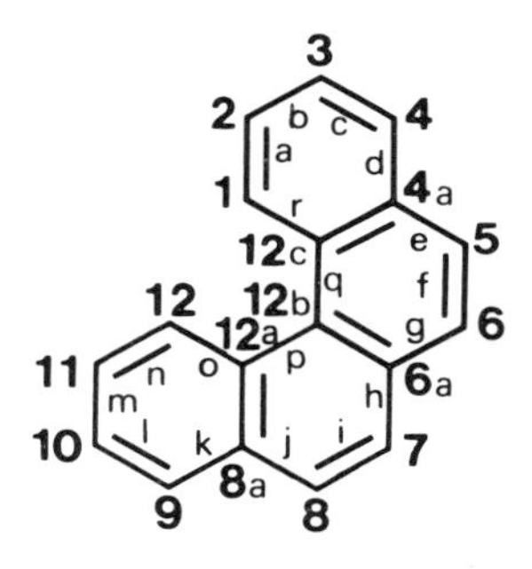

$C_{18}H_{12}$ Mol. wt: 228.3

1.3 Chemical and physical properties of the pure substance

From Weast (1975), unless otherwise specified

(*a*) *Description*: Needles or plates (recrystallized from ethanol or ligroin)

(*b*) *Melting-point*: 66.1°C (Karcher *et al.*, 1983)

(*c*) *Spectroscopy data*: λ_{max} 219, 228, 245, 271, 281, 302, 313, 326, 346, 370 nm (in cyclohexane) (Karcher *et al.*, 1983). Mass and nuclear magnetic resonance spectra have been tabulated (NIH/EPA, Chemical Information System, 1982; Karcher *et al.*, 1983).

(*d*) *Solubility*: Virtually insoluble in water; slightly soluble in ethanol and ligroin

(*e*) *Stability*: No data were available.

(*f*) *Reactivity*: 5-bromo-, 5-nitro- and 5-acetyl-3-benzo[*c*]phenanthrene can be prepared by direct substitution (Clar, 1964).

2. Production, Use, Occurrence and Analysis

2.1 Production and use

There is no commercial production or known use of this compound.

2.2 Occurrence and analysis

Data on occurrence and methods of analysis are summarized in the 'General Remarks on the Substances Considered', p. 35.

Benzo[*c*]phenanthrene occurs ubiquitously in products of incomplete combustion. It has been identified in mainstream cigarette smoke (Van Duuren, 1958; Snook *et al.*, 1977); exhaust emissions from gasoline engines (Grimmer *et al.*, 1977); surface water (1.0-9.1 ng/l) (Grimmer *et al.*, 1981); and waste water (0.042-0.699 μg/l) (Grimmer *et al.*, 1981).

3. Biological Data Relevant to the Evaluation of Carcinogenic Risk to Humans

3.1 Carcinogenicity studies in animals

(*a*) *Skin application*

Mouse: A group of 40 mice received twice weekly skin applications of a 0.3% solution of benzo[*c*]phenanthrene in benzene. The last mouse had died by 580 days. One papilloma and four epitheliomas [squamous-cell carcinomas] were reported (Badger *et al.*, 1940).

Two groups of 30 female CD-1 mice, seven to eight weeks old, received a single skin application of a solution of 0.4 or 2.0 μmol benzo[*c*]phenanthrene in a mixture of acetone and dimethyl sulphoxide (95:5) on the dorsal skin, followed seven days later by twice weekly applications of 16 nmol 12-*O*-tetradecanoylphorbol-13-acetate (TPA) in 200 μl acetone for 20 weeks. Papillomas were induced in 5/30 (17%) and 11/30 animals (38%) in the two groups, respectively. No skin tumour occurred in 30 controls receiving the solvent and TPA (Levin *et al.*, 1980).

(*b*) *Subcutaneous and/or intramuscular administration*

Mouse: No injection-site tumours occurred in 10 mice injected s.c. with 5 mg benzo[*c*]phenanthrene in 0.2% sesame oil at intervals of several weeks. The last mouse died at 445 days (Badger *et al.*, 1940).

Rat: Of six rats given several repeated s.c. injections of 5 mg benzo[*c*]phenanthrene in sesame oil, one developed a sarcoma at the injection site after 533 days (Badger *et al.*, 1940).

3.2 Other relevant biological data[1]

(a) *Experimental systems*

No data were available to the Working Group on toxic effects or on effects on reproduction and prenatal toxicity.

Metabolism and activation

Benzo[*c*]phenanthrene was metabolized to the 3,4- and 5,6-dihydrodiols and unidentified monohydroxy derivatives in the presence of rat-liver preparations (Ittah *et al.*, 1983). The 3,4-dihydrodiol was metabolized by rat-liver microsomes to products that were mutagenic to bacterial and mammalian cells (Wood *et al.*, 1980). This dihydrodiol was a more potent tumour initiator on mouse skin than benzo[c]phenanthrene (Levin *et al.*, 1980).

Mutagenicity and other short-term tests

Benzo[*c*]phenanthrene induced mutations in *Salmonella typhimurium* strain TA100 (his^-/his^+) when added at a concentration of 25 nmol/plate in the presence of an exogenous metabolic system (postmitochondrial supernatant from Aroclor-induced rat liver) (Wood *et al.*, 1980).

(*b*) *Humans*

No data were available to the Working Group.

3.3 Case reports and epidemiological studies of carcinogenicity to humans

No data were available to the Working Group.

---

[1] See also 'General Remarks on the Substances Considered', p. 53.

4. Summary of Data Reported and Evaluation[1]

4.1 Experimental data

Benzo[*c*]phenanthrene was tested in a mouse-skin initiation-promotion assay and was active as an initiator. Other experiments in mice by skin application and in mice and rats by subcutaneous injection were considered inadequate for evaluation.

No data on the teratogenicity of this chemical were available.

In the one study evaluated, benzo[*c*]phenanthrene was mutagenic to *Salmonella typhimurium* in the presence of an exogenous metabolic system.

There is *inadequate evidence* that benzo[*c*]phenanthrene is active in short-term tests.

4.2 Human data[2]

Benzo[*c*]phenanthrene is present as a minor component of the total content of polynuclear aromatic compounds in the environment. Human exposure to benzo[*c*]phenanthene occurs primarily through the smoking of tobacco, inhalation of polluted air and by ingestion of food and water contaminated by combustion effluents (for details, see 'General Remarks on the Substances Considered', p. 35).

4.3 Evaluation

The available data are inadequate to permit an evaluation of the carcinogenicity of benzo[*c*]phenanthrene to experimental animals.

5. References

Badger, G.M., Cook, J.W., Hewett, C.L., Kennaway, E.L., Kennaway, N.M., Martin, R.H., & Robinson, A.M. (1940) The production of cancer by pure hydrocarbon. V. *Proc. R. Soc. London*, *129*, 439-467

Clar, E., ed. (1964) *Polycyclic Hydrocarbons*, Vol. 2, New York, Academic Press, pp. 256-264

Grimmer, G., Böhnke, H. & Glaser, A. (1977) Investigation on the carcinogenic burden by air pollution in man. XV. Polycyclic aromatic hydrocarbons in automobile exhaust gas - An inventory. *Zbl. Bakt. Hyg., 1 Abt., Orig. Bl64*, 218-234

---

[1] For definitions of the italicized terms, see Preamble, p. 19.

[2] Studies on occupational exposure to polynuclear aromatic compounds will be considered in future *IARC Monographs*.

Grimmer, G., Schneider, D. & Dettbarn, G. (1981) The load of different rivers in the Federal Republic of Germany by PAH (PAH - profiles of surface water) (Ger.). *Wasser*, *56*, 131-144

Ittah, Y., Thakker, D.R., Levin, W., Croisy-Delcey, M., Ryan, D.E., Thomas, P.E., Conney, A.H. & Jerina, D.M. (1983) Metabolism of benzo[*c*]phenanthrene by rat liver microsomes and by a purified monooxygenase system reconstituted with different forms of cytochrome P-450. *hem.-biol. Interactions*, *45*, 15-28

Karcher, W., Fordham, R., Dubois, J. & Gloude, P. (1983) *Spectral Atlas of Polycyclic Aromatic Compounds*, Dordrecht, The Netherlands, D. Reidel (in press)

Levin, W., Wood, A.W., Chang, R.L., Ittah, Y., Croisy-Delcey, M., Yagi, H., Jerina, D.M. & Conney, A.H. (1980) Exceptionally high tumor-initiating activity of benzo(*c*)phenanthrene bay-region diol-epoxides on mouse skin. *Cancer Res.*, *40*, 3910-3914

NIH/EPA Chemical Information System (1982) *Mass Spectral Search System*, Washington DC, CIS Project, Information Services Corporation

Snook, M.E., Severson, R.F., Arrendale, R.F., Higman, H.C. & Chortyk, O.T. (1977) The identification of high molecular weight polynuclear aromatic hydrocarbons in a biologically active fraction of cigarette smoke condensate. *Beitr. Tabakforsch.*, *9*, 79-101

Van Duuren, B.L. (1958) The polynuclear aromatic hydrocarbons in cigarette-smoke condensate. II. *J. natl Cancer Inst.*, *21*, 623-630

Weast, A.C., ed. (1975) *CRC Handbook of Chemistry and Physics*, 56th ed., Cleveland, OH, Chemical Rubber Co., p. C-194

Wood, A.W., Chang, R.L., Levin, W., Ryan, D.E., Thomas, P.E., Croisy-Delcey, M., Ittah, Y., Yagi, H., Jerina, D.M. & Conney, A.H. (1980) Mutagenicity of the dihydrodiols and bay-region diol-epoxides of benzo(*c*)phenanthrene in bacterial and mammalian cells. *Cancer Res.*, *40*, 2876-2883

BENZO[*a*]PYRENE

This compound was considered by earlier working groups, in December 1972 (IARC, 1973) and in February 1982 (IARC, 1982).

1. Chemical and Physical Data

1.1 Synonyms and trade names

Chem. Abstr. Services Reg. No.: 50-32-8

Chem. Abstr. Name: Benzo(a)pyrene

IUPAC Systematic Name: Benzo[*a*]pyrene

Synonyms: Benzo[*def*]chrysene; 1,2-benzopyrene[1] ; 3,4-benzopyrene ; 6,7-benzopyrene; 3,4-benzpyrene; 3,4-benz(a)pyrene; benz(a)pyrene; BP; B(a)P

1.2 Structural and molecular formulae and molecular weight

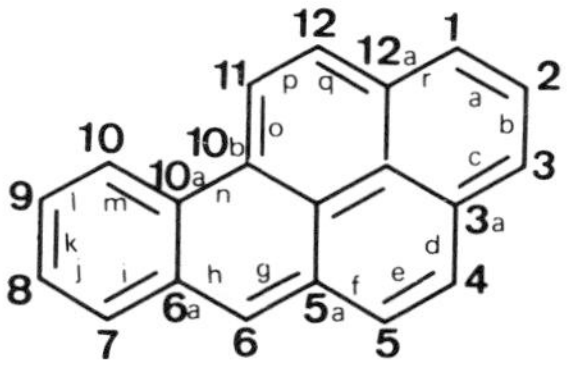

$C_{20}H_{12}$

Mol. wt: 252.3

1.3 Chemical and physical properties of the pure substance

From National Library of Medicine (1982), unless otherwise specified

(*a*) *Description*: Pale-yellow needles (recrystallized from benzene/methanol); crystals may be monoclinic or orthorhombic

(*b*) *Boiling-point*: 310-312°C at 10 mm Hg

(*c*) *Melting-point*: 178.1°C (Karcher *et al.*, 1983)

---

[1] Until about 1936

(*d*) *Spectroscopy data*: λ_{max} 219, 226, 254, 265, 272, 283, 296, 330, 345, 363, 379, 383, 393, 402 nm (in cyclohexane) (Karcher *et al.*, 1983). Infrared spectra (Pouchert, 1970), mass spectra (Strömberg & Widmark, 1970) and nuclear magnetic resonance spectra (Karcher *et al.*, 1983) have been reported; mass spectra have been tabulated (NIH/EPA Chemical Information System, 1982).

(*e*) *Solubility*: Practically insoluble in water (0.5-6 μg/l) (Davis *et al.*, 1942; Weil-Malherbe, 1946; Eisenbrand, 1971); sparingly soluble in ethanol and methanol; soluble in benzene, toluene and xylene

(*f*) *Stability*: Undergoes photo-oxidation after irradiation in indoor sunlight or by fluorescent light in organic solvents (Karatsune & Hirohata, 1962)

(*g*) *Reactivity*: Oxidation with ozone yields benzo[*a*]pyrene-(1,6 or 3,6)-quinone, and, on further oxidation, benzanthrone dicarboxylic anhydride (Moriconi *et al.*, 1961). Readily undergoes nitration and halogenation (Clar, 1964). Reacts with NO and NO_2 to form nitro derivatives (Butler & Crossley, 1981)

2. Production, Use, Occurrence and Analysis

2.1 Production and use

There is no commercial production or known use of this compound. A reference material of certified high purity is available (Karcher *et al.*, 1980; Community Bureau of Reference, 1982).

2.2 Occurrence and analysis

Data on occurrence and methods of analysis are summarized in the 'General Remarks on the Substances Considered', p. 35.

Benzo[*a*]pyrene occurs ubiquitously in products of incomplete combustion; it also occurs in fossil fuels. It has been identified in mainstream cigarette smoke (20-40 μg/cigarette) (US Department of Health & Human Services, 1982), (4.66 μg/100 cigarettes) (Sforzolini & Mariani, 1961), (4.7-7.8 μg/cigarette) (Hoffmann *et al.*, 1963, 1974), (3.1 μg/100 cigarettes) (Hoffmann *et al.*, 1967), (1.4-1.5 g/100 cigarettes) (Borisyuk, 1966), (3.4 μg/100 g of burnt material) (Masuda & Kuratsune, 1972), (2.5 μg/100 cigarettes) (Wynder & Hoffmann, 1963), (12.9-35.2 ng/cigarette) (Rathkamp *et al.*, 1973), (1.7-2.3 μg/100 cigarettes) (Severson *et al.*, 1979), (0.99-1.3 μg/g condensate) (Müller *et al.*, 1964), (1.8-8 μg/1000 cigarettes) (Kiryu & Karatsune, 1966); sidestream cigarette smoke (135 μg/cigarette) (Kotin & Falk, 1960), (4.7 mg/kg cigarette smoke condensate) (Wynder & Hoffmann, 1961), (3.2-20.8 μg/500 g material) (Bentley & Burgan, 1960); smoke of cigars (5.1 μg/100 g) (Hoffmann *et al.*, 1963), (3.4 μg/100 g) (Campbell & Lindsey, 1957), (1.8-3 μg/cigar) (Hoffmann & Wynder, 1972), (2.2-4 μg/cigar) (Schmeltz *et al.*, 1976), (1.8-2.1 g/100 cigars) (Borisyuk, 1966); the air of restaurants (0.25-10.1 ng/m^3) (Just *et al.*, 1972); at public gatherings (7.1-21.7 ng/m^3) (Elliott & Rowe, 1975); in smoke-filled rooms (22 ng/m^3) (Grimmer *et al.*, 1977b); mainstream smoke of marijuana cigarettes (2.9 μg/100 cigarettes) (Lee *et al.*, 1976), (3.1 μg/cigarette) (Hoffmann *et al.*, 1975); urban air (0.05-74 ng/m^3) (Hoffmann & Wynder, 1976); gasoline engine exhaust (50-81 μg/l fuel) (Grimmer *et al.*, 1977a), (72.6 mg/kg tar) (Hoffmann & Wynder, 1963); diesel engine exhaust (2-170 μg/mg extract) (National Research Council, 1981); various crude oils (0.1-3.6 mg/kg) (Grimmer *et al.*, 1983); various fresh (0.008-0.266 mg/kg) (Grimmer *et al.*, 1981b) and used motor oils (5.2-35.1 mg/kg) (Grimmer *et al.*, 1981c); gasolines (1-2 mg/kg) (Begeman & Colucci, 1968); charcoal-broiled steaks (8 μg/kg) (Lijinsky & Shubik, 1964); various processed (refined, broiled, smoked) foods (0.01-33.5 μg/kg) (Fretheim, 1976; Lintas *et al.*, 1979; Gray & Morton, 1981), (50.4 μg/kg) (Lijinsky & Ross, 1967), (up to 289 μg/kg) (Thorsteinsson, 1969); various oils, margarine, butter, fats (up to 62 μg/kg) (Biernoth & Rost, 1968); fruit, vegetables and cereals (up to 48.1 μg/kg) (Hettche, 1971); roasted coffee (0.1-16.5 μg/kg) (Fritz, 1972); tea (3.9-21.3 μg/kg) (Gray & Morton, 1981); surface water (0.2-13 000 ng/l) (Iljnitsky & Rozhkova, 1970; Grimmer *et al.*, 1981a); tap water (0.2-1000 ng/l) (Trakhtman & Manita, 1966; Melchiorri *et al.*, 1973); rain water (2.2-7.3 ng/l) (Woidich *et al.*, 1976); subterranean water (0.4-7 ng/l) (Borneff & Kunte, 1969); waste water (0.001-6000 μg/l) (Bliokh, 1965; Andelman & Suess, 1970); sludge (3-1330 μg/kg) (Borneff & Kunte, 1967; Iljnitsky & Rozhkova, 1970); freeze-dried sewage sludge (540-13 300 μg/kg) (Grimmer *et al.*, 1980); and dried sediments from lakes (1-2380 μg/kg) (Grimmer & Böhnke, 1975, 1977).

3. Biological Data Relevant to the Evaluation of Carcinogenic Risk to Humans

3.1 Carcinogenicity studies in animals

There is *sufficient evidence* that benzo[*a*]pyrene is carcinogenic to experimental animals (see General Remarks on the Substances Considered, p. 33).

3.2 Other relevant biological data[1]

(*a*) *Experimental systems*

Toxic effects

The LD_{50} for the mouse (i.p.) is about 250 mg/kg (Salamone, 1981). The ID_{50} (skin irritant activity) for the mouse is 5.6×10^{-5} mmol/ear (Brune *et al.*, 1978).

[1] See also 'General Remarks on the Substances Considered', p. 53.

The growth rate of mouse ascites sarcoma cells in culture was slightly (6%) inhibited when benzo[*a*]pyrene was added at a concentration of 1 μmol/ml dissolved in dimethyl sulphoxide (Pilotti *et al.*, 1975).

Marked differences in the toxic effects of benzo[*a*]pyrene have been reported in different strains of mice depending on their genetic constitutions. The *Ah* locus, which determines the inducibility of aryl hydrocarbon hydroxylase, appears to be of particular importance. For example, oral administration of about 120 mg/kg bw benzo[*a*]pyrene per day with the diet induced aplastic anaemia in nonresponsive (poorly inducible) AKR/N mice (Ah^d/Ah^d type) and death within four weeks, whereas responsive (markedly inducible) mice (Ah^b/Ah^b type) remained healthy for at least six months. In the former, bone marrow was hypocellular with myeloid precursors and promegakaryocytes. When benzo[*a*]pyrene was injected i.p. (500 mg/kg bw) instead of being given orally (120 mg/kg bw), survival time of responsive mice (Ah^b/Ah^b type) was found to be significantly shorter than that of nonresponsive mice (Ah^d/Ah^d type) (Robinson *et al.*, 1975). These differences may be explained in part by the greater capacity of the bowel and liver of responsive mice to detoxify an orally administered dose of benzo[*a*]pyrene metabolically. However, if the hydrocarbon reaches bone marrow and other distal tissues in responsive mice, it is metabolized to toxic metabolites to a greater extent (Nebert *et al.*, 1977).

A detailed description of the liver damage induced by benzo[a]pyrene in mice has been reported (Meiss *et al.*, 1982).

I.p. administration of single doses of 10 mg benzo[*a*]pyrene produced an immediate and sustained reduction in the growth rate of young rats (Haddow *et al.*, 1937).

The effects of benzo[*a*]pyrene on organ cultures of rat trachea (Crocker *et al.*, 1965), stratification of epithelium (Dirksen & Crocker, 1968) and human fetal lung (Lasnitzki, 1956) have been reported. Consistent findings in rat trachea were suppression of mesenchyme, stimulation of basal-cell replication and induction of metaplasia; epithelial hyperplasia and inhibition of connective tissue growth were seen in human lung cultures. When benzo[*a*]pyrene (in a beeswax pellet) was implanted into isogenically transplanted rat trachea, persistent hyperplasia and metaplasia were observed (Topping *et al.*, 1978).

A single topical application of 0.05 ml of a 1% solution of benzo[*a*]pyrene in acetone to the interscapular area of hairless mice (hr/hr) induced an increase in the mitotic rate of epidermal cells (Elgjo, 1968).

Effects on reproduction and prenatal toxicity

Benzo[*a*]pyrene crosses the placenta in mice and rats (Shendrikova & Aleksandrova, 1974; Shendrikova *et al.*, 1974; Takahashi, 1974; Baranova *et al.*, 1976), and, at a dose of 5 mg/day s.c. to pregnant rats, can cause death of all fetuses (Wolfe & Bryan, 1939).

Benzo[a]pyrene is embryotoxic to mice. Quantitative effects are in part dependent on the genetically-determined ability of the cytochrome-P-450 monooxygenase system of the mother and fetus to be induced by polynuclear aromatic hydrocarbons. When the mother is 'inducible', the genotype of the fetus is relatively unimportant, since it appears that the active metabolites formed in the mother can cross the placenta, causing fetal death or malformation. When the mother is a 'non-inducible', however, the genotype of the fetus is important in that individual fetuses within one litter may or may not be inducible. Such observations help to explain why some fetuses may be more sensitive to the action of teratogens and carcinogens than others within the same litter (Shum *et al.*, 1979). An 'inducible' genotype is not the only factor involved in benzo[*a*]pyrene reproductive toxicity, however, since 'non-inducible' mice may sometimes be more sensitive to effects on fetal weight reduction and congenital anomalies due to benzo[*a*]pyrene; additionally, the incidence of external malformations may not be the same in mice of different genotypes after treatment with benzo[*a*]pyrene, even if both dams and fetuses are inducible (Hoshino *et al.*, 1981).

Oral administration of 10 mg/kg bw benzo[*a*]pyrene to CD-1 mice during pregnancy resulted in a marked and specific reduction of gonadal weight but had no effect on body weight of either male or female offspring. Reduction in fertility and reproductive capacity occurred. With 40 mg/kg bw/day benzo[*a*]pyrene, almost complete sterility was observed in the offspring of both sexes (Mackenzie & Angevine, 1981).

I.p. administration of 100-150 mg/kg bw benzo[a]pyrene to C3H/Anf mice in mid- or late pregnancy results in a marked and persistent suppression of the immune system in the offspring (Urso & Gengozian, 1980).

Application of benzo[*a*]pyrene to the skin of pregnant mice [strain unspecified] over four generations resulted in sensitization of the offspring to the effects of benzo[*a*]pyrene, so that an increase in the rate of appearance of papillomas and carcinomas was observed following skin application to offspring over that in control mice that had not been treated *in utero* (Andrianova, 1971).

Benzo[*a*]pyrene has been shown to act as a transplacental carcinogen in mice of the Ha/ICR strain (Bulay, 1970; Bulay & Wattenberg, 1971) and of the A and C57BL strains (Nikonova, 1977): lung tumours were induced in the Ha/ICR strain and liver tumours in all three strains. Transplacental carcinogenesis has also been shown in rabbits (Beniashvili, 1978).

Metabolism and activation

Benzo[*a*]pyrene is metabolized to approximately 20 primary and secondary oxidized metabolites and to a variety of conjugates. Several metabolites can induce mutations, transform cells and/or bind to cellular macromolecules; however only a 7,8-diol-9,10-epoxide is presently considered to be an ultimate carcinogenic metabolite. The reader is referred to 'General Remarks on the Substances Considered' (Section 4) of this monograph and to several reviews (DePierre & Ernster, 1978; Pelkonen & Nebert, 1982) for access to the very extensive literature on the metabolism and activation of benzo[*a*]pyrene.

Mutagenicity and other short-term tests

Benzo[*a*]pyrene, an indirect carcinogen which undergoes metabolism to reactive electrophiles capable of binding covalently to DNA (Lutz, 1979), has been used extensively as a positive control in a variety of short-term tests. It was active in assays for bacterial DNA repair, bacteriophage induction and bacterial mutation; mutation in *Drosophila melanogaster*; DNA binding, DNA repair, sister chromatid exchange, chromosomal aberration, point mutation and transformation in mammalian cells in culture; and tests in mammals *in vivo*, including DNA binding, sister chromatid exchange, chromosomal aberration, sperm abnormality and the somatic specific locus (spot) test (Hollstein *et al.*, 1979; de Serres & Ashby, 1981).

(*b*) *Humans*

Toxic effects

A 1% solution of benzo[*a*]pyrene in benzene was applied daily to protected and unprotected surfaces of the skin of 26 patients suffering from pemphigus vulgaris, mycosis fungoides, prokeratosis, xeroderma pigmentosum, basal-cell cancer, squamous-cell cancer, lupus erythematosis, psoriasis, syphilis in various stages or ringworm. The period of application did not exceed four months, and the diameter of the treated area was 2 cm. A progressive series of alterations developed in normal skin (chronologically): erythema, pigmentation, desquamation, formation of verrucae[1], and infiltration. The manifestations regressed completely within two to three months of cessation of treatment. Clinically, perceptible erythema occurred in only two patients with basal-cell cancer. Pigmentation, which occurred in all patients, consisted of an increase in melanin in the basal-cell layer of the epidermis and was more evident in the exposed skin (e.g., hand, face). It developed more readily in the skin of senile individuals than in younger patients. Rarely, small masses of pigment granules were found in the more superficial layers. Desquamation was proportional in extent to the erythema of the first stage. The formation of verrucae was the most constant manifestation caused by the treatment. The skin of the patient with xeroderma pigmentosum did not react differently to benzo[*a*]pyrene application from that of other patients (Cottini & Mazzone, 1939).

Human bronchial mucosa treated with benzo[*a*]pyrene in organ culture showed destruction of all cell types and distortion of columnar cell morphology but not of regenerative epithelium (Crocker *et al*., 1973).

No data were available to the Working Group on effects on reproduction and prenatal toxicity, absorption, distribution, excretion and metabolism or mutagenicity and chromosomal effects.

3.3 Case reports and epidemiological studies of carcinogenicity to humans

No data were available to the Working Group.

4. Summary of Data Reported and Evaluation[2]

4.1 Experimental data

Benzo[*a*]pyrene has been shown to be carcinogenic to experimental animals (see 'General Remarks on the Substances Considered', p. 33).

[1] Statement of the authors: 'Although clinically the lesions were not true verrucae, they are for convenience considered as such. It would be more accurate to describe the alteration as an accentuation of the relief pattern of the skin secondary to a deepening of the sulci.'

[2] For definitions of the italicized terms, see Preamble, p. 19.

Benzo[*a*]pyrene is embryotoxic and teratogenic in mice; the inducibility of aryl hydrocarbon hydroxylase activity in dams and fetuses is an important factor in determining these effects. A reduction in fertility in both male and female offspring was observed in mice following exposure to benzo[*a*]pyrene *in utero*.

Benzo[*a*]pyrene undergoes metabolism to reactive electrophiles capable of binding covalently to DNA. It was active in assays for bacterial DNA repair, bacteriophage induction and bacterial mutation; mutation in *Drosophila melanogaster*; DNA binding, DNA repair, sister chromatid exchange, chromosomal aberrations, point mutation and transformation in mammalian cells in culture; and in tests in mammals *in vivo*, including DNA binding, sister chromatid exchange, chromosomal aberration, sperm abnormality and the somatic specific locus (spot) test.

There is *sufficient evidence* that benzo[*a*]pyrene is active in short-term tests.

4.2 Human data[1]

Benzo[*a*]pyrene is present as a component of the total content of polynuclear aromatic compounds in the environment. Human exposure to benzo[*a*]pyrene occurs primarily through the smoking of tobacco, inhalation of polluted air and by ingestion of food and water contaminated by combustion effluents (for details, see 'General Remarks on the Substances Considered', p. 35).

4.3 Evaluation[2]

There is *sufficient evidence* that benzo[*a*]pyrene is carcinogenic to experimental animals.

5. References

Andelman, J.B. & Suess, M.J. (1970) Polynuclear aromatic hydrocarbons in the water environment. *Bull. World Health Organ.*, *43*, 479-508

Andrianova, M.M. (1971) Transplacental action of 3-methylcholanthrene and benz[α]pyrene on four generations of mice. *Bull. exp. Biol. Med.*, *71*, 677-680

Baranova, L.N., Shendrikova, I.A., Aleksandrov, V.A., Likhachev, A.Ya., Ivanov-Galitsin, M.N., Dikun, P.P. & Napalkov, N.P. (1976) Problem of the mechanism of penetration of carcinogenic polycyclic hydrocarbons through the placenta of rats and mice. *Dokl. Akad. Nauk. SSR*, *228*, 733-736

---

[1] Studies on occupational exposure to polynuclear aromatic compounds will be considered in future *IARC Monographs*.

[2] In the absence of adequate data on humans, it is reasonable, for practical purposes, to consider chemicals for which there is *sufficient evidence* of carcinogenicity in animals as if they presented a carcinogenic risk to humans (see also Preamble, p. 20).

Begeman, C.R. & Colucci, J.M. (1968) Benzo[*a*]pyrene in gasoline partially persists in automobile exhaust. *Science*, *161*, 271

Beniashvili, D.Sh. (1978) A comparative study on the action of some carcinogens inducing transplacental blastomogenesis in rabbits. *Vopr. Onkol.*, *24*, 77-83

Bentley, H.R. & Burgan, J.G. (1960) Polynuclear hydrocarbons in tobacco and tobacco smoke. II. The origin of 3:4 benzopyrene found in tobacco and tobacco smoke. *Analyst*, *85*, 723-727

Biernoth, G. & Rost, H.E. (1968) Occurrence of polycyclic aromatic hydrocarbons in edible oils and their refining (Ger.). *Arch. Hyg.*, *152*, 238-250

Bliokh, S.S. (1965) The pollution of water by carcinogenic substances. *Hyg. Sanit.*, *30*, 100-104

Borisyuk, Y.P. (1966) Content of 3,4-benzopyrene in smoking products of some tobacco brands (Russ.). *Vopr. eksp. Onkol.*, *2*, 93-97 [*Chem. Abstr.*, *67*, 18634c]

Borneff, J. & Kunte, H. (1967) Carcinogenic substances in water and soil. XIX. Action of sewage purification on polycylic hydrocarbons (Ger.). *Arch. Hyg. Bakt.*, *151*, 202-210

Borneff, J. & Kunte, H. (1969) Carcinogenic substances in water and soil. 25. A routine method for the determination of polynuclear aromatic hydrocarbons in water (Ger.). *Arch. Hyg. Bakt.*, *153*, 220-229

Brune, K., Kalin, H., Schmidt, R. & Hecker, E. (1978) Inflammatory, tumor initiating and promoting activities of polycyclic aromatic hydrocarbons and diterpene esters in mouse skin as compared with their prostaglandin releasing potency *in vitro*. *Cancer Lett.*, *4*, 333-342

Bulay, O.M. (1970) The study of development of lung and skin tumors in mice exposed *in utero* to polycyclic hydrocarbons. *Acta med. turc.*, *7*, 3-38

Bulay, O.M. & Wattenberg, L.W. (1971) Carcinogenic effects of polycyclic hydrocarbon carcinogen administration to mice during pregnancy on the progeny. *J. natl Cancer Inst.*, *46*, 397-402

Butler, J.D. & Crossley, P. (1981) Reactivity of polycyclic aromatic hydrocarbons adsorbed on soot particles. *Atmos. Environ.*, *15*, 91-94

Campbell, J.M. & Lindsey, A.J. (1957) Polycyclic hydrocarbons in cigar smoke. *Br. J. Cancer*, *11*, 192-195

Clar, E., ed. (1964) *Polycyclic Hydrocarbons*, vol. 2, New York, Academic Press, pp. 130-140

Community Bureau of Reference (1982) *Polycyclic Aromatic Hydrocarbon Reference Materials of Certified Purity* (*Information Handout, No. 22*), Brussels, Commission of the European Communities

Cottini, G.B. & Mazzone, G.B. (1939) The effects of 3:4-benzpyrene on human skin. *Am. J. Cancer*, *37*, 186-195

Crocker, T.T., Nielsen, B.I. & Lasnitzki, I. (1965) Carcinogenic hydrocarbons: Effects on suckling rat trachea in organ culture. *Arch. environ. Health*, *10*, 240-250

Crocker, T.T., O'Donnell, T.V & Nunes, L.L. (1973) Toxicity of benzo(*a*)pyrene and air pollution composite for adult human bronchial mucosa in organ culture. *Cancer Res.*, *33*, 88-93

Davis, W.W., Krahl, M.E. & Clowes, G.H.A. (1942) Solubility of carcinogenic and related hydrocarbons in water. *J. Am. chem. Soc.*, *64*, 108-110

DePierre, J.W. & Ernster, L. (1978) The metabolism of polycyclic hydrocarbons and its relationship to cancer. *Biochim. biophys. Acta*, *473*, 149-186

Dirksen, E.R. & Crocker, T.T. (1968) Ultrastructural alterations produced by polycyclic aromatic hydrocarbons on rat tracheal epithelium in organ culture. *Cancer Res.*, *28*, 906-923

Eisenbrand, J. (1971) On the water solubility of 3,4-benzopyrene and other aromatic hydrocarbons and its increase by solubilizers (Ger.). *Dtsch. Lebensm. Rundsch.*, *67*, 435-444

Elgjo, K. (1968) Growth kinetics of the mouse epidermis after a single application of 3,4-benzopyrene, croton oil or 1,2-benzopyrene. *Acta pathol. microbiol. scand.*, *73*, 183-190

Elliott, L.P. & Rowe, D.R. (1975) Air quality during public gatherings. *J. Air Pollut. Control Assoc.*, *25*, 635-636

Fretheim, K. (1976) Carcinogenic polycyclic aromatic hydrocarbons in Norwegian smoked meat sausages. *J. agric. Food Chem.*, *24*, 976-979

Fritz, W. (1972) A contribution to the contamination of foods with carcinogenic hydrocarbons during processing and cooking (Ger.). *Arch. Geschwulstforsch.*, *40*, 81-90

Gray, J.I. & Morton, I.D. (1981) Some toxic compounds produced in food by cooking and processing. *J. human Nutr.*, *35*, 5-23

Grimmer, G. & Böhnke, H. (1975) Profile analysis of polycyclic aromatic hydrocarbons and metal content in sediment layers of a lake. *Cancer Lett.*, *1*, 75-84

Grimmer, G. & Böhnke, H. (1977) Investigation on drilling cores of sediments of Lake Constance. I. Profiles of the polycyclic aromatic hydrocarbons. *Z. Naturforsch.*, *32c*, 703-711

Grimmer, G., Böhnke, H. & Glaser, A. (1977a) Investigation on the carcinogenic burden by air pollution in man. XV. Polycyclic aromatic hydrocarbons in automobile exhaust gas - An inventory. *Zbl. Bakt. Hyg., 1 Abt., Orig. B164*, 218-234

Grimmer, G., Böhnke, H. & Harke, H.-P. (1977b) Passive smoking: Intake of polycyclic aromatic hydrocarbons by breathing of cigarette smoke containing air (Ger.). *Int. Arch. occup. environ. Health*, *40*, 93-99

Grimmer, G., Hilge, G. & Niemitz, W. (1980) Comparison of the profiile of polycyclic aromatic hydrocarbons in sewage sludge samples from 25 sewage treatment works (Ger.). *Wasser*, *54*, 255-272

Grimmer, G., Schneider, D. & Dettbarn, G. (1981a) The load of different rivers in the Federal Republic of Germany by PAH (PAH-profiles of surface water) (Ger.). *Wasser*, *56*, 131-144

Grimmer, G., Jacob, J. & Naujack, K.-W. (1981b) Profile of the polycyclic aromatic compounds from lubricating oils. Inventory by GCGC/MS - PAH in environmental materials, Part 1. *Fresenius Z. anal. Chem.*, *306*, 345-355

Grimmer, G., Jacob, J., Naujack, K.-W. & Dettbarn, G. (1981c) Profile of the polycyclic aromatic compounds from used engine oil - Inventory by GCGC/MS - PAH in environmental materials, Part 2. *Fresenius Z. anal. Chem.*, *309*, 13-19

Grimmer, G., Jacob, J. & Naujack, K.-W. (1983) Profile of the polycyclic aromatic compounds from crude oils. Inventory by GCGC/MS - PAH in environmental materials. Part 3. *Fresenius Z. anal. Chem.*, *314*, 29-36

Haddow, A., Scott, C.M. & Scott, J.D. (1937) The influence of certain carcinogenic and other hydrocarbons on body growth in the rat. *Proc. R. Soc. London Ser. B*, *122*, 477-507

Hettche, H.O. (1971) Plant waxes as collectors of polycyclic aromatics in the air of residential areas. *Staub-Reinhalt. Luft*, *31*, 34-41

Hoffmann, D. & Wynder, E.L. (1963) Studies on gasoline engine exhaust. *J. Air Pollut. Control Assoc.*, *13*, 322-327

Hoffmann, D. & Wynder, E.L. (1972) Smoke of cigarettes and little cigars: An analytical comparison. *Science*, *178*, 1197-1199

Hoffmann, D. & Wynder, E. (1976) *Environmental respiratory carcinogenesis*. In: Searle, C.E., ed., *Chemical Carcinogens* (*ACS Monograph 173*), Washington DC, American Chemical Society, p. 341

Hoffmann, D., Rathkamp, G. & Wynder, E.L. (1963) Comparison of the yields of several selected components in the smoke from different tobacco products. *J. natl Cancer Inst.*, *31*, 627-637

Hoffmann, D., Rathkamp. G. & Rubin, J. (1967) Chemical studies on tobacco smoke. II. Comparison of the yields of several selected components in the smoke from five major Turkish tobacco varieties. *Food Cosmet. Toxicol.*, *5*, 37-38

Hoffmann, D., Sanghvi, L.D. & Wynder, E.L. (1974) Comparative chemical analysis of Indian bidi and American cigarette smoke. *Int. J. Cancer*, *14*, 49-53

Hoffmann, D., Brunnemann, K.D., Gori, G.B. & Wynder, E.L. (1975) *On the carcinogenicity of marijuana smoke*. In: Runeckles, V.C., ed., *Recent Advances in Phytochemistry*, Vol. 9, New York, Plenum Press, pp. 63-81

Hollstein, M., McCann, J., Angelosanto, F.A. & Nichols, W.W. (1979) Short-term tests for carcinogens and mutagens. *Mutat. Res.*, *65*, 133-226

Hoshino, K., Hayashi, Y., Takehira, Y. & Kameyama, Y. (1981) Influences of genetic factors on the teratogenicity of environmental pollutants: Teratogenic susceptibility to benzo[*a*]pyrene and *Ah* locus in mice. *Congenital Anomalies*, *21*, 97-103

IARC (1973) *IARC Monographs on the Evaluation of Carcinogenic Risk of Chemicals to Man*, Vol. 3, *Certain Polycyclic Aromatic Hydrocarbons and Heterocyclic Compounds*, Lyon, pp. 91-136

IARC (1982) *IARC Monographs on the Evaluation of the Carcinogenic Risk of Chemicals to Humans*, Supplement 4, *Chemicals, Industrial Processes and Industries Associated with Cancer in Humans*, Lyon, pp. 15, 227-228

Iljnitsky, A.P. & Rozhkova, L.G. (1970) Pollution of resources by carcinogenic hydrocarbons (Russ.). *Vopr. Onkol.*, *16*, 78-82

Just, J., Borkowska, M. & Maziarka, S. (1972) Air pollution by tobacco smoke in Warsaw coffee houses (Pol.). *Rocz. Panstw. Zakł. Hyg.*, *23*, 129-135

Karcher, W., Jacob, J. & Haemers, L. (1980) *The Certification of Eight Polycyclic Aromatic Hydrocarbon Materials (PAH) (BCR Reference Materials Nos. 46, 47, 48, 49, 50, 51, 52 and 53)* (*EUR 6967EN*), Luxembourg, Commission of the European Communities

Karcher, W., Fordham, R., Dubois, J. & Gloude, P. (1983) *Spectral Atlas of Polycyclic Aromatic Compounds*, Dordrecht, The Netherlands, D. Reidel (in press)

Kiryu, S. & Kuratsune, M. (1966) Polycyclic aromatic hydrocarbons in the cigarette tar produced by human smoking. *Gann*, *57*, 317-322

Kotin, P. & Falk, H.L. (1960) The role and action of environmental agents in the pathogenesis of lung cancer. II. Cigarette smoke. *Cancer*, *13*, 250-262

Kuratsune, M. & Hirohata, T. (1962) Decomposition of polycyclic aromatic hydrocarbons under laboratory illuminations. *Natl Cancer Inst. Monogr.*, *9*, 117-125

Lasnitzki, I. (1956) The effect of 3,4-benzpyrene on human foetal lung grown *in vitro*. *Br. J. Cancer.*, *10*, 510-516

Lee, M.L., Novotny, M. & Bartle, K.D. (1976) Gas chromatography/mass spectrometry and nuclear magnetic resonance spectrometric studies of carcinogenic polynuclear aromatic hydrocarbons in tobacco and marijuana smoke condensates. *Anal. Chem.*, *48*, 405-416

Lijinsky, W. & Shubik, P. (1964) Benzo(*a*)pyrene and other polynuclear hydrocarbons in charcoal-broiled meat. *Science*, *145*, 53-55

Lijinsky, W. & Ross, A.E. (1967) Production of carcinogenic polynuclear hydrocarbons in the cooking of food. *Food Cosmet. Toxicol.*, *5*, 343-347

Lintas, C., De Matthaeis, M.C. & Merli, F. (1979) Determination of benzo[*a*]pyrene in smoked, cooked and toasted food products. *Food Cosmet. Toxicol.*, *17*, 325-328

Lutz, W.K. (1979) *In vivo* covalent binding of organic chemicals to DNA as a quantitative indicator in the process of chemical carcinogenesis. *Mutat. Res.*, *65*, 289-356

MacKenzie, K.M. & Angevine, D.M. (1981) Infertility in mice exposed *in utero* to benzo[*a*]pyrene. *Biol. Reprod.*, *24*, 183-191

Masuda, Y. & Kuratsune, M. (1972) Comparison of the yield of polycyclic aromatic hydrocarbons in smoke from Japanese tobacco. *Jpn. J. Hyg.*, *27*, 339-341

Meiss, R., Heinrich, U., Offermann, M. & Themann, H. (1982) Long-term study of liver damage following subcutaneous injection of airborne particle extracts and polycyclic aromatic hydrocarbon fractions. *Int. Arch. occup. environ. Health*, *49*, 305-314

Melchiorri, C., Chiacchiarini, L., Grella, A. & D'Arca, S.U. (1973) Identification and determination of polycyclic aromatic hydrocarbons in some tap waters of Roma city (Ital.). *Nuovi Ann. Ig. Microbiol.*, *24*, 279-301

Moriconi, E.J., Rakoczy, B. & O'Connor, W.F. (1961) Ozonolysis of polycyclic aromatics. VIII. Benzo(a)pyrene. *J. Am. chem. Soc.*, *83*, 4618-4622

Müller, K.H., Neurath, G. & Horstmann, H. (1964) Influence of air permeability of cigarette paper on the yield of smoke and on its composition (Ger.). *Beitr. Tabakforsch.*, *2*, 271-281

National Library of Medicine (1982) *Toxicology Data Bank*, Bethesda, MD, National Library of Medicine Specialized Information Services, Toxicology Information Program

National Research Council (1981) *Health Effects of Exposure to Diesel Exhaust*, Washington DC, National Academy Press

Nebert, D.W., Levitt, R.C., Jensen, N.M., Lambert, G.S. & Felton, J.S. (1977) Birth defects and aplastic anemia: Differences in polycyclic hydrocarbon toxicity associated with the *Ah* locus. *Arch. Toxicol.*, *39*, 109-132

NIH/EPA Chemical Information System (1982) *Mass Spectral Search System*, Washington DC, CIS Project, Information Services Corporation

Nikonova, T.V. (1977) Transplacental action of benzo(*a*)pyrene and pyrene. *Bull. exp. Biol. Med.*, *84*, 1025-1027

Pelkonen, O. & Nebert, D.W. (1982) Metabolism of polycyclic aromatic hydrocarbons: Etiologic role in carcinogenesis. *Pharmacol. Rev.*, *34*, 189-222

Pilotti, A., Ancker, K., Arrhenius, E. & Enzell, C. (1975) Effects of tobacco and tobacco smoke constituents on cell multiplication *in vitro*. *Toxicology*, *5*, 49-62

Pouchert, C.J., ed. (1970) *The Aldrich Library of Infrared Spectra*, Milwaukee, WI, Aldrich Chemical Co., p. 439

Rathkamp, G., Tso, T.C. & Hoffmann, D. (1973) Chemical analysis on tobacco smoke. XX. Smoke analysis of cigarettes made from Bright tobaccos differing in variety and stalk position. *Beitr. Tabakforsch.*, *7*, 179-189

Robinson, J.R., Felton, J.S., Levitt, R.C., Thorgeirsson, S.S. & Nebert, D.W. (1975) Relationship between 'aromatic hydrocarbon responsiveness' and the survival times in mice treated with various drugs and environmental compounds. *Mol. Pharmacol.*, *11*, 850-865

Salamone, M.F. (1981) *Toxicity of 41 carcinogens and noncarcinogenic analogs*. In: de Serres, F.J. & Ashby, J., eds, *Evaluation of Short-Term Tests for Carcinogens. Report of the*

International Collaborative Program. Progress in Mutation Research, Vol. 1, Amsterdam, Elsevier/North-Holland, pp. 682-685

Schmeltz, I., Brunnemann, K.D., Hoffmann, D. & Cornell, A. (1976) On the chemistry of cigar smoke. Comparisons between experimental little and large cigars. *Beitr. Tabakforsch.*, *8*, 367-377

de Serres, F.J. & Ashby, J. (1981) *Evaluation of Short-Term Tests for Carcinogens. Report of the International Collaborative Program. Progress in Mutation Research*, Vol. 1, Amsterdam, Elsevier/North-Holland

Severson, R.F., Arrendale, R.F., Chaplin, J.F. & Williamson, R.E. (1979) Use of pale-yellow tobacco to reduce smoke polynuclear aromatic hydrocarbons. *J. agric. Food Chem.*, *27*, 896-900

Sforzolini, G.S. & Mariani, A. (1961) Occurrence of 3:4-benzopyrene and other polycyclic hydrocarbons in the smoke of Italian cigarettes (Ital.). *Ric. Sci.*, *1*, 98-117 [*Chem. Abstr.*, *55*, 13178b]

Shendrikova, I.A. & Aleksandrov, V.A. (1974) Comparative characteristics of penetration of polycyclic hydrocarbons through the placenta into the fetus in rats (Russ.). *Biull. eksp. Biol. Med.*, *77*, 77-79

Shendrikova, I.A., Ivanov-Golitsyn, M.N. & Likhachev, A.Ya. (1974) The transplacental penetration of benzo(a)pyrene in mice (Russ.). *Vopr. Onkol.*, *20*, 53-56

Shum, S., Jensen, N.M. & Nebert, D.W. (1979) The murine *Ah* locus: *In utero* toxicity and teratogenesis with genetic differences in benzo[*a*]pyrene metabolism. *Teratology*, *20*, 365-376

Strömberg, L.E. & Widmark, G. (1970) Qualitative determination of polyaromatic hydrocarbons in the air near gas-works retorts. *J. Chromatogr.*, *47*, 27-44

Takahashi, G. (1974) Distribution and metabolism of benzo(*a*)pyrene in fetal mouse. *Bull. Chest Dis. Res. Inst. Kyoto Univ.*, *7*, 155-160

Thorsteinsson, T. (1969) Polycyclic hydrocarbons in commercially- and home-smoked food in Iceland. *Cancer*, *23*, 455-457

Topping, D.C., Pal, B.C., Martin, D.H., Nelson, F.R. & Nettesheim, P. (1978) Pathologic changes induced in respiratory tract mucosa by polycyclic hydrocarbons of differing carcinogenic activity. *Am. J. Pathol.*, *93*, 311-324

Trakhtman, N.N. & Manita, M.D. (1966) Effect of chlorination of water on pollution by 3,4-benzpyrene. *Hyg. Sanit.*, *31*, 316-320

Urso, P. & Gengozian, N. (1980) Depressed humoral immunity and increased tumor incidence in mice following *in utero* exposure to benzo[*a*]pyrene. *J. Toxicol. environ. Health*, *6*, 569-576

US Department of Health & Human Services (1982) *The Health Consequences of Smoking. Cancer. A Report of the Surgeon General* (*DHHS (PHS) 82-50179*), Washington DC, US Government Printing Office, p. 240

Weil-Malherbe, H. (1946) The solubilization of polycyclic aromatic hydrocarbons by purines. *Biochem. J.*, *40*, 351-363

Woidich, H., Pfannhauser, W., Blaicher, G. & Tiefenbacher, K. (1976) Analysis of polycyclic aromatic hydrocarbons in drinking and industrial water (Ger.). *Lebensmittelchem. gerichtl. Chem.*, *30*, 141-145

Wolfe, J.M. & Bryan, W.R. (1939) Effects induced in pregnant rats by injection of chemically pure carcinogenic agents. *Am. J. Cancer*, *36*, 359-368

Wynder, E.L. & Hoffmann, D. (1961) Present status of laboratory studies on tobacco carcinogenesis. *Acta pathol. microbiol. scand.*, *52*, 119-132

Wynder, E.L. & Hoffmann, D. (1963) Experimental contribution to carcinogenicity of tobacco smoke (Ger.). *Dtsch. med. Wochenschr.*, *88*, 623-628

BENZO[*e*]PYRENE

This compound was considered by a previous working group, in December 1972 (IARC, 1973). Data that have become available since that time have been incorporated in the present monograph and taken into consideration in the evaluation.

1. Chemical and Physical Data

1.1 Synonyms and trade names

Chem. Abstr. Services Reg. No.: 192-97-2

Chem. Abstr. Name: Benzo(e)pyrene

IUPAC Systematic Name: Benzo[*e*]pyrene

Synonyms: 1,2-Benzopyrene; 4,5-benzopyrene; 1,2-benzpyrene; 4,5-benzpyrene; B(e)P

1.2 Structural and molecular formulae and molecular weight

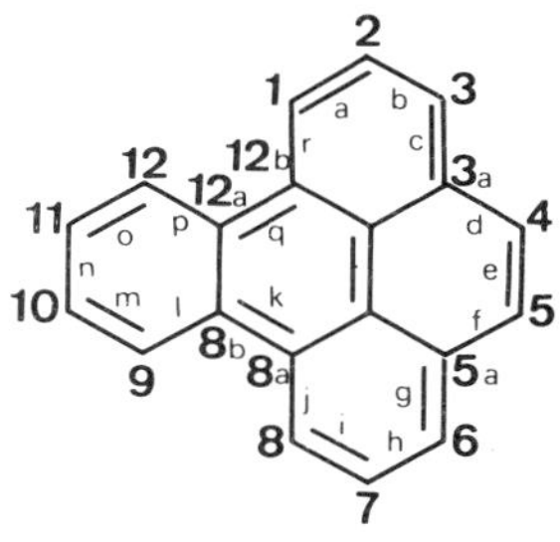

$C_{20}H_{12}$

Mol. wt: 252.3

1.3 Chemical and physical properties of the pure substance

(*a*) *Description*: Prisms or plates (recrystallized from benzene) (National Library of Medicine, 1982) (*b*) *Boiling-point*: 310-312°C (10 mm Hg) (Rappoport, 1967)

(*c*) *Melting-point*: 178.7°C (Karcher *et al.*, 1983)

(*d*) *Spectroscopy data*: λ_{max} 222, 236, 256, 266, 277, 288, 302, 316, 327, 330 nm (in cyclohexane) (Karcher *et al.*, 1983). Infrared spectra have been reported (API Research Project 44, 1960). Mass and nuclear magnetic resonance spectra have been tabulated (NIH/EPA Chemical Information System, 1982; Karcher *et al.*, 1983).

(*e*) *Solubility*: Very slightly soluble in water; soluble in acetone (Wynder & Hoffmann, 1959)

(*f*) *Stability*: No data were available.

(*g*) *Reactivity*: Reacts with NO and NO_2 to form nitro derivatives (Butler & Crossley, 1981)

2. Production, Use, Occurrence and Analysis

2.1 Production and use

There is no commercial production or known use of this compound. A reference material of certified high purity is available (Karcher *et al.*, 1980; Community Bureau of Reference, 1982).

2.2 Occurrence and analysis

Data on occurrence and methods of analysis are summarized in the 'General Remarks on the Substances Considered', p. 35.

Benzo[e]pyrene occurs ubiquitously in products of incomplete combustion; it is also found in fossil fuels. It has been identified in mainstream cigarette smoke (0.1 mg/kg cigarette smoke condensate) (Hoffmann & Wynder, 1960), (0.5 μg/100 g burnt material) (Masuda & Kuratsune, 1972), (1.3 μg/100 cigarettes) (Lee *et al.*, 1976), (2.5 μg/cigarette) (Grimmer *et al.*, 1977a), (1.9 μg/1000 cigarettes) (Kiryu & Kuratsune, 1966); sidestream cigarette smoke (13.5 μg/cigarette) (Grimmer *et al.*, 1977a); smoke-filled rooms (18 ng/m^3) (Grimmer *et al.*, 1977b); the air of restaurants (3.0-23.4 ng/m^3) (Just *et al.*, 1972); mainstream smoke of marijuana cigarettes (1.8 μg/100 cigarettes) (Lee *et al.*, 1976); the atmosphere of cities (1-37 ng/m^3) (Hoffmann & Wynder,

1976); emissions from the burning of various types of coal (0.02-10.3 mg/kg) (Brockhaus & Tomingas, 1976); crude oils (1.2-28.9 mg/kg oil) (Grimmer *et al.*, 1983); fresh (0.030-0.402 mg/kg) (Grimmer *et al.*, 1981b) and used motor oils (0.23-48.9 mg/kg) (Grimmer *et al.*, 1981c); gasolines (0.18-1.82 mg/kg) (Müller & Meyer, 1974); charcoal-broiled steaks (6 μg/kg) (Lijinsky & Shubik, 1964); smoked and cooked food (up to 74 μg/kg) (Thorsteinsson, 1969); vegetables (up to 67.2 μg/kg) (Hettche, 1971); vegetable oils (0.0-32.7 μg/kg) (Grimmer & Hildebrandt, 1967); tap-water (0.2 ng/l) (Olufsen, 1980); surface water (3.4-30.8 ng/l) (Grimmer *et al.*, 1981a); waste water (0.323-2.928 μg/l) (Grimmer *et al.*, 1981a); freeze-dried sewage sludge (530-12400 μg/kg) (Grimmer *et al.*, 1980); dried sediments from lakes (2-2010 μg/kg) (Grimmer & Böhnke, 1975, 1977).

3. Biological Data Relevant to the Evaluation of Carcinogenic Risk to Humans

3.1 Carcinogenicity studies in animals[1]

(*a*) *Skin application*

Mouse: A group of 20 female Swiss mice [age unspecified] received applications of a 0.1% solution of benzo[*e*]pyrene [purity unspecified] in acetone thrice weekly by skin painting for life. Of five animals that survived 13 months after the start of treatment, two had skin papillomas and three had carcinomas. There was no survivor after 14 months. In the same study, similar or lower concentrations of benzo[*a*]pyrene and dibenz[*a,h*]anthracene given under the same experimental conditions produced more tumours: 80% papillomas and 20% carcinomas at six months with 0.1% dibenz[*a,h*]anthracene and 95% papillomas and 95% carcinomas at 11 months with 0.01% benzo[*a*]pyrene (Wynder & Hoffmann, 1959).

A single application of 1.0 mg benzo[*e*]pyrene in 0.1 ml acetone to the skin of 20 female ICR/Ha Swiss mice, eight weeks old, induced no tumour in an experiment lasting 64 weeks. In a similar group of mice, the same treatment was followed by repeated paintings with 25 μg croton resin in 0.1 ml acetone, 2/20 animals developed a papilloma (Van Duuren *et al.*, 1968).

A group of 20 female CD-1 mice, eight weeks of age, received a single topical application of 10 μmol benzo[*e*]pyrene (purified by preparative thin-layer chromatography) in acetone, followed

---

[1] The Working Group was aware of a study in progress by skin application in mice and of a recently completed study by s.c. injection in mice (IARC, 1983).

one week later by twice-weekly applications of 5 μmol 12-*O*-tetradecanoylphorbol-13-acetate (TPA) for 34 weeks. A control group of 30 mice received TPA only (10 μmol). Tumour incidence was 85% in the treated group, with 5.77 papillomas/mouse, whereas a 3% tumour incidence was found in control mice; this difference was statistically significant (Scribner, 1973). [The Working Group noted that the dose of TPA was probably 5 μg (Scribner & Süss, 1978).]

Groups of 30 female Charles River CD-1 mice, seven to nine weeks of age, received 100 μg benzo[*e*]pyrene (purity, >99%) in 0.2 ml acetone by skin application twice weekly for 30 weeks. At 30 weeks, 68% of the mice had papillomas (2.1 papillomas/mouse); and at 40 weeks, 24% of mice had carcinomas. In a group of 30 female mice of the same strain twice-weekly administration for 30 weeks of 10 μg TPA by skin application resulted in papillomas in 14% of mice (0.2 papillomas/mouse); at 40 weeks, no carcinoma was observed. In two other groups, 30 female mice of the same strain received a single topical application of 100 or 252 μg benzo[*e*]pyrene followed one week later by twice-weekly skin application of 10 μg TPA for 30 weeks. With the high dose of benzo[*e*]pyrene 19% of mice had papillomas (0.4 papillomas/mouse) at 30 weeks and no carcinoma at 40 weeks; no activity was detected with the low dose of benzo[*e*]pyrene. In another group of 30 mice of the same strain, benzo[*e*]pyrene (100 μg twice weekly for 30 weeks) was administered one week after a single topical application of 51 μg 7,12-dimethylbenz[*a*]anthracene (DMBA); at 30 weeks, 92% of mice had papillomas (4.5 papillomas/mouse), and at 40 weeks 45% of mice had carcinomas. No skin tumour was observed in 30 corresponding control mice treated with a single dose of 51 μg DMBA (Slaga *et al.*, 1979).

Groups of 30 female CD-1 mice, seven to eight weeks old, received single applications of 1.0, 2.5 or 6.0 μmol benzo[*e*]pyrene (purity, 99%) in 200 μl acetone containing 10% dimethyl sulphoxide, followed seven days later by twice-weekly applications of 16 nmol TPA in 200 μl acetone for 25 weeks. The percentages of mice with papillomas (and the numbers of papillomas/mouse) were 15 (0.15), 11 (0.11) and 14 (0.14) in the three benzo[*e*]pyrene-treated groups, respectively. In the control group, which received only TPA, 7% of mice had tumours with 0.07 papillomas/mouse (Buening *et al.*, 1980).

A group of 30 female Sencar mice, seven to nine weeks old, received a single topical application of 2 μmol benzo[*e*]pyrene (purity, >95%) in acetone, followed one week later by twice-weekly applications of 2 μg TPA for 14 weeks. A control group of 30 animals received TPA alone. At 15 weeks, the percentages of mice with skin tumours (and the numbers of papillomas/mouse) were 17 (0.2) in the treated group and 10 (0.1) in the control group (Slaga *et al.*, 1980).

(*b*) *Other experimental systems*

Intrapulmonary injection: Three groups of 30-35 female Osborne-Mendel *rats*, three months old, body weight 245 g, were injected directly into the lung after thoracotomy (Stanton *et al.*, 1972) [with 0.05 ml] of a mixture of beeswax + tricaprylin (1:1) containing 0.2, 1.0 or 5.0 mg benzo[*e*]pyrene (purity, 99.7%) [0.8, 4.2 or 20 mg/kg bw]. Three similar groups of 35 rats received 0.1, 3.0 or 1.0 mg benzo[*a*]pyrene (positive controls); and an untreated group of 35 rats and a group of 35 rats receiving only the vehicle were also available. The rats were observed until spontaneous death; mean survival times were 117, 111, 104, 111, 77, 54, 118 and 104 weeks for the 0.2, 1.0, 5.0 mg benzo[*e*]pyrene-treated, the 0.1, 0.3, 1.0 mg benzo[*a*]pyrene-treated, the untreated and the vehicle-control groups, respectively. Pulmonary squamous-cell carcinomas were found in one of the high-dose animals treated with benzo[*e*]pyrene, and a pulmonary sarcoma was seen in the mid-dose group. The incidences of pulmonary carcinomas in the positive-control groups were 4/35, 21/35 and 33/35 in the low-, mid- and high-dose groups, respectively; six low-dose rats and two mid-dose rats of the positive-control groups also developed a pulmonary fibrosarcoma. No lung tumour was observed in the vehicle or untreated controls (Deutsch-Wenzel *et al.*, 1983).

Perinatal exposure: Newborn Swiss-Webster BLU:Ha (ICR) *mice* of both sexes were given i.p. injections on day 1, 8 and 15 of life, of 0.4, 0.8 and 1.6 μmol benzo[*e*]pyrene (purity, 99%) (total dose, 2.8 μmol) in dimethyl sulphoxide, respectively. Other newborn mice of both sexes were given i.p. injections of 0.8, 1.6 and 3.2 μmol benzo[*e*]pyrene on the same days (total dose, 5.6 μmol). At the end of the experiment (62-66 weeks), 30 male and 30 female mice from the low-dose group and 25 male and 23 female mice from the high-dose group were still alive. No tumour related to the treatment was seen in any organ, including the lung and liver, in any animals from either dose group (Buening *et al.*, 1980).

Pellet implantation: Groups of 20 female Fischer 344 *rats* [age unspecified] received implants of beeswax pellets containing either 1 mg benzo[*a*]pyrene, 0.5 mg benzo[*a*]pyrene, 1 mg benzo[*e*]pyrene [purity unspecified], 0.5 mg benzo[*a*]pyrene + 1 mg benzo[*e*]pyrene, or 1 mg benzo[*a*]pyrene + 1 mg benzo[*e*]pyrene in tracheas from isogenic donors transplanted subcutaneously in the retroscapular region (two tracheas/animal). All surviving animals were killed 28 months after the start of exposure. Benzo[*e*]pyrene did not induce tumours in tracheal explants, while 1 mg benzo[*a*]pyrene induced carcinomas in 65% of the grafts. Benzo[*e*]pyrene appeared to reduce the incidence of carcinomas from 65% (benzo[*a*]pyrene alone) to 40% (benzo[*a*]pyrene plus benzo[*e*]pyrene). However, the incidence of sarcoma in tracheal and peritracheal explants was enhanced two- to three-fold by benzo[*e*]pyrene given with benz[*a*]pyrene compared with benzo[*a*]pyrene alone (Topping *et al.*, 1981).

(*c*) *Effects of combinations*

Benzo[*e*]pyrene, when administered to mice by skin application together with 7,12-dimethylbenz[*a*]anthracene (DMBA) or benzo[*a*]pyrene, resulted in fewer skin tumours than with DMBA alone (Slaga *et al.*, 1979) and in more skin tumours than with benzo[*a*]pyrene alone (Van Duuren *et al.*, 1973; Van Duuren & Goldschmidt, 1976; Slaga *et al.*, 1979).

3.2 Other relevant biological data[1]

(*a*) *Experimental systems*

No data were available to the Working Group on toxic effects or on effects on reproduction and prenatal toxicity.

Metabolism and activation

The 4,5-dihydrodiol is the major metabolite of benzo[*e*]pyrene, but the 9,10-dihydrodiol has also been detected as a minor product following incubation of benzo[e]pyrene with rat-liver preparations (MacLeod *et al.*, 1980). Glucuronic acid conjugates of 3-hydroxy-benzo[*e*]pyrene and of the 4,5-dihydrodiol have been reported in hamster embryo cells (MacLeod *et al.*, 1979, 1982). The formation of the 4,5,9,10-tetraol and the 9,10,11,12-tetraol as well as unidentified

---

[1] See also 'General Remarks on the Substances Considered', p. 53.

phenols from the precursor 9,10-dihydrodiol have also been reported using hepatic microsomal preparations from various species, including humans (Wood *et al.*, 1979; Thakker *et al.*, 1981).

Benzo[*e*]pyrene-9,10-diol-11,12-epoxide was weakly mutagenic in bacterial and mammalian cells (Wood *et al.*, 1980).

Benzo[*e*]pyrene-9,10-dihydrodiol is not significantly active as a tumour initiator on mouse skin (Buening *et al.*, 1980; Slaga *et al.*, 1980) and does not induce pulmonary tumours following its i.p. injection into newborn mice (Buening *et al.*, 1980). It did induce hepatic tumours in newborn male mice (Buening *et al.*, 1980). Also in newborn male mice, the 9,10-diol-11,12-epoxide diastereomer, in which the benzylic 9-hydroxyl and the epoxide are in the *trans* position, induced a significant incidence of hepatic tumours; but only the *cis* diastereoisomer increased the incidence of pulmonary tumours (Chang *et al.*, 1981). The *cis* diastereoisomer has not been tested for tumour-initiating activity on mouse skin, but the *trans* diastereoisomer is not significantly active in that system (Slaga *et al.*, 1980).

Nucleic acid binding has been observed in hamster embryo cells treated with benzo[*e*]pyrene (MacLeod *et al.*, 1979).

Mutagenicity and other short-term tests

Results from short-term tests are given in Table 1.

Table 1. Results from short-term tests: Benzo[e]pyrene[a]

Test	Organism/assay[b]	Exogenous metabolic system[b]	Reported result	Comments	References
PROKARYOTES					
DNA damage	*Escherichia coli* (polA$^+$/polA$^-$)	UI-R-PMS	Negative	Tested at up to 250 μg/ml [The Working Group noted that UI-PMS was used]	Rosenkranz & Poirier (1979)
Mutation	*Salmonella typhimurium* (his$^-$/his$^+$)	Aro-R-PMS	Positive	At 10 μg/plate in strain TA100	McCann *et al.* (1975); LaVoie *et al.* (1979)
				At 10 μg/plate in strain TA98	Simmon (1979a); Hermann (1981)
	Salmonella typhimurium (8AG^S/8AG^R)	Aro-R-PMS	Positive	At 90 nmol/ml in strain TM677	Kaden *et al.* (1979)
FUNGI					
Mutation	*Saccharomyces cerevisiae* D3 (mitotic recombination)	Aro-R-PMS	Negative	Tested at up to 5% w/v [*sic*]	Simmon (1979b)
MAMMALIAN CELLS *IN VITRO*					
DNA damage	Human foreskin epithelial cells (unscheduled DNA synthesis)	--	Inconclusive	Tested at up to 400 μg/ml	Lake *et al.* (1978)
	Primary rat hepatocytes (unscheduled DNA synthesis)	--	Negative	At up to 1 μmol/ml	Probst *et al.* (1981); Tong *et al.* (1981a)
	HeLa cells (unscheduled DNA synthesis)	3MC-R-PMS	Positive	At 1 nmol/ml	Martin *et al.* (1978)
Mutation	Chinese hamster V79 cells (OUA^S/OUA^R)	SHE feeder layer	Negative	Tested at 1 μg/ml	Hubermann (1978)

Test	Organism/assay[b]	Exogenous metabolic system[b]	Reported result	Comments	References
	Rat liver epithelial ARL 18 cells ($6TG^S/6TG^R$)	--	Negative	Tested at up to 100 nmol/ml	Tong *et al.* (1981a)
	Mouse C3H 10T1/2 cells (OUA^S/OUA^R)	--	Negative	At up to 15 nmol/ml	Gehly *et al.* (1982)
Chromosome effects	Rat liver epithelial ARL 18 cells (sister chromatid exchange)	--	Negative	At up to 1 μmol/ml	Tong *et al.* (1981b)
	Mouse C3H 10T1/2 cells (aberrations; sister chromatid exchange)	--	Negative	At up to 10 nmol/ml	Gehly *et al.* (1982)
Cell transformation	Mouse C3H 10T1/2 cells (morphological)	--	Negative	At up to 15 nmol/ml	Gehly *et al.* (1982)
	Syrian hamster embryo cells (morphological)	--	Negative	Tested at up to 100 μg/ml	Pienta *et al.* (1977)
		--	Positive	At 10 μg/ml [The Working Group noted that only one transformed focus was induced.]	DiPaolo *et al.* (1969)
MAMMALIAN CELLS *IN VIVO*					
Chromosome effects	Chinese hamster bone-marrow cells: sister chromatid exchange	--	Positive	At 2 x 450 mg/kg bw i.p.	Roszinsky-Köcher *et al.* (1979)
	aberrations	--	Negative	Treated i.p. with 2 x 450 mg/kg bw	

[a] This table comprises selected assays and references and is not intended to be a complete review of the literature.
[b] For an explanation of the abbreviations, see the Appendix, p. 452.

(*b*) *Humans*

No data were available to the Working Group.

3.3 Case reports and epidemiological studies of carcinogenicity to humans

No data were available to the Working Group.

4. Summary of Data Reported and Evaluation[1]

4.1 Experimental data

Benzo[*e*]pyrene was tested for carcinogenicity in mice by skin application in two studies, and skin tumours were observed in one experiment. Benzo[*e*]pyrene was also tested in various studies in mice in the mouse-skin initiation-promotion assay; promoting activity was detected after initiation with 9,10-dimethylbenz[*a*]anthracene in one study; the initiating activity was not clearly evident in other studies in which 12-*O*-tetradecanoylphorbol-13-acetate was used as a promoting agent.

Multiple intraperitoneal administrations of benzo[*e*]pyrene to newborn mice did not result in a significant increase of tumours in two studies.

In rats, pulmonary injection of benzo[*e*]pyrene at various dose levels resulted in one squamous-cell carcinoma of the lung at the highest dose level and one pulmonary sarcoma at the mid-dose level.

No data on the teratogenicity of this chemical were available.

Benzo[*e*]pyrene was mutagenic to *Salmonella typhimurium* in the presence of an exogenous metabolic system. It did not induce mitotic recombination in yeast. It did not induce mutations or sister chromatid exchange in cultured mammalian cells and was negative in assays for morphological transformation. It induced unscheduled DNA synthesis in HeLa cells in the presence of an exogenous metabolic system, but not in primary rat hepatocytes. In the one available report, it did not induce chromosomal aberrations *in vitro*. In the one available in-vivo study, it induced sister chromatid exchange, but not chromosomal aberrations in hamster bone marrow.

There is *limited evidence* that benzo[*e*]pyrene is active in short-term tests.

4.2 Human data[2]

Benzo[*e*]pyrene is present as a major component of the total content of polynuclear aromatic compounds in the environment. Human exposure to benzo[*e*]pyrene occurs primarily through the smoking of tobacco, inhalation of polluted air and by ingestion of food and water contaminated by combustion effluents (for details, see 'General Remarks on the Substances Considered', p. 35).

4.3 Evaluation

The available data are inadequate to permit an evaluation of the carcinogenicity of benzo[*e*]pyrene to experimental animals.

---

[1] For definitions of italicized terms, see Preamble, p. 19.

[2] Studies on occupational exposure to polynuclear aromatic compounds will be considered in future *IARC Monographs*.

5. References

API Research Project 44 (1960) *Selected Infrared Spectral Data*, Vol. VI, No. 2249, Washington DC, American Petroleum Institute

Brockhaus, A. & Tomingas, R. (1976) Emission of polycyclic hydrocarbons during burning processes in small heating installations and their concentration in the atmosphere (Ger.). *Staub-Reinhalt. Luft*, *36*, 96-101

Buening, M.K., Levin, W., Wood, A.W., Chang, R.L., Lehr, R.E., Taylor, C.W., Yagi, H., Jerina, D.M. & Conney, A.H. (1980) Tumorigenic activity of benzo(*e*)pyrene derivatives on mouse skin and in newborn mice. *Cancer Res.*, *40*, 203-206

Butler, J.D. & Crossley, P. (1981) Reactivity of polycyclic aromatic hydrocarbons adsorbed on soot particles. *Atmos. Environ.*, *15*, 91-94

Chang, R.L., Levin, W., Wood, A.W., Lehr, R.E., Kumar, S., Yagi, H., Jerina, D.M. & Conney, A.H. (1981) Tumorigenicity of the diastereomeric bay-region benzo(*e*)pyrene 9,10-diol-11,12-epoxides in newborn mice. *Cancer Res.*, *41*, 915-918

Community Bureau of Reference (1982) *Polycyclic Aromatic Hydrocarbon Reference Materials of Certified Purity* (*Information Handout, No. 22*), Brussels, Commission of the European Communities

Deutsch-Wenzel, R.P., Brune, H., Grimmer, G., Dettbarn, G. & Misfeld, J. (1983) Experimental studies in rat lungs on the carcinogenicity and dose-response relationships of eight frequently occurring environmental polycyclic aromatic hydrocarbons. *J. natl Cancer Inst.*, *71*, 539-544

DiPaolo, J.A., Donovan, J.P. & Nelson, R. (1969) Quantitative studies of *in vitro* transformation by chemical carcinogens. *J. natl Cancer Inst.*, *42*, 867-876

Gehly, E.B., Landolph, J.R., Heidelberger, C., Nagasawa, H. & Little, J.B. (1982) Induction of cytotoxicity, mutation, cytogenetic changes and neoplastic transformation by benzo(*a*)pyrene and derivatives in C3H/10T 1/2 clone 8 mouse fibroblasts. *Cancer Res.*, *42*, 1866-1875

Grimmer, G. & Böhnke, H. (1975) Profile analysis of polycyclic aromatic hydrocarbons and metal content in sediment layers of a lake. *Cancer Lett.*, *1*, 75-84

Grimmer, G. & Böhnke, H. (1977) Investigation on drilling cores of sediments of Lake Constance. I. Profiles of the polycyclic aromatic hydrocarbons. *Z. Naturforsch.*, *32c*, 703-711

Grimmer, G. & Hildebrandt, A. (1967) Content of polycyclic hydrocarbons in crude vegetable oils. *Chem. Ind.*, 25 November, 2000-2002

Grimmer, G., Böhnke, H. & Harke, H.-P. (1977a) Passive smoking: Measuring of concentrations of polycyclic aromatic hydrocarbons in rooms after machine smoking of cigarettes (Ger.). *Int. Arch. occup. environ. Health*, *40*, 83-92

Grimmer, G., Böhnke, H. & Harke, H.-P. (1977b) Passive smoking: Intake of polycyclic aromatic hydrocarbons by breathing of cigarette smoke containing air (Ger.). *Int. Arch. occup. environ. Health*, *40*, 93-99

Grimmer, G., Hilge, G. & Niemitz, W. (1980) Comparison of the profile of polycyclic aromatic hydrocarbons in sewage sludge samples from 25 sewage treatment works (Ger.). *Wasser*, *54*, 255-272

Grimmer, G., Schneider, D. & Dettbarn, G. (1981a) The load of different rivers in the Federal Republic of Germany by PAH (PAH-profiles of surface water) (Ger.). *Wasser*, *56*, 131-144

Grimmer, G., Jacob, J. & Naujack, K.-W. (1981b) Profile of the polycyclic aromatic compounds from lubricating oils. Inventory by GCGC/MS - PAH in environmental materials, Part 1. *Fresenius Z. anal. Chem.*, *306*, 347-355

Grimmer, G., Jacob, J., Naujack, K.-W. & Dettbarn, G. (1981c) Profile of the polycyclic aromatic compounds from used engine oil - Inventory by GCGC/MS - PAH in environmental materials, Part 2. *Fresenius Z. anal. Chem.*, *309*, 13-19

Grimmer, G., Jacob, J. & Naujack, K.-W. (1983) Profile of the polycyclic aromatic compounds from crude oils. Inventory by GCGC/MS - PAH in environmental materials, Part 3. *Fresenius Z. anal. Chem.*, *314*, 29-36

Hermann, M. (1981) Synergistic effects of individual polycyclic aromatic hydrocarbons on the mutagenicity of their mixtures. *Mutat. Res.*, *90*, 399-409

Hettche, H.O. (1971) Plant waxes as collectors of polycyclic aromatics in the air of residential areas. *Staub-Reinhalt. Luft*, *31*, 34-41

Hoffmann, D. & Wynder, E.L. (1960) On the isolation and identification of polycyclic aromatic hydrocarbons. *Cancer*, *13*, 1062-1073

Hoffmann, D. & Wynder, E. (1976) *Environmental respiratory carcinogenesis*. In: Searle, C.E., ed., *Chemical Carcinogens* (*ACS Monograph 173*), Washington DC, American Chemical Society, p. 341

Huberman, E. (1978) *Cell transformation and mutability of different genetic loci in mammalian cells by metabolically activated carcinogenic polycyclic hydrocarbons*. In: Gelboin, H.V. & Ts'o, P.O.P., eds, *Polycyclic Hydrocarbons and Cancer*, Vol. 2, *Molecular and Cell Biology*, New York, Academic Press, pp. 161-174

IARC (1973) *IARC Monographs on the Evaluation of Carcinogenic Risk of Chemicals to Man*, Vol. 3, *Certain polycyclic aromatic hydrocarbons and heterocyclic compounds*, Lyon, pp. 137-158

IARC (1983) *Information Bulletin on the Survey of Chemicals Being Tested for Carcinogenicity*, No. 10, Lyon, pp. 15, 24

Just, J., Borkowska, M. & Maziarka, S. (1972) Air pollution by tobacco smoke in Warsaw coffee houses (Pol.). *Rocz. Panstw. Zakł. Hyg.*, *23*, 129-135

Kaden, D.A., Hites, R.A. & Thilly, W.G. (1979) Mutagenicity of soot and associated polycyclic aromatic hydrocarbons to *Salmonella typhimurium*. *Cancer Res.*, *39*, 4152-4159

Karcher, W., Jacob, J. & Haemers, L. (1980) *The Certification of Eight Polycyclic Aromatic Hydrocarbon Materials (PAH) (BCR Reference Materials Nos. 46, 47, 48, 49, 50, 51, 52 and 53) (EUR 6967 EN)*, Luxembourg, Commission of the European Communities

Karcher, W., Fordham, R., Dubois, J. & Gloude, P. (1983) *Spectral Atlas of Polycyclic Aromatic Compounds*, Dordrecht, The Netherlands, D. Reidel (in press)

Kiryu, S. & Kuratsune, M. (1966) Polycyclic aromatic hydrocarbons in the cigarette tar produced by human smoking. *Gann*, *57*, 317-322

Lake, R.S., Kropko, M.L., Pezzutti, M.R., Shoemaker, R.H. & Igel, H.J. (1978) Chemical induction of unscheduled DNA synthesis in human skin epithelial cell cultures. *Cancer Res.*, *38*, 2091-2098

LaVoie, E., Bedenko, V., Hirota, N., Hecht., S.S. & Hoffmann, D. (1979) *A comparison of the mutagenicity, tumor-initiating activity and complete carcinogenicity of polynuclear aromatic hydrocarbons*. In: Jones, P.W. & Leber, P., eds, *Polynuclear Aromatic Hydrocarbons*, Ann Arbor, MI, Ann Arbor Science Publishers, Inc., pp. 705-721

Lee, M.L., Novotny, M. & Bartle, K.D. (1976) Gas chromatography/mass spectrometry and nuclear magnetic resonance spectrometric studies of carcinogenic polynuclear aromatic hydrocarbons in tobacco and marijuana smoke condensates. *Anal. Chem.*, *48*, 405-416

Lijinsky, W. & Shubik, P. (1964) Benzo(*a*)pyrene and other polynuclear hydrocarbons in charcoal-broiled meat. *Science*, *145*, 53-55

MacLeod, M.C., Cohen, G.M. & Selkirk, J.K. (1979) Metabolism and macromolecular binding of the carcinogen benzo[*a*]pyrene and its relatively inert isomer benzo[*e*]pyrene by hamster embryo cells. *Cancer Res.*, *39*, 3463-3470

MacLeod, M.C., Levin, W., Conney, A.H., Lehr, R.E., Mansfield, B.K., Jerina, D.M. & Selkirk, J.K. (1980) Metabolism of benzo(*e*)pyrene by rat liver microsomal enzymes. *Carcinogenesis*, *1*, 165-173

MacLeod, M.C., Mansfield, B.K. & Selkirk, J.K. (1982) Time course of metabolism of benzo[*e*]pyrene by hamster embryo cells and the effect of chemical modifiers. *Chem.-biol. Interactions*, *40*, 275-285

Martin, C.N., McDermid, A.C. & Garner, R.C. (1978) Testing of known carcinogens and noncarcinogens for their ability to induce unscheduled DNA synthesis in HeLa cells. *Cancer Res.*, *38*, 2621-2627

Masuda, Y. & Kuratsune, M. (1972) Comparison of the yield of polycyclic aromatic hydrocarbons in smoke from Japanese tobacco. *Jpn. J. Hyg.*, *27*, 339-341

McCann, J., Choi, E., Yamasaki, E. & Ames, B.N. (1975) Detection of carcinogens as mutagens in the *Salmonella*/microsome test: Assay of 300 chemicals. *Proc. natl Acad. Sci. USA*, *72*, 5135-5139

Müller, K. & Meyer, J.P. (1974) *Einfluss von Ottokraftstoffen auf die Emission von Polynuklearen aromatischen Kohlenwasserstoffen in Automobilabgasen im Europa-Test* (Effect of gasoline components on emission of polynuclear aromatic hydrocarbons in car exhaust in the Europa test) (*Forschungsbericht 4568*), Hamburg, Deutsche Gesellschaft für Mineralölwissenschaft und Kohlechemie e.V.

National Library of Medicine (1982) *Toxicology Data Bank*, Bethesda, MD, National Library of Medicine Specialized Information Services, Toxicology Information Program

NIH/EPA Chemical Information System (1982) *Mass Spectral Search System*, Washington DC, CIS Project, Information Services Corporation

Olufsen, B. (1980) *Polynuclear aromatic hydrocarbons in Norwegian drinking water resources.* In: Bjørseth, A. & Dennis, A.J., eds, *Polynuclear Aromatic Hydrocarbons: Chemistry and Biological Effects*, *4th Int. Symposium*, Columbus, OH, Battelle Press, pp. 333-343

Pienta, R.J., Poiley, J.A. & Lebherz III, W.B. (1977) Morphological transformation of early passage golden Syrian hamster embryo cells derived from cryopreserved primary cultures as a reliable *in vitro* bioassay for identifying diverse carcinogens. *Int. J. Cancer*, *19*, 642-655

Probst, G.S., McMahon, R.E., Hill, L.E., Thompson, C.Z., Epp, J.K. & Neal, S.B. (1981) Chemically-induced unscheduled DNA synthesis in primary rat hepatocyte cultures: A comparison with bacterial mutagenicity using 218 compounds. *Environ. Mutagenesis*, *3*, 11-32

Rappoport, Z., ed. (1967) *CRC Handbook of Tables for Organic Compound Identification*, 3rd ed., Boca Raton, FL, Chemical Rubber Co., p. 50

Rosenkranz, H.S. & Poirier, L.A. (1979) Evaluation of the mutagenicity and DNA-modifying activity of carcinogens and noncarcinogens in microbial systems. *J. natl Cancer Inst.*, *62*, 873-892

Roszinsky-Köcher, G., Basler, A. & Röhrborn, G. (1979) Mutagenicity of polycyclic hydrocarbons. V. Induction of sister-chromatid exchanges *in vivo*. *Mutat. Res.*, *66*, 65-67

Scribner, J.D. (1973) Brief communication: Tumor initiation by apparently non carcinogenic polycyclic aromatic hydrocarbons. *J. natl Cancer Inst.*, *50*, 1717-1719

Scribner, J.D. & Süss, R. (1978) Tumour initiation and promotion. *Int. Rev. exp. Pathol.*, *18*, 137-198

Simmon, V.F. (1979a) *In vitro* mutagenicity assays of chemical carcinogens and related compounds with *Salmonella typhimurium*. *J. natl Cancer Inst.*, *62*, 893-899

Simmon, V.F. (1979b) *In vitro* assays for recombinogenic activity of chemical carcinogens and related compounds with *Saccharomyces cerevisiae D3*. *J. natl Cancer Inst.*, *62*, 901-909

Slaga, T.J., Jecker, L., Bracken, W.M. & Weeks, C.E. (1979) The effects of weak or non-carcinogenic polycyclic hydrocarbons on 7,12-dimethylbenz[*a*]anthracene and benzo[*a*]pyrene skin tumor-initiation. *Cancer Lett.*, *7*, 51-59

Slaga, T.J., Gleason, G.L., Mills, G., Ewald, L., Fu, P.P., Lee, H.M. & Harvey, R.G. (1980) Comparison of the skin tumor-initiating activities of dihydrodiols and diol-epoxides of various polycyclic aromatic hydrocarbons. *Cancer Res.*, *40*, 1981-1984

Stanton, M.F., Miller, E., Wrench, C. & Blackwell, R. (1972) Experimental induction of epidermoid carcinoma in the lungs of rats by cigarette smoke condensate. *J. natl Cancer Inst.*, *49*, 867-877

Thakker, D.R., Levin, W., Buening, M., Yagi, H., Lehr, R.E., Wood, A.W., Conney, A.H. & Jerina, D.M. (1981) Species-specific enhancement by 7,8-benzoflavone of hepatic microsomal metabolism of benzo[*e*]pyrene 9,10-dihydrodiol to bay-region diol epoxides. *Cancer Res.*, *41*, 1389-1396

Thorsteinsson, T. (1969) Polycyclic hydrocarbons in commercially and home-smoked food in Iceland. *Cancer Lett.*, *23*, 455-457

Tong, C., Laspia, M.F., Telang, S. & Williams, G.M. (1981a) The use of adult rat liver cultures in the detection of the genotoxicity of various polycyclic aromatic hydrocarbons. *Environ. Mutagenesis*, *3*, 477-487

Tong, C., Brat, S.V. & Williams, G.M. (1981b) Sister-chromatid exchange induction by polycyclic aromatic hydrocarbons in an intact cell system of adult rat-liver epithelial cells. *Mutat. Res.*, *91*, 467-473

Topping, D.C., Martin, D.H. & Nettesheim, P. (1981) Determination of cocarcinogenic activity of benzo[*e*]pyrene for respiratory tract mucosa. *Cancer Lett.*, *11*, 315-321

Van Duuren, B.L. & Goldschmidt, B.M. (1976) Cocarcinogenic and tumor-promoting agents in tobacco carcinogenesis. *J. natl Cancer Inst.*, *56*, 1237-1242

Van Duuren, B.L., Sivak, A., Langseth, L., Goldschmidt, B.M. & Segal, A. (1968) Initiators and promoters in tobacco carcinogenesis. *Natl Cancer Inst. Monogr.*, *28*, 173-180

Van Duuren, B.L., Katz, C. & Goldschmidt, B.M. (1973) Brief communication: Cocarcinogenic agents in tobacco carcinogenesis. *J. natl Cancer Inst.*, *51*, 703-705

Wood, A.W., Levin, W., Thakker, D.R., Yagi, H., Chang, R.L., Ryan, D.E., Thomas, P.E., Dansette, P.M., Whittaker, N., Turujman, S., Lehr, R.E., Kumar, S., Jerina, D.M. & Conney, A.H. (1979) Biological activity of benzo[*e*]pyrene. An assessment based on mutagenic activities and metabolic profiles of the polycyclic hydrocarbon and its derivatives. *J. biol. Chem.*, *254*, 4408-4415

Wood, A.W., Chang, R.L., Huang, M.-T., Levin, W., Lehr, R.E., Kumar, S., Thakker, D.R., Yagi, H., Jerina, D.M. & Conney, A.H. (1980) Mutagenicity of benzo(*e*)pyrene and triphenylene tetrahydroepoxides and diol-epoxides in bacterial and mammalian cells. *Cancer Res.*, *40*, 1985-1989

Wynder, E.L. & Hoffmann, D. (1959) A study of tobacco carcinogenesis. VII. The role of higher polycyclic hydrocarbons. *Cancer*, *12*, 1079-1086

CARBAZOLE

1. Chemical and Physical Data

1.1 Synonyms and trade names

Chem. Abstr. Services Reg. No.: 86-74-8

Chem. Abstr. Name: 9H-Carbazole

IUPAC Systematic Name: Carbazole

Synonyms: 9-Azafluorene; dibenzopyrrole; dibenzo(b,d)pyrrole; diphenylenimine

1.2 Structural and molecular formulae and molecular weight

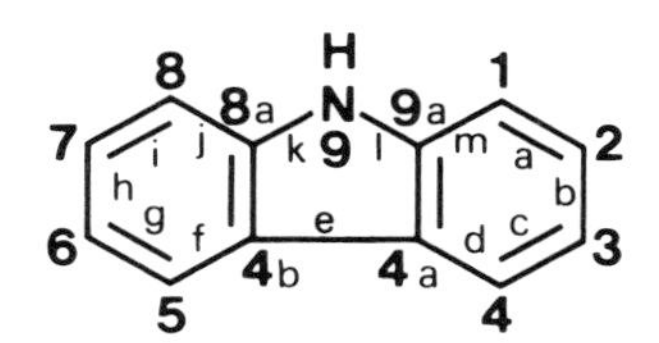

$C_{12}H_9N$ Mol. wt: 167.2

1.3 Chemical and physical properties of the pure substance

(*a*) *Description*: White crystals, plates or leaves (National Library of Medicine, 1982)

(*b*) *Boiling-point*: 355°C ((Weast, 1975)

(*c*) *Melting-point*: 247-248°C (Weast, 1975)

(*d*) *Spectroscopy data*: λ_{max} 233, 245, 256, 292, 323, 337 nm (in ethanol) (Friedel & Orchin, 1951). Mass spectra and infrared spectra have been tabulated (NIH/EPA Chemical Information System, 1982).

(*e*) *Solubility*: Practically insoluble in water; slightly soluble in acetic acid, chlorinated hydrocarbons and petroleum ether; one gram dissolves in the following volumes of organic solvents: 9 ml acetone, 2 ml acetone at 50°C, 120 ml benzene, 35 ml diethyl ether, 135 ml absolute ethanol, 6 ml pyridine and 3 ml quinoline (Windholz, 1976)

(*f*) *Volatility*: Vapour pressure, 400 mm Hg at 323°C (National Library of Medicine, 1982)

(*g*) *Stability*: No data were available.

(*h*) *Reactivity*: Extremely weak base. Reaction with potassium hydroxide yields the *N*-potassium salt (Windholz, 1976). Reacts with nitrogen oxides (Tokiwa *et al.*, 1981)

2. Production, Use, Occurrence and Analysis

2.1 Production and use

There is no known commercial production of this compound. However, it has been reported to be an important dye intermediate; to be used in making photographic plates sensitive to ultraviolet light; and as a reagent for detecting lignin, carbohydrates and formaldehyde (Windholz, 1976).

2.2 Occurrence and analysis

Data on occurrence and methods of analysis are summarized in the 'General Remarks on the Substances Considered', p. 35.

Carbazole occurs in the products of incomplete combustion of nitrogen-containing organic matter. It has been identified in mainstream cigarette smoke (100 μg/100 cigarettes) (Hoffmann *et al.*, 1968), crude oils (Grimmer *et al.*, 1983) and coal-tar (Lang & Eigen, 1967).

3. Biological Data Relevant to the Evaluation of Carcinogenic Risk to Humans

3.1 Carcinogenic studies in animals

(*a*) *Oral administration*

Mouse: Groups of 50 male and 50 female B6C3 F_1 mice, six weeks of age, were fed a pellet diet containing technical-grade carbazole (purity, 96%) at concentrations of 0.6, 0.3 or 0.15% or none (control group). The treatment was continued for 96 weeks; the animals were then fed a basal diet until killed in week 104. Neoplastic lesions were found in the liver and in the forestomach. The lesions in liver were classified as neoplastic nodules and hepatocellular carcinomas. The incidences of both types of lesion in livers of all the groups fed carbazole were significantly ($p < 0.05$) greater than that in the control group. The incidences of neoplastic nodules and hepatocellular carcinomas were, respectively: in the high-dose group, females: 16/46 (34.8%) and 30/46 (65.2%) with 3 (10%) lung metastases; males: 10/48 (20.9%) and 37/48 (77.1%) with 11 (29.7%) lung metastases; mid-dose group, females: 21/43 (48.8%) and 24/43 (55.8%) with 3 (12.5%) lung metastases; males: 22/42 (52.4%) and 20/42 (47.4%) with 7 (35%) lung metastases; low-dose group, females: 13/49 (26.5%) and 35/49 (71.4%) with 5 (14.3%) lung metastases, males: 30/42 (71.4%) and 12/42 (28.6%) with 2 (16.7%) lung metastases. In control animals, with a mean survival time of about 100 weeks, the incidences of neoplastic nodules and hepatocellular carcinomas were, respectively, 4.4% (2/45) and 4.4% (2/45) in females and 28.2% (13/46) and 19.6% (9/46) in males. The numbers of papillomas in the forestomach in groups of mice given 0.6% carbazole were 4/46 in females ($p < 0.05$) and 4/48 in males ($p < 0.05$); in mice given 0.3%, 7/43 in females ($p < 0.01$) and 1/42 in males; in mice given 0.15%, 5/49 in females ($p < 0.05$) and 0/42 in males, whereas no such tumour was observed in the respective control groups (female 0/45, male 0/46). Squamous-cell carcinoma incidence was increased significantly ($p < 0.01$) in males fed 0.6% carbazole (7/48, 14.6%). No squamous-cell carcinoma was observed in the forestomachs of male or female controls (Tsuda *et al*., 1982).

(*b*) *Skin application*

Mouse: Two skin application experiments in mice were published in 1924 and 1927, with inconclusive results (Kennaway, 1924; Maisin *et al*., 1927).

A solution of 0.5% carbazole in benzene was applied 120 times to the skin of 50 mice. Epilation of the treated area was the only reaction observed after 276 days (Schürch & Winterstein, 1935).

(*c*) *Subcutaneous and/or intramuscular administration*

Mouse: A group of ten male A strain mice, three to four months old, received six s.c. injections, in the left flank, of 10 mg crystallized carbazole moistened with glycerol. All ten mice were still alive after one year and four after 19 months. No tumour was reported at the injection site (Shear & Leiter, 1941). [The Working Group noted the small number of animals used.]

3.2 Other relevant biological data[1]

(*a*) *Experimental systems*

Toxic effects

Widely disparate LD_{50}s (oral) have been reported in rats: > 500 mg/kg bw (determined on two rats) (Dieke *et al*., 1947) and > 5000 mg/kg bw (Eagle & Carlson, 1950).

---

[1] See also 'General Remarks on the Substances Considered', p. 53.

The growth rate of mouse ascites sarcoma cells in culture was slightly inhibited (8%) when carbazole was added at a concentration of 1 μmol/mol dissolved in dimethyl sulphoxide (Pilotti *et al.*, 1975).

No data were available to the Working Group on effects on reproduction and prenatal toxicity.

Metabolism and activation

3-Hydroxycarbazole has been reported to be a urinary metabolite of carbazole in rats and rabbits (Johns & Wright, 1964).

Mutagenicity and other short-term tests

Results of short-term tests are given in Table 1.

Table 1. Results from short-term tests: Carbazole

Test	Organism/assay[a]	Exogenous metabolic system[a]	Reported result	Comments	References
PROKARYOTES					
Mutation	*Salmonella typhimurium* (his⁻/his⁺)	Aro-R-PMS	Negative	Tested at up to 2500 μg/plate in strains TA1535, TA1538, TA98 and TA100	Anderson & Styles (1978)
		Aro-R-PMS	Negative	Tested at up to 0.50 mg/plate in strain TA98	Ho *et al.* (1981)
		Aro-R-PMS	Negative	Tested at up to 250 μg/plate in strains TA98 and TA100	LaVoie *et al.*, 1982
	Salmonella typhimurium ($8AG^S/8AG^R$)	Aro-R-PMS, PB-R-PMS	Negative	Tested at up to 3 μmol/ml in strain TM677	Kaden *et al.* (1979)

[a] For an explanation of the abbreviations, see the Appendix, p. 452.

(*b*) *Humans*

No data were available to the Working Group.

3.3 Case reports and epidemiological studies of carcinogenicity to humans

No data were available to the Working Group.

4. Summary of Data Reported and Evaluation[1]

4.1 Experimental data

Carbazole was tested for carcinogenicity in mice by administration in the diet, by skin application and by subcutaneous injection. In the study by oral administration, a dose-dependent increase in the incidence of liver neoplastic nodules and hepatocellular carcinomas was observed. Papillomas and carcinomas of the forestomach occurred in animals receiving the high-dose level. The other studies in mice were considered inadequate for evaluation.

No data on the teratogenicity of this compound were available.

Carbazole was not mutagenic to *Salmonella typhimurium*.

There is *inadequate evidence* that carbazole is active in short-term tests.

4.2 Human data[2]

Carbazole is present as a major component of the total content of polynuclear aromatic compounds in the environment, arising primarily from the combustion of tobacco and coal (for details, see General Remarks on the Substances Considered, p. 35).

4.3 Evaluation

There is *limited evidence* that carbazole is carcinogenic to experimental animals.

---

[1] For definitions of the italicized terms, see Preamble, p. 19.

[2] Studies on occupational exposure to polynuclear aromatic compounds will be considered in future *IARC Monographs*.

5. References

Anderson, D. & Styles, J.A. (1978) The bacterial mutation test. *Br. J. Cancer*, *37*, 924-930

Dieke, S.H., Allen, G.S. & Richter, C.P. (1947) The acute toxicity of thioureas and related compounds to wild and domestic Norway rats. *J. Pharmacol. exp. Ther.*, *90*, 260-270

Eagle, E. & Carlson, A.J. (1950) Toxicity, antipyretic and analgesic studies on 39 compounds including aspirin, phenacetin and 27 derivatives of carbazole and tetrahydrocarbazole. *J. Pharmacol. exp. Ther.*, *99*, 450-457

Friedel, R.A. & Orchin, M., eds (1951) *Ultraviolet Spectra of Aromatic Compounds*, New York, John Wiley & Sons, No. 338

Grimmer, G., Jacob, J. & Naujack, K.-W. (1983) Profile of the polycyclic aromatic compounds from crude oils. Inventory by GCGC/MS - PAH in environmental materials. Part 3. *Fresenius Z. anal. Chem.*, *314*, 29-36

Ho, C.-H., Clark, B.R., Guerin, M.R., Barkenbus, B.D., Rao, T.K. & Epler, J.L. (1981) Analytical and biological analyses of test materials from the synthetic fuel technologies. IV. Studies of chemical structure-mutagenic activity relationships of aromatic nitrogen compounds relevant to synfuels. *Mutat. Res.*, *85*, 335-345

Hoffmann, D., Rathkamp, G. & Woziwodzki, H. (1968) Chemical studies on tobacco smoke. VI. The determination of carbazoles in cigarette smoke. *Beitr. Tabakforsch.*, *4*, 253-263

Johns, S.R. & Wright, S.E. (1964) The metabolism of carbazole in rats and rabbits. *J. med. Chem.*, *7*, 158-161

Kaden, D.A., Hites, R.A. & Thilly, W.G. (1979) Mutagenicity of soot and associated polycyclic aromatic hydrocarbons to *Salmonella typhimurium*. *Cancer Res.*, *39*, 4152-4159

Kennaway, E.L. (1924) On the cancer-producing factor in tar. *Br. med. J.*, *i*, 564-567

Lang, K.F. & Eigen, I. (1967) *Organic compounds in coal tar* (Ger.) In: Heilbronner, E., Hofmann, U., Schäfer, K. & Wittig, G., eds, *Fortschrifte der chemischer Forschung* (Advances in chemical research), Vol. 8, Berlin, Springer, pp. 91-170

LaVoie, E.J., Briggs, G., Bedenko, V. & Hoffmann, D. (1982) Mutagenicity of substituted carbazoles in *Salmonella typhimurium*. *Mutat. Res.*, *101*, 141-150

Maisin, J., Desmedt, P. & Jacqmin, L. (1927) Carcinogenic action of carbazole (Fr.). *C.R. Soc. Biol. Paris*, *96*, 1056-1058

National Library of Medicine (1982) *Toxicology Data Bank*, Bethesda, MD, National Library of Medicine Specialized Information Services, Toxicology Information Program

NIH/EPA Chemical Information System (1982) *Mass Spectral Search System*, Washington DC, CIS Project, Information Services Corporation

Pilotti, A., Ancker, K., Arrhenius, E. & Enzell, C. (1975) Effects of tobacco and tobacco smoke constituents on cell multiplication *in vitro*. *Toxicology*, *5*, 49-62

Schürch, O. & Winterstein, A. (1935) Carcinogenic action of aromatic hydrocarbons (Ger.). *Hoppe-Seyler's Z. physiol. Chem.*, *236*, 79-91

Shear, M.J. & Leiter, J. (1941) Studies in carcinogenesis. XVI. Production of subcutaneous tumors in mice by miscellaneous polycyclic compounds. *J. natl Cancer Inst.*, *2*, 241-258

Tokiwa, H., Nakagawa, R., Morita, K. & Ohnishi, Y. (1981) Mutagenicity of nitro derivatives induced by exposure of aromatic compounds to nitrogen dioxide. *Mutat. Res.*, *85*, 195-205

Tsuda, H., Hagiwara, A., Shibata, M., Oshima, M. & Ito, N. (1982) Carcinogenic effect of carbazole in the liver of (C57BL/6NxC3H/HeN)F_1 mice. *J. natl Cancer Inst.*, *69*, 1383-1387

Weast, R.C., ed. (1975) *CRC Handbook of Chemistry and Physics*, 56th ed., Cleveland, OH, Chemical Rubber Co., p. C-233

Windholz, M., ed. (1976) *The Merck Index*, 9th ed., Rahway, NJ, Merck & Co., p. 227

CHRYSENE

This compound was considered by a previous working group, in December 1972 (IARC, 1973). Data that have become available since that time have been incorporated in the present monograph and taken into consideration in the evaluation.

1. Chemical and Physical Data

1.1 Synonyms and trade names

Chem. Abstr. Services Reg. No.: 218-01-9

Chem. Abstr. Name: Chrysene

IUPAC Systematic Name: Chrysene

Synonyms: 1,2-Benzophenanthrene; benzo(a)phenanthrene; 1,2-benzphenanthrene; benz(a)phenanthrene; 1,2,5,6-dibenzonaphthalene

1.2 Structural and molecular formulae and molecular weight

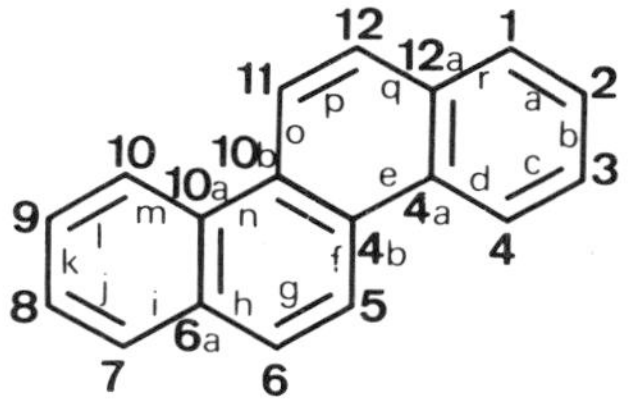

$C_{18}H_{12}$ Mol. wt: 228.3

1.3 Chemical and physical properties of the pure substance

From National Library of Medicine (1982), unless otherwise specified

(*a*) *Description*: Orthorhombic bipyramidal plates (recrystallized from benzene); colourless platelets with blue fluorescence

(*b*) *Boiling-point*: 448°C

(*c*) *Melting-point*: 255-256°C

(*d*) *Spectroscopy data*: λ_{max} 220, 240, 250, 257, 293, 305, 318, 342, 350, 359 nm (in cyclohexane) (Karcher *et al.*, 1983). Infrared spectra have been reported (API Research Project 44, 1960). Mass and nuclear magnetic resonance spectra have been tabulated (NIH/EPA Chemical Information System, 1982; Karcher *et al.*, 1983).

(*e*) *Solubility*: Virtually insoluble in water (1.5-2.2 μg/l) (Davis *et al.*, 1942; May *et al.*, 1978); slightly soluble in acetone, carbon disulphide, diethyl ether, ethanol (1 g/1300 ml absolute ethanol), glacial acetic acid, toluene (0.24 parts/100 parts at 18°C; 1 g/480 ml at 25°C; 5% at 100°C) and hot xylene. Soluble in benzene

(*f*) *Stability*: Does not undergo photo-oxidation in organic solvents under fluorescent light or indoor sunlight (Kuratsune & Hirohata, 1962)

(*g*) *Reactivity*: Oxidized by osmium tetroxide in pyridine to the 11,12-dihydrodiol; 11,12-chrysoquinone is formed by treatment with a boiling solution of chromic acid in glacial acetic acid (Clar, 1964); reacts with NO and NO_2 to form nitro derivatives (Butler & Crossley, 1981; Tokiwa *et al.*, 1981)

2. Production, Use, Occurrence and Analysis

2.1 Production and use

There is no commercial production or known use of this compound.

2.2 Occurrence and analysis

Data on occurrence and methods of analysis are summarized in the 'General Remarks on the Substances Considered', p. 35.

Chrysene occurs ubiquitously and in about the same concentration as benzo[*a*]pyrene in products of incomplete combustion. In addition, chrysene and, preferentially, related stuctures (methyl-substituted and partially hydrogenated chrysenes) occur in higher concentrations than most of the other polynuclear aromatic hydrocarbons in fossil fuels such as crude oil and lignite.

Chrysene has been identified in mainstream cigarette smoke (9.6 μg/100 cigarettes) (Ayres & Thornton, 1965), (3.6 μg/cigarette) (Hecht *et al.*, 1974), (7.2 μg/100 g burnt material) (Masuda & Kuratsune, 1972), (5.1 μg/100 cigarettes) (Lee *et al.*, 1976), (6.6-8.2 μg/100 cigarettes) (Ellington *et al.*, 1978), (6.3-15 μg/1000 cigarettes) (Kiryu & Kuratsune, 1966); mainstream smoke of marijuana cigarettes (5.5 μg/100 cigarettes) (Lee *et al.*, 1976); the air of cities (1.3-13.3 ng/m^3) (Hoffmann & Wynder, 1976); exhaust emissions from gasoline engines (85-123 μg/l fuel) (Grimmer *et al.*, 1977); with triphenylene in exhaust of burnt coals (0.07-21.8 mg/kg) (Brockhaus & Tomingas, 1976); gasolines (0.23-2.96 mg/kg) (Müller & Meyer, 1974); crude oils (Grimmer *et al.*, 1983); charcoal-broiled steaks (1.4 μg/kg) (Lijinsky & Shubik, 1964); broiled and smoked meat (up to 25.4 μg/kg) (Lijinsky & Ross, 1967), smoked fish (10-173 μg/kg) (Masuda & Kuratsune, 1971), vegetables (up to 395 μg/kg) (Hettche, 1971); edible oils and fats (up to 200 μg/kg) (Biernoth & Rost, 1968); roasted and soluble coffee (3.0-19.1 μg/kg) (Fritz, 1969); surface water (7.9-62.0 ng/l) (Grimmer *et al.*, 1981); waste water (0.732-6.44 μg/l) (Grimmer *et al.*, 1981); freeze-dried sewage sludge (780-23 700 μg/kg) (Grimmer *et al.*, 1980); and sediments (40-240 μg/kg) (Giger & Blumer, 1974).

3. Biological Data Relevant to the Evaluation of Carcinogenic Risk to Humans

In earlier bioassays, the chrysene used contained some impurities (e.g., methylchrysenes) that were not detectable with the instrumentation available at that time.

13.1 Carcinogenicity studies in animals[1]

(a) Skin application

Mouse: Groups of 20, 50 and 100 mice were painted twice weekly with different samples of a 0.3% solution of chrysene in benzene for over 440 days. A sample of chrysene of 'doubtful purity' produced two papillomas in 50 mice (30 of which were still alive at six months); another sample produced one papilloma and one epithelioma [squamous-cell carcinoma] in 100 mice (74 of which were still alive at six months); and a sample of synthetic chrysene (not dissolved in benzene) did not induce tumours in 20 mice (11 of which were still alive at six months) (Barry *et al.*, 1935).

In another study, a 7.5% solution of chrysene in liquid paraffin or oleic acid was painted on the skin of mice for 50-85 weeks and induced a few malignant skin tumours (Bottomley & Twort, 1934). However, skin application of a 1% solution of chrysene in 90% benzene induced no tumour after nine months (Kennaway, 1924). A 0.3% solution in mouse fat or a 7.5% solution in oleic acid produced no tumour in 120 mice (Barry *et al.*, 1935).

[1] The Working Group was aware of a study in progress by skin application on mice (IARC, 1983).

A group of 50 mice were painted twice weekly with 'pure' chrysene [concentration unspecified] in benzene; 11 animals were still alive after 276 days and no tumour was reported (Schürch & Winterstein, 1935).

One epithelioma [squamous-cell carcinoma] was induced in 15 CD-1 mice still alive after 31 weeks of twice weekly skin applications with a 0.2% solution of chrysene in acetone (Riegel *et al.*, 1951).

A group of 20 female Swiss mice [age unspecified] received skin applications of a 1% solution of chrysene in acetone thrice weekly for life. The first skin tumour was observed after seven months. At 12 months, six animals were still alive. Of the original 20 mice, nine developed skin papillomas and eight, carcinomas (Wynder & Hoffmann, 1959). [No solvent-treated control was used.]

Groups of 20 C3H male mice, two months of age, were given repeated topical applications of 0.15% chrysene (specially purified) in either decahydronaphthalene (decalin) or 50:50 decalin:*n*-dodecane. When applied in decalin alone, chrysene produced a papilloma in 1/12 mice at 76 weeks; when the combination of 50:50 decalin:*n*-dodecane was used as the vehicle, papillomas were produced in 5/19 mice and carcinomas in 12/19 animals at 49 weeks (Horton & Christian, 1974). [The Working Group noted that *n*-dodecane has a cocarcinogenic effect when tested simultaneously with polynuclear aromatic hydrocarbons (Bingham & Falk, 1969)].

In an initiation-promotion experiment, a single dose of 1 mg chrysene in 0.4 ml acetone, followed by 25 μg croton resin in 0.1 ml acetone, was applied three times weekly to the skin of 20 female ICR/Ha Swiss mice, eight weeks of age, beginning 13-21 days after the initial treatment; another group of 20 mice received applications of croton resin solution only (promoter controls). Papillomas developed in 5/20 promoter controls and in 16/20 of the mice receiving both chrysene and croton resin solution; two carcinomas were also induced in the latter group and one in the controls (Van Duuren *et al.*, 1966).

In another study, 30 female CD-1 mice, eight weeks old, received a single skin application of a 4.4 μmol solution of chrysene in acetone, followed one week later by twice-weekly applications of 10 μmol 12-*O*-tetradecanoylphorbol-13-acetate (TPA); papillomas developed in 73% of the mice at 35 weeks. In a group of mice treated with TPA alone, 3% tumours were observed (Scribner, 1973). [The Working Group noted that the dose of TPA was probably 10 μg (Scribner & Süss, 1978).]

A group of 20 female Swiss albino mice (Ha/ICR/Mil) received ten daily applications of 100 μg chrysene [high purity] in 0.1 ml acetone, followed ten days after the last initiator dose by applications of 2.5 μg TPA in 0.1 ml acetone thrice weekly for 20 weeks. An increased incidence of tumours (papillomas and carcinomas) was seen by the end of the promoter application: in 11/18 mice, compared with 4/11 mice with carcinomas after 72 weeks in a group receiving chrysene only (Hecht *et al.*, 1974).

A group of 30 female CD-1 mice seven to eight weeks old, received a single application of a 0.4, 1.25 or 4.0 μmol solution of chrysene in 200 μl tetrahydrofuran:dimethyl sulphoxide (95:5) followed by applications of a 16 nmol solution of TPA in 200 μl acetone or applications of TPA alone twice weekly for 25 weeks; 25, 43 and 52% of mice receiving both chrysene (low, mid- and high dose) and TPA developed skin papillomas (0.32, 0.97 and 1.45 tumours/mouse, respectively), whereas such tumours were induced in only 7% of mice (0.07 tumours/mouse) receiving TPA alone (Levin *et al.*, 1978).

A group of 30 female Charles River CD-1 mice, eight weeks of age, received single topical applications of 2 μmol chrysene (purity, 95%; melting-point, 250-252°C) twice weekly in 200 μl acetone:dimethyl sulphoxide:ammonium hydroxide (1000:100:1) on the shaved dorsal skin. A group of 30 controls received acetone alone. One week later, both groups received 16 nmol TPA in 200 μl acetone twice weekly for 25 weeks. By that time, 21/30 (67%) treated mice had developed skin papillomas, compared with 1/30 of controls (Wood *et al.*, 1979).

In a further study, a group of 30 female Charles River CD-1 mice, eight weeks of age, received single topical applications of 2.5 μmol chrysene (purity, 98%) in 200 μl acetone. A group of 30 controls received acetone alone. One week later, the mice received twice weekly skin applications of 16 nmol TPA in 200 μl acetone for 26 weeks. In this experiment, 80% of mice developed skin papillomas (2.16 tumours/mouse) compared with 4% in controls (0.04 tumours/mouse) (Wood *et al.*, 1980).

A group of 30 female Sencar mice, seven to nine weeks old, received a single application of 2 μmol chrysene (purity, >95%) in acetone, followed one week later by applications of 2 μg TPA twice weekly. Papillomas were observed in 21/29 (73%) mice in the experimental group, whereas only 3/30 (10%) promoter-control animals receiving TPA alone developed such tumours (Slaga *et al.*, 1980).

(*b*) *Subcutaneous and/or intramuscular administration*

Mouse: No tumour was reported in 50 mice that received one s.c. injection of 'purified' chrysene and were observed for 45 weeks (Bottomley & Twort, 1934), nor did any tumour appear in ten mice receiving weekly injections of 1 or 2 mg chrysene [purity unspecified] in lard and observed for 350 days (Barry & Cook, 1934). Similarly, two s.c. injections of 10 mg chrysene to 30 Jackson A mice gave negative results at 15 months (Shear & Leiter, 1941).

A group of 50 male and female C57Bl mice, three to four months old, received one s.c. injection of 5 mg chrysene [purity unspecified] in tricaprylin. Four sarcomas were observed in 39 animals still alive after four months, with an average induction time of 401 days; the experiment was terminated at 22 months (Steiner & Falk, 1951). A group of 40 or 50 male and female C57Bl mice received injections of 5 mg chrysene [purity unspecified] in tricaprylin. Of 22 mice still alive at 150 days, five had developed sarcomas, with an average induction time of 271 days; the experiment was terminated after 22-28 months (Steiner, 1955). In another experiment in which male C57Bl mice, 120 days old, received 10 weekly injections of 1 mg chrysene [purity unspecified] in arachis oil, two out of 20 animals still alive at the time of appearance of the first tumour developed injection-site tumours 60-80 weeks after the start of treatment; no tumour appeared in controls receiving the solvent alone (Boyland & Sims, 1967).

Rat: Groups of 10 rats received repeated injections of 2-6 mg chrysene [purity unspecified] in lard, or the solvent only; four tumours were observed in treated animals, and sarcomas were found in two solvent controls (Barry & Cook, 1934). Weekly injections of 2 ml of a 0.05% aqueous-acetonic suspension of purified chrysene to a group of 10 rats induced no tumour in the four animals still alive at 18 months (Boyland & Burrows, 1935). Repeated injection of 5 mg chrysene [purity unspecified] in water or sesame oil (total number of treatments, 6-11) to 14 Wistar rats, six to eight weeks of age, produced no tumour after ten months (Pollia, 1941).

(*c*) *Perinatal exposure*

Mouse: A group of 104 male and female Swiss mice received s.c. injections of 100 μg chrysene [purity unspecified] in polyethylene glycol (PEG 400) on days 0, 1 and 2 after birth (total dose, 300 μg) and were observed for 70-75 weeks. In the 27 males and 21 females still alive at 70 weeks, 13 liver tumours were observed in males, and three lung tumours were found in two males and one female. In controls treated with PEG 400 once, five liver tumours and two lung tumours occurred in 20 surviving males, and three liver tumours and three lung tumours in 21 surviving females. In controls treated with PEG 400 three times, ten liver tumours and four lung tumours occurred in 30 surviving males, and one lung tumour in 15 surviving females (Grover *et al.*, 1975).

A group of 100 Swiss-Webster BLU:Ha (ICR) mice received i.p. injections of chrysene (purity checked by liquid chromatography) in dimethyl sulphoxide on days 1, 8 and 15 after birth (total dose, 1.4 μmol) and were observed for 38-42 weeks. Of the animals still alive at 38-42 weeks, pulmonary tumours were found in 5/24 (21%) males and 2/11 (9%) females; hepatic tumours were found in 6/24 (25%) males and lymphosarcomas in 1/24 (4%). Among controls given dimethyl sulphoxide and still alive at 38-42 weeks, lung tumours occurred in 2/21 (10%) males and 7/38 (18%) females; no liver tumour or lymphosarcoma was observed (Buening *et al.*, 1979).

A group of 80 male and female newborn Swiss-Webster BLU:Ha (ICR) mice received three i.p. injections of 0.2, 0.4 and 0.8 μmol chrysene (repurified; melting-point, 256°C) in dimethyl sulphoxide on days 1, 8 and 15 after birth. A total of 56 mice survived until weaning, and 27 males and 11 females were killed at weeks 39-41. Hepatic tumours [not further specified] developed in 6/27 (22%) male mice compared with 0/52 in controls given dimethyl sulphoxide. No increase in the incidence of pulmonary tumours was observed (Chang *et al.*, 1983).

3.2 Other relevant biological data[1]

(*a*) *Experimental systems*

Toxic effects

The LD_{50} (i.p.) for the mouse is >320 mg/kg bw (Simmon *et al.*, 1979). The growth rate of mouse ascites sarcoma cells in culture was slightly (9%) inhibited when chrysene was added at a concentration of 1 μmol/ml dissolved in dimethyl sulphoxide (Pilotti *et al.*, 1975).

I.p. administration of 30 mg chrysene in sesame oil did not reduce the growth rate of young rats as did administration of 10 mg benzo[*a*]pyrene (Haddow *et al.*, 1937).

Effects on reproduction and prenatal toxicity

Application of small amounts of chrysene (about 0.1% in 10 μl of a petroleum hydrocarbon mixture of relatively low embryotoxicity) to the egg-shell of Mallard ducks resulted in embryotoxic and teratogenic effects in the ducklings (Hoffman & Gay, 1981).

Chrysene, in a dose of 60 mg/kg bw given orally to pregnant rats on day 19 of gestation, can induce cytochrome P-450 enzymes, such as benzo[*a*]pyrene hydroxylase, in fetal liver (Welch *et al.*, 1972). Benzo[*a*]pyrene hydroxylase activity of the rat placenta can also be induced (Welch *et al.*, 1969).

---

[1] See also 'General Remarks on the Substances Considered', p. 53.

Metabolism and activation

The 1,2-, 3,4- and 5,6-dihydrodiols and some monohydroxy derivatives, including the 1- and 3-phenols, have been reported to be metabolites of chrysene following incubation of this compound with rat-liver preparations (Sims, 1970; Nordquist *et al.*, 1981; Jacob *et al.*, 1982) and with mouse-skin cultures (MacNicoll *et al.*, 1980). The formation of the 1,2-diol-3,4-epoxide and of the 3,4-diol-1,2-epoxide from the precursor dihydrodiols has also been reported in studies using rat-liver preparations (Chou *et al.*, 1981; Nordquist *et al.*, 1981; Vyas *et al.*, 1982).

The 1,2-dihydrodiol, in the presence of an exogenous metabolic system, and the 1,2-diol-3,4-epoxide are mutagenic in bacterial and mammalian cells (Wood *et al.*, 1977, 1979); and the 1,2-dihydrodiol is active as a tumour initiating agent on mouse skin (Levin *et al.*, 1978; Slaga *et al*, 1980; Chang *et al.*, 1983). The 1,2-dihydrodiol and the 1,2-diol-3,4-epoxide are also active in inducing pulmonary adenomas in newborn mice (Buening *et al.*, 1979; Chang *et al.*, 1983).

Nucleic acid adducts formed in hamster cells treated with chrysene appear to arise from reactions of the 1,2-diol-3,4-epoxide with DNA (Hodgson *et al.*, 1982; Vigny *et al.*, 1982).

Mutagenicity and other short-term tests

Results from short-term tests are given in Table 1.

Table 1. Results from short-term tests: Chrysene[a]

Test	Organism/assay[b]	Exogenous metabolic system[b]	Reported result	Comments	References
PROKARYOTES					
DNA damage	*Escherichia coli* ($polA^+/polA^-$)	UI-R-PMS	Negative	Tested at up to 250 μg/ml [The Working Group noted that UI-PMS was used.]	Rosenkranz & Poirier (1979)
Mutation	*Salmonella typhimurium* (his^-/his^+)	Aro-R-PMS	Positive	At 10 μg/plate in strain TA100	McCann *et al.* (1975); LaVoie *et al.* (1979)
		Aro-R-Micr	Positive	At 125 nmol/plate in strain TA100; dose-response seen with increasing concentrations of Micr	Wood *et al.* (1977)
	Salmonella typhimurium ($8AG^S/8AG^R$)	Aro-R-PMS	Positive	At 45 nmol/ml in strain TM677	Kaden *et al.* (1979)
FUNGI					
Mutation	*Saccharomyces cerevisiae* D3 (mitotic recombination)	Aro-R-PMS	Negative	Tested at up to 5% w/v (*sic*)	Simmon (1979)
MAMMALIAN CELLS *IN VITRO*					
DNA damage	Primary rat hepatocytes (unscheduled DNA synthesis)	--	Negative	Tested at up to 100 nmol/ml	Tong *et al.* (1981)

Test	Organism/assay[b]	Exogenous metabolic system[b]	Reported result	Comments	References
Mutation	Chinese hamster V79 cells (OUA^S/OUA^R; $8AG^S/8AG^R$)	SHE feeder layer	Negative	Tested at up to 10 μg/ml	Huberman & Sachs (1976)
Cell transformation	Syrian hamster embryo cells (morphological)	--	Positive	At 10 μg/ml	Pienta *et al.*, 1977
	Mouse prostate C3HG23 cells (morphological)	--	Negative	Tested at up to 10 μg/ml	Marquardt *et al.* (1972)
MAMMALIAN CELLS *IN VIVO*					
Chromosome effects	Chinese hamster bone-marrow cells: sister chromatid exchange	--	Positive	At 2 x 450 mg/kg bw i.p.	Roszinsky-Köcher *et al.* (1979)
	aberrations	--	Negative	Treated i.p. with 2 x 450 mg/kg bw	
	NMRI mice (metaphase II oocytes) (aberrations)	--	Positive	At 450 mg/kg bw orally	Basler *et al.* (1977)

[a] This table comprises selected assays and references and is not intended to be a complete review of the literature.

[b] For an explanation of the abbreviations, see the Appendix, p. 452.

(*b*) *Humans*

No data were available to the Working Group.

3.3 Case reports and epidemiological studies of carcinogenicity to humans

No data were available to the Working Group.

4. Summary of Data Reported and Evaluation[1]

4.1 Experimental data

Chrysene was tested for carcinogenicity in several studies by skin application to mice and produced skin tumours; in one study, an enhancing effect was observed when chrysene was tested simultaneously with *n*-dodecane. Chrysene was also tested in the mouse-skin

[1] For definitions of the italicized terms, see Preamble, p. 19.

initiation-promotion assay and was active as an initiator. Local tumours were observed following its subcutaneous injection in mice. Perinatal administration of chrysene to mice by subcutaneous or intraperitoneal injection increased the incidences of liver tumours.

No relevant data on the teratogenicity of this chemical were available.

Chrysene was mutagenic to *Salmonella typhimurium* in the presence of an exogenous metabolic system. It did not induce mitotic recombination in yeast, unscheduled DNA synthesis in primary rat hepatocytes, or mutations in Chinese hamster V79 cells. However, in one study each in mice and hamsters it induced sister chromatid exchange and chromosomal aberrations, respectively. It was positive in one of two reported studies of morphological transformation in mammalian cells.

There is *limited evidence* that chrysene is active in short-term tests.

4.2 Human data[1]

Chrysene is present as a major component of the total content of polynuclear aromatic compounds in the environment. Human exposure to chrysene occurs primarily through the smoking of tobacco, inhalation of polluted air and by ingestion of food and water contaminated by combustion effluents (for details, see 'General Remarks on the Substances Considered', p. 35).

4.3 Evaluation

There is *limited evidence* that chrysene is carcinogenic to experimental animals.

5. References

API Research Project 44 (1960) *Selected Infrared Spectral Data*, Vol. VI, No. 2241, Washington DC, American Petroleum Institute

Ayres, C.I. & Thornton, R.E. (1965) Determination of benzo(*a*)pyrene and related compounds in cigarette smoke. *Beitr. Tabakforsch.*, *3*, 285-290

Barry, G. & Cook, J.W. (1934) A comparison of the action of some polycyclic aromatic hydrocarbons in producing tumours of connective tissue. *Am. J. Cancer*, *20*, 58-69

---

[1] Studies on occupational exposure to polynuclear aromatic compounds will be considered in future *IARC Monographs*.

Barry G., Cook, J.W., Haslewood, G.A.D., Hewett, C.L., Hieger, I. & Kennaway, E.L. (1935) The production of cancer by pure hydrocarbons. Part III. *Proc. R. Soc. London Ser. B*, *117*, 318-351

Basler, A., Herbold, B., Peter, S. & Röhrborn, G. (1977) Mutagenicity of polycyclic hydrocarbons. II. Monitoring genetical hazards of chrysene *in vitro* and *in vivo*. *Mutat. Res.*, *48*, 249-254

Biernoth, G. & Rost, H.E. (1968) Occurrence of polycyclic aromatic hydrocarbons in edible oils and their refining (Ger.). *Arch. Hyg.*, *152*, 238-250

Bingham, E. & Falk, H.L. (1969) Environmental carcinogens. The modifying effect of cocarcinogens on the threshold response. *Arch. environ. Health*, *19*, 779-783

Bottomley, A.C. & Twort, C.C. (1934) The carcinogenicity of chrysene and oleic acid. *Am. J. Cancer*, *21*, 781-786

Boyland, E. & Burrows, H. (1935) The experimental production of sarcoma in rats and mice by a colloidal aqueous solution of l:2:5:6-dibenzanthracene. *J. Pathol. Bacteriol.*, *41*, 231-238

Boyland, E. & Sims. P. (1967) The carcinogenic activities in mice of compounds related to benzo[*a*]anthracene. *Int. J. Cancer*, *2*, 500-504

Brockhaus, A. & Tomingas, R. (1976) Emission of polycyclic hydrocarbons during burning processes in small heating installations and their concentration in the atmosphere (Ger.). *Staub-Reinhalt. Luft*, *36*, 96-101

Buening, M.K., Levin, W., Karle, J.M., Yagi, H., Jerina, D.M. & Conney, A.H. (1979) Tumorigenicity of bay-region epoxides and other derivatives of chrysene and phenanthrene in newborn mice. *Cancer Res.*, *39*, 5063-5068

Butler, J.D. & Crossley, P. (1981) Reactivity of polycyclic aromatic hydrocarbons adsorbed on soot particles. *Atmos. Environ.*, *15*, 91-94

Chang, R.L., Levin, W., Wood, A.W., Yagi, H., Tada, M., Vyas, K.P., Jerina, D.M. & Conney, A.H. (1983) Tumorigenicity of enantiomers of chrysene 1,2-dihydrodiol and of the diastereomeric bay-region chrysene 1,2-diol-3,4-epoxides on mouse skin and in newborn mice. *Cancer Res.*, *43*, 192-196

Chou, M.W., Fu, P.P. & Yang, S.K. (1981) Metabolic conversion of dibenz[*a*,*h*]anthracene (±) *trans*-1,2-dihydrodiol and chrysene (±) *trans*-3,4-dihydrodiol to vicinal dihydrodiol epoxides. *Proc. natl Acad. Sci. USA*, *78*, 4270-4273

Clar, E. (1964) *Polycyclic Hydrocarbons*, Vol. 1, London, Academic Press, p. 243

Davis, W.W., Krahl, M.E. & Clowes, G.H.A. (1942) Solubility of carcinogenic and related hydrocarbons in water. *J. Am. chem. Soc.*, *64*, 108-110

Ellington, J.J., Schlotzhauer, P.F. & Schepartz, A.I. (1978) Quantitation of hexane-extractable lipids in serial samples of flue-cured tobaccos. *J. Food agric. Chem.*, *26*, 270-273

Fritz, W. (1969) Solubility of polyaromatic compounds after boiling up coffee substrates and pure coffee (Ger.). *Dtsch. Lebensm. Rundsch.*, *65*, 83-85

Giger, W. & Blumer, M. (1974) Polycyclic aromatic hydrocarbons in the environment: Isolation and characterization by chromatography, visible, ultraviolet, and mass spectrometry. *Anal. Chem.*, *46*, 1663-1671

Grimmer, G., Böhnke, H. & Glaser, A. (1977) Investigation on the carcinogenic burden by air pollution in man. XV. Polycyclic aromatic hydrocarbons in automobile exhaust gas - An inventory. *Zbl. Bakt. Hyg., 1 Abt., Orig. B164*, 218-234

Grimmer, G., Hilge, G. & Niemitz, W. (1980) Comparison of the profile of polycyclic aromatic hydrocarbons in sewage sludge samples from 25 sewage treatment works (Ger.). *Wasser*, *54*, 255-272

Grimmer, G., Schneider, D. & Dettbarn, G. (1981) The load of different rivers in the Federal Republic of Germany by PAH (PAH-profiles of surface water) (Ger.). *Wasser*, *56*, 131-144

Grimmer, G., Jacob, J. & Naujack, K.-W. (1983) Profile of the polycyclic aromatic compounds from crude oils. Inventory by GCGC/MS - PAH in environmental materials, Part 3. *Fresenius Z. anal. Chem.*, *314*, 29-36

Grover, P.L., Sims, P., Mitchley, B.C.V. & Roe, F.J.C. (1975) The carcinogenicity of polycyclic hydrocarbon epoxides in newborn mice. *Br. J. Cancer*, *31*, 182-188

Haddow, A., Scott, C.M. & Scott, J.D. (1937) The influence of certain carcinogenic and other hydrocarbons on body growth in the rat. *Proc. R. Soc. London Ser. B*, *122*, 477-507

Hecht, S.S., Bondinell, W.E. & Hoffmann, D. (1974) Chrysene and methylchrysenes: Presence in tobacco smoke and carcinogenicity. *J. natl Cancer Inst.*, *53*, 1121-1133

Hettche, H.O. (1971) Plant waxes as collectors of polycyclic aromatics in the air of residential areas. *Staub-Reinhalt. Luft*, *31*, 34-41

Hodgson, R.M., Pal, K., Grover, P.L. & Sims, P. (1982) The metabolic activation of chrysene by hamster embryo cells. *Carcinogenesis*, *3*, 1051-1056

Hoffmann, D.J. & Gay, M.L. (1981) Embryotoxic effects of benzo[*a*]pyrene, chrysene, and 7,12-dimethylbenz[*a*]anthracene in petroleum hydrocarbon mixtures in Mallard ducks. *J. Toxicol. environ. Health*, *7*, 775-787

Hoffmann, D. & Wynder, E. (1976) *Environmental respiratory carcinogenesis*. In: Searle, C.E., ed., *Chemical Carcinogens* (*ACS Monograph 173*), Washington DC, American Chemical Society, p. 341

Horton, A.W. & Christian, G.M. (1974) Cocarcinogenic versus incomplete carcinogenic activity among aromatic hydrocarbons: Contrast between chrysene and benzo[*b*]triphenylene. *J. natl Cancer Inst.*, *53*, 1017-1020

Huberman, E. & Sachs, L. (1976) Mutability of different genetic loci in mammalian cells by metabolically activated carcinogenic polycyclic hydrocarbons. *Proc. natl Acad. Sci. USA*, *73*, 188-192

IARC (1973) *IARC Monographs on the Evaluation of Carcinogenic Risk of Chemicals to Man*, Vol. 3, *Certain Polycyclic Aromatic Hydrocarbons and Heterocyclic Compounds*, Lyon, pp. 159-177

IARC (1983) *Information Bulletin on the Survey of Chemicals Being Tested for Carcinogenicity*, No. 10, Lyon, p. 16

Jacob, J., Schmoldt, A. & Grimmer, G. (1982) Formation of carcinogenic and inactive chrysene metabolites by rat liver microsomes of various monooxygenase activities. *Arch. Toxicol.*, *51*, 255-265

Kaden, D.A., Hites, R.A. & Thilly, W.G. (1979) Mutagenicity of soot and associated polycyclic aromatic hydrocarbons to *Salmonella typhimurium*. *Cancer Res.*, *39*, 4152-4159

Karcher, W., Fordham, R., Dubois, J. & Gloude, P. (1983) *Spectral Atlas of Polycyclic Aromatic Compounds*, Dordrecht, The Netherlands, D. Reidel (in press)

Kennaway, E.L. (1924) On the cancer-producing factor in tar. *Brit. med. J.*, *i*, 564-567

Kiryu, S. & Kuratsune, M. (1966) Polycyclic aromatic hydrocarbons in the cigarette tar produced by human smoking. *Gann*, *57*, 317-322

Kuratsune, M. & Hirohata, T. (1962) Decomposition of polycyclic aromatic hydrocarbons under laboratory illuminations. *Natl Cancer Inst. Monogr.*, *9*, 117-125

LaVoie, E., Bedenko, V., Hirota, N., Hecht, S.S, & Hoffmann, D. (1979) *A comparison of the mutagenicity, tumor-initiating activity and complete carcinogenicity of polynuclear aromatic hydrocarbons*. In: Jones, P.W. & Leber, P., eds, *Polynuclear Aromatic Hydrocarbons*, Ann Arbor, MI, Ann Arbor Science Publishers, pp. 705-721

Lee, M.L., Novotny, M. & Bartle, K.D. (1976) Gas chromatography/mass spectrometry and nuclear magnetic resonance spectrometric studies of carcinogenic polynuclear aromatic hydrocarbons in tobacco and marijuana smoke condensates. *Anal. Chem.*, *48*, 405-416

Levin, W., Wood, A.W., Chang, R.L., Yagi, H., Mah, H.D., Jerina, D.M. & Conney, A.H. (1978) Evidence for bay region activation of chrysene l,2-dihydrodiol to an ultimate carcinogen. *Cancer Res.*, *38*, 1831-1834

Lijinsky, W. & Ross, A.E. (1967) Production of carcinogenic polynuclear hydrocarbons in the cooking of food. *Food cosmet. Toxicol.*, *5*, 343-347

Lijinsky, W. & Shubik, P. (1964) Benzo(*a*)pyrene and other polynuclear hydrocarbons in charcoal-broiled meat. *Science*, *145*, 53-55

MacNicoll, A.D., Grover, P.L. & Sims, P. (1980) The metabolism of a series of polycyclic hydrocarbons by mouse skin maintained in short-term organ culture. *Chem.-biol. Interactions*, *29*, 169-188

Marquardt, H., Kuroki, T., Huberman, E., Selkirk, J.K., Heidelberger, C., Grover, P.L. & Sims, P. (1972) Malignant transformation of cells derived from mouse prostate by epoxides and other derivatives of polycyclic hydrocarbons. *Cancer Res.*, *32*, 716-720

Masuda, Y. & Kuratsune, M. (1971) Polycyclic aromatic hydrocarbons in smoked fish, 'katsuobushi'. *Gann*, *62*, 27-30

Masuda, Y. & Kuratsune, M. (1972) Comparison of the yield of polycyclic aromatic hydrocarbons in smoke from Japanese tobacco. *Jpn. J. Hyg.*, *27*, 339-341

May, W.E., Wasik, S.P. & Freeman, D.H. (1978) Determination of the solubility behaviour of some polycyclic aromatic hydrocarbons in water. *Anal. Chem.*, *50*, 997-1000

McCann, J., Choi, E., Yamasaki, E. & Ames, B.N. (1975) Detection of carcinogens as mutagens in the *Salmonella*/microsome test: Assay of 300 chemicals. *Proc. natl Acad. Sci. USA*, *72*, 5135-5139

Müller, K. & Meyer, J.P. (1974) *Einfluss von Ottokraftstoffen auf die Emission von polynuklearen aromatischen Kohlenwasserstoffen in Automobilabgasen im Europa-test* (Effect of gasoline components on emission of polynuclear aromatic hydrocarbons in car exhaust in the Europa test) (*Forschungsbericht 4568*), Hamburg, Deutsche Gesellschaft für Mineralölwissenschaft und Kohlechemie e.V.

National Library of Medicine (1982) *Toxicology Data Bank*, Bethesda, MD, National Library of Medicine Specialized Information Services, Toxicology Information Program

NIH/EPA Chemical Information System (1982) *Mass Spectral Search System*, Washington DC, CIS Project, Information Services Corporation

Nordqvist, M., Thakker, D.R., Vyas, K.P., Yagi, H., Levin, W., Ryan, D.E., Thomas, P.E., Conney, A.H. & Jerina, D.M. (1981) Metabolism of chrysene and phenanthrene to bay-region diol epoxides by rat liver enzymes. *Mol. Pharmacol.*, *19*, 168-178

Pienta, R.J., Poiley, J.A. & Lebherz III, W.B. (1977) Morphological transformation of early passage golden Syrian hamster embryo cells derived from cryopreserved primary cultures as a reliable *in vitro* bioassay for identifying diverse carcinogens. *Int. J. Cancer*, *19*, 642-655

Pilotti, A., Ancker, K., Arrhenius, E. & Enzell, C. (1975) Effects of tobacco and tobacco smoke constituents on cell multiplication *in vitro*. *Toxicology*, *5*, 49-62

Pollia, J.A. (1941) Investigations on the possible carcinogenic effect of anthracene and chrysene and some of their compounds. II. The effect of subcutaneous injection in rats *J. ind. Hyg. Toxicol.*, *23*, 449-451

Riegel, B., Watman, W.B., Hill, W.T., Reeb, B.B., Shubik, P. & Stanger, D.W. (1951) Delay of methylcholanthrene skin carcinogenesis in mice by 1,2,5,6-dibenzofluorene. *Cancer Res.*, *11*, 301-306

Rosenkranz, H.S. & Poirier, L.A. (1979) Evaluation of the mutagenicity and DNA-modifying activity of carcinogens and noncarcinogens in microbial systems. *J. natl Cancer Inst.*, *62*, 873-892

Roszinsky-Köcher, G., Basler, A. & Röhrborn, G. (1979) Mutagenicity of polycyclic hydrocarbons. V. Induction of sister chromatid exchanges *in vivo*. *Mutat. Res.*, *66*, 65-67

Schürch, O. & Winterstein, A. (1935) On the carcinogenic action of aromatic hydrocarbons (Ger.). *Hoppe-Seylers Z. Physiol. Chem.*, *236*, 79-91

Scribner, J.D. (1973) Brief communication: Tumor initiation by apparently noncarcinogenic polycyclic aromatic hydrocarbons. *J. natl Cancer Inst.*, *50*, 1717-1719

Scribner, J.D. & Süss, R. (1978) Tumour initiation and promotion. *Int. Rev. exp. Pathol.*, *18*, 137-198

Shear, M.J. & Leiter, J. (1941) Studies in carcinogenesis. XVI. Production of subcutaneous tumors in mice by miscellaneous polycyclic compounds. *J. natl Cancer Inst.*, *2*, 241-258

Simmon, V.F. (1979) *In vitro* assays for recombinogenic activity of chemical carcinogens and related compounds with *Saccharomyces cerevisiae* D3. *J. natl Cancer Inst.*, *62*, 901-909

Simmon, V.F., Rosenkranz, H.S., Zeiger, E. & Poirier, L.A. (1979) Mutagenic activity of chemical carcinogens and related compounds in the intraperitoneal host-mediated assay. *J. natl Cancer Inst.*, *62*, 911-918

Sims, P. (1970) Qualitative and quantitative studies on the metabolism of a series of aromatic hydrocarbons by rat-liver preparation. *Biochem. Pharmacol.*, *19*, 795-818

Slaga, T.J., Gleason, G.L., Mills, G., Ewald, L., Fu, P.P., Lee, H.M. & Harvey, R.G. (1980) Comparison of the skin tumor-initiating activities of dihydrodiols and diol-epoxides of various polycyclic aromatic hydrocarbons. *Cancer Res.*, *40*, 1981-1984

Steiner, P.E. (1955) Carcinogenicity of multiple chemicals simultaneously administered. *Cancer Res.*, *15*, 632-635

Steiner, P.E. & Falk, H.L. (1951) Summation and inhibition effects of weak and strong carcinogenic hydrocarbons: 1:2-benzanthracene, chrysene, 1:2:5:6-dibenzanthracene, and 20-methylcholanthrene. *Cancer Res.*, *11*, 56-63

Tokiwa, H., Nakagawa, R., Morita, K. & Ohnishi, Y. (1981) Mutagenicity of nitro derivatives induced by exposure of aromatic compounds to nitrogen dioxide. *Mutat. Res.*, *85*, 195-205

Tong, C., Laspia, M.F., Telang, S. & Williams, G.M. (1981) The use of adult rat liver cultures in the detection of the genotoxicity of various polycyclic aromatic hydrocarbons. *Environ. Mutagenesis*, *3*, 477-487

Van Duuren, B.L., Sivak, A., Segal, A., Orris, L. & Langseth, L. (1966) The tumor-promoting agents of tobacco leaf and tobacco smoke condensate. *J. natl Cancer Inst.*, *37*, 519-526

Vigny, P., Spiro, M., Hodgson, R.M., Grover, P.L. & Sims, P. (1982) Fluorescence spectral studies on the metabolic activation of chrysene by hamster embryo cells. *Carcinogenesis*, *3*, 1491-1493

Vyas, K.P., Levin, W., Yagi, H., Thakker, D.R., Ryan, D.E., Thomas, P.E., Conney, A.H. & Jerina, D.M. (1982) Stereoselective metabolism of the (+) and (-)-enantiomers of *trans*-1,2-dihydroxy-1,2-dihydrochrysene to bay-region 1,2-diol-3,4-epoxide diastereomers by rat liver enzymes. *Mol. Pharmacol.*, *22*, 182-189

Welch, R.M., Harrison, Y.E., Gommi, B.W., Poppers, P.J., Finster, M. & Conney, A.H. (1969) Stimulatory effect of cigarette smoking on the hydroxylation of 3,4-benzpyrene and the *N*-demethylation of 3-methyl-4-monomethylaminoazobenzene by enzymes in human placenta. *Clin. Pharmacol.*, *10*, 100-109

Welch, R.M., Gommi, B., Alvares, A.P. & Conney, A.H. (1972) Effect of enzyme induction on the metabolism of benzo(*a*)pyrene and 3′-methyl-4-monomethylaminoazobenzene in the pregnant and fetal rat. *Cancer Res.*, *32*, 973-978

Wood, A.W., Levin, W., Ryan, D., Thomas, P.E., Yagi, H., Mah, H.D., Thakker, D.R., Jerina, D.M. & Conney, A.H. (1977) High mutagenicity of metabolically activated chrysene 1,2-dihydrodiol: Evidence for bay region activation of chrysene. *Biochem. biophys. Res. Commun.*, *78*, 847-854

Wood, A.W., Chang, R.L., Levin, W., Ryan, D.E., Thomas, P.E., Mah, H.D., Karle, J.M., Yagi, H., Jerina, D.M. & Conney, A.H. (1979) Mutagenicity and tumorigenicity of phenanthrene and chrysene epoxides and diol epoxides. *Cancer Res.*, *39*, 4069-4077

Wood, A.W., Levin, W., Chang, R.L., Huang, M.-T., Ryan, D.E., Thomas, P.E., Lehr, R.E., Kumar, S., Koreeda, M., Akagi, H., Ittah, Y., Dansette, P., Yagi, H., Jerina, D.M. & Conney, A.H. (1980) Mutagenicity and tumor-initiating activity of cyclopenta(*c,d*)pyrene and structurally related compounds. *Cancer Res.*, *40*, 642-649

Wynder, E.L. & Hoffmann, D. (1959) A study of tobacco carcinogenesis. VII. The role of higher polycyclic hydrocarbons. *Cancer*, *12*, 1079-1086

CORONENE

1. Chemical and Physical Data

1.1 Synonyms and trade names

Chem. Abstr. Services Reg. No.: 191-07-1

Chem. Abstr. Name: Coronene

IUPAC Systematic Name: Coronene

Synonym: Hexabenzobenzene

1.2 Structural and molecular formulae and molecular weight

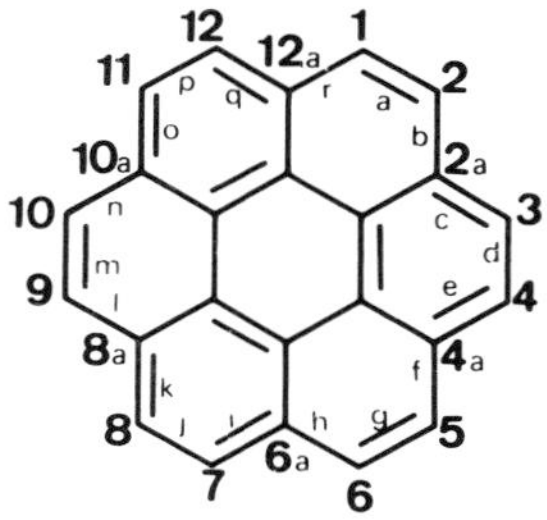

$C_{24}H_{12}$

Mol. wt: 300.4

1.3 Chemical and physical properties of the pure substance

From Weast (1975), unless otherwise specified

(*a*) *Description*: Yellow needles (recrystallized from benzene)

(*b*) *Boiling-point*: 525°C

(*c*) *Melting-point*: 438-440°C

(*d*) *Spectroscopy data*: λ_{max} 228, 252, 290 nm (in ethanol) (Clar, 1964); 305, 316.5, 319.5, 325.5, 336, 341.5, 347.5, 368.5, 378, 381.5, 388, 296.5, 402, 410, 420, 428 (in benzene) (Clar, 1964). Mass spectra have been tabulated (NIH/EPA, Chemical Information System, 1982).

(*e*) *Solubility*: Insoluble in water and concentrated sulphuric acid; slightly soluble in benzene

(*f*) *Stability*: No data were available.

(*g*) *Reactivity*: Hydrogenation under pressure gives perhydrocoronene. Reacts with nitric acid to various mono-, di-, tri- and hexanitro products. Can be halogenated. Reacts with benzoyl chloride and aluminium chloride to form benzoylcoronene. It also reacts with one molecule of succinic anhydride or phthalic anhydride to form mono substitution products (Clar, 1964).

2. Production, Use, Occurrence and Analysis

2.1 Production and use

There is no commercial production or known use of this compound. Coronene can be used in the preparation of sulpha dyes (Clar, 1964).

2.2 Occurrence and analysis

Data on occurrence and methods of analysis are summarized in the 'General Remarks on the Substances Considered', p. 35.

Coronene occurs ubiquitously in products of incomplete combustion; it also occurs in fossil fuels. It has been identified in mainstream cigarette smoke (Lyons, 1962; Snook *et al.*, 1977); the air of coffee houses (0.5-1.2 ng/m^3) (Just *et al.*, 1972); gasoline engine exhaust (106-271 μg/l fuel burned) (Grimmer *et al.*, 1977); various fresh (0.02-0.65 mg/kg) and used gasoline motor oils (2.8-29.4 mg/kg) (Grimmer *et al.*, 1981); gasolines (0.06-1.11 mg/kg) (Müller & Meyer, 1974); coal-tar (Lang & Eigen, 1967); charcoal-broiled steaks (2.3 μg/kg) (Lijinsky & Shubik, 1964); edible oils (0-2.8 μg/kg) (Grimmer & Hildebrandt, 1967); and dried sediment of lakes (9-810 μg/kg) (Grimmer & Böhnke, 1975, 1977).

3. Biological Data Relevant to the Evaluation of Carcinogenic Risk to Humans

3.1 Carcinogenicity studies in animals

Skin application

Mouse: Groups of 40 female NMRI mice, 10 weeks old, received coronene (purity, >96%) at doses of 5 or 15 μg/animal in 0.05 ml dimethyl sulphoxide applied four times per week for 104 weeks; 2/40 animals in the high-dose and 1/39 animals in the low-dose group developed local tumours (sarcomas in the high-dose group and a carcinoma in the low-dose group) at the site of application. No tumour developed in 36 control animals given dimethyl sulphoxide [$p > 0.09$] (Habs *et al.*, 1980).

In an initiation-promotion experiment, 20 female ICR/Ha Swiss mice, eight weeks of age, received a total dose of 0.5 mg coronene (purity checked by thin-layer chromatography) in benzene (applied as five consecutive doses of 0.1 mg), followed two weeks later by thrice weekly applications of 25 μg croton resin in acetone for 62 weeks. Papillomas developed in six mice; none occurred in mice receiving coronene alone; 5/20 occurred in mice receiving croton resin alone (one carcinoma was also seen in this last group) and 1/20 in another croton resin control group (Van Duuren *et al.*, 1968).

3.2 Other relevant biological data[1]

(*a*) *Experimental systems*

Toxic effects

The growth rate of mouse ascites sarcoma cells in culture was not inhibited when coronene was added at a concentration of 1 μmol/ml in dimethyl sulphoxide (Pilotti *et al.*, 1975).

No data were available to the Working Group on effects on reproduction and prenatal toxicity or on metabolism and activation.

Mutagenicity and other short-term tests

Results from short-term tests are given in Table 1.

[1] See also 'General Remarks on the Substances Considered', p. 53.

Table 1. Results from short-term tests: Coronene

Test	Organism/assay[a]	Exogenous metabolic system	Reported result	Comments	References
PROKARYOTES					
Mutation	*Salmonella typhimurium* (his$^-$/his$^+$)	3MC-R-PMS	Positive	At 0.1 μmol/plate in strain TA98	Florin *et al.* (1980)
		Aro-R-PMS	Positive	At 2 nmol/plate in strain TA98	Salamone *et al.* (1979); Hermann (1981)
	Salmonella typhimurium (8AG^S/8AG^R)	Aro-R-PMS, PB-R-PMS	Negative	Tested at up to 170 nmol/ml in strain TM677	Kaden *et al.* (1979)

[a] For an explanation of the abbreviations, see the Appendix, p. 452.

(*b*) *Humans*

No data were available to the Working Group.

3.3 Case reports and epidemiological studies of carcinogenicity to humans

No data were available to the Working Group.

4. Summary of Data Reported and Evaluation[1]

4.1 Experimental data

Coronene was tested for carcinogenicity in one experiment in mice by skin application. No significant increase in the incidence of skin tumours was observed. It was also tested in the mouse-skin initiation-promotion assay and was active as an initiator.

No data on the teratogenicity of this compound were available.

Coronene was mutagenic to *Salmonella typhimurium* in the presence of an exogenous metabolic system.

There is *inadequate evidence* that coronene is active in short-term tests.

[1] For definitions of the italicized terms, see Preamble, p. 19.

4.2 Human data[1]

Coronene is present as a minor component of the total content of polynuclear aromatic compounds in the environment; it is a major component of the polynuclear aromatic compound content of gasoline engine exhaust.

4.3 Evaluation

The available data are inadequate to permit an evaluation of the carcinogenicity of coronene to experimental animals.

5. References

Clar, E., ed. (1964) *Polycyclic Hydrocarbons*, Vol. 2, New York, Academic Press, pp. 79-86

Florin, I., Rutberg, L., Curvall, M. & Enzell. C.R. (1980) Screening of tobacco smoke constituents for mutagenicity using the Ames' test. *Toxicology*, *18*, 219-232

Grimmer, G. & Böhnke, H. (1975) Possible analysis of polycyclic aromatic hydrocarbons and metal content in sediment layers of a lake. *Cancer Lett.*, *1*, 75-84

Grimmer, G. & Böhnke, H. (1977) Investigation on drilling cores of sediments of Lake Constance. I. Profiles of the polycyclic aromatic hydrocarbons (Ger.). *Z. Naturforsch.*, *32c*, 703-711

Grimmer, G. & Hildebrandt, A. (1967) Content of polycylic hydrocarbons in crude vegetable oils. *Chem. Ind.*, 25 November, 2000-2002

Grimmer, G., Böhnke, H. & Glaser, A. (1977) Investigation on the carcinogenic burden by air pollution in man. XV. Polycyclic aromatic hydrocarbons in automobile exhaust gas - An inventory. *Zbl. Bakt. Hyg., I Abt., Orig. B 164*, 218-234

Grimmer, G., Jacob, J., Naujack, K.-W. & Dettbarn, G. (1981) Profile of the polycyclic aromatic hydrocarbons from used engine oil - Inventory by GCGC/MS - PAH in environmental materials, Part 2. *Fresenius Z. anal. Chem.*, *309*, 13-19

---

[1] Studies on occupational exposure to polynuclear aromatic compounds will be considered in future *IARC Monographs*.

Habs, M., Schmähl, D. & Misfeld, J. (1980) Local carcinogenicity of some environmentally relevant polycyclic aromatic hydrocarbons after lifelong topical application to mouse skin. *Arch. Geschwulstforsch.*, *50*, 266-274

Hermann, M. (1981) Synergistic effects of individual polycyclic aromatic hydrocarbons on the mutagenicity of their mixtures. *Mutat. Res.*, *90*, 399-409

Just, J., Borkowska, M. & Maziarka, S. (1972) Air pollution by (Pol.). *Rocz. Panstw. Zakł Hyg.*, *23*, 129-135

Kaden, D.A., Hites, R.A. & Thilly, W.G. (1979) Mutagenicity of soot and associated polycyclic aromatic hydrocarbons to *Salmonella typhimurium*. *Cancer Res*, *39*, 4152-4159

Lang, K.F. & Eigen, I. (1967) *Organic compounds in coal tar* (Ger.). In: Heilbronner, E., Hofmann, U., Schäfer, K. & Wittig, G., eds, *Fortschifte der chemischer Forschung* (Advances in chemical research), Vol. 8, Berlin, Springer, pp. 91-170

Lijinsky, W. & Shubik, P. (1964) Benzo(*a*)pyrene and other polynuclear hydrocarbons in charcoal-broiled meat. *Science*, *145*, 53-55

Lyons, M.J. (1962) Comparison of aromatic polycyclic hydrocarbons from gasoline engine and diesel engine exhaust, general atmospheric dust, and cigarette-smoke condensate. *Natl Cancer Inst. Monogr.*, *9*, 193-199

Müller, K. & Meyer, J.P. (1974) *Einfluss von Ottokraftstoffen auf die Emission von polynukleären aromatischen Kohlenwasserstoffen in Automobilabgasen im Europa-Test* (Effect of gasoline components on emission of polynuclear aromatic hydrocarbons in car exhaust in the Europa test) (*Forschungsbericht 4568*), Hamburg, Deutsche Gesellschaft für Mineralölwissenschaft und Kohlechemie e.V.

NIH/EPA Chemical Information System (1982) *Mass Spectral Search System*, Washington DC, CIS Project, Information Services Corporation

Pilotti, A., Ancker, K., Arrhenius, E. & Enzell, C. (1975) Effects of tobacco and tobacco smoke constituents on cell multiplication *in vitro*. *Toxicology*, *5*, 49-62

Salamone, M.F., Heddle, J.A. & Katz, M. (1979) The mutagenic activity of thirty polycyclic aromatic hydrocarbons (PAH) and oxides in urban airborne particulates. *Environ. Int.*, *2*, 37-43

Snook, M.E., Severson, R.F., Arrendale, R.F., Higman, H.C. & Chortyk, O.T. (1977) The identification of high molecular weight polynuclear aromatic hydrocarbons in a biologically active fraction of cigarette smoke condensate. *Beitr. Tabakforsch.*, *9*, 79-101

Van Duuren, B.L., Sivak, A., Langseth, L., Goldschmidt, B.M. & Segal, A. (1968) Initiators and promotors in tobacco carcinogenesis. *Natl Cancer Inst. Monogr.*, *28*, 173-180

Weast, R.C., ed. (1975) *CRC Handbook of Chemistry and Physics*, 56th ed., Cleveland, OH, Chemical Rubber Co., p. C-248

CYCLOPENTA[*cd*]PYRENE

1. Chemical and Physical Data

1.1 Synonyms and trade names

Chem. Abstr. Services Reg. No.: 27208-37-3

Chem. Abstr. Name: Cyclopenta(cd)pyrene

IUPAC Systematic Name: Cyclopenta[*cd*]pyrene

Synonyms: Acepyrene; cyclopenteno[*cd*]pyrene

1.2 Structural and molecular formula and molecular weight

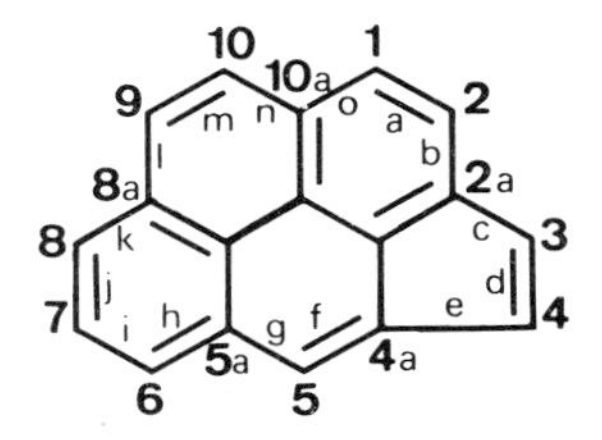

$C_{18}H_{10}$

Mol. wt: 226.3

1.3 Chemical and physical properties of the pure substance

(*a*) *Melting-point*: 170°C (Karcher *et al.*, 1983)

(*b*) *Spectroscopy data*: λ_{max} 238, 248, 286, 308, 338, 354, 366, 374, 386 nm (in cyclohexane) (Karcher *et al.*, 1983)

(*c*) *Solubility*: Soluble in acetone (Habs *et al.*, 1980) (*d*) *Stability*: No data were available.

(*e*) *Reactivity*: No data were available.

2. Production, Use, Occurrence and Analysis

2.1 Production and use

There is no commercial production or known use of this compound.

2.2 Occurrence and analysis

Data on occurrence and methods of analysis are summarized in the 'General Remarks on the Substances Considered', p. 35.

Cyclopenta[*cd*]pyrene occurs ubiquitously in products of incomplete combustion; it also occurs in fossil fuels and in particularly high levels in gasoline engine exhaust and carbon black. It has been identified in polluted air (Grimmer *et al.*, 1980); exhaust from gasoline engines (750-987 μg/l burned fuel) (Grimmer *et al.*, 1977); emissions from burned coals (0.00-3.8 mg/kg) (Brockhaus & Tomingas, 1978); combustion particles (Lee *et al.*, 1977); and carbon black (Gold, 1975; Wallcave *et al.*, 1975; Lee & Hites, 1976; Locati *et al.*, 1979).

3. Biological Data Relevant to the Evaluation of Carcinogenic Risk to Humans

3.1 Carcinogenicity studies in animals

Skin application

Mouse: A group of 40 female NRMI mice, 10 weeks old, received twice weekly skin applications of 27.2 μg cyclopenta[*cd*]pyrene (purity, >96%) in acetone for life (112 weeks). In 3/38 surviving mice two skin carcinomas and one sarcoma were observed. No skin tumour occurred when doses of 1.7 μg or 6.8 μg/animal were administered twice weekly for life or in controls (Habs *et al.*, 1980).

Groups of 30 female Swiss albino mice, nine weeks old, received skin applications of 1.8, 0.6 or 0.2 μmol [300, 100 or 30 μg] cyclopenta[*cd*]pyrene (purity, >99.9%) in acetone twice weekly on the dorsal skin for 30 weeks. The numbers of mice with skin tumours in the high-, mid- and low-dose groups 57 weeks after the start of treatment, when all the animals were killed, were 7, 11 and 17, respectively. Tumours were predominantly malignant neoplasms: squamous cell carcinomas, keratoacanthomas and basal-cell carcinomas. No skin tumour occurred in 30 solvent-treated controls (Cavalieri *et al.*, 1981).

Groups of 30 female Charles River CD-1 mice, eight weeks of age, received a single initiating application of 2.5, 1.0, 0.4 or 0.1 μmol cyclopenta[*cd*]pyrene (purity, >98%) in acetone, followed one week later by 16 nmol 12-*O*-tetradecanoylphorbol-13-acetate (TPA) twice a week for 26 weeks. The incidences of papillomas in the dosed groups 27 weeks after the start of treatment were 37, 30, 21 and 10%, respectively (Wood *et al.*, 1980).

Groups of 30 female CD-1 mice, nine weeks old, received applications of 0.18, 0.06 or 0.02 μmol cyclopenta[*cd*]pyrene in acetone on the dorsal skin every other day for 20 days, followed one week later by applications of 0.017 μmol TPA twice weekly for 40 weeks. The incidences of skin papilloma in the high-, mid- and low-dose groups at 44 weeks after the start of treatment were 6/29 (21%), 9/29 (31%) and 1/30 (3%), respectively. Skin papillomas developed in 3/29 mice treated with TPA alone (Cavalieri *et al.*, 1981).

Cyclopenta[*cd*]pyrene was tested for initiating activity in groups of 30 female Sencar mice, seven to nine weeks old, at single doses of 10, 100 and 200 μg/mouse followed one week later by twice weekly skin applications of 2 μg TPA for 25 weeks. At that time, at least 28 mice in each group had survived, and the mean numbers of papillomas per mouse (and the percentages of mice with papillomas) were: 0.1 (11%) in the low-dose group, 0.4% (39%) in the mid-dose group and 0.9 (57%) in the high-dose group. In control mice treated with TPA only, the respective values were 0.2 (10%) (Raveh *et al.*, 1982).

3.2 Other relevant biological data[1]

(*a*) *Experimental systems*

No data were available to the Working Group on toxic effects or on effects on reproduction and prenatal toxicity.

Metabolism and activation

The 3,4- and 9,10-dihydrodiols have been detected as metabolites of cyclopenta[*cd*]pyrene following incubation of this compound with rat-liver preparations (Gold & Eisenstadt 1980; Eisenstadt *et al.*, 1981).

---

[1] See also 'General Remarks on the Substances Considered', p. 53.

Cyclopenta[*cd*]pyrene 3,4-oxide, the presumed intermediate metabolic precursor of the 3,4-dihydrodiols, is a direct-acting mutagen in bacteria and can transform mammalian cells morphologically (Gold *et al.*, 1980).

The synthetic compound, 3,4-dihydrocyclopenta[*cd*]pyrene, was tested for carcinogenicity by both skin-application studies and in the mouse-skin initiation-promotion model. It was inactive in both systems (Cavalieri *et al.*, 1981).

Mutagenicity and other short-term tests

Results from short-term tests are given in Table 1.

Table 1. Results from short-term tests: Cyclopenta[*cd*]pyrene

Test	Organism/assay[a]	Exogenous metabolic system	Reported result	Comments	References
PROKARYOTES					
Mutation	*Salmonella typhimurium* (his⁻/his⁺)	Aro-R-PMS	Positive	At 1 μg/plate in strain TA1537	Eisenstadt & Gold (1978)
		Aro-R- Cyt P450, Aro-R-Micr	Positive	At 4 nmol/plate in strain TA98	Wood *et al.* (1980)
	Salmonella typhimurium ($8AG^S/8AG^R$)	Aro-R-PMS	Positive	At 7 nmol/ml in strain TM677	Kaden *et al.* (1979)
MAMMALIAN CELLS *IN VITRO*					
Mutation	Human lymphoblastoid HH-4 cells ($6TG^S/6TG^R$)	Aro-R-PMS	Positive	At 4 μg/ml	Skopek *et al.* (1979)
	Mouse lymphoma L5178Y cells (TFT^S/TFT^R)	Aro-R-PMS	Positive	At 1.6 μg/ml	Gold *et al.* (1980)
Cell transformation	Mouse C3H 10T1/2 CL8 cells (morphological)	--	Positive	At 1 μg/ml	Gold *et al.* (1980)

[a] For an explanation of the abbreviations, see the Appendix, p. 452.

(*b*) *Humans*

No data were available to the Working Group.

3.3 Case reports and epidemiological studies of carcinogenicity to humans

No data were available to the Working Group.

4. Summary of Data Reported and Evaluation[1]

4.1 Experimental data[2]

Cyclopenta[*cd*]pyrene was tested for carcinogenicity in two studies in female mice by skin application; it produced skin tumours in one study. It was also tested in the mouse-skin initiation-promotion assay in three studies and was active as an initiator.

No data were available on the teratogenicity of this chemical.

Cyclopenta[*cd*]pyrene was mutagenic to *Salmonella typhimurium* and to mammalian cells *in vitro* in the presence of an exogenous metabolic system. It induced morphological transformation in mammalian cells.

There is *sufficient evidence* that cyclopenta[*cd*]pyrene is active in short-term tests.

4.2 Human data[3]

Cyclopenta[*cd*]pyrene is present as a minor component of the total content of polynuclear aromatic compounds in the environment but occurs as a major polynuclear aromatic component of gasoline engine exhaust.

4.3 Evaluation

There is *limited evidence* that cyclopenta[*cd*]pyrene is carcinogenic to experimental animals.

---

[1] For definitions of the italicized terms, see Preamble, p. 19.

[2] Subsequent to the meeting, the Secretariat became aware of a study (Cavalieri *et al.*, 1983) in which cyclopenta[*cd*]pyrene was tested alone or in combination with benzo[*a*]pyrene by repeated application to the skin of female Swiss mice. At the end of the experiment, analysis of the incidence of malignant skin tumours showed that the two compounds acted synergistically.

[3] Studies on occupational exposure to polynuclear aromatic compounds will be considered in future *IARC Monographs*.

5. References

Brockhaus, A. & Tomingas, R. (1976) Emission of polycylic hydrocarbons during burning processes in small heating installations and their concentration in the atomosphere (Ger.). *Staub-Reinhalt. Luft*, *36*, 96-101

Cavalieri; E., Rogan, E., Toth, B. & Munhall, A. (1981) Carcinogenicity of the environmental pollutants cyclopenteno[*c,d*]pyrene and cyclopentano[*c,d*]pyrene in mouse skin. *Carcinogenesis*, *2*, 277-281

Cavalieri, E., Munhall, A., Rogan, E., Salmasi, S. & Patil, K. (1983) Syncarcinogenic effect of the environmental pollutants cyclopenteno[*cd*]pyrene and benzo[*a*]pyrene in mouse skin. *Carcinogenesis*, *4*, 393-397

Eisenstadt, E. & Gold, A. (1978) Cyclopenta(*c,d*)pyrene: A highly mutagenic polycyclic aromatic hydrocarbon. *Proc. natl, Acad. Sci. USA*, *75*, 1667-1669

Eisenstadt, E., Shpizner, B. & Gold, A. (1981) Metabolism of cyclopenta(*cd*)pyrene at the K-region by microsomes and a reconstituted cytochrome P-450 system from rat liver. *Biochem. biophys. Res. Commun.*, *100*, 965-971

Gold, A. (1975) Carbon black adsorbates: Separation and identification of a carcinogen and some oxygenated polyaromatics. *Anal. Chem.*, *47*, 1469-1472

Gold, A. & Eisenstadt, E. (1980) Metabolic activation of cyclopenta(*cd*)pyrene to 3,4-epoxycyclopenta(*cd*)pyrene by rat liver microsomes. *Cancer Res.*, *40*, 3940-3944

Gold, A., Nesnow, S., Moore, M., Garland, H., Curtis, G., Howard, B., Graham, D. & Eisenstadt, E. (1980) Mutagenesis and morphological transformation of mammalian cells by a non-bay-region polycyclic cyclopenta(*cd*)pyrene and its 3,4-oxide. *Cancer Res.*, *40*, 4482-4484

Grimmer, G., Böhnke, H. & Glaser, A. (1977) Investigation on the carcinogenic burden of air pollution in man. XV. Polycyclic aromatic hydrocarbons in automobile exhaust gas - An inventory. *Zbl. Bakt. Hyg., 1 Abt., Orig. B164*, 218-234

Grimmer, G., Naujack, K.-W. & Schneider, D. (1980) *Changes in PAH-profiles in different areas of a city during the year.* In: Bjørseth, A. & Dennis, A.J., eds, *Polynuclear Aromatic Hydrocarbons: Chemistry and Biological Effects*, *4th Int. Symposium*, Columbus, OH, Battelle Press, pp. 107-125

Habs, M., Schmähl, D. & Misfeld, J. (1980) Local carcinogenicity of some environmentally relevant polycyclic aromatic hydrocarbons after lifelong topical application mouse skin. *Arch. Geschwulstforsch.*, *50*, 266-274

Kaden, D.A., Hites, R.A. & Thilly, W.G. (1979) Mutagenicity of soot and associated polycyclic aromatic hydrocarbons to *Salmonella typhimurium*. *Cancer Res.*, *39*, 4152-4159

Karcher, W., Fordham, R., Dubois, J. & Gloude, P. (1983) *Spectral Atlas of Polycyclic Aromatic Compounds*, Dordrecht, The Netherlands, D. Reidel (in press)

Lee, M.L. & Hites, R.A. (1976) Characterization of sulfur-containing polycyclic aromatic compounds in carbon blacks. *Anal. Chem.*, *48*, 1890-1893

Lee, M.L., Prado, G.P., Howard, J.B. & Hites, R.A. (1977) Source identificaton of urban airborne polycyclic aromatic hydrocarbons by gas chromatographic mass spectrometry and high resolution mass spectrometry. *Biomed. Mass Spectrom.*, *4*, 182-187

Locati, G., Fantuzzi, A., Consonni, G., Li Gotti, I. & Bonomi, G. (1979) Identification of polycyclic aromatic hydrocarbons in carbon black with reference to cancerogenic risk in fire production. *Am. ind. Hyg. Assoc. J.*, *40*, 644-652

Raveh, D., Slaga, T.J. & Huberman, E. (1982) Cell-mediated mutagenesis and tumor-initiating activity of the ubiquitous polycyclic hydrocarbon, cyclopenta[*c,d*]pyrene. *Carcinogenesis*, *3*, 763-766

Skopek, T.R., Liber, H.L., Kaden, D.A., Hites, R.A. & Thilly, W.G. (1979) Mutation of human cells by kerosene soot. *J. natl Cancer Inst.*, *63*, 309-312

Wallcave, L., Nagel, D.L., Smith, J.W. & Waniska, R.D. (1975) Two pyrene derivatives of widespread environmental distribution: Cyclopenta(*cd*)pyrene and acepyrene. *Environ. Sci. Technol.*, *9*, 143-345

Wood, A.W., Levin, W., Chang, R.L., Huang, M.-T., Ryan, D.E., Thomas, P.E., Lehr, R.E., Kumar, S., Koreeda, M., Akagi, H., Ittah, Y., Dansette, P., Yagi, H., Jerina, D.M. & Conney, A.H. (1980) Mutagenicity and tumor-initiating activity of cyclopenta(*c,d*)pyrene and structurally related compounds. *Cancer Res.*, *40*, 642-649

DIBENZ[*a,h*]ACRIDINE

This compound was considered by a previous working group, in December 1972 (IARC, 1973).

1. Chemical and Physical Data

1.1 Synonyms and trade-names

Chem. Abstr. Services Reg. No.: 226-36-8

Chem. Abstr. Name: Dibenz(a,h)acridine

IUPAC Systematic Name: Dibenz[*a,h*]acridine

Synonyms: 7-Azadibenz(a,h)anthracene; DB(a,h)AC; 1,2:5,6-dibenzacridine; 1,2,5,6-dibenzacridne; dibenz(a,d)acridine; 1,2,5,6-dibenzoacridine; 1,2,5,6-dinaphthacridine

1.2 Structural and molecular formulae and molecular weight

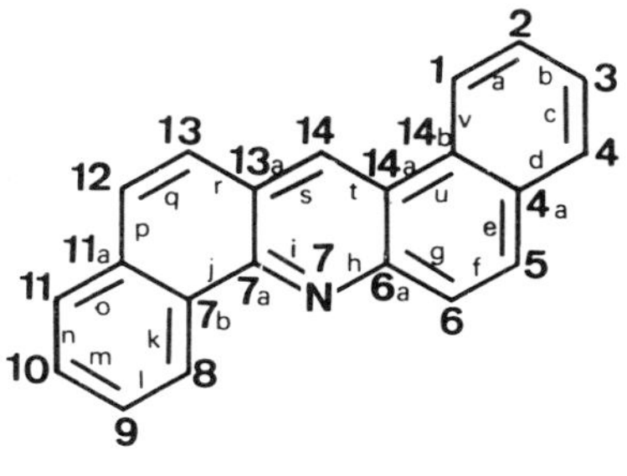

$C_{21}H_{13}N$

Mol. wt: 279.4

1.3 Chemical and physical properties of the pure substance

From National Library of Medicine (1982), unless otherwise specified

(*a*) *Description*: Yellow crystals

(*b*) *Melting-point*: 226°C (Karcher *et al.*, 1983)

(*c*) *Spectroscopy data*: λ_{max} 219, 224, 246, 258, 267, 287, 295, 316, 330, 344, 353, 363, 372, 382, 392 (in cyclohexane) (Karcher *et al.*, 1983); infrared and nuclear magnetic resonance spectra have also been reported (Pouchert, 1970; Karcher *et al.*, 1983).

(*d*) *Solubility*: Sparingly soluble in ethanol; soluble in benzene, acetone (Lacassagne *et al.*, 1956) and cyclohexane

(*e*) *Stability*: No data were available.

(*f*) *Reactivity*: Reacts in general as an acridine (Raulins, 1973; Selby, 1973)

2. Production, Use, Occurrence and Analysis

2.1 Production and use

There is no commercial production or known use of this compound.

2.2 Occurrence and analysis

Data on occurrence and methods of analysis are summarized in the 'General Remarks on the Substances Considered', p. 35.

Dibenz[*a,h*]acridine has been identified in cigarette smoke (0.01 μg/100 cigarettes) (Van Duuren *et al.*, 1960); air pollution from coal heating (<0.12-17 μg/m^3), industrial effluent (0.7 μg/m^3), a coal-tar-pitch polluted air sample (0.01 μg/m^3) (Sawicki *et al.*, 1965b); and gasoline engine exhaust (<300 μg/kg benzene-soluble fraction) (Sawicki *et al.*, 1965a).

3. Biological Data Relevant to the Evaluation of Carcinogenic Risk to Humans

3.1 Carcinogenicity studies in animals

There is *sufficient evidence* that dibenz[*a,h*]acridine is carcinogenic to experimental animals (see 'General Remarks on the Substances Considered', p. 33).

3.2 Other relevant biological data[1]

(*a*) *Experimental systems*

Toxic effects

The i.p. administration of a single dose per animal of 10 mg dibenz[*a,h*]acridine in sesame oil produced an immediate and persistent reduction in the growth rate of young rats (Haddow *et al.*, 1937).

No data were available to the Working Group on reproduction and prenatal toxicity or on metabolism.

Mutagenicity and other short-term tests

Results from short-term tests are given in Table 1.

Table 1. Results from short-term tests: Dibenz[*a,h*]acridine

Test	Organism/assay[a]	Exogenous metabolic system[a]	Reported	Comments	References
PROKARYOTES					
Mutation	*Salmonella typhimurium* (his⁻/his⁺)	Aro-R-PMS	Negative	Tested at up to 1000 μg/plate in strains TA1535, TA1537, TA1538, TA98 and TA100	Salamone *et al.* (1979)
		PCB-R-PMS	Positive	At 4 μg/plate in strain TA100	Kitahara *et al.* (1978)

[a] For an explanation of the abbreviations, see the Appendix, p. 452.

(*b*) *Humans*

No data were available to the Working Group.

3.3 Case reports and epidemiological studies of carcinogenicity to humans

No data were available to the Working Group.

[1] See also 'General Remarks on the Substances Considered', p. 53.

4. Summary of Data Reported and Evaluation[1]

4.1 Experimental data

Dibenz[*a*,*h*]acridine has been shown to be carcinogenic to experimental animals (see 'General Remarks on the Substances Considered', p. 33, and IARC, 1973).

No data on the teratogenicity of this compound were available.

Both positive and negative results were obtained in tests of mutagenicity in *Salmonella typhimurium*.

There is *inadequate evidence* that dibenz[*a*,*h*]acridine is active in short-term tests.

4.2 Human data[2]

Dibenz[*a*,*h*]acridine is present as a minor component of the total content of polynuclear aromatic compounds in tobacco smoke and urban pollutants.

4.3 Evaluation[3]

There is *sufficient evidence* that dibenz[*a*,*h*]acridine is carcinogenic to experimental animals.

---

[1] For definitions of the italicized terms, see Preamble, p. 19.

[2] Studies on occupational exposure to polynuclear aromatic compounds will be considered in future *IARC Monographs*.

[3] In the absence of adequate data on humans it is reasonable, for practical purposes, to regard chemicals for which there is *sufficient evidence* of carcinogenicity in animals as if they presented a carcinogenic risk to humans (see also Preamble, p. 20).

5. References

Haddow, A., Scott, C.M. & Scott, J.D. (1937) The influence of certain carcinogenic and other hydrocarbons on body growth in the rat. *Proc. R. Soc. London Ser. B*, *122*, 477-507

IARC (1973) *IARC Monographs on the Evaluation of Carcinogenic Risk of Chemicals to Man*, Vol. 3, *Certain polycyclic aromatic hydrocarbons and heterocyclic compounds*, Lyon, pp. 247-253

Karcher, W., Fordham, R., Dubois, J. & Gloude, P. (1983) *Spectral Atlas of Polycyclic Aromatic Compounds*, Dordrecht, The Netherlands, D. Reidel (in press)

Kitahara, Y., Okuda, H., Shudo, K., Okamoto, T., Nagao, M., Seino, Y. & Sugimura, T. (1978) Synthesis and mutagenicity of 10-azabenzo[*a*]pyrene-4,5-oxide and other pentacyclic aza-arene oxides. *Chem. pharm. Bull.*, *26*, 1950-1953

Lacassagne, A., Buu-Hoï, N.P., Daudel, R. & Zajdela, F. (1956) *The relation between carcinogenic activity and the physical and chemical properties of angular benzoacridines.* In: Greenstein, J.P. & Haddow, A., eds, *Advances in Cancer Research*, Vol. 4, New York, Academic Press, pp. 315-360

National Library of Medicine (1982) *Toxicology Data Bank*, Bethesda, MD, National Library of Medicine Specialized Information Services, Toxicology Information Program

Pouchert, C.J., ed. (1970) *The Aldrich Library of Infrared Spectra*, Milwaukee, WI, Aldrich Chemical Company, p. 1011

Raulins, N.R. (1973) *Acridines*. In: Acheson, R.M., ed., *Acridines*, 2nd ed., New York, Interscience, pp. 63-67

Salamone, M.F., Heddle, J.A. & Katz, M. (1979) The mutagenic activity of thirty polycyclic aromatic hydrocarbons (PAH) and oxides in urban airborne particulates. *Environ. Int.*, *2*, 37-43

Sawicki, E., Meeker, J.E. & Morgan, M.J. (1965a) Polynuclear aza compounds in automotive exhaust. *Arch. environ. Health*, *11*, 773-775

Sawicki, E., Meeker, J.E. & Morgan, M.J. (1965b) The quantitative comparison of air pollution source effluents in terms of aza heterocyclic compounds and polynuclear aromatic hydrocarbons. *Int. J. Air Water Pollut.*, *9*, 291-298

Selby, I.A. (1973) *Acridinium salts and reduced acridines*. In: Acheson, R.M., ed., *Acridines*, 2nd ed., New York, Interscience, pp. 434-445

Van Duuren, B.L., Bilbao, J.A. & Joseph, C.A. (1960) The carcinogenic nitrogen heterocyclics in cigarette-smoke condensate. *J. natl Cancer Inst.*, *25*, 53-61

DIBENZ[*a,j*]ACRIDINE

This compound was considered by a previous working group, in December 1972 (IARC, 1973).

1. Chemical and Physical Data

1.1 Synonyms and trade names

Chem. Abstr. Services Reg. No.: 224-42-0

Chem. Abstr. Name: Dibenz(a,j)acridine

IUPAC Systematic Name: Dibenz[*a,j*]acridine

Synonyms: 7-Azadibenz(a,j)anthracene; DB(a,j)AC; 1,2:7,8-dibenzacridine; 1,2,7,8-dibenzacridine; 3,4,5,6-dibenzacridine; dibenz(a,f)acridine; dibenzo(a,j)acridine; 3,4,6,7-dinaphthacridine

1.2 Structural and molecular formulae and molecular weight

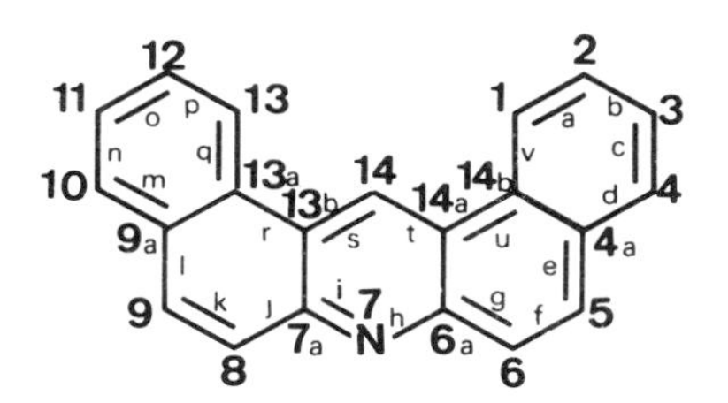

$C_{21}H_{13}N$

Mol. wt: 279.3

1.3 Chemical and physical properties of the pure substance

From National Library of Medicine (1982), unless otherwise specified

(*a*) *Description*: Yellow needles or prisms

(*b*) *Melting-point*: 216°C (Karcher *et al.*, 1983)

(*c*) *Spectroscopy data*: λ_{max} 213, 224, 236, 255, 270, 293, 300, 320, 334, 352, 362, 371, 381, 391 nm (in cyclohexane) (Karcher *et al.*, 1983); infrared spectra have also been reported (Pouchert, 1970). Mass and nuclear magnetic resonance spectra have been tabulated (NIH/EPA Chemical Information System, 1982; Karcher *et al.*, 1983).

(*d*) *Solubility*: Sparingly soluble in benzene; soluble in ethanol and acetone (Lacassagne *et al.*, 1956)

(*e*) *Stability*: Stable to oxidation by air (Blout & Corley, 1947)

(*f*) *Reactivity*: Reacts in general as an acridine (Raulins, 1973; Selby, 1973). Catalytic reduction yields Morgan's base (a molecular compound with 7,14-dihydro-dibenz[*a,j*]acridine) (Blout & Corley, 1947).

2. Production, Use, Occurrence and Analysis

2.1 Production and use

There is no commercial production or known use of this compound.

2.2 Occurrence and analysis

Data on occurrence and methods of analysis are summarized in the 'General Remarks on the Substances Considered', p. 35.

Dibenz[*a,j*]acridine has been identified in cigarette smoke (0.27-1 μg/100 cigarettes) (Van Duuren *et al.*, 1960; Wynder & Hoffmann, 1963); polluted air of coal heating (<0.15-2 μg/m³), industrial effluent (1.8 μg/m³), a coal-tar-pitch-polluted air sample (0.001 μg/m³) (Sawicki *et al.*, 1965b); and gasoline engine exhaust (<300 μg/kg benzene-soluble fraction) (Sawicki *et al.*, 1965a).

3. Biological Data Relevant to the Evaluation of Carcinogenic Risk to Humans

3.1 Carcinogenicity studies in animals

There is *sufficient evidence* that dibenzo[*a,j*]acridine is carcinogenic to experimental animals (see 'General Remarks on the Substances Considered', p. 33).

3.2 Other relevant biological data[1]

(*a*) *Experimental systems*

No data were available to the Working Group on toxic effects or on metabolism.

Effects on reproduction and prenatal toxicity

The only study available was reported in the form of an abstract; it suggested that placental transfer of dibenz[*a,j*]acridine or its metabolites occurs (Reno, 1969).

Mutagenicity and other short-term test

Results from short-term tests are given in Table 1.

Table 1. Results from short-term tests: Dibenz[*a,j*]acridine

Test	Organism/assay[a]	Exogenous metabolic system[a]	Reported result	Comments	References
PROKARYOTES					
Mutation	*Salmonella typhimurium* (his⁻/his⁺)	PCB-R-PMS, Aro-R-PMS	Positive	At 5 μg/plate in strain TA100	Kitahara *et al.* (1978); McCann *et al.* (1975)
		Aro-R-PMS	Positive	At 5 μg/plate in strain TA98	Ho *et al.* (1981)
		Aro-R-PMS, 3MC-GP-PMS	Positive	At 25 μg/plate in strain TA100	Baker *et al.* (1980)
MAMMALIAN CELLS *IN VITRO*					
DNA damage	Primary rat hepatocytes (unscheduled DNA synthesis)	--	Negative	Tested at up to 1 μmol/ml	Probst *et al.* (1981)

[a] For an explanation of the abbreviations, see the Appendix, p. 452.

(*b*) *Humans*

No data were available to the Working group.

[1] See also 'General Remarks on the Substances Considered', p. 53.

3.3 Case reports and epidemiological studies of carcinogenicity to humans

No data were available to the Working Group.

4. Summary of Data Reported and Evaluation[1]

4.1 Experimental data

Dibenz(*a,j*]acridine has been shown to be carcinogenic to experimental animals (see 'General Remarks on the Substances Considered', p. 33, and IARC, 1973).

No data on the teratogenicity of this compound were available.

Dibenz[*a,j*]acridine was mutagenic to *Salmonella typhimurium* in the presence of an exogenous metabolic system. It did not induce unscheduled DNA synthesis in rat hepatocytes.

There is *inadequate evidence* that dibenz[*a,j*]acridine is active in short-term tests.

4.2 Human data[2]

Dibenz[*a,j*]acridine occurs in tobacco smoke and urban pollutants.

4.3 Evaluation[3]

There is *sufficient evidence* that dibenz[*a,j*]acridine is carcinogenic to experimental animals.

---

[1] For definitions of the italicized terms, see Preamble, p. 19.

[2] Studies on occupational exposure to polynuclear aromatic compounds will be considered in future *IARC Monographs*.

[3] In the absence of adequate data on humans it is reasonable, for practical purposes, to regard chemicals for which there is *sufficient evidence* of carcinogenicity in animals as if they presented a carcinogenic risk to humans (see also Preamble, p. 20).

5. References

Blout, E.R. & Corley, R.S. (1947) The reaction of β-naphthol, β-naphthylamine and formaldehyde. III. The dibenzacridine products. *J. Am. chem. Soc.*, *69*, 763-769

Ho, C.H., Clark, B.R., Guerin, M.R., Barkenbus, B.D., Rao, T.K. & Epler, J.L. (1981) Analytical and biological analyses of test materials from the synthetic fuel technologies. IV. Studies of chemical structure-mutagenic activity relationships of aromatic nitrogen compounds relevant to synfuels. *Mutat. Res.*, *85*, 335-345

IARC (1973) *IARC Monographs on the Evaluation of Carcinogenic Risk to Man*, Vol. 3, *Certain polycyclic aromatic hydrocarbons and heterocyclic compounds*, Lyon, pp. 254-259

Karcher, W., Fordham, R., Dubois, J. & Gloude, P. (1983) *Spectral Atlas of Polycyclic Aromatic Compounds*, Dordrecht, The Netherlands, D., Reidel (in press)

Kitahara, Y., Okuda, H., Shudo, K., Okamoto, T., Nagao, M., Seino, Y. & Sugimura, T. (1978) Synthesis and mutagenicity of 10-azabenzo[*a*]pyrene-4,5-oxide and other pentacyclic aza-arene oxides. *Chem. pharm. Bull.*, *26*, 1950-1953

Lacassagne, A., Buu-Hoï, N.P., Daudel, R. & Zajdela, F. (1956) *The relation between carcinogenic activity and physical and chemical properties of angular benzoacridines*. In: Greenstein, J.P. & Haddow, A., eds, *Advances in Cancer Research*, Vol. 4, New York, Academic Press, pp. 315-360

McCann, J., Choi, E., Yamasaki, E. & Ames, B.N. (1975) Detection of carcinogens as mutagens in the *Salmonella*/microsome test: Assay of 300 chemicals. *Proc. natl Acad. Sci. USA*, *72*, 5135-5139

National Library of Medicine (1982) *Toxicology Data Bank*, Bethesda, MD, National Library of Medicine Specialized Information Services, Toxicology Information Program

NIH/EPA Chemical Information System (1982) *Mass Spectral Search System*, Washington DC, CIS Project, Information Services Corporation

Pouchert, C.J., ed. (1970) *The Aldrich Library of Infrared Spectra*, Milwaukee, WI, Aldrich Chemical Company, p. 1011

Probst, G.S., McMahon, R.E., Hill, L.E., Thompson, C.Z., Epp, J.K. & Neal, S.B. (1981) Chemically-induced unscheduled DNA synthesis in primary rat hepatocyte cultures: A comparison with bacterial mutagenicity using 218 compounds. *Environ. Mutagenesis*, *3*, 11-32

Raulins, N.R. (1973) *Acridines*. In: Acheson, R.M., ed., *Acridines*, 2nd ed., New York, Interscience, pp. 63-67

Reno, F.E. (1969) Toxic effects of polynuclear aromatic hydrocarbons in chicken embryos and mice. *Diss. Abstr. Int.*, *B29*, 4777

Sawicki, E., Meeker, J.E. & Morgan, M.J. (1965a) Polynuclear aza compounds in automotive exhaust. *Arch. environ. Health*, *11*, 773-775

Sawicki, E., Meeker, J.E. & Morgan, M.J. (1965b) The quantitative comparison of air pollution source effluents in terms of aza heterocyclic compounds and polynuclear aromatic hydrocarbons. *Int. J. Air Water. Pollut.*, *9*, 291-298

Selby, I.A. (1973) *Acridinium salts and reduced acridines*. In: Acheson, R.M., ed., *Acridines*, 2nd ed., New York, Interscience, pp. 434-445

Van Duuren, B.L., Bilbao, J.A. & Joseph, C.A. (1960) The carcinogenic nitrogen heterocyclics in cigarette-smoke condensate. *J. natl Cancer Inst.*, *25*, 53-61

Wynder, E.L. & Hoffmann, D. (1963) Experimental contribution to tobacco smoke carcinogenesis (Ger.). *Dtsch. med. Wochenschr.*, *88*, 623-628

DIBENZ[*a*,*c*]ANTHRACENE

1. Chemical and Physical Data

1.1 Synonyms and trade names

Chem. Abstr. Services Reg. No.: 215-58-7

Chem. Abstr. Name: Benzo(b)triphenylene

IUPAC Systematic Name: Benzo[*b*]triphenylene

Synonyms: Dibenzo[*a*,*c*]anthracene; 1,2:3,4-dibenzanthracene; 1,2:3,4-dibenzoanthracene

1.2 Structural and molecular formulae and molecular weight

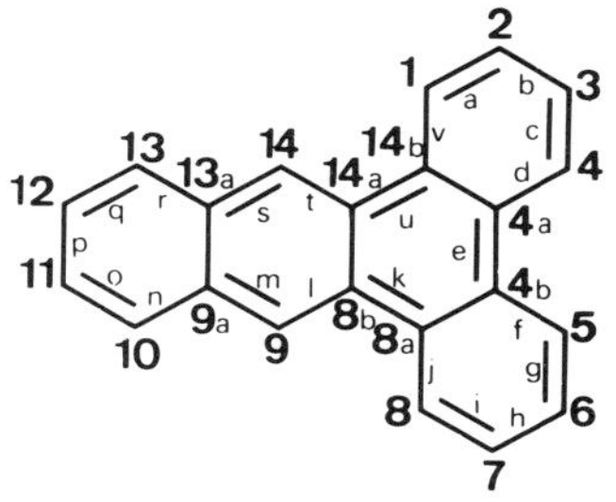

$C_{22}H_{14}$

Mol. wt: 278.4

1.3 Chemical and physical properties of the pure substance

(*a*) *Description*: Needles (recrystallized from acetic acid or ethanol) (Weast, 1975)

(*b*) *Melting-point*: 205.7% (Karcher *et al.*, 1983)

(*c*) *Spectroscopy data*: λ_{max} 217, 241, 247, 264, 275, 286, 305, 322, 332, 349, 374 nm (in cyclohexane). Nuclear magnetic resonance spectra have been reported (Karcher *et al.*, 1983).

(*d*) *Solubility*: Practically insoluble in water; slightly soluble in petroleum ether (a low-boiling petroleum fraction); soluble in hot acetic acid and benzene

(*e*) *Stability*: No data were available.

(*f*) *Reactivity*: No data were available.

2. Production, Use, Occurrence and Analysis

2.1 Production and use

There is no commercial production or known use of this compound. A reference material of certified high purity is available (Community Bureau of Reference, 1982).

2.2 Occurrence and analysis

Data on occurrence and methods of analysis are summarized in the 'General Remarks on the Substances Considered', p. 35.

Dibenz[*a*,*c*]anthracene occurs ubiquitously in products of incomplete combustion; it also occurs in fossil fuels. It has been identified in mainstream cigarette smoke (Lee *et al.*, 1976; Severson *et al.*, 1977; Snook *et al.*, 1977) and smoke of marijuana cigarettes (Lee *et al.*, 1976); coal-tar (Lang & Eigen, 1967); lubricating (0.08 mg/kg) (Grimmer *et al.*, 1981a) and used motor oils (2.86 mg/kg) (Grimmer *et al.*, 1981b); crude oil (Guerin *et al.*, 1978); and waste water (2.2 μg/l) (Olufsen, 1980).

3. Biological Data Relevant to the Evaluation of Carcinogenic Risk to Humans

3.1 Carcinogenicity studies in animals[1]

Skin application

Mouse: A group of 30 female Swiss mice, eight to ten weeks old, received skin applications of 85 μg dibenzo[*a*,*c*]anthracene (purity, >99%) in acetone twice weekly for 65 weeks. The

[1] The Working Group was aware of two studies by s.c. injection in mice completed and one in mice by skin application in progress (IARC, 1983).

first tumour was observed 60 weeks after the start of treatment, at which time 16 animals were still alive; by 100 weeks, 8/30 mice had developed carcinomas and 1/30 papilloma were observed. No skin tumour occurred in 20 acetone-treated controls, 14 of which survived 60 or more weeks (Lijinsky *et al.*, 1970).

Single applications of 1.0 mg dibenz[*a,c*]anthracene ('plate-chromatographically pure') in 100 μl benzene given to a group of 20 female ICR/Ha Swiss mice, six to eight weeks old, failed to induce tumours during an observation period of 58-60 weeks. In a similar group, the same treatment was followed two weeks later by thrice-weekly applications of 2.5 μg phorbol myristyl acetate per 0.1 ml acetone; 19/20 and 4/20 animals developed papillomas (30) and carcinomas, respectively. Of 20 controls treated with phorbol myristyl acetate, one mouse developed two papillomas (Van Duuren *et al.*, 1970).

A group of 30 female CD-1 mice, eight weeks old, received 2.5 μmol [700 μg] dibenz[*a,c*]anthracene ('thin-layer chromatographically pure') in benzene. One week later the mice were painted with 5 μmol 12-*O*-tetradecanoylphorbol-13-acetate (TPA) twice weekly for up to 34 weeks. Papillomas developed in 63% of treated mice (2.3 tumours/mouse) and in 3% of controls treated only with TPA (Scribner, 1973). [The Working Group noted that the dose of TPA was probably 5 μg (Scribner & Süss, 1978).]

A group of 30 female Sencar mice, seven to nine weeks old, received single applications of 2 μmol dibenz[*a,c*]anthracene (purity, >95%) in acetone followed one week later by twice-weekly applications of 2 μg TPA. At 15 weeks, 27/28 surviving mice had tumours (0.5 papillomas/mouse), compared with 3/30 controls treated with TPA alone (0.1 papillomas/mouse) (Slaga *et al.*, 1980).

3.2 Other relevant biological data[1]

(*a*) *Experimental systems*

Toxic effects

Hyperplasia and cases of so-called 'transitional hyperplasia' were observed when dibenz[*a,c*]anthracene was implanted (in a beeswax pellet) into isogenetically transplanted rat tracheas (Topping *et al.*, 1978).

Effects on reproduction and prenatal toxicity

No data on reproductive or prenatal toxicity were available to the Working Group. Dibenz[*a,c*]anthracene (20 mg/kg bw) given i.p. on days 18-21 of gestation to Sprague-Dawley rats was a more effective inducer of cytochrome P-450 and aminopyrine *N*-demethylase activity in fetal liver than either 3-methylcholanthrene or benzo[*a*]pyrene (Guenthner & Mannering, 1977).

---

[1] See also 'General Remarks on the Substances Considered', p. 53.

Metabolism and activation

The 10,11-dihydrodiol is the major metabolite of dibenz[*a,c*]anthracene, but the 1,2- and 3,4-diols have also been detected following incubation of dibenz[*a,c*]anthracene with rat-liver preparations (Sims, 1970; MacNicoll *et al.*, 1979) and cultures of mouse skin (MacNicoll *et al.*, 1980). The formation of the 10,11-diol-12,13-epoxide from the 10,11-dihydrodiol has been described (Hewer *et al.*, 1981).

The 10,11-dihydrodiol was found to be mutagenic to bacteria in the presence of an exogenous metabolic system (Malaveille *et al.*, 1980) and was active as a tumour-initiating agent (Chouroulinkov *et al.*, 1983). [In an earlier study of shorter duration, no tumour-initiating activity was observed (Slaga *et al.*, 1980).] DNA binding was detected in mouse skin and in mouse embryo cells treated with dibenz[*a,c*]anthracene (Goshman & Heidelberger, 1967; Duncan *et al.*, 1969).

Some of the hydrocarbon-dioxyribonucleoside adducts formed in hamster embryo cells treated with the compound appear to arise from reactions of the 10,11-diol-12,13-epoxide (Hewer *et al.*, 1981)

Mutagenicity and other short-term tests

Results from short-term tests are given in Table 1.

Table 1. Results from short-term tests: Dibenz[*a,c*]anthracene[a]

Test	Organism/assay[b]	Exogenous metabolic system[b]	Reported result	Comments	References
PROKARYOTES					
DNA damage	*Bacillus subtilis* (rec^{+}/rec^{-})	Aro-R-PMS	Positive	Minimal inhibitory concentrations: rec^{+} at 50 μg/well *versus* rec^{-} at 6 μg/well	McCarroll *et al.* (1981)
Mutation	*Salmonella typhimurium* (his^{-}/his^{+})	Aro-R-PMS, 3MC-R-PMS Aro-R-PMS	Positive Positive	At 5 μg/plate in strain TA100 At 3 nmol/plate in strain TA98	McCann *et al.* (1975); Salamone *et al.* (1979); Malaveille *et al.* (1980); Hermann (1981)
	Salmonella typhimurium ($8AG^{S}/8AG^{R}$)	Aro-R-PMS	Positive	At 13 nmol/ml in strain TM677	Kaden *et al.* (1979)
MAMMALIAN CELLS *IN VITRO*					
DNA damage	HeLa cells (unscheduled DNA synthesis)	3MC-R-PMS	Positive	At 100 pmol/ml	Martin *et al.* (1978)
	Primary rat hepatocytes (unscheduled DNA synthesis)		Positive	At 50 nmol/ml	Probst *et al.* (1981)
Mutation	Chinese hamster V79 cells ($8AG^{S}/8AG^{R}$; OUA^{S}/OUA^{R})	SHE feeder layer	Positive	At 1 μg/ml	Huberman & Sachs (1976); Huberman (1978)
	Chinese hamster V79 cells ($6TG^{S}/6TG^{R}$)	3MC-R-PMS	Positive	At 25 nmol/ml	Krahn & Heidelberger (1977)
Chromosome effects	Chinese hamster ovary cells (sister chromatid exchange)	None	Negative	Tested at up to 4 μg/ml	Pal (1981)

Test	Organism/assay[b]	Exogenous metabolic system[b]	Reported result	Comments	References
Cell transformation	Syrian hamster embryo cells (morphological)	--	Positive	At 10 μg/ml At 1.0 μg/ml [Negative data were reported in an earlier study (Pienta *et al.*, 1977).]	DiPaolo *et al.* (1969) Pienta (1980)
	Mouse prostate C3H cells (morphological)	MEF feeder layer	Negative	Tested at up to 5 μg/ml	Chen & Heidelberger (1969)

[a] This table comprises selected assays and references and is not intended to be a complete review of the literature.
[b] For an explanation of the abbreviations, see the Appendix, p. 452.

(*b*) *Humans*

No data were available to the Working Group.

3.3 Case reports and epidemiological studies of carcinogenicity to humans

No data were available to the Working Group.

4. Summary of Data Reported and Evaluation[1]

4.1 Experimental data

Dibenz[*a,c*]anthracene was tested for carcinogenicity in one experiment by skin application to female mice and produced skin tumours. It was also tested in three experiments using the mouse-skin intiation-promotion assay and showed initiating activity.

No data on the teratogenicity of this chemical were available.

In the presence of an exogenous metabolic system, dibenz[*a,c*]anthracene was mutagenic to *Salmonella typhimurium* and mammalian cells in culture. It was positive in an assay for differential survival, using DNA repair-proficient/-deficient strains of *Bacillus subtilis*, and induced unscheduled DNA repair synthesis in cultured mammalian cells. In two of three studies it induced morphological transformation in mammalian cells.

There is *sufficient evidence* that dibenz[*a,c*]anthracene is active in short-term tests.

---

[1] For definitions of the italicized terms, see Preamble, p. 19.

4.2 Human data[1]

Dibenz[*a,c*]anthracene is present as a minor component of the total content of polynuclear aromatic compounds in the environment. Human exposure to dibenz[*a,c*]anthracene occurs primarily through the smoking of tobacco, inhalation of polluted air and by ingestion of food and water contaminated by combustion effluents (for details, see 'General Remarks on the Substances Considered', p. 35).

4.3 Evaluation

There is *limited evidence* that dibenz[*a,c*]anthracene is carcinogenic to experimental animals.

5. References

Chen, T.T. & Heidelberger, C. (1969) Quantitative studies on the malignant transformation of mouse prostate cells by carcinogenic hydrocarbons *in vitro*. *Int. J. Cancer*, *4*, 166-178

Chouroulinkov, I., Coulomb, H., MacNicoll, A.D., Grover, P.L. & Sims, P. (1983) Tumour-initiating activities of dihydrodiols of dibenz[*a,c*]anthracene. *Cancer Lett.*, *19*, 21-26

Community Bureau of Reference (1982) *Polycyclic Aromatic Hydrocarbon Reference Materials of Certified Purity* (*Information Handout, No. 22*), Brussels, Commission of the European Communities

DiPaolo, J.A., Donovan, P. & Nelson, R. (1969) Quantitative studies of in vitro transformation by chemical carcinogens. *J. natl Cancer Inst.*, *42*, 867-876

Duncan, M., Brookes, P. & Dipple, A. (1969) Metabolism and binding to cellular macromolecules of a series of hydrocarbons by mouse embryo cells in culture. *Int. J. Cancer*, *4*, 813-819

Goshman, L.M. & Heidelberger, C. (1967) Binding of tritium-labeled polycyclic hydrocarbons to DNA of mouse skin. *Cancer Res.*, *27*, 1678-1688

Grimmer, G., Jacob, J. & Naujack, K.-W. (1981a) Profile of the polycyclic aromatic compounds from lubricating oils. Inventory by GCGC/MS - PAH in environmental materials, Part 1. *Fresenius Z. anal. Chem.*, *306*, 345-355

Grimmer, G., Jacob, J., Naujack, K.-W. & Dettbarn, G. (1981b) Profile of the polycyclic aromatic compounds from used engine oil - Inventory by GCGC/MS - PAH in environmental materials, Part 2. *Fresenius Z. anal. Chem.*, *309*, 13-19

---

[1] Studies on occupational exposure to polynuclear aromatic compounds will be considered in future *IARC Monographs*.

Guenthner, T.M. & Mannering, G.J. (1977) Induction of hepatic mono-oxygenase systems in fetal and neonatal rats with phenobarbital, polycyclic hydrocarbons and other xenobiotics. *Biochem. Pharmacol.*, *26*, 567-575

Guerin, M.R., Epler, J.L., Griest, W.H., Clark, B.R. & Rao, T.K. (1978) *Polycyclic aromatic hydrocarbons from fossil fuel conversion processes*. In: Jones, P.W. & Freudenthal, R.I., eds, *Carcinogenesis*, Vol. 3, *Polynuclear Aromatic Hydrocarbons*, New York, Raven Press, pp. 21-33

Hermann, M. (1981) Synergistic effects of individual polycyclic aromatic hydrocarbons on the mutagenicity of their mixtures. *Mutat. Res.*, *90*, 399-409

Hewer, A., Cooper, C.S., Ribeiro, O., Pal, K., Grover, P.L. & Sims, P. (1981) The metabolic activation of dibenz[*a,c*]anthracene. *Carcinogenesis*, *2*, 1345-1352

Huberman, E. (1978) Cell transformation and mutability of different genetic loci in mammalian cells by metabolically activated carcinogenic polycyclic hydrocarbons. In: Gelboin, H.V. & Ts'O, P.O.P., eds, *Polycyclic Hydrocarbons and Cancer*, Vol. 2, *Molecular and Cell Biology*, New York, Academic Press, pp. 161-174

Huberman, E. & Sachs, L. (1976) Mutability of different genetic loci in mammalian cells by metabolically activated carcinogenic polycyclic hydrocarbons. *Proc. natl Acad. Sci. USA*, *73*, 188-192

IARC (1983) *Information Bulletin on the Survey of Chemicals Being Tested for Carcinogenicity*, No. 10, Lyon, pp. 6, 17, 25

Kaden, D.A., Hites, R.A. & Thilly, W.G. (1979) Mutagenicity of soot and associated polycyclic aromatic hydrocarbons to *Salmonella typhimurium*. *Cancer Res.*, *39*, 4152-4159

Karcher, W., Fordham, R., Dubois, J. & Gloude, P. (1983) *Spectral Atlas of Polycyclic Aromatic Compounds*, Dordrecht, The Netherlands, D. Reidel (in press)

Krahn, D.F. & Heidelberger, C. (1977) Liver homogenate-mediated mutagenesis in Chinese hamster V79 cells by polycyclic aromatic hydrocarbons and aflatoxins. *Mutat. Res.*, *46*, 27-44

Lang, K.F. & Eigen, I. (1967) *Organic compounds in coal tar*. In: Heilbronner, E., Hofmann, U., Schäfer, K. & Wittig, G., eds, *Fortschifte der chemischer Forschung* (Advances in chemical research), Vol. 8, Berlin, Springer, pp. 91-170

Lee, M.L., Novotny, M. & Bartle, K.D. (1976) Gas chromatography/mass spectrometry and nuclear magnetic resonance spectrometric studies of carcinogenic polynuclear aromatic hydrocarbons in tobacco and marijuana smoke condensates. *Anal. Chem.*, *48*, 405-416

Lijinsky, W., Garcia, H. & Saffiotti, U. (1970) Structure activity relationships among some polynuclear hydrocarbons and their hydrogenated derivatives. *J. natl Cancer Inst.*, *44*, 641-649

MacNicoll, A.D., Burden, P.M., Rattle, H., Grover, P.L. & Sims, P. (1979) The formation of dihydrodiols in the chemical or enzymic oxidation of dibenz[*a,c*]anthracene, dibenz[*a,h*]anthracene and chrysene. *Chem.-biol. Interactions*, *27*, 365-379

MacNicoll, A.D., Grover, P.L. & Sims, P. (1980) The metabolism of a series of polycyclic hydrocarbons by mouse skin maintained in short-term organ culture. *Chem.-biol. Interactions*, *29*, 169-188

Malaveille, C., Hautefeuille, A., Bartsch, H., MacNicoll, A.D., Grover, P.L. & Sims, P. (1980) Liver microsome-mediated mutagenicity of dihydrodiols derived from dibenz(*a*,*c*)anthracene in *S. typhimurium TA 100*. *Carcinogenesis*, *1*, 287-289

Martin, C.N., McDermid, A.C. & Garner, R.C. (1978) Testing of known carcinogens and noncarcinogens for their ability to induce unscheduled DNA synthesis in HeLa cells. *Cancer Res.*, *38*, 2621-2627

McCann, J., Choi, E., Yamasaki, E. & Ames, B.N. (1975) Detection of carcinogens as mutagens in the *Salmonella*/microsome test: Assay of 300 chemicals. *Proc. natl Acad. Sci. USA*, *72*, 5135-5139

McCarroll, N.E., Keech, B.H. & Piper, C.E. (1981) A microsuspension adaptation of the *Bacillus subtilis* 'rec' assay. *Environ. Mutagenesis*, *3*, 607-616

Olufsen, B. (1980) *Polynuclear aromatic hydrocarbons in Norwegian drinking water resources*. In: Bjørseth. A. & Dennis, A.J., eds, *Polynuclear Aromatic Hydrocarbons: Chemistry and Biological Effects*, *4th Int. Symposium,* Columbus, OH, Battelle Press, pp. 333-343

Pal., K. (1981) The induction of sister-chromatid exchanges in Chinese hamster ovary cells by K-region epoxides and some dihydrodiols derived from benz[*a*]anthracene, dibenz[*a*,*c*]anthracene and dibenz[*a*,*h*]anthracene. *Mutat. Res.*, *84*, 389-398

Pienta, R.J. (1980) *A transformation bioassay employing cryopreserved hamster embryo cells*. In: Mishra, N., Dunkel, V. & Mehlman, M., eds, *Advances in Modern Environmental Toxicology*, Vol. 1, *Mammalian Cell Transformation by Chemical Carcinogens*, Princeton Junction, NJ, Senate Press, pp. 47-83

Pienta, R.J., Poiley, J.A. & Lebherz III, W.B. (1977) Morphological transformation of early passage golden Syrian hamster embryo cells derived from cryopreserved primary cultures as a reliable *in vitro* bioassay for identifying diverse carcinogens. *Int. J. Cancer*, *19*, 642-655

Probst, G.S., McMahon, R.E., Hill, L.E., Thompson, C.Z., Epp, J.K. & Neal, S.B. (1981) Chemically-induced unscheduled DNA synthesis in primary rat hepatocyte cultures: A comparison with bacterial mutagenicity using 218 compounds. *Environ. Mutagenesis*, *3*, 11-32

Salamone, M.F., Heddle, J.A. & Katz, M. (1979) The mutagenic activity of thirty polycyclic aromatic hydrocarbons (PAH) and oxides in urban airborne particulates. *Environ. Int.*, *2*, 37-43

Scribner, J.D. (1973) Brief communication: Tumor initiation by apparently noncarcinogenic polycyclic aromatic hydrocarbons. *J. natl Cancer Inst.*, *50*, 1717-1719

Scribner, J.D. & Süss, R. (1978) *Tumor initiation and promotion*. In: Richter, G.W. & Epstein, M.A., eds, *International Review of Experimental Pathology*, Vol. 18, New York, Academic Press, pp. 166-167

Severson, R.F., Schlotzhauer, W.S., Arrendale, R.F., Snook, M.E. & Higman, H.C. (1977) Correlation of polynuclear aromatic hydrocarbon formation between pyrolysis and smoking. *Beitr. Tabakforsch.*, *9*, 23-37

Sims, P. (1970) Qualitative and quantitative studies on the metabolism of a series of aromatic hydrocarbons by rat-liver preparations. *Biochem. Pharmacol.*, *19*, 795-818

Slaga, T.J., Gleason, G.L., Mills, G., Ewald, L., Fu, P.P., Lee, H.M. & Harvey, R.G. (1980) Comparison of the skin tumor-initiating activities of dihydrodiols and diol-epoxides of various polycyclic aromatic hydrocarbons. *Cancer Res.*, *40*, 1981-1984

Snook, M.E., Severson, R.F., Arrendale, R.F., Higman, H.C. & Chortyk, O.T. (1977) The identification of high molecular weight polynuclear aromatic hydrocarbons in a biologically active fraction of cigarette smoke condensate. *Beitr. Tabakforsch.*, *9*, 79-101

Topping, D.C., Pal, B.C., Martin, D.H., Nelson, F.R. & Nettesheim, P. (1978) Pathologic changes induced in respiratory tract mucosa by polycyclic hydrocarbons of differing carcinogenic activity. *Am. J. Pathol.*, *93*, 311-324

Van Duuren, B.L., Sivak, A., Goldschmidt, B.M., Katz, C. & Melchionne, S. (1970) Initiating activity of aromatic hydrocarbons in two-stage carcinogenesis. *J. natl Cancer Inst.*, *44*, 1167-1173

Weast, R.C., ed. (1975) *CRC Handbook of Chemistry and Physics*, 56th ed., Cleveland, OH, Chemical Rubber Co., p. C-267

DIBENZ[*a*,*h*]ANTHRACENE

This compound was considered by a previous working group, in December 1972 (IARC, 1973).

1. Chemical and Physical Data

1.1 Synonyms and trade names

Chem. Abstr. Services Reg. No.: 53-70-3

Chem. Abstr. Name: Dibenz(a,h)anthracene

IUPAC Systematic Name: Dibenz[*a*,*h*]anthracene

Synonyms: 1,2:5,6-Benzanthracene; DBA; DB(a,h)A; 1,2:5,6-dibenzanthracene; 1,2,5,6-dibenzanthracene; 1,2:5,6-dibenz(a)anthracene; 1,2,7,8-dibenzanthracene; 1,2:5,6-dibenzoanthracene; dibenzo[*a*,*h*]anthracene

1.2 Structural and molecular formulae and molecular weight

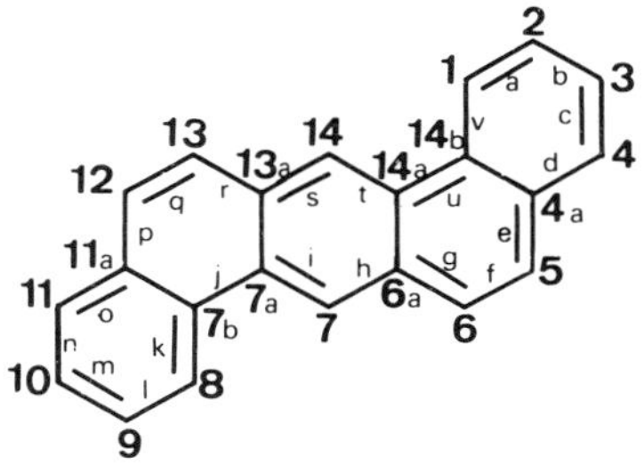

$C_{22}H_{14}$

Mol. wt: 278.4

1.3 Chemical and physical properties of the pure substance

From National Library of Medicine (1982), unless otherwise specified

(*a*) *Description*: Colourless plates or leaflets (recrystallized from acetic acid); crystals may be monoclinic or orthorhombic

(*b*) *Melting-point*: 266.6°C (Karcher *et al.*, 1983)

(*c*) *Spectroscopy data*: λ_{max} 215, 221, 231, 272, 277, 285, 297, 319, 331, 347 nm (Karcher *et al.*, 1983). Mass spectra have been tabulated (NIH/EPA Chemical Information System, 1982). Infrared spectra have been reported (Pouchert, 1981). Nuclear magnetic resonance spectra are available (Karcher *et al.*, 1983).

(*d*) *Solubility*: Virtually insoluble in water (0.5 μg/l) (Davis *et al.*, 1942); slightly soluble in diethyl ether and ethanol; soluble in benzene, toluene, xylene and most other organic solvents and oils

(*e*) *Stability*: Undergoes photo-oxidation under indoor sunlight in solution (Kuratsune & Hirohata, 1962)

(*f*) *Reactivity*: Can be hydrogenated to the octadecahydro derivative (Clar, 1952); oxidized by chromic acid to dibenz[*a*,*h*]anthra-7,14-quinone and anthraquinone-1,2,5,6-tetracarboxylic acid; oxidized by osmium tetroxide to 5,6-dihydroxy-5,6-dihydro-dibenz[*a*,*h*]anthracene and to dibenz[*a*,*h*]anthra-5,6-quinone

2. Production, Use, Occurrence and Analysis

2.1 Production and use

There is no commercial production or known use of this compound.

2.2 Occurrence and analysis

Data on occurrence and methods of analysis are summarized in the 'General Remarks on the Substances Considered', p. 35.

Dibenz[*a*,*h*]anthracene occurs ubiquitously as a product of incomplete combustion; it also occurs in fossil fuels. It has been identified in mainstream cigarette smoke (0.1-0.15 mg/kg cigarette smoke condensate) (Hoffmann & Wynder, 1960); mainstream smoke of marijuana cigarettes (Lee *et al.*, 1976); gasoline engine exhaust tar (2.5 mg/kg tar) (Hoffmann & Wynder, 1962); urban air (3.2-32 ng/m^3) (Hoffmann & Wynder, 1976); coal-heating exhaust (Brockhaus & Tomingas, 1976); used engine oil (14.32 mg/kg) (Grimmer *et al.*, 1981); coal-tar (Lijinsky *et al.*, 1963; Lang & Eigen, 1967); charcoal-broiled steaks (0.2 μg/kg) (Lijinsky & Shubik, 1964), vegetables (0.5-2.6 μg/kg) (Hettche, 1971), edible oils (0-1.9 μg/kg) (Grimmer & Hildebrandt, 1967); and waste water (Olufsen, 1980).

3. Biological Data Relevant to the Evaluation of Carcinogenic Risk to Humans

3.1 Carcinogenicity studies in animals

There is *sufficient evidence* that dibenz[*a,h*]anthracene is carcinogenic to experimental animals (see 'General Remarks on the Substances considered', p. 33).

3.2 Other relevant biological data[1]

(*a*) *Experimental systems*

Toxic effects

Acute i.p. administration of 3-90 mg/kg bw dibenz[*a,h*]anthracene in sesame oil produced a reduction in the growth rate of young rats that persisted for at least 15 weeks (Haddow *et al.*, 1937).

Effects on reproduction and prenatal toxicity

Dibenz[*a,h*]anthracene at a dose of 5 mg/rat given s.c. daily from the first day of pregnancy resulted in fetal death and resorption and may also have affected the subsequent fertility of the dams (Wolfe & Bryan, 1939). [The Working Group noted the small number of animals used.] It can also induce benzo[*a*]pyrene hydroxylase activity in the rat placenta (Welch *et al.*, 1969).

In a study reported as an abstract, it was suggested that dibenz[*a,h*]anthracene or its metabolites cross the placenta (Reno, 1969).

Metabolism and activation

The 5,6-oxide (Selkirk *et al.*, 1971) and the 1,2-, 3,4- and 5,6-dihydrodiols have been detected as metabolites of dibenz[*a,h*]anthracene following incubation of this compound with rat-liver preparations (MacNicoll *et al.*, 1979; Nordqvist *et al.*, 1979) and mouse skin in organ culture (MacNicoll *et al.*, 1980).

The 3,4-dihydrodiol was active as a tumour initiator in mouse skin (Buening *et al.*, 1979; Slaga *et al.*, 1980) and induced pulmonary tumours in newborn mice (Buening *et al.*, 1979).

---

[1] See also 'General Remarks on the Substances Considered', p. 53.

The 3,4-dihydrodiol was the most mutagenic of the three dihydrodiols to bacteria in the presence of rat-liver preparations (Wood *et al.*, 1978). The 5,6-oxide transformed hamster embryo cells (Huberman *et al.*, 1972; Marquardt *et al.*, 1972). The 5,6-oxide was found to bind to cellular macromolecules in mammalian cells (Kuroki *et al.*, 1971/72), but was a poor tumour initiator on mouse skin (Van Duuren *et al.*, 1967). Nucleoside adducts have been detected in mouse skin following topical application of the parent hydrocarbon, but they have not been characterized (Phillips *et al.*, 1979).

Mutagenicity and other short-term tests

Results from short-term tests are given in Table 1.

Table 1. Results from short-term tests: Dibenz[*a*,*h*]anthracene[a]

Test	Organism/assay[b]	Exogenous metabolic system[b]	Reported result	Comments	References
PROKARYOTES					
DNA damage	*Escherichia coli* (rec^+/rec^-)	Aro-R-PMS	Positive	At 25 μg/well	Ichinotsubo *et al.* (1977)
	Bacillus subtilis (rec^+/rec^-)	Aro-R-PMS	Positive	Minimal inhibitory concentration: rec^+ at 50 μg/well *versus* rec^- at 12 μg/well	McCarroll *et al.* (1981)
Mutation	*Salmonella typhimurium* (his^-/his^+)	Aro-R-PMS Aro-GP-PMS 3MC-GP-PMS	Positive	At 5 μg/plate in strain TA100	McCann *et al.* (1975); Andrews *et al.* (1978);
		Aro-R-PMS	Positive	At 12 nmol/plate in strain TA98	Baker *et al.* (1980); Hermann (1981)
	Salmonella typhimurium ($8AG^S/8AG^R$)	Aro-R-PMS	Positive	At 75 nmol/ml in strain TM677	Kaden *et al.* (1979)
MAMMALIAN CELLS *IN VITRO*					
DNA damage	Human foreskin epithelial cells (unscheduled DNA synthesis)	--	Positive	In the range 1-100 μg/ml	Lake *et al.* (1978)
	HeLa cells (unscheduled DNA synthesis)	3MC-R-PMS	Positive	At 100 pmol/ml	Martin *et al.* (1978)
	Syrian hamster embryo cells (unscheduled DNA synthesis)	--	Negative	Tested at up to 20 μg/ml	Casto (1979)
	Primary rat hepatocytes (unscheduled DNA synthesis)	--	Negative	Tested at up to 100 nmol/ml	Probst *et al.* (1981)
Mutation	Chinese hamster V79 cells (OUA^S/OUA^R; $8AG^S/8AG^R$)	SHE feeder layer	Positive	At 1 μg/ml	Huberman d Sachs (1976); Huberman (1978)
	Chinese hamster V79 cells ($6TG^S/6TG^R$)	3MC-R-PMS	Positive	At 56 nmol/ml	Krahn & Heidelberger (1977)
Chromosome effects	Chinese hamster ovary cells (sister chromatid exchange)	None	Positive	At 8 μg/ml [The Working Group noted that the effect did not appear to be significant.]	Pal (1981)

Test	Organism/assay[b]	Exogenous metabolic system[b]	Reported result	Comments	References
Cell transformation	Syrian hamster embryo cells (morphological)	--	Positive	At 0.5-10 μg/ml	DiPaolo *et al.* (1969); Pienta *et al.* (1977); Casto *et al.* (1977); Casto (1979)
	Mouse C3H 10T1/2 cells (morphological)	--	Positive	At 20 μg/ml	Reznikoff *et al.* (1973)
	Mouse prostate C3H cells (morphological)	--	Positive	At 10 μg/ml	Chen & Heidelberger (1969)
	Mouse prostate C3HG23 cells (morphological)	None	Negative	Tested at up to 10 μg/ml	Marquardt *et al.* (1972)
MAMMALIAN CELLS *IN VIVO*					
Chromosome effects	Chinese hamster bone-marrow cells: sister chromatid exchange	--	Positive	At 2 x 450 mg/kg bw i.p.	Roszinsky-Köcher *et al.* (1979)
	chromosome breaks	--	Negative	Treated i.p. with 2 x 450 mg/kg bw	

[a] This table comprises selected assays and references and is not intended to be a complete review of the literature.

[b] For an explanation of the abbreviations, see the Appendix, p. 452.

(*b*) *Humans*

No data were available to the Working Group.

3.3 Case reports and epidemiological studies of carcinogenicity to humans

No data were available to the Working Group.

4. Summary of Data Reported and Evaluation[1]

4.1 Experimental data

Dibenz[*a,h*]anthracene has been shown to be carcinogenic to experimental animals (see 'General Remarks on the Substances Considered', p. 33, and IARC, 1973).

[1] For definitions of the italicized terms, see Preamble, p. 19.

Dibenz[*a,h*]anthracene is embryotoxic to rats when given at high doses. The available data on teratogenicity were inadequate for evaluation.

Dibenz[*a,h*]anthracene was positive in differential survival assays using DNA-repair-proficient/-deficient strains of bacteria and was mutagenic to *Salmonella typhimurium* in the presence of an exogenous metabolic system. In cultured mammalian cells, dibenz[*a,h*]anthracene was mutagenic and induced unscheduled DNA synthesis in the presence of an exogenous metabolic system. It was positive in assays for morphological transformation. In the one available study, it induced sister chromatid exchange but not chromosomal aberrations *in vivo*.

There is *sufficient evidence* that dibenz[*a,h*]anthracene is active in short-term tests.

4.2 Human data[1]

Dibenz[*a,h*]anthracene is present as a minor component of the total content of polynuclear aromatic compounds in the environment. Human exposure to dibenz[*a,h*]anthracene occurs primarily through the smoking of tobacco, inhalation of polluted air and by ingestion of food and water contaminated with combustion products (for details, see 'General Remarks on the Substances Considered', p. 35).

4.3 Evaluation[2]

There is *sufficient evidence* that dibenz[*a,h*]anthracene is carcinogenic to experimental animals.

5. References

Andrews, A.W., Thibault, L.H. & Lijinsky, W. (1978) The relationship between carcinogenicity and mutagenicity of some polynuclear hydrocarbons. *Mutat. Res.*, *51*, 311-318

Baker, R.S.U., Bonin, A.M., Stupans, I. & Holder, G.M. (1980) Comparison of rat and guinea pig as sources of the S9 fraction in the *Salmonella*/mammalian microsome mutagenicity test. *Mutat. Res.*, *71*, 43-52

---

[1] Studies on occupational exposure to polynuclear aromatic compounds will be considered in future *IARC Monographs*.

[2] In the absence of adequate data on humans, it is reasonable, for practical purposes, to regard chemicals for which there is *sufficient evidence* of carcinogenicity in animals as if they presented a carcinogenic risk to humans (see also Preamble, p. 20).

Brockhaus, A. & Tomingas, R. (1976) Emission of polycyclic hydrocarbons during burning processes in small heating installations and their concentration in the atmosphere (Ger.). *Staub-Reinhalt. Luft*, *36*, 96-101

Buening, M.K., Levin, W., Wood, A.W., Chang, R.L., Yagi, H., Karle, J.M., Jerina, D.M. & Conney, A.H. (1979) Tumorigenicity of the dihydrodiols of dibenzo(*a,h*)anthracene on mouse skin and in newborn mice. *Cancer Res.*, *39*, 1310-1314

Casto, B.C. (1979) *Polycyclic hydrocarbons and Syrian hamster embryo cells: Cell transformation, enhancement of viral transformation and analysis of DNA damage.* In: Jones, P.W. & Leber, P., eds, *Polynuclear Aromatic Hydrocarbons*, Ann Arbor, MI, Ann Arbor Science Publishers, pp. 51-66

Casto, B.C., Janosko, N. & DiPaolo, J.A. (1977) Development of a focus assay model for transformation of hamster cells *in vitro* by chemical carcinogens. *Cancer Res.*, *37*, 3508-3515

Chen, T.T. & Heidelberger, C. (1969) Quantitative studies on the malignant transformation of mouse prostate cells by carcinogenic hydrocarbons *in vitro*. *Int. J. Cancer*, *4*, 166-178

Clar, E. (1952) *Aromatische Kohlenwasserstoffe. Polycyclische Systeme* (Aromatic hydrocarbons. Polycyclic system), Berlin, Springer, pp. 200-201

Davis, W.W., Krahl, M.E. & Clowes, G.H.A. (1942) Solubility of carcinogenic and related hydrocarbons in water. *J. Am. chem. Soc.*, *64*, 108-110

Di Paolo, J.A., Donovan, P. & Nelson, R.L. (1969) Quantitative studies of *in vitro* transformation by chemical carcinogens. *J. natl Cancer Inst.*, *42*, 867-876

Grimmer, G. & Hildebrandt, A. (1967) Content of polycyclic hydrocarbons in crude vegetable oils. *Chem. Ind.*, 25 November, 2000-2002

Grimmer, G., Jacob, J., Naujack, K.-W. & Dettbarn, G. (1981) Profile of the polycyclic aromatic compounds from used engine oil - Inventory by GCGC/MS - PAH in environmental materials, Part 2. *Fresenius Z. anal. Chem.*, *309*, 13-19

Haddow, A., Scott, C.M. & Scott, J.D. (1937) The influence of certain carcinogenic and other hydrocarbons on body growth in the rat. *Proc. R. Soc. London Ser. B*, *122*, 477-507

Hermann, M. (1981) Synergistic effects of individual polycyclic aromatic hydrocarbons on the mutagenicity of their mixtures. *Mutat. Res.*, *90*, 399-409

Hettche, H.O. (1971) Plant waxes as collectors of polycyclic aromatics in the air of residential areas. *Staub-Reinhalt. Luft*, *31*, 34-41

Hoffmann, D. & Wynder, E.L. (1960) On the isolation and identification of polycyclic aromatic hydrocarbons. *Cancer*, *13*, 1062-1073

Hoffmann, D. & Wynder, E.L. (1962) A study of air pollution carcinogenesis. II. The isolation and identification of polynuclear aromatic hydrocarbons from gasoline engine exhaust condensate. *Cancer*, *15*, 93-102

Hoffmann, D. & Wynder, E.L. (1976) *Environmental respiratory carcinogenesis*. In: Searle, C.E., ed., *Chemical Carcinogens* (*ACS Monograph 173*), Washington DC, American Chemical Society, p. 341

Huberman, E. (1978) *Cell transformation and mutability of different genetic loci in mammalian cells by metabolically activated carcinogenic polycyclic hydrocarbons*. In: Gelboin, H.V. and Ts'O, P.O.P., eds, *Polycyclic Hydrocarbons and Cancer*, Vol. 2, *Molecular and Cell Biology*, New York, Academic Press, pp. 161-174

Huberman, E. & Sachs, L. (1976) Mutability of different genetic loci in mammalian cells by metabolically activated carcinogenic polycyclic hydrocarbons. *Proc. natl Acad. Sci. USA*, *73*, 188-192

Huberman, E., Kuroki, T., Marquardt, H., Selkirk, J.K., Heidelberger, C., Grover, P.L. & Sims, P. (1972) Transformation of hamster embryo cells by epoxides and other derivatives of polycyclic hydrocarbons. *Cancer Res.*, *32*, 1391-1396

IARC (1973) *IARC Monographs on the Evaluation of Carcinogenic Risk of Chemicals to Man*, Vol. 3, *Certain polycyclic aromatic hydrocarbons and heterocyclic compounds*, Lyon, pp. 178-196

Ichinotsubo, D., Mower, H.F., Setliff, J. & Mandel, M. (1977) The use of *rec*- bacteria for testing of carcinogenic substances. *Mutat. Res.*, *46*, 53-62

Kaden, D.A., Hites, R.A. & Thilly, W.G. (1979) Mutagenicity of soot and associated polycyclic aromatic hydrocarbons to *Salmonella typhimurium*. *Cancer Res.*, *39*, 4152-4159

Karcher, W., Fordham, R., Dubois, J. & Gloude, P. (1983) *Spectral Atlas of Polycyclic Aromatic Compounds*, Dordrecht, The Netherlands, D. Reidel (in press)

Krahn, D.F. & Heidelberger, C. (1977) Liver homogenate-mediated mutagenesis in Chinese hamster V79 cells by polycyclic aromatic hydrocarbons and aflatoxins. *Mutat. Res.*, *46*, 27-44

Kuratsune, M. & Hirohata, T. (1962) Decomposition of polycyclic aromatic hydrocarbons under laboratory illuminations. *Natl Cancer Inst. Monogr.*, *9*, 117-125

Kuroki, T., Huberman, E., Marquardt, H., Selkirk, J.K., Heidelberger, C., Grover, P.L. & Sims, P. (1971/72) Binding of K-region epoxides and other derivatives of benz[*a*]anthracene and dibenz[*a*,*h*]anthracene to DNA, RNA and proteins of transformable cells. *Chem.-biol. Interactions*, *4*, 389-397

Lake, R.S., Kropko, M.L., Pezzutti, M.R., Shoemaker, R.H. & Igel, H.J. (1978) Chemical induction of unscheduled DNA synthesis in human skin epithelial cell cultures. *Cancer Res.*, *38*, 2091-2098

Lang, K.F. & Eigen, I. (1967) *Organic compounds in coal tar* (Ger.). In: Heilbronner, E., Hofmann, U., Schäfer, K. & Wittig, G., eds, *Fortschifte der chemischer Forschung* (Advances in chemical research), Vol. 8, Berlin, Springer, pp. 91-170

Lee, M.L., Novotny, M. & Bartle, K.D. (1976) Gas chromatography/mass spectrometry and nuclear magnetic resonance spectrometric studies of carcinogenic polynuclear aromatic hydrocarbons in tobacco and marijuana smoke condensates. *Anal. Chem.*, *48*, 405-416

Lijinsky, W. & Shubik, P. (1964) Benzo(*a*)pyrene and other polynuclear hydrocarbons in charcoal-broiled meat. *Science*, *145*, 53-55

Lijinsky, W., Domsky, I., Mason, G., Ramahi, H.Y. & Safavi, T. (1963) The chromatographic determination of trace amounts of polynuclear hydrocarbons in petrolatum, mineral oil, and coal tar. *Anal. Chem.*, *35*, 952-956

MacNicoll, A.D., Burden, P.M., Rattle, H., Grover, P.L. & Sims, P. (1979) The formation of dihydrodiols in the chemical or enzymic oxidation of dibenz[*a,c*]anthracene, dibenz[*a,h*]anthracene and chrysene. *Chem.-biol. Interactions*, *27*, 365-379

MacNicoll, A.D., Grover, P.L. & Sims, P. (1980) The metabolism of a series of polycyclic hydrocarbons by mouse skin maintained in short-term organ culture. *Chem.-biol. Interactions*, *29*, 169-188

Marquardt, H., Kuroki, T., Huberman, E., Selkirk, J.K., Heidelberger, C., Grover, P.L. & Sims, P. (1972) Malignant transformation of cells derived from mouse prostate by epoxides and other derivatives of polycyclic hydrocarbons. *Cancer Res.*, *32*, 716-720

Martin, C.N., McDermid, A.C. & Garner, R.C. (1978) Testing of known carcinogens and noncarcinogens for their ability to induce unscheduled DNA synthesis in HeLa cells. *Cancer Res.*, *38*, 2621-2627

McCann, J., Choi, E., Yamasaki, E. & Ames, B.N. (1975) Detection of carcinogens as mutagens in the *Salmonella*/microsome test: Assay of 300 chemicals. *Proc. natl Acad. Sci. USA*, *72*, 5135-5139

McCarroll, N.E., Keech, B.H. & Piper, C.E. (1981) A microsuspension adaptation of the *Bacillus subtilis* 'rec' assay. *Environ. Mutagenesis*, *3*, 607-616

National Library of Medicine (1982) *Toxicology Data Bank*, Bethesda, MD, National Library of Medicine, Specialized Information Services, Toxicology Information Program

NIH/EPA Chemical Information System (1982) *Mass Spectral Search System*, Washington DC, CIS Project, Information Services Corporation

Nordqvist, M., Thakker, D.R., Levin, W., Yagi, H., Ryan, D.E., Thomas, P.E., Conney, A.H. & Jerina, D.M. (1979) The highly tumorigenic 3,4-dihydrodiol is a principal metabolite formed from dibenzo[*a,h*]anthracene by liver enzymes. *Mol. Pharmacol.*, *16*, 643-655

Olufsen, B. (1980) *Polynuclear aromatic hydrocarbons in Norwegian drinking water resources.* In: Bjørseth, A. & Dennis, A.J., eds, *Polynuclear Aromatic Hydrocarbons: Chemistry and Biological Effects*, *4th Int. Symposium*, Columbus, OH, Battelle Press, pp. 333-343

Pal, K. (1981) The induction of sister-chromatid exchanges in Chinese hamster ovary cells by K-region epoxides and some dihydrodiols derived from benz[*a*]anthracene, dibenz[*a,c*]anthracene and dibenz[*a,h*]anthracene. *Mutat. Res.*, *84*, 389-398

Phillips, D.H., Grover, P.L. & Sims, P. (1979) A quantitative determination of the covalent binding of a series of polycyclic hydrocarbons to DNA in mouse skin. *int. J. Cancer*, *23*, 201-208

Pouchert, C.J. (1981) *The Aldrich Library of Infrared Spectra*, 2nd ed., Milwaukee, WI, Aldrich Chemical Co., p. 519

Probst, G.S., McMahon, R.E., Hill, L.E., Thompson, C.Z., Epp, J.K. & Neal, S.B. (1981) Chemically-induced unscheduled DNA synthesis in primary rat hepatocyte cultures: A comparison with bacterial mutagenicity using 218 compounds. *Environ. Mutagenesis*, *3*, 11-32

Reno, F.E. (1969) Toxic effects of polynuclear aromatic hydrocarbons in chicken embryos and mice. *Diss. Abstr. Int., B*, *29*, 4777

Reznikoff, C.A., Bertram, J.S., Brankow, D.W. & Heidelberger, C. (1973) Quantitative and qualitative studies of chemical transformation of cloned C3H mouse embryo cells sensitive to postconfluence inhibition of cell division. *Cancer Res.*, *33*, 3239-3249

Roszinsky-Köcher, G., Basler, A. & Röhrborn, G. (1979) Mutagenicity of polycyclic hydrocarbons. V. Induction of sister chromatid exchanges *in vivo*. *Mutat. Res.*, *66*, 65-67

Selkirk, J.K., Huberman, E. & Heidelberger, C. (1971) An epoxide is an intermediate in the microsomal metabolism of the chemical carcinogen dibenz[*a*,*h*]anthracene. *Biochem. biophys. Res. Commun.*, *43*, 1010-1016

Slaga, T.J., Gleason, G.L., Mills, G., Ewald, L., Fu, P.P., Lee, H.M. & Harvey, R.G. (1980) Comparison of the skin tumor-initiating activities of dihydrodiols and diol-epoxides of various polycyclic aromatic hydrocarbons. *Cancer Res.*, *40*, 1981-1984

Van Duuren, B.L., Langseth, L., Goldschmidt, B.M. & Orris, L. (1967) Carcinogenicity of epoxides, lactones, and peroxy compounds. VI. Structure and carcinogenic activity. *J. natl Cancer Inst.*, *39*, 1217-1228

Welch, R.M., Harrison, Y.E., Gommi, B.W., Poppers, P.J., Finster, M. & Conney, A.H. (1969) Stimulatory effect of cigarette smoking on the hydroxylation of 3,4-benzpyrene and the *N*-demethylation of 3-methyl-4-monomethylaminoazobenzene by enzymes in human placenta. *Clin. Pharmacol. Ther.*, *10*, 100-109

Wolfe, J.M. & Bryan, W.R. (1939) Effects induced in pregnant rats by injection of chemically pure carcinogenic agents. *Am. J. Cancer*, *36*, 359-368

Wood, A.W., Levin, W., Thomas, P.E., Ryan, D., Karle, J.M., Yagi, H., Jerina, D.M. & Conney, A.H. (1978) Metabolic activation of dibenzo[*a*,*h*]anthracene and its dihydrodiols to bacterial mutagens. *Cancer Res.*, *38*, 1967-1973

DIBENZ[*a,j*]ANTHRACENE

1. Chemical and Physical Data

1.1 Synonyms and trade names

Chem. Abstr. Services Reg. No.: 224-41-9

Chem. Abstr. Name: Dibenz(a,j)anthracene

IUPAC Systematic Name: Dibenz[*a,j*]anthracene

Synonyms: 1,2:7,8-Dibenzanthracene; 3,4,5,6-dibenzanthracene; dibenzo-1,2,7,8-anthracene; dibenzo[*a,j*]anthracene

1.2 Structural and molecular formulae and molecular weight

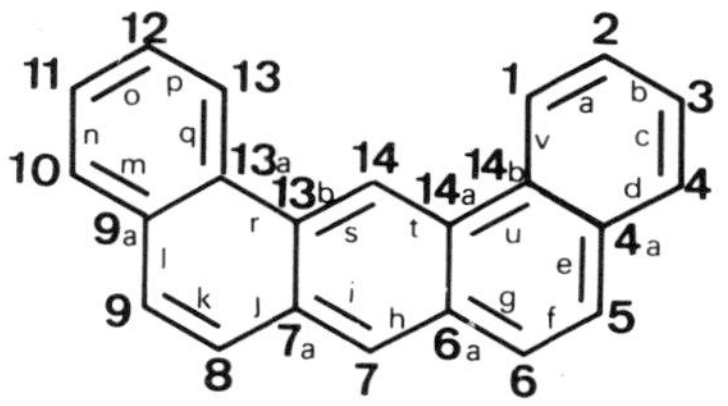

$C_{22}H_{14}$

Mol. wt: 278.4

1.3 Chemical and physical properties of the pure substance

(*a*) *Description*: Orange leaves or needles (recrystallized from benzene) (Weast, 1975); bluish-green fluorescence in solution (Rappoport, 1967)

(*b*) *Melting-point*: 197.3°C (Karcher *et al.*, 1983)

(*c*) *Spectroscopy data*: λ_{max} 214, 224, 248, 257, 274, 285, 296.5, 300, 321, 333, 336, 348 nm (in cyclohexane) (Karcher *et al*, 1983) (*d*) *Solubility*: Virtually insoluble in water (12 μg/l) (Davis *et al.*, 1942) and acetic acid; slightly soluble in benzene, diethyl ether and ethanol; soluble in petroleum ether (a low-boiling coal-tar fraction)

(*e*) *Stability*: No data were available.

(*f*) *Reactivity:* Oxidized by osmium tetroxide to 5,6-dihydrodiol, which can be oxidized first to a quinone and then to a diquinone. Ozonolysis gives the quinone and the dicarboxylic acid (Clar, 1964).

2. Production, Use, Occurrence and Analysis

2.1 Production and use

There is no commercial production or known use of this compound. A reference material of certified high purity is available (Community Bureau of Reference, 1982).

2.2 Occurrence and analysis

Data on occurrence and methods of analysis are summarized in the 'General Remarks on the Substances Considered', p. 35.

Dibenz[*a,j*]anthracene occurs ubiquitously in products of incomplete combustion; it also occurs in fossil fuels. It has been identified in dried sediment from lakes (1-309 μg/kg) (Grimmer & Böhnke, 1977); mainstream cigarette smoke (1.1 μg/kg/cigarette); sidestream cigarette smoke (4.1 μg/cigarette) (Grimmer *et al.*, 1977a); smoke-filled rooms (6 ng/m^3) (Grimmer *et al.*, 1977b); coal-tar (Lang & Eigen, 1967); sewage sludge (0.23-0.30 mg/kg) (Grimmer *et al.*, 1978); and used engine oil (22-86 mg/kg) (Grimmer *et al.*, 1981).

3. Biological Data Relevant to the Evaluation of Carcinogenic Risk to Humans

3.1 Carcinogenicity studies in animals[1]

(*a*) *Skin application*

Mouse: Two groups of 30 female Swiss mice, eight to ten weeks old, were painted on the skin twice weekly with 78 or 39 μg dibenz[*a,j*]anthracene (purity, >99%) in acetone for 60-81

[1] The Working Group was aware of a study in progress in mice by skin application (IARC, 1983).

weeks; 20 controls received acetone alone. The first skin tumours were observed at 64 and 66 weeks, respectively. Among the animals still alive at those times, 6/20 developed carcinomas and 2/20, papillomas in the high-dose group; 2/9 mice developed carcinomas and 2/9, papillomas in the low-dose group. No skin tumour occurred in 20 controls, 14 of which survived 60 weeks and six of which survived 100 weeks after treatment with acetone (Lijinsky *et al.*, 1970).

(*b*) *Subcutaneous and/or intramuscular administration*

Mouse: A group of 25 female Swiss mice, eight to ten weeks of age, received 0.4 mg dibenz[*a,j*]anthracene (purity, >99%) in 0.2 ml olive oil injected s.c. and were observed for life. Sarcomas at the injection site developed in 3/21 animals surviving 60 or more weeks; the average latent period was 73 weeks. No local tumour occurred in 25 controls injected with olive oil, 16 of which survived 60 or more weeks (Lijinsky *et al.*, 1970).

3.2 Other relevant biological data[1]

(*a*) *Experimental systems*

No data were available to the Working Group on toxic effects, on effects on reproduction and prenatal toxicity or on metabolism and activation.

Mutagenicity and other short-term tests

Dibenz[*a,j*]anthracene was reported to induce mutations in *Salmonella typhimurium* strain TA100 (*his*$^-$/*his*$^+$) at a concentration of 10 μg/plate in the presence of an exogenous metabolic system (postmitochondrial supernatant from Aroclor-induced rat liver) (Andrews *et al.*, 1978).

(*b*) *Humans*

No data were available to the Working Group.

3.3 Case reports and epidemiological studies of carcinogenicity to humans

No data were available to the Working Group.

---

[1] See also 'General Remarks on the Substances Considered', p. 53.

4. Summary of Data Reported and Evaluation[1]

4.1 Experimental data

Dibenz[*a,j*]anthracene was tested for carcinogenicity in female mice by skin application and produced skin tumours. It was also tested in mice by subcutaneous injection and produced a few tumours at the site of injection.

No data were available on the teratogenicity of this compound.

In the one available study, dibenz[*a,j*]anthracene was mutagenic to *Salmonella typhimurium* in the presence of an exogenous metabolic system.

There is *inadequate evidence* that dibenz[*a,j*]anthracene is active in short-term tests.

4.2 Human data[2]

Dibenz[*a,j*]anthracene is present as a minor component of the total content of polynuclear aromatic compounds in the environment. Human exposure to dibenz[*a,j*]anthracene occurs primarily through the smoking of tobacco, inhalation of polluted air and by ingestion of food and water contaminated by combustion effluents (for details, see 'General Remarks on the Substances Considered', p. 35).

4.3 Evaluation

There is *limited evidence* that dibenz[*a,j*]anthracene is carcinogenic to experimental animals.

5. References

Andrews, A.W., Thibault, L.H. & Lijinsky, W. (1978) The relationship between carcinogenicity and mutagenicity of some polynuclear hydrocarbons. *Mutat. Res.*, *51*, 311-318

Clar, E., ed. (1964) *Polycyclic Hydrocarbons*, Vol. 1, New York, Academic Press, pp. 337-341

---

[1] For definitions of the italicized terms, see Preamble, p. 19.

[2] Studies on occupational exposure to polynuclear aromatic compounds will be considered in future *IARC Monographs*.

Community Bureau of Reference (1982) *Polycyclic Aromatic Hydrocarbon Reference Materials of Certified Purity* (*Information Handout, No. 22*), Brussels, Commission of the European Communities

Davis, W.W., Krahl, M.E. & Clowes, G.H.A. (1942) Solubility of carcinogenic and related hydrocarbons in water. *J. Am. Chem. Soc.*, *64*, 108-110

Grimmer, G. & Böhnke, H. (1977) Investigation on drilling cores of sediments of Lake Constance. I. Profiles of the polycyclic aromatic hydrocarbons (Ger.). *Z. Naturforsch.*, *32c*, 703-711

Grimmer, G., Böhnke, H. & Harke, H.-P. (1977a) Passive smoking. Measurement of concentrations of polycyclic aromatic hydrocarbons in rooms after machine smoking of cigarettes (Ger.). *Int. Arch. occup. environ. Health*, *40*, 83-92

Grimmer, G., Böhnke, H. & Harke, H.-P. (1977b) Passive smoking. Intake of polycyclic aromatic hydrocarbons by breathing of cigarette smoke containing air (Ger.). *Int. Arch. occup. environ. Health*, *40*, 93-99

Grimmer, G., Böhnke, H. & Borwitzky, H. (1978) Gas-chromatographic profile analysis of polycyclic aromatic hydrocarbons in sewage sludge (Ger.). *Fresenius Z. anal. Chem.*, *289*, 91-95

Grimmer, G., Jacob, J., Naujack, K.-W. & Dettbarn, G. (1981) Profile of the polycyclic aromatic hydrocarbons from used engine oil - Inventory by GCGC/MS - PAH in environmental materials, part 2. *Fresenius Z. anal. Chem.*, *309*, 13-19

IARC (1983) *Information Bulletin on the Survey of Chemicals Being Tested for Carcinogenicity*, No. 10, Lyon, p. 17

Karcher, W., Fordham, R., Dubois, J. & Gloude, P. (1983) *Spectral Atlas of Polycyclic Aromatic Compounds*, Dordrecht, The Netherlands, D. Reidel (in press)

Lang, K.F. & Eigen, I. (1967) *Organic compounds in coal tar* (Ger.). In: Heilbronner, E., Hofmann, U., Schäfer, K. & Wittig, G., eds, *Fortschrifte der chemischer Forschung* (Advances in chemical research, Vol. 8, Berlin, Springer, pp. 91-170

Lijinsky, W., Garcia, H. & Saffiotti, U. (1970) Structure-activity relationships among some polynuclear hydrocarbons and their hydrogenated derivatives. *J. natl Cancer Inst.*, *44*, 641-649

Rappoport, Z., ed. (1967) *CRC Handbook of Tables for Organic Compound Identification*, 3rd ed., Boca Raton, FL, Chemical Rubber Co., p. 50

Weast, R.C., ed. (1975) *CRC Handbook of Chemistry and Physics*, 56th ed., Cleveland, OH, Chemical Rubber Co., p. C-267

7*H*-DIBENZO[*c*,*g*]CARBAZOLE

This compound was considered by a previous working group, in December 1972 (IARC, 1973).

1. Chemical and Physical Data

1.1 Synonyms and trade names

Chem. Abstr. Services Reg. No.: 194-59-2

Chem. Abstr. Name: 7H-Dibenzo(c,g)carbazole

IUPAC Systematic Name: 7*H*-Dibenzo[*c*,*g*]carbazole

Synonyms: 7-Aza-7H-dibenzo(c,g)fluorene; 3,4,5,6-dibenzocarbazole

1.2 Structural and molecular formulae and molecular weight

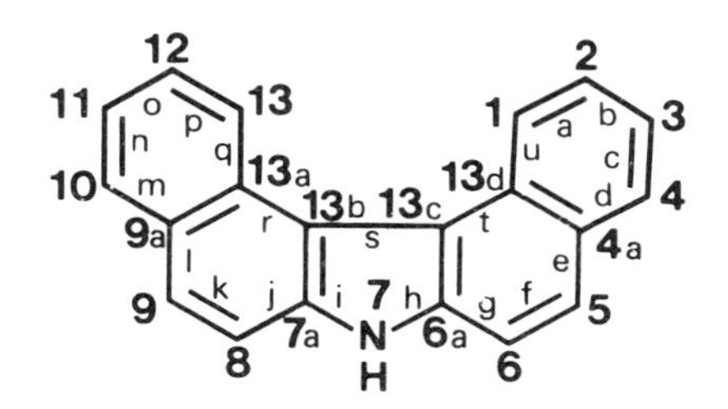

$C_{20}H_{13}N$

Mol. wt: 267.3

1.3 Chemical and physical properties of the pure substance

From National Library of Medicine (1982), unless otherwise specified

(*a*) *Description*: Needles (recrystallized from ethanol)

(*b*) *Melting-point*: 158°C

(*c*) *Solubility*: Insoluble in petroleum ether; soluble in common organic solvents

(*d*) *Stability*: No data were available.

(*e*) *Reactivity*: Reacts generally as a carbazole (Sumpter & Miller, 1954); can be acylated and nitrosated (Buckingham, 1982)

2. Production, Use, Occurrence and Analysis

2.1 Production and use

There is no commercial production or known use of this compound.

2.2 Occurrence and analysis

Data on occurrence and methods of analysis are summarized in the 'General Remarks on the Substances Considered', p. 35.

7*H*-Dibenzo[*c,g*]carbazole has been identified in cigarette smoke (0.07 μg/100 cigarettes) (Van Duuren *et al*., 1960).

3. Biological Data Relevant to the Evaluation of Carcinogenic Risk to Humans

3.1 Carcinogenicity studies in animals

There is *sufficient evidence* that 7*H*-dibenzo[*c,g*]carbazole is carcinogenic to experimental animals (see 'General Remarks on the Substances Considered', p. 33).

3.2 Other relevant biological data[1]

(*a*) *Experimental systems*

Toxic effects

Male Syrian hamsters, six to seven weeks old, received five weekly instillations into the trachea of a suspension of 3 mg 7*H*-dibenzo[*c,g*]carbazole in 0.2 ml saline or water. In animals

---

[1] See also General Remarks on the Substances Considered', p. 53.

killed one week after the final instillation, hyperplastic epithelium and squamous metaplasia were observed at the site of application (Nagel *et al.*, 1976).

No data were available on effects on reproduction and prenatal toxicity or on metabolism and activation of this compound.

Distribution

^{3}H-7*H*-Dibenzo[*c,g*]carbazole (3 mg), administered intratracheally in saline or water to hamsters, was cleared from the respiratory tract with a half-life of one to three hours, which was several times faster than the rate observed for benzo[*a*]pyrene (Nagel *et al.*, 1976). 7*H*-Dibenzo[*c,g*]carbazole was distributed to the liver, kidney, brain and fat, with highest concentrations observed in the intestine; six hours after administration, five times as much radioactivity was observed in the faeces (16%) as in the urine (3%).

Mutagenicity and other short-term tests

Results from short-term tests are given in Table 1.

Table 1. Results from short-term tests: 7*H*-Dibenzo[*c,g*]carbazole

Test	Organism/assay[a]	Exogenous metabolic system[a]	Reported result	Comments	References
PROKARYOTES					
Mutation	*Salmonella typhimurium* (his⁻/his⁺)	Aro-R-PMS	Negative	Tested at up to 0.1 mg/plate in strain TA98	Ho *et al.* (1981)
		Aro-R-PMS	Inconclusive	Tested at up to 1000 μg/plate in strain TA98	Salamone *et al.* (1979)

[a] For an explanation of the abbreviations, see the Appendix, p. 452.

(*b*) *Humans*

No data were available to the Working Group.

3.3 Case reports and epidemiological studies of carcinogenicity to humans

No data were available to the Working Group.

4. Summary of Data Reported and Evaluation[1]

4.1 Experimental data

7*H*-Dibenzo[*c,g*]carbazole has been shown to be carcinogenic to experimental animals. (See 'General Remarks on the Substances Considered', p. 33, and IARC, 1973.)

No data on the teratogenicity of this compound were available.

There were insufficient data available to evaluate the mutagenicity of 7*H*-dibenzo-[*c,g*]carbazole to *Salmonella typhimurium*

There is *inadequate evidence* that 7*H*-dibenzo[*c,g*]carbazole is active in short-term tests.

4.2 Human data[2]

7*H*-Dibenzo[*c,g*]carbazole occurs in tobacco smoke.

4.3 Evaluation[3]

There is *sufficient evidence* that 7*H*-dibenzo[*c,g*]carbazole is carcinogenic to experimental animals.

5. References

Buckingham, J., ed. (1982) *Dictionary of Organic Compounds*, 5th ed., Vol. 2, New York, Chapman & Hall, p. 1595

Ho, C.H., Clark, B.R., Guerin, M.R., Barkenbus, B.D., Rao, T.K. & Epler, J.L. (1981) Analytical and biological analyses of test materials from the synthetic fuel technologies. IV. Studies of chemical structure-mutagenic activity relationships of aromatic nitrogen compounds relevant to synfuels. *Mutat. Res.*, *85*, 335-345

---

[1] For definitions of the italicized terms, see Preamble, p. 19.

[2] Studies on occupational exposure to polynuclear aromatic compounds will be considered in future *IARC Monographs*.

[3] In the absence of adequate data on humans it is reasonable, for practical purposes, to regard chemicals for which there is *sufficient evidence* of carcinogenicity in animals as if they presented a carcinogenic risk to humans (see also Preamble, p. 20).

IARC (1973) *IARC Monographs on the Evaluation of Carcinogenic Risk of Chemicals to Man*, Vol. 3, *Certain polycyclic aromatic hydrocarbons and heterocyclic compounds*, Lyon, pp. 260-268

Nagel, D.L., Stenbäck, F., Clayson, D.B. & Wallcave, L. (1976) Intratracheal installation studies with 7*H*-dibenzo[*c,g*]carbazole in the Syrian hamster. *J. natl Cancer Inst.*, *57*, 119-123

National Library of Medicine (1982) *Toxicology Data Bank*, Bethesda, MD, National Library of Medicine Specialized Information Services, Toxicology Information Program

Salamone, M.F., Heddle, J.A. & Katz, M. (1979) The mutagenic activity of thirty polycyclic aromatic hydrocarbons (PAH) and oxides in urban airborne particulates. *Environ. Int.*, *2*, 37-43

Sumpter, W.C. & Miller, F.M. (1954) *The Chemistry of Heterocyclic Compounds*, Vol. 8, *Heterocyclic Compounds with Indole and Carbazole Systems*, New York, Interscience, pp. 70-109

Van Duuren, B.L., Bilbao, J.A. & Joseph, C.A. (1960) The carcinogenic nitrogen heterocyclics in cigarette-smoke condensate. *J. natl Cancer Inst.*, *25*, 53-61

DIBENZO[*a,e*]FLUORANTHENE[1]

1. Chemical and Physical Data

1.1 Synonyms and trade names

Chem. Abstr. Services Reg. No.: 5385-75-1

Chem. Abstr. Name: Dibenz(a,e)aceanthrylene

IUPAC Systematic Name: Dibenz[*a,e*]aceanthrylene

Synonym: 2,3,5,6-Dibenzofluoranthene (Buckingham, 1982)

1.2 Structural and molecular formulae and molecular weight

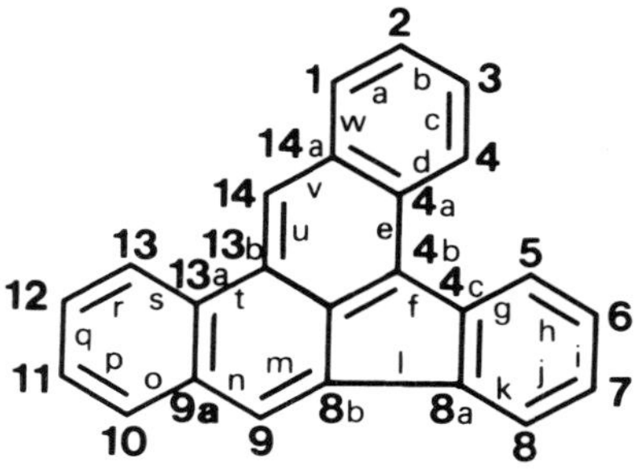

$C_{24}H_{14}$

Mol. wt: 302.4

1.3 Chemical and physical properties of the pure substance

From Buckingham (1982), unless otherwise specified

(*a*) *Description*: Yellow needles (recrystallized from benzene)

---

[1] Before 1968, dibenzo[*a,e*]fluoranthene was considered erroneously to be dibenzo[*a,l*]pyrene, because of an incorrect structural assignment (Lacassagne *et al.*, 1968).

(*b*) *Melting-point*: 232°C

(*c*) *Solubility*: Soluble in 1,4-dioxane (Hoffmann & Wynder, 1966)

(*d*) *Stability*: No data were available.

(*e*) *Reactivity*: No data were available.

2. Production, Use, Occurrence and Analysis

2.1 Production and use

There is no commercial production or known use of this compound.

2.2 Occurrence and analysis

Data on occurrence and methods of analysis are summarized in the 'General Remarks on the Substances Considered', p. 35.

Dibenzo[*a,e*]fluoranthene has been reported sporadically to occur in products of incomplete combustion. It has been identified in tobacco smoke (Snook *et al.*, 1977).

3. Biological Data Relevant to the Evaluation of Carcinogenic Risk to Humans

3.1 Carcinogenicity studies in animals

The Working Group noted that earlier reports of studies carried out with a compound called 'dibenzo[*a,l*]pyrene' were in fact done with dibenzo[*a,e*]fluoranthene (Lacassagne *et al.*, 1968). For this reason, papers published prior to 1966 on 'dibenzo[*a,l*]pyrene' have been reviewed under the assumption that the compound tested was, in fact, dibenzo[*a,e*]fluoranthene.

(*a*) *Skin application*

Mouse: Two groups of 20 female Swiss albino (Ha/ICR/Mil) mice, seven to eight weeks of age, received thrice weekly skin paintings of either a 0.1% or a 0.05% solution of dibenzo[*a,e*]fluoranthene (purified by chromatography) in dioxane [1,4-dioxane] for 12 months. In the high-dose group, the first papilloma appeared after five months, at which time 18 animals were still alive. The first epitheliomas [squamous-cell carcinomas] appeared after six months, at which time l5 mice were still alive. The last animal died at 14 months, at which time l8 mice had both papillomas and epitheliomas [squamous-cell carcinomas]. In the low-dose group, the last animal had died by 11 months, by which time 17 mice had developed papillomas (the first at five months) and 17 had epitheliomas [squamous-cell carcinomas], the first of which developed at six months. No skin tumour occurred in a solvent-control group of 20 mice (Hoffmann & Wynder, 1966).

A group of 30 female Swiss albino (Ha/ICR/Mil) mice, seven to eight weeks of age, received ten skin applications of 0.025 ml of a 0.1% solution of dibenzo[*a,e*]fluoranthene in 1,4-dioxane every two days (total dose, 0.25 mg). On day 28, skin applications were begun of a 2.5% solution of croton oil in acetone. The experiment was terminated after six months of promotion, at which time 28 mice were still alive. Papillomas were recorded in l8 mice (total, 52), and one animal developed an epithelioma [squamous-cell carcinoma]. In a control group of 30 mice receiving only the 2.5% solution of croton oil in acetone, two animals developed a papilloma (Hoffmann & Wynder, 1966).

(*b*) *Subcutaneous and/or intramuscular administration*

Mouse: A group of 16 male and 14 female XVII nc/Z strain mice, seven and six to seven months of age, respectively, received three s.c. injections of 0.6 mg dibenzo[*a,e*]fluoranthene in 0.2 ml olive oil; 10/16 males developed a sarcoma at the injection site, with the first tumour appearing after 115 days and the last after 170 days, with an average latent period of 144 days; 12/14 females developed sarcomas at the injection site, with the first tumour appearing after 120 days, the last after 272 days and an average latent period of 179 days. Of six surviving male mice from another group of eight mice that received single injections of 0.6 mg dibenzo[*a,e*]fluoranthene, four developed sarcomas at the injection site, with the first occurring at 202 days, the last at 209 days and an average latent period of 206 days (Lacassagne *et al.*, 1963). [The Working Group noted that no vehicle control was used in this study.]

3.2 Other relevant biological data[1]

(*a*) *Experimental systems*

No data were available to the Working Group on toxic effects or on effects on reproduction and prenatal toxicity.

Metabolism

A comprehensive study of the in-vitro metabolism of dibenzo[*a,e*]fluoranthene is available (Saguem *et al.*, 1983a,b).

7-, 10-, and 11-Monohydroxy derivatives of dibenzo[*a,e*]fluoranthene and the 1,2- and 10,11-dihydrodiols have been detected as metabolites following the incubation of dibenzo[*a,e*]fluoranthene with mouse- and rat-liver preparations (Périn-Roussel *et al.*, 1980).

---

[1] See also 'General Remarks on the Substances Considered', p. 53.

Nucleic acid adducts were formed when the compound was metabolized in a mouse- and rat-liver preparation in the presence of DNA (Périn-Roussel *et al.*, 1978), and some evidence has been obtained that these may arise from either the 1,2-diol-3,4-epoxide or the 10,11-diol-12,13-epoxide (Périn-Roussel *et al.*, 1983).

No data were available to the Working Group on mutagenicity and other short-term tests.

(*b*) *Humans*

No data were available to the Working Group.

3.3 Case reports and epidemiological studies of carcinogenicity to humans

No data were available to the Working Group.

4. Summary of Data Reported and Evaluation[1]

4.1 Experimental data

Dibenzo[*a,e*]fluoranthene was tested for carcinogenicity in one study by skin application to mice and produced skin tumours. It was also tested in the mouse-skin initiation-promotion assay and was active as an initiator.

Dibenzo[*a,e*]fluoranthene was tested in mice by subcutaneous administration and produced sarcomas at the injection site.

No data on the teratogenicity of this chemical were available.

No data were available to evaluate the activity of dibenzo[*a,e*]fluoranthene in short-term tests.

4.2 Human data[2]

Dibenzo[*a,e*]fluoranthene is present as a minor component of the total content of polynuclear aromatic compounds in the environment. Human exposure to dibenzo[*a,e*]fluoranthene occurs primarily through the smoking of tobacco; it may also occur through inhalation of polluted air and by ingestion of food and water contaminated by combustion effluents (for details, see 'General Remarks on the Substances Considered', p. 35).

---

[1] For definitions of the italicized terms, see Preamble, p. 19.

[2] Studies on occupational exposure to polynuclear aromatic compounds will be considered in future *IARC Monographs*.

4.3 Evaluation

There is *limited evidence* that dibenzo[a,e]fluoranthene is carcinogenic to experimental animals.

5. References

Buckingham, J., ed. (1982) *Dictionary of Organic Compounds*, 5th ed., Vol. 2, New York, Chapman & Hall, p. 1599

Hoffmann, D. & Wynder, F.L. (1966) Contribution to the carcinogenic action of dibenzopyrenes (Ger.). *Z. Krebsforsch.*, *68*, 137-149

Lacassagne, A., Buu-Hoï, N.P., Zajdela, F. & Lavit-Lamy, D. (1963) High carcinogenic activity of 1.2:3.4-dibenzopyrene and l.2:4.5-dibenzopyrene (Fr.). *C.R. Acad. Sci. Paris*, *256*, 2728-2730

Lacassagne, A., Buu-Hoï, N.P., Zajdela, F. & Vingiello, A. (1968) The true dibenzo[*a,l*]pyrene, a new, potent carcinogen. *Naturwissenschaften*, *55*, 43

Périn-Roussel, O., Ekert, B., Zajdela F. & Jacquignon, P. (1978) Binding of dibenzo(*a,e*)fluoranthene, a carcinogenic, polycyclic hydrocarbon without K-region, to nucleic acids in a sub-cellular microsomal system. *Cancer Res.*, *38*, 3499-3504

Périn-Roussel, O., Croisy-Delcey, M., Mispelter, J., Saguem, S., Chalvet, O., Ekert, B., Fouquet, J., Jacquigon, P., Lhoste, J.M., Muel, B. & Zajdela, F.E. (1980) Metabolic activation of dibenzo[*a,e*]fluoranthene, a nonalternant carcinogenic polycyclic hydrocarbon, in liver homogenates. *Cancer Res.*, *40*, 1742-1749

Périn-Roussel, O., Saguem, S., Ekert, B. & Zajdela, F. (1983) Binding to DNA of bay-region and pseudo-bay region diol-epoxides of dibenzo[*a,e*]fluoranthene and comparison with adducts obtained with dibenzo[*a,e*]fluoranthene or its dihydrodiols in the presence of microsomes. *Carcinogenesis*, *4*, 27-32

Saguem, S., Mispelter, J., Périn-Roussel, O., Lhoste, J.M. & Zajdela, F. (1983a) Multi-step metabolism of the carcinogen dibenzo[*a,e*]fluoranthene. I. Identification of the metabolites from rat microsomes. *Carcinogenesis*, *4*, 827-835

Saguem, S., Périn-Roussel, O., Mispelter, J., Lhoste, J.M. & Zajdela, F. (1983b) Multi-step metabolism of the carcinogen dibenzo[*a,e*]fluoranthene. II. Metabolic pathways. *Carcinogenesis*, *4*, 837-842

Snook, M.E., Severson, R.S., Arrendale, R.F, Highman, H.C. & Chortyk, O.T. (1977) The identification of high molecular weight polynuclear aromatic hydrocarbons in a biologically active fraction of cigarette smoke condensate. *Beitr. Tabakforsch.*, *9*, 79-101

DIBENZO[*a,e*]PYRENE

This compound was considered by a previous working group, in December 1972 (IARC, 1973).

1. Chemical and Physical Data

1.1 Synonyms and trade names

Chem. Abstr. Services Reg. No.: 192-65-4

Chem. Abstr. Name: Naphtho(1,2,3,4-def)chrysene

IUPAC Systematic Name: Naphtho[1,2,3,4-*def*]chrysene

Synonyms: DB(a,e)P; 1,2:4,5-dibenzopyrene; 1,2,4,5-dibenzopyrene; naphtho-(1,2,3,4,def)chrysene

1.2 Structural and molecular formulae and molecular weight

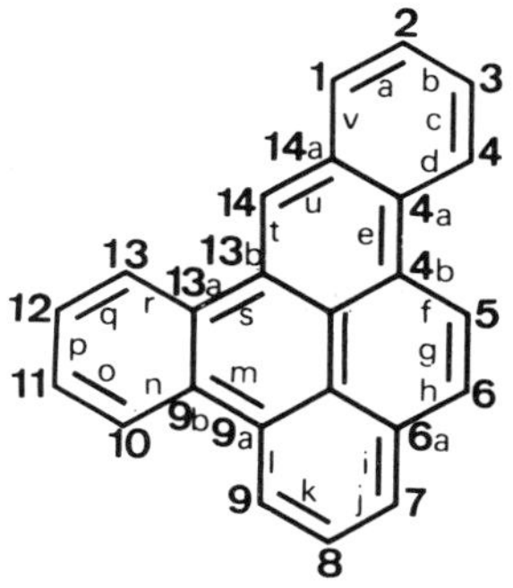

$C_{24}H_{14}$

Mol. wt: 302.4

1.3 Chemical and physical properties of the pure substance

From National Library of Medicine (1982), unless otherwise specified

(*a*) *Description:* Pale-yellow needles (recrystallized from xylene)

(*b*) *Melting-point*: 233°C (Karcher *et al.*, 1983)

(*c*) *Spectroscopy data*: λ_{max} 221, 241, 247, 271, 290, 302, 323, 338, 354, 373, 384, 394 nm (in cyclohexane) (Karcher *et al.*, 1983)

(*d*) *Solubility*: Slightly soluble in acetic acid, acetone, benzene and ethanol; soluble in hot toluene

(*e*) *Stability*: No data were available.

(*f*) *Reactivity*: Oxidized by sodium dichromate in glacial acetic acid, first to dibenzo[*a,e*]pyrene-5,6-quinone and then to 9,14-dioxo-9,14-dihydrodibenz[*a,c*]anthracene carboxylic acid (Zinke & Zimmer, 1951; Ott, 1955). Reacts with bromine or acetyl chloride to give the mono substitution products, 7-bromodibenzo[*a,e*]pyrene or 7-acetodibenzo[*a,e*]pyrene (Lang & Zander, 1965)

2. Production, Use, Occurrence and Analysis

2.1 Production and use

There is no commercial production or known use of this compound.

2.2 Occurrence and analysis

Data on occurrence and methods of analysis are summarized in the 'General Remarks on the Substances Considered', p. 35.

Dibenzo[*a,e*]pyrene occurs ubiquitously in products of incomplete combustion; it also occurs in fossil fuels. It has been identified in tobacco smoke (Snook *et al.*, 1977) and gasoline engine exhaust (Lyons, 1962).

3. Biological Data Relevant to the Evaluation of Carcinogenic Risk to Humans

3.1 Carcinogenicity studies in animals

There is *sufficient evidence* that dibenzo[*a,e*]pyrene is carcinogenic to experimental animals (see 'General Remarks on the Substances Considered', p. 33).

3.2 Other relevant biological data[1]

(*a*) *Experimental systems*

No data were available to the Working Group on toxic effects, on effects on reproduction and prenatal toxicity or on metabolism and activation.

Mutagenicity and other short-term tests

Dibenzo[*a,e*]pyrene was mutagenic to *Salmonella typhimurium* strain TA100 (*his*$^-$/*his*$^+$) when added at a concentration of 100 μg/plate in the presence of an exogenous metabolic system (post-mitochondrial supernatant from Aroclor-induced rat liver) (LaVoie *et al.*, 1979).

(*b*) *Humans*

No data were available to the Working Group.

3.3 Case reports and epidemiological studies of carcinogenicity to humans

No data were available to the Working Group.

4. Summary of Data Reported and Evaluation[1]

4.1 Experimental data

Dibenzo[*a,e*]pyrene has been shown to be carcinogenic to experimental animals (see 'General Remarks on the Substances Considered', p. 33, and IARC, 1973).

No data on the teratogenicity of this compound were available.

In the one study evaluated, dibenzo[*a,e*]pyrene was mutagenic to *Salmonella typhimurium* in the presence of an exogenous metabolic system.

There is *inadequate evidence* that dibenzo[*a,e*]pyrene is active in short-term tests.

4.2 Human data[2]

Dibenzo[*a,e*]pyrene is present as a minor component of the total content of polynuclear aromatic compounds in the environment. Human exposure to dibenzo[*a,e*]pyrene occurs

---

[1] See also 'General Remarks on the Substances Considered', p. 53.

[1] For definitions of the italicized terms, see Preamble, p. 19.

[2] Studies on occupational exposure to polynuclear aromatic compounds will be considered in future *IARC Monographs*.

primarily through the smoking of tobacco, inhalation of polluted air and by ingestion of food and water contaminated with combustion products (for details, see 'General Remarks on the Substances Considered', p. 35).

4.3 Evaluation[1]

There is *sufficient evidence* that dibenzo[*a*,*e*]pyrene is carcinogenic to experimental animals.

5. References

IARC (1973) *IARC Monographs on the Evaluation of Carcinogenic Risk of Chemicals to Man*, Vol. 3, *Certain polycyclic aromatic hydrocarbons and heterocyclic compounds*, Lyon, pp. 201-206

Karcher, W., Fordham, R., Dubois, J. & Gloude, P. (1983) *Spectral Atlas of Polycyclic Aromatic Compounds*, Dordrecht, The Netherlands, D. Reidel (in press)

Lang, K.F. & Zander, M. (1965) On knowledge of 1,2,4,5-dibenzopyrene (Ger.). *Chem. Ber.*, *98*, 597-600

LaVoie, E., Bedenko, V., Hirota, N., Hecht, S.S. & Hoffmann, D. (1979) *A comparison of the mutagenicity, tumor-initiating activity and complete carcinogenicity of polynuclear aromatic hydrocarbons*. In: Jones, P.W. & Leber, P., eds, *Polynuclear Aromatic Hydrocarbons*, Ann Arbor, MI, Ann Arbor Science Publishers, pp. 705-721

Lyons, M.J. (1962) Comparison of aromatic polycyclic hydrocarbons from gasoline engine and diesel engine exhausts, general atmospheric dust, and cigarette-smoke condensate. *Natl Cancer Inst. Monogr.*, *9*, 193-199

National Library of Medicine (1982) *Toxicology Data Bank*, Bethesda, MD, National Library of Medicine Specialized Information Services, Toxicology Information Program

Ott, R. (1955) Synthesis of higher condensed ring system through intermolecular deshydration of different molecules by combination and ring closing. IX. On the 1,2,4,5-dibenzopyrenequinone with observations on amphi- and *o*-quinone (Ger.). *Monatsch. Chem.*, *86*, 622-636

Snook, M.E., Severson, R.F., Arrendale, R.F., Higman, H.C. & Chortyk, O.T. (1977) The identification of high molecular weight polynuclear aromatic hydrocarbons in a biologically active fraction of cigarette smoke condensate. *Beitr. Tabakforsch.*, *9*, 79-101

Zinke, A. & Zimmer, W. (1951) On the formation of benzopyrenes from chrysenes. II. On the 1,2,4,5-dibenzopyrene (Ger.). *Monatsch. Chem.*, *82*, 348-358

---

[1] In the absence of adequate data on humans, it is reasonable, for practical purposes, to regard chemicals for which there is *sufficient evidence* of carcinogenicity in animals as if they presented a carcinogenic risk to humans (see also Preamble, p. 20).

DIBENZO[*a,h*]PYRENE

This compound was considered by a previous working group in December 1972 (IARC, 1973).

1. Chemical and Physical Data

1.1 Synonyms and trade names

Chem. Abstr. Services Reg. No.: 189-64-0

Chem. Abstr. Name: Dibenzo(b,def)chrysene

IUPAC Systematic Name: Dibenzo[*b,def*]chrysene

Synonyms: DB(a,h)P; 1,2,6,7-dibenzopyrene; 3,4:8,9-dibenzopyrene; 3,4,8,9-dibenzopyrene

1.2 Structural and molecular formulae and molecular weight

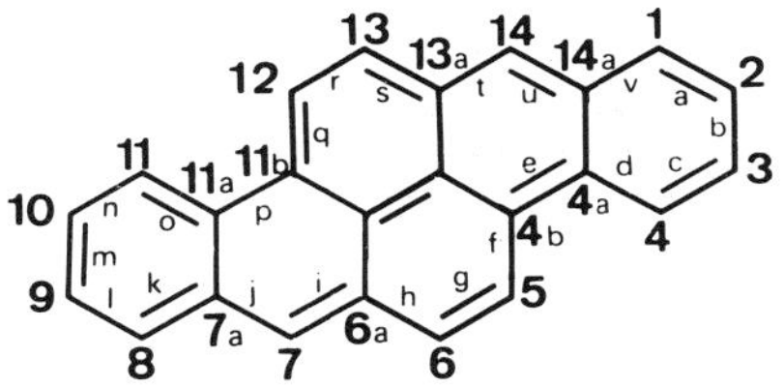

$C_{24}H_{14}$

Mol. wt: 302.4

1.3 Chemical and physical properties of the pure substance

From National Library of Medicine (1982), unless otherwise specified

(*a*) *Description*: Golden-yellow plates (recrystallized from xylene or trichlorobenzene)

(*b*) *Melting-point*: 317°C (Karcher *et al.*, 1983)

(*c*) *Spectroscopy data*: λ_{max} 216, 237, 246, 254, 267, 285, 297, 309, 385, 419, 441, 445 nm (in cyclohexane) (Karcher *et al.*, 1983). Mass (Thomas & Las, 1978) and nuclear magnetic resonance spectra (Karcher *et al.*, 1983) have been reported.

(*d*) *Solubility*: Soluble in 1,4-dioxane (Hoffmann & Wynder, 1966)

(*e*) *Stability*: No data were available.

(*f*) *Reactivity*: The 7-mono- and 7,14-dinitro-derivatives are formed by nitration (Ioffe & Efross, 1946; Clar, 1964).

2. Production, Use, Occurrence and Analysis

2.1 Production and use

There is no commercial production or known use of this compound.

2.2 Occurrence and analysis

Data on occurrence and methods of analysis are summarized in the 'General Remarks on the Substances Considered', p. 35.

Dibenzo[*a*,*h*]pyrene occurs ubiquitously in products of incomplete combustion; it is also found in fossil fuels. It is a minor constituent of coal-tar (Buu-Hoï, 1958; Lang & Eigen, 1967) and automobile engine exhaust (Lyons, 1962); it has been identified in tobacco smoke (Snook *et al.*, 1977) and possibly in urban atmospheres (Hoffmann & Wynder, 1976).

3. Biological Data Relevant to the Evaluation of Carcinogenic Risk to Humans

3.1 Carcinogenicity studies in animals

There is *sufficient evidence* that dibenzo[*a*,*h*]pyrene is carcinogenic to experimental animals (see 'General Remarks on the Substances Considered', p. 33).

3.2 Other relevant biological data[1]

(*a*) *Experimental systems*

No data were available to the Working Group on toxic effects or on effects on reproduction and prenatal toxicity.

Metabolism

The 1,2- and 3,4-dihydrodiols have been identified as metabolites of dibenzo[*a,h*]pyrene following incubation of this compound with rat-liver preparations (Hecht *et al.*, 1981).

The 1,2-dihydrodiol is mutagenic to bacteria in the presence of an exogenous metabolic system (Wood *et al.*, 1981) and is a tumour initiator on mouse skin and tumorigenic in newborn mice (Chang *et al.*, 1982).

Mutagenicity and other short-term tests

Results from short-term tests are given in Table 1.

Table 1. Results from short-term tests: Dibenz[*a,h*]pyrene

Test	Organism/assay[a]	Exogenous metabolic system[a]	Reported result	Comments	References
PROKARYOTES					
Mutation	*Salmonella typhimurium* (his⁻/his⁺)	Aro-R-Micr	Positive	At 6 nmol/plate in strains TA98 and TA100	Wood *et al.* (1981)
		Aro-R-PM	Inconclusive	Tested at up to 100 μg/plate in strain TA100	LaVoie *et al.* (1979)

[a] For an explanation of the abbreviations see the Appendix, p. 452.

(*b*) *Humans*

No data were available to the Working Group.

3.3 Case reports and epidemiological studies of carcinogenicity to humans

No data were available to the Working Group.

[1] See also 'General Remarks on the Substances Considered', p. 53.

4. Summary of Data Reported and Evaluation[1]

4.1 Experimental data

Dibenzo[*a*,*h*]pyrene has been shown to be carcinogenic to experimental animals. (See 'General Remarks on the Substances Considered', p. 33, and IARC, 1973).

No data on the teratogenicity of this compound were available.

In one study, dibenzo[*a*,*h*]pyrene was mutagenic to *Salmonella typhimurium* in the presence of an exogenous metabolic system.

There is *inadequate evidence* that dibenzo[*a*,*h*]pyrene is active in short-term tests.

4.2 Human data[2]

Dibenzo[*a*,*h*]pyrene is present as a minor component of the total content of polynuclear aromatic compounds in the environment. Human exposure to dibenzo[*a*,*h*]pyrene occurs primarily through the smoking of tobacco, inhalation of polluted air and by ingestion of food and water contaminated with combustion products (for details, see 'General Remarks on the Substances Considered', p. 35).

4.3 Evaluation[3]

There is *sufficient evidence* that dibenzo[*a*,*h*]pyrene is carcinogenic to experimental animals.

5. References

Buu-Hoï, N.P. (1958) Presence of 3:4-8:9-dibenzpyrene in coal-tar. *Nature*, *182*, 1158-1159

Chang, R.L., Levin, W., Wood, A.W., Lehr, R.E., Kumar, S., Yagi, H., Jerina, D.M. & Conney, A.H. (1982) Tumorigenicity of bay-region diol-epoxides and other benzo-ring derivatives of dibenzo(*a*,*h*)pyrene and dibenzo(*a*,*i*)pyrene on mouse skin and in newborn mice. *Cancer Res.*, *42*, 25-29

---

[1] For definitions of the italicized terms, see 'Preamble', p. 19.

[2] Studies on occupational exposure to polynuclear aromatic compounds will be considered in future *IARC Monographs*.

[3] In the absence of adequate data on humans, it is reasonable, for practical purposes, to regard chemicals for which there is *sufficient evidence* of carcinogenicity in animals as if they presented a carcinogenic risk to humans (see also, Preamble, p. 20).

Clar, E., ed. (1964) *Polycyclic Hydrocarbons*, Vol. 2, New York, Academic Press, pp. 151-153

Hecht, S.S., LaVoie, E.J., Bedenko, V., Hoffmann, D., Sardella, D.J., Boger, E. & Lehr, R.E. (1981) *On the metabolic activation of dibenzo[*a,i*]pyrene and dibenzo[*a,h*]pyrene*. In: Cooke, M. & Dennis, A.J., eds, *Polynuclear Aromatic Hydrocarbons, Chemical Analysis and Biological Fate, 5th Int. Symposium*, Columbus, OH, Battelle Press, pp. 43-54

Hoffmann, D. & Wynder, E.L. (1966) Contribution to the carcinogenic properties of dibenzopyrenes (Ger.). *Z. Krebsforsch.*, *68*, 137-149

Hoffmann, D. & Wynder, E. (1976) *Environmental respiratory carcinogenesis*. In: Searle, C.E., ed., *Chemical Carcinogens* (*ACS Monograph 173*), Washington DC, American Chemical Society, p. 341

IARC (1973) *IARC Monographs on the Evaluation of Carcinogenic Risk of Chemicals to Man*, Vol. 3, *Certain polycyclic aromatic hydrocarbons and heterocyclic compounds*, Lyon, pp. 207-214

Ioffe, I.S. & Efross, L.S. (1946) Nitro and amino derivatives of dibenzopyrene (Russ.). *J. gen. Chem. USSR*, *16*, 111-116

Karcher, W., Fordham, R., Dubois, J. & Gloude, P. (1983) *Spectral Atlas of Polycyclic Aromatic Compounds*, Dordrecht, The Netherlands, D.Reidel (in press)

Lang, K.F. & Eigen, I. (1967) *Organic compounds in coal tar* (Ger.). In: Heilbronner, E., Hofmann, U., Schäfer, K. & Wittig, G., eds, *Fortschrifte der chemischer Forschung* (Advances in chemical research), Vol. 8, Berlin, Springer, pp. 91-170

LaVoie, E., Bedenko, V., Hirota, N., Hecht, S.S. & Hoffmann, D. (1979) *A comparison of the mutagenicity, tumor-initiating activity and complete carcinogenicity of polynuclear aromatic hydrocarbons*. In: Jones, P.W. & Leber, P., eds, *Polynuclear Aromatic Hydrocarbons*, Ann Arbor, MI, Ann Arbor Science Publishers, pp. 705-721

Lyons, M.J. (1962) Comparison of aromatic polycyclic hydrocarbons from gasoline engine and diesel engine exhausts, general atmospheric dust, and cigarette-smoke condensate. *Natl Cancer Inst. Monogr.*, *9*, 193-199

National Library of Medicine (1982) *Toxicology Data Bank*, Bethesda, MD, National Library of Medicine Specialized Information Services, Toxicology Information Program

Snook, M.E., Severson, R.F., Arrendale, R.F., Higman, H.C. & Chortyk, O.T. (1977) The identification of high molecular weight polynuclear aromatic hydrocarbons in a biologically active fraction of cigarette smoke condensate. *Beitr. Tabakforsch.*, *9*, 79-101

Thomas, R.S. & Las, R.C. (1978) Mass spectra of isomeric dibenzopyrenes. *Adv. Mass Spectrom.*, *7B*, 1709-1712

Wood, A.W., Chang, R.L., Levin, W., Ryan, D.E., Thomas, P.E., Lehr, R.E., Kumar, S., Sardella, D.J., Boger, E., Yagi, H., Sayer, J.M., Jerina, D.M. & Conney, A.H. (1981) Mutagenicity of the bay-region diol-epoxides and other benzo-ring derivatives of dibenzo(*a,h*)pyrene and dibenzo(*a,i*)pyrene. *Cancer Res.*, *41*, 2585-2597

DIBENZO[*a,i*]PYRENE

This compound was considered by a previous working group, in December 1972 (IARC, 1973).

1. Chemical and Physical Data

1.1 Synonyms and trade names

Chem. Abstr. Services Reg. No.: 189-55-9

Chem. Abstr. Name: Benzo(rst)pentaphene

IUPAC Systematic Name: Benzo[*rst*]pentaphene

Synonyms: DB(a,i)P; 1,2,7,8-dibenzopyrene; 3,4:9,10-dibenzopyrene; dibenzo(b,h)pyrene; 1,2:7,8-dibenzpyrene; 3,4,9,10-dibenzpyrene

1.2 Structural and molecular formulae and molecular weight

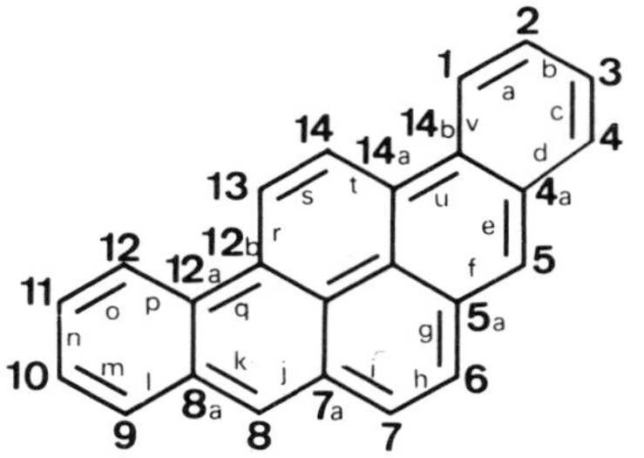

$C_{24}H_{14}$

Mol. wt: 302.4

1.3 Chemical and physical properties of the pure substance

From National Library of Medicine (1982), unless otherwise specified

(*a*) *Description:* Greenish-yellow needles, prisms or lamellae

(*b*) *Boiling-point:* 275°C at 0.05 mm Hg (Buckingham, 1982) (*c*) *Melting-point*: 281.5-282.5°C (Buckingham, 1982); 280°C

(*d*) *Spectroscopy data:* λ_{max} 222, 242, 272 nm (in ethanol); 285, 297, 317, 332, 356, 375, 397, 419, 433 nm (in benzene) (Clar, 1964). Infrared spectra have been reported (Cannon & Sutherland, 1951).

(*e*) *Solubility:* Soluble in 1,4-dioxane (Hoffmann & Wynder, 1966), in boiling glacial acetic acid (2 g/l) and boiling benzene (5 g/l); almost insoluble in diethyl ether and ethanol

(*f*) *Stability*: No data were available.

(*g*) *Reactivity:* Attacked by nitric acid and other reagents, mainly at the 5- and 8-positions; slowly oxidized by sulphuric acid to a quinone; oxidized by lead tetraacetate to the 5,8-diacetoxy derivative and by chromic oxide or selenium dioxide to the 5,8-quinone (Ünseren & Fieser, 1962)

2. Production, Use, Occurrence and Analysis

2.1 Production and use

There is no commercial production or known use of this compound.

2.2 Occurrence and analysis

Data on occurrence and methods of analysis are summarized in the 'General Remarks on the Substances Considered', p. 35.

Dibenzo[*a,i*]pyrene occurs ubiquitously in products of incomplete combustion; it also occurs in fossil fuels. It has been identified in mainstream cigarette smoke (about 1 μg/100 cigarettes) (Bonnet & Neukomm, 1956), (0.52 μg/300 cigarettes) (Müller *et al.*, 1967); and coal-tar (Schoental, 1957).

3. Biological Data Relevant to the Evaluation of Carcinogenic Risk to Humans

3.1 Carcinogenicity studies in animals

There is *sufficient evidence* that dibenzo[*a,i*]pyrene is carcinogenic to experimental animals (see 'General Remarks on the Substances Considered', p. 33).

3.2 Other relevant biological data[1]

(*a*) *Experimental systems*

No data were available to the Working Group on toxic effects or on effects on reproduction and prenatal toxicity.

Metabolism and activation

The 1,2- and 3,4-dihydrodiols have been reported to be metabolites of dibenzo[*a,i*]pyrene following incubation of this compound with rat-liver preparations (Hecht *et al.*, 1981a,b).

The 3,4-dihydrodiol has been reported to be mutagenic to bacteria in the presence of an exogenous metabolic system (Wood *et al.*, 1981); it is a tumour initiator on mouse skin and tumorigenic in newborn mice (Chang *et al.*, 1982).

Mutagenicity and other short-term tests

Results from short-term tests are given in Table 1.

Table 1. Results from short-term tests: Dibenzo[*a,i*]pyrene

Test	Organism/assay[a]	Exogenous metabolic system[a]	Reported result	Comments	References
PROKARYOTES					
DNA damage	*Escherichia coli* (rec^+/rec^-)	Aro-R-PMS	Positive	At 600 μg/well	Ichinotsubo *et al.* (1977)
	Bacillus subtilis (rec^+/rec^-)	Aro-R-PMS	Positive	Minimum inhibitory concentration: rec^+ at 500 μg/well *versus* rec^- at 125 μg/well	McCarroll *et al.* (1981)
Mutation	*Salmonella typhimurium* (his^-/his^+)	Aro-R-Micr	Positive	At 3 nmol/plate in strain TA98	Wood *et al.* (1981)
		3MC-GP-PMS Aro-R-PMS Aro-GP-PMS	Positive	At 5 μg/plate in strain TA100	Baker *et al.* (1980)
		Aro-R-PMS	Positive	Tested at 20 μg/plate in strain TA100	McCann *et al.* (1975)
MAMMALIAN CELLS *IN VITRO*					
DNA damage	Primary rat hepatocytes (unscheduled DNA synthesis)	--	Negative	Tested at up to 1 μmol/ml	Probst *et al.* (1981)

[a] For an explanation of the abbreviations see the Appendix, p. 452.

[1] See also 'General Remarks on the Substances Considered', p. 53.

(*b*) *Humans*

No data were available to the Working Group.

3.3 Case reports and epidemiological studies of carcinogenicity to humans

No data were available to the Working Group.

4. Summary of Data Reported and Evaluation[1]

4.1 Experimental data

Dibenzo[*a,i*]pyrene has been shown to be carcinogenic to experimental animals. (See 'General Remarks on the Substances Considered', p. 33, and IARC, 1973.)

No data on the teratogenicity of this compound were available.

Dibenzo[*a,i*]pyrene was mutagenic to *Salmonella typhimurium* in the presence of an exogenous metabolic system. It was positive in assays for differential killing in strains of DNA-repair-proficient/-deficient bacteria. It did not induce unscheduled DNA synthesis in rat hepatocytes.

There is *inadequate evidence* that dibenzo[*a,i*]pyrene is active in short-term tests.

4.2 Human data[2]

Dibenzo[*a,i*]pyrene is present as a minor component of the total content of polynuclear aromatic compounds in the environment. Human exposure to dibenzo[*a,i*]pyrene occurs primarily through the smoking of tobacco, inhalation of polluted air and by ingestion of food and water contaminated with combustion products (for details, see 'General Remarks on the Substances Considered', p. 35).

---

[1] For definitions of the italicized terms, see Preamble, p. 19.

[2] Studies on occupational exposure to polynuclear aromatic compounds will be considered in future *IARC Monographs*.

4.3 Evaluation[1]

There is *sufficient evidence* that dibenzo[*a,i*]pyrene is carcinogenic to experimental animals.

5. References

Baker, R.S.U., Bonin, A.M., Stupans, I. & Holder, G.M. (1980) Comparison of rat and guinea pig as sources of the S9 fraction in the *Salmonella*/mammalian microsome mutagenicity test. *Mutat. Res.*, *71*, 43-52

Bonnet, J. & Neukomm, S. (1956) On the chemical composition of tobacco smoke. I. Analysis of the neutral fraction (Fr.). *Helv. chim. Acta*, *39*, 1724-1733

Buckingham, J., ed. (1982) *Dictionary of Organic Compounds,* 5th ed., Vol. 1, New York, Chapman & Hall, p. 567

Canon, C.G. & Sutherland, G.B.B.M. (1951) The infra-red absorption spectra of some aromatic compounds. *Spectrochim. Acta*, *4*, 373-395

Chang, R.L., Levin, W., Wood, A.W., Lehr, R.E., Kumar, S., Yagi, H., Jerina, D.M. & Conney,A.H. (1982) Tumorigenicity of bay-region diol-epoxides and other benzo-ring derivatives of dibenzo(*a,h*)pyrene and dibenzo(*a,i*)pyrene on mouse skin and in newborn mice. *Cancer Res.*, *42*, 25-29

Clar, E., ed. (1964) *Polycyclic Hydrocarbons,* Vol. 2, New York, Academic Press, pp. 153-157

Hecht, S.S., LaVoie, E.J., Bedenko, V., Hoffmann, D., Sardella, D.J., Boger, E. & Lehr, R.E. (1981a) *On the metabolic activation of dibenzo[*a,i*]pyrene and dibenzo[*a,h*]pyrene*. In: Cooke, M. & Dennis, A.J., eds, *Polynuclear Aromatic Hydrocarbons, Chemical Analysis and Biological Fate, 5th Int. Symposium*, Columbus, OH, Battelle Press, pp. 43-54

Hecht, S.S., LaVoie, E.J., Bedenko, V., Pingaro, L., Katayama, S., Hoffmann, D., Sardella, D.J., Boger, E. & Lehr, R.E. (1981b) Reduction of tumorigenicity and of dihydrodiol formation by fluorine substitution in the angular rings of dibenzo(*a,i*)pyrene. *Cancer Res.*, *41*, 4341-4345

Hoffmann, D. & Wynder, E.L. (1966) Contribution to the carcinogenic properties of dibenzopyrenes (Ger.). *Z. Krebsforsch.*, *68*, 137-149

[1] In the absence of adequate data on humans, it is reasonable, for practical purposes, to regard chemicals for which there is *sufficient evidence* of carcinogenicity in animals as if they presented a carcinogenic risk to humans (see also, Preamble, p. 20).

IARC (1973) *IARC Monographs on the Evaluation of Carcinogenic Risk of Chemicals to Man*, Vol. 3, *Certain polycyclic aromatic hydrocarbons and heterocyclic compounds*, Lyon, pp. 215-223

Ichinotsubo, D., Mower, H.F., Setliff, J. & Mandel, M. (1977) The use of *rec-* bacteria for testing of carcinogenic substances. *Mutat. Res.*, *46*, 53-62

McCann, J., Choi, E., Yamasaki, E. & Ames, B.N. (1975) Detection of carcinogens as mutagens in the *Salmonella*/microsome test: Assay of 300 chemicals. *Proc. natl Acad. Sci. USA*, *72*, 5135-5139

McCarroll, N.E., Keech, B.H. & Piper, C.E. (1981) A microsuspension adaptation of the *Bacillus subtilis* 'rec' assay. *Environ. Mutagenesis*, *3*, 607-616

Müller, R., Moldenhauer, W. & Schlemmer, P. (1967) Experiences with quantitative determination of polycyclic hydrocarbons in tobacco smoke (Ger.). *Ber. Inst. Tabakforsch. Dresden*, *14*, 159-173

National Library of Medicine (1982) *Toxicology Data Bank,* Bethesda, MD, National Library of Medicine Specialized Information Services, Toxicology Information Program

Probst, G.S., McMahon, R.E., Hill, L.E., Thompson, C.Z., Epp, J.K. & Neal, S.B. (1981) Chemically-induced unscheduled DNA synthesis in primary rat hepatocyte cultures: A comparison with bacterial mutagenicity using 218 compounds. *Environ. Mutagenesis*, *3*, 11-32

Schoental, R. (1957) Isolation of 3:4-9:10-dibenzopyrene from coal-tar. *Nature*, *180*, 606

Ünseren, E. & Fieser, L.F. (1962) Investigation of the metabolism of 3,4,9,10-dibenzpyrene. *J. org. Chem.*, *27*, 1386-1389

Wood, A.W., Chang, R.L., Levin, W., Ryan, D.E., Thomas, P.E., Lehr, R.E., Kumar, S., Sardella, D.J., Boger, E., Yagi, H., Sayer, J.M., Jerina, D.M. & Conney, A.H. (1981) Mutagenicity of the bay-region diol-epoxides and other benzo-ring derivatives of dibenzo(*a,h*)pyrene and dibenzo(*a,i*)pyrene. *Cancer, Res.*, *41*, 2589-2597

DIBENZO[*a,l*]PYRENE[1]

This compound was considered by a previous working group, in December 1972 (IARC, 1973). Data that have become available since that time have been incorporated in the present monograph and taken into consideration in the evaluation.

1. Chemical and Physical Data

1.1 Synonyms and trade names

Chem. Abstr. Services Reg. No.: 191-30-0

Chem. Abstr. Name: Dibenzo(def,p)chrysene

IUPAC Systematic Name: Dibenzo[*def,p*]chrysene

Synonyms: Ba 51-090462; DB(a,l)P; 1,2:3,4-dibenzopyrene; 1,2,3,4-dibenzopyrene; 1,2,9,10-dibenzopyrene; 2,3:4,5-dibenzopyrene; 4,5,6,7-dibenzpyrene

1.2 Structural and molecular formulae and molecular weight

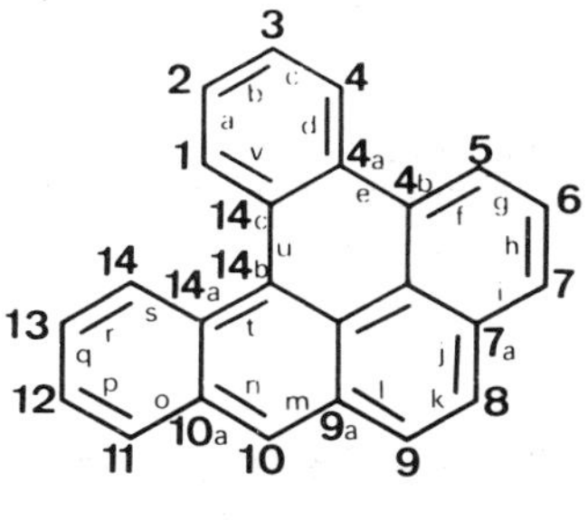

$C_{24}H_{14}$

Mol. wt: 302.4

[1] The compound referred to in the literature before 1968 as 1,2,3,4-dibenzopyrene (dibenz[*a,l*]pyrene) was in fact dibenzo[*a,e*]fluoranthene (see monograph, p. 321). Dibenzo[*a,l*]pyrene was synthesized by Buu-Hoi *et al.* (1969), who discuss the earlier confusion. Only data obtained since 1968 are reviewed here.

1.3 Chemical and physical properties of the pure substance

(*a*) *Description:* Pale-yellow plates (recrystallized from ethanol or ethanol/benzene) (National Library of Medicine, 1982)

(*b*) *Melting-point:* 162.4°C (Karcher *et al.*, 1983)

(*c*) *Spectroscopy data:* λ_{max} 237, 268, 291, 302, 315, 334, 353, 370, 389 nm in cyclohexane (Karcher *et al.*, 1983); mass and nuclear magnetic resonance spectra have been tabulated (NIH/EPA Chemical Information System, 1982; Karcher *et al.*, 1983).

(*d*) *Solubility:* Soluble in olive oil (Lacassagne *et al.*, 1968)

(*e*) *Stability*: No data were available.

(*f*) *Reactivity*: No data were available.

2. Production, Use, Occurrence and Analysis

2.1 Production and use

There is no commercial production or known use of this compound. A reference sample of certified high purity is available (Community Bureau of Reference, 1982).

2.2 Occurrence and analysis

Data on occurrence and methods of analysis are summarized in the 'General Remarks on the Substances Considered', p. 35.

Dibenzo[*a*,*l*]pyrene occurs in some products of incomplete combustion; it also occurs in fossil fuels. It has been identified in mainstream cigarette smoke (Snook *et al.*, 1977) and products of coal gasification (Bridbord & French, 1978; Young *et al.*, 1978).

3. Biological Data Relevant to the Evaluation of Carcinogenic Risk to Humans

3.1 Carcinogenicity studies in animals

The Working Group noted that earlier studies reported to have been carried out with a compound called dibenzo[*a*,*l*]pyrene were in fact done with dibenzo[*a*,*e*]fluoranthene (Lacassagne *et al.*, 1968). For this reason, only papers published since from 1968 are reviewed here.

(*a*) *Skin application*

Mouse: Groups of 19-21 female ICR Swiss albino mice [age unspecified] received skin applications of 0.1%, 0.05%, 0.01%, 0.005% or 0.001% dibenzo[*a,l*]pyrene [solvent unspecified] (total doses, 700, 350, 240, 200 and 55 μg, respectively, and given in 7, 7, 24, 40 and 55 applications, respectively) [dose schedule unspecified]. Skin tumours [types unspecified] occurred in all groups beginning two months after the start of treatment. Tumours developed in 16, 19, 21, 19 and 20 animals in these groups, respectively, by six months, at which time all animals in the three highest dose groups had died. No solvent control group was included in the study, but positive control groups receiving applications of 0.1 and 0.05% benzo[*a*]pyrene (total doses, 6000 μg and 3000 μg, respectively, over 60 applications) were available; skin tumour yield six months after the start of treatment in these groups was similar to that of dibenzo[*a,l*]pyrene-treated groups (18 and 13 tumour-bearing mice in the high- and low-dose benzo[*a*]pyrene groups, respectively) (Masuda & Kagawa, 1972).

(*b*) *Subcutaneous and/or intramuscular administration*

Mouse: A group of 12 male and 12 female XVII nc/ZE mice, three to four months of age, received two s.c. injections of 0.6 mg dibenzo[*a,l*]pyrene in 0.2 ml olive oil at one month intervals. A third injection was given two months later to those mice that had not developed a strong fibrous reaction at the injection site. All 24 mice developed sarcomas at the injection site within a mean latent period of 130 days for males and 113 days for females. (The solvent alone was reported to have been inactive in over 500 control mice used in that laboratory.) (Lacassagne *et al.*, 1968).

3.2 Other relevant biological data[1]

No data were available to the Working Group.

3.3 Case reports and epidemiological studies of carcinogenicity to humans

No data were available to the Working Group.

---

[1] See also 'General Remarks on the Substances Considered', p. 53.

4. Summary of Data Reported and Evaluation[1]

4.1 Experimental data

Dibenzo[*a,l*]pyrene was tested for carcinogenicity in mice in one study by skin application and in one study by subcutaneous administration. It induced tumours at the sites of application.

No data on the teratogenicity of this compound were available.

No data were available to evaluate the activity of dibenzo[*a,l*]pyrene in short-term tests.

4.2 Human data[2]

Dibenzo[*a,l*]pyrene is present as a minor component of the total content of polynuclear aromatic compounds in the environment. Human exposure to dibenzo[*a,l*]pyrene occurs mainly through the smoking of tobacco, inhalation of polluted air and by ingestion of food and water contaminated with combustion products (for details, see 'General Remarks on the Substances Considered', p. 35).

4.3 Evaluation[3]

There is *sufficient evidence* that dibenzo[*a,l*]pyrene is carcinogenic to experimental animals.

---

[1] For definitions of the italicized terms, see Preamble, p. 19.

[2] Studies on occupational exposure to polynuclear aromatic compounds will be considered in future *IARC Monographs*.

[3] In the absence of adequate data on human, it is reasonable, for practical purposes, to regard chemicals for which there is *sufficient evidence* of carcinogenicity in animals as if they presented a carcinogenic risk to humans (see also Preamble, p. 20).

5. References

Bridbord, K. & French, J.G. (1978) *Carcinogenic and mutagenic risks associated with fossil fuels*. In: Jones, P.W. & Freudenthal, R.I., eds, *Carcinogenesis*, Vol. 3, *Polynuclear Aromatic Hydrocarbons*, New York, Raven Press, pp. 451-463

Buu-Hoï, N.P., Périn-Roussel, O. & Jacquignon, P. (1969) On the new synthesis of dibenzo[*a,l*]pyrene (Fr.). *Bull. Soc. chim. Fr., Ser. 5*, *36*, 3566-3568

Community Bureau of Reference (1982) *Polycyclic Aromatic Hydrocarbon Reference Materials of Certified Purity* (*Information Handout, No. 22*), Brussels, Commission of the European Communities

IARC (1973) *IARC Monographs on the Evaluation of Carcinogenic Risk of Chemicals to Man*, Vol. 3, *Certain polycyclic aromatic hydrocarbons and heterocyclic compounds*, Lyon, pp. 224-228

Karcher, W., Fordham, R., Dubois, J. & Gloude, P. (1983) *Spectral Atlas of Polycyclic Aromatic Compounds*, Dordrecht, The Netherlands, D. Reidel (in press)

Lacassagne, A., Buu-Hoï, N.P., Zajdela, F & Vingiello, F.A. (1968) The true dibenzo[*a,l*]pyrene, a new, potent carcinogen. *Naturwissenschaften*, *55*, 43

Masuda, Y. & Kagawa, R. (1972) A novel synthesis and carcinogenicity of dibenzo[*a,l*]pyrene. *Chem. pharm. Bull.*, *20*, 2736-2737

National Library of Medicine (1982) *Toxicology Data Bank,* Bethesda, MD, National Library of Medicine Specialized Information Services, Toxicology Information Program

NIH/EPA Chemical Information System (1982) *Mass Spectral Search System*, Washington DC, CIS Project, Information Services Corporation

Snook, M.E., Severson, R.F., Arrendale, R.F., Higman, H.C. & Chortyk, O.T. (1977) The identification of high molecular weight polynuclear aromatic hydrocarbons in a biologically active fraction of cigarette smoke condensate. *Beitr. Tabakforsch.*, *9*, 79-101

Young, R.J., McKay, W.J. & Evans, J.M. (1978) Coal gasification and occupational health. *Am. ind. Hyg. Assoc. J.*, *39*, 985-997

1,4-DIMETHYLPHENANTHRENE

1. Chemical and Physical Data

1.1 Synonyms and trade names

Chem. Abstr. Services Reg. No.: 22349-59-3

Chem. Abstr. Name: Phenanthrene, 1,4-dimethyl-

IUPAC Systematic Name: 1,4-Dimethylphenanthrene

1.2 Structural and molecular formulae and molecular weight

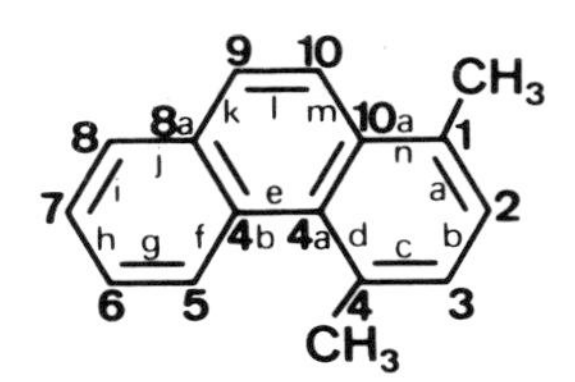

$C_{16}H_{14}$ Mol. wt: 206.3

1.3 Chemical and physical properties of the pure substance

(*a*) *Description*: Needles (recrystallized from methanol) (Buckingham, 1982)

(*b*) *Melting-point*: 50-51°C (Rappoport, 1967)

(*c*) *Solubility*: Soluble in acetone (La Voie *et al.*, 1981)

(*d*) *Stability*: No data were available.

(*e*) *Reactivity*: No data were available.

2. Production, Use, Occurrence and Analysis

2.1 Production and use

There is no commercial production or known use of this compound.

2.2 Occurrence and analysis

Data on occurrence and methods of analysis are summarized in the 'General Remarks on the Substances Considered', p. 35.

Several dimethylphenanthrenes have been found in tobacco smoke (Snook *et al.*, 1977), gasoline engine exhaust gas (Grimmer *et al.*, 1977) and used engine oil (Grimmer *et al.*, 1981); however, individual dimethylphenanthrenes have not been identified.

3. Biological Data Relevant to the Evaluation of Carcinogenic Risk to Humans

3.1 Carcinogenicity studies in animals

Skin application

Mouse: A group of 20 female Swiss albino Ha/ICR mice, seven to eight weeks of age, received applications of a 0.1% solution of 1,4-dimethylphenanthrene (purity, >99.5%) in 100 μl acetone ten times (total dose, 1.0 mg) on alternate days to the shaven back. Starting ten days after the final dose, applications of 2.5 μg 12-*O*-tetradecanoylphorbol-13-acetate (TPA) in 100 μl acetone were given thrice weekly for 20 weeks. A positive control group of 20 mice received a total of 0.3 mg benzo[*a*]pyrene thrice weekly, and a vehicle control group of 20 mice received acetone alone. Of the 1,4-dimethylphenanthrene-treated mice, 95% developed skin tumours, with an average of 8.2 tumours/animal. No tumour was seen in the acetone-control group. Of the benzo[*a*]pyrene-treated animals, 93% developed tumours, with an average of 3.7 tumours/animal. In a footnote to this report, the authors recorded that 1,4-dimethylphenanthrene was active as a tumour initiator when the concentration of the promoter was reduced to 0.0025% and the initiating dose was kept the same, or when the initiating dose was 300 μg (LaVoie *et al.*, 1981).

Groups of 20 outbred albino female Crl: CD-1(ICR)Br mice, 50-55 days old, received skin applications of either 30 or 100 μg 1,4-dimethylphenanthrene (purity, >99.5%) in 100 μl acetone ten times on alternate days (total dose, 0.3 mg or 1.0 mg). Control mice received acetone alone. Starting ten days after the final dose, applications of 2.5 μg TPA in 100 μg acetone were given thrice weekly for 20 weeks. Of the low-dose 1,4-dimethylphenanthrene-treated mice, 80% developed skin tumours, with an average of 3.25 tumours/mouse; in the high-dose 1,4-dimethylphenanthrene-treated group, 100% developed skin tumours, with an average of 5.3 tumours/mouse. No tumour was seen in the acetone-control group (LaVoie *et al.*, 1982).

3.2 Other relevant biological data[1]

(*a*) *Experimental systems*

No data were available to the Working Group on toxic effects, on effects on reproduction and prenatal toxicity or on metabolism and activation.

Mutagenicity and other short-term tests

Results from short-term tests are given in Table 1.

Table 1. Results from short-term tests: 1,4-Dimethylphenanthrene

Test	Organism/assay[a]	Exogenous metabolic system[a]	Reported result	Comments	References
PROKARYOTES					
Mutation	*Salmonella typhimurium* (his^-/his^+)	Aro-R-PMS	Positive	At 50 μg/plate in strain TA100	LaVoie *et al.* (1981)
MAMMALIAN CELLS *IN VITRO*					
DNA damage	Primary rat hepatocytes (unscheduled DNA synthesis)	--	Positive	At 10 nmol/ml	Tong *et al.* (1981)

[a] For an explanation of the abbreviations, see the Appendix, p. 452.

(*b*) *Humans*

No data were available to the Working Group.

3.3 Case reports and epidemiological studies of carcinogenicity to humans

No data were available to the Working Group.

[1] See also 'General Remarks on the Substances Considered', p. 53.

4. Summary of Data Reported and Evaluation[1]

4.1 Experimental data

1,4-Dimethylphenanthrene was tested only for tumour-initiating activity in the mouse-skin initiation-promotion assay. It was active as an initiator.

No data on the teratogenicity of this compound were available.

1,4-Dimethylphenanthrene was mutagenic to *Salmonella typhimurium* in the presence of an exogenous metabolic system and induced unscheduled DNA synthesis in cultured primary rat hepatocytes.

There is *limited evidence* that 1,4-dimethylphenanthrene is active in short-term tests.

4.2 Human data[2]

Humans are exposed to dimethylphenanthrenes in tobacco smoke and urban air; however, 1,4-dimethylphenanthrene has not been specifically identified.

4.3 Evaluation

The available data were inadequate to permit an evaluation of the carcinogenicity of 1,4-dimethylphenanthrene *per se* to experimental animals.

5. References

Buckingham, J., ed. (1982) *Dictionary of Organic Compounds*, 5th ed., Vol. 2, New York, Chapman & Hall, p. 2I98

Grimmer, G., Böhnke, H. & Glaser, A. (1977) Investigation on the carcinogenic burden by air pollution in man. XV. Polycyclic aromatic hydrocarbons in automobile exhaust gas - An inventory. *Zbl. Bakt. Hyg., 1 Abt., Orig. B164*, 218-234

---

[1] For definitions of the italicized terms, see Preamble, p. 19.

[2] Studies on occupational exposure to polynuclear aromatic compounds will be considered in future *IARC Monographs*.

Grimmer, G., Jacob, J., Naujack, K.-W. & Dettbarn, G. (1981) Profile of the polycyclic aromatic compounds from used engine oil - Inventory by GCGC/MS - PAH in environmental materials, Part 2. *Fresenius Z. anal. Chem.*, *309*, 13-19

LaVoie, E.J., Tulley-Freiler, L., Bedenko, V. & Hoffmann, D. (1981) Mutagenicity, tumor-initiating activity, and metabolism of methylphenanthrenes. *Cancer Res.*, *41*, 3441-3447

LaVoie, E.J., Bedenko, V., Tulley-Freiler, L. & Hoffmann, D. (1982) Tumor-initiating activity and metabolism of polymethylated phenanthrenes. *Cancer Res.*, *42*, 4045-4048

Rappoport, Z., ed. (1967) *CRC Handbook of Tables for Organic Compound Identification*, 3rd ed., Boca Raton, FL, Chemical Rubber Co., p. 44

Snook, M.E., Severson, R.F., Arrendale, R.F., Higman, H.C. & Chortyk, O.T. (1977) The identification of high molecular weight polynuclear aromatic hydrocarbons in a biologically active fraction of cigarette smoke condensate. *Beitr. Tabakforsch.*, *9*, 79-101

Tong C., Laspia, M.F., Telang, S. & Williams, G.M. (1981) The use of adult rat liver cultures in the detection of the genotoxicity of various polycyclic aromatic hydrocarbons. *Environ. Mutagenesis*, *3*, 477-487

FLUORANTHENE

1. Chemical and Physical Data

1.1 Synonyms and trade names

Chem. Abstr. Services Reg. No.: 206-44-0

Chem. Abstr. Name: Fluoranthene

IUPAC Systematic Name: Fluoranthene

Synonyms: 1,2-Benzacenaphthene; benzo(jk)fluorene; idryl; 1,2-(1,8-naphthalenediyl)benzene; 1,2-(1,8-naphthylene)benzene

1.2 Structural and molecular formulae and molecular weight

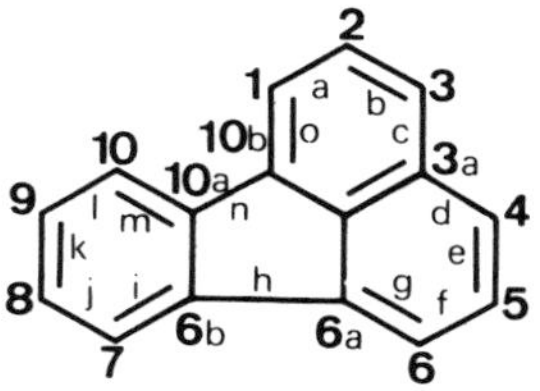

$C_{16}H_{10}$

Mol. wt: 202.3

1.3 Chemical and physical properties of the pure substance

(*a*) *Description:* Pale-yellow needles or plates (recrystallized from ethanol) (Weast, 1975)

(*b*) *Boiling-point:* 250-251°C at 60 mm Hg (Buckingham, 1982); ca. 375°C (Weast, 1975) (*c*) *Melting-point:* 111°C (Karcher *et al.*, 1983)

(*d*) *Spectroscopy data:* λ_{max} 235, 252, 262, 270, 275, 281, 286, 306, 322, 339, 357 nm (in cyclohexane) (Karcher *et al.*, 1983) Mass and nuclear magnetic resonance spectra have been tabulated (NIH/EPA Chemical Information System, 1982; Karcher *et al.*, 1983).

(*e*) *Solubility*: Virtually insoluble (0.20-0.26 mg/l) in water (Davis *et al.*, 1942; May *et al.*, 1978); soluble in acetic acid, benzene, carbon disulphide, chloroform, diethyl ether and ethanol (Weast, 1975)

(*f*) *Stability*: Does not undergo photo-oxidation in organic solvents under fluorescent light or sunlight (Kuratsune & Hirohata, 1962); resistant to photodecomposition (Korfmacher *et al.*, 1980)

(*g*) *Reactivity*: Most reactive at positions 3 and 8; reduction with sodium in ethanol gives tetrahydrofluoranthene; can be chlorinated, brominated and nitrated relatively easily. Concentrated sulphuric acid gives a mono- and a disulphonic acid. Numerous Friedel-Crafts reactions have been carried out with fluoranthene (Clar, 1964); reacts with NO and NO_2 to form nitro derivatives (Butler & Crossley, 1981; Tokiwa *et al.*, 1981)

2. Production, Use, Occurrence and Analysis

2.1 Production and use

There is no commercial production or known use of this compound.

2.2 Occurrence and analysis

Data on occurrence and methods of analysis are summarized in the 'General Remarks on the Substances Considered', p. 35.

Fluoranthene occurs ubiquitously in products of incomplete combustion; it also occurs in fossil fuels. It has been identified in mainstream cigarette smoke (26.3 μg/cigarette) (Hoffmann *et al.*, 1972), (8.3 μg/100 cigarettes) (Lee *et al.*, 1976), (9.8-13.2 μg/100 cigarettes) (Severson *et al.*, 1979), (10.5-11.3 μg/100 cigarettes) (Ellington *et al.*, 1978), (27.2 μg/cigarette) (Grimmer *et al.*, 1977b), (17-53 μg/1000 cigarettes) (Kiryu & Kuratsune, 1966); sidestream cigarette smoke (125.5 μg/cigarette) (Grimmer *et al.*, 1977b); smoke-filled rooms (99 ng/m^3) (Grimmer *et al.*, 1977c); mainstream smoke from marijuana cigarettes (8.9 μg/100 cigarettes) (Lee *et al.*, 1976); urban air (0.9-15.0 ng/m^3) (Hoffmann & Wynder, 1976); gasoline engine exhaust (1060-1662 μg/l fuel) (Grimmer *et al.*, 1977a); coals (0.13-29.4 mg/kg) (Brockhaus & Tomingas, 1976); various lubricating oils (0.008-2.75 mg/kg) (Grimmer *et al.*, 1981a), used gasoline motor oils (3.4-109.0 mg/kg) and used diesel engine oils (0.18-58.9 mg/kg) (Grimmer *et al.*, 1981b); gasolines (0.70-10.10 mg/kg) (Müller & Meyer, 1974); crude oils (1.7 mg/kg) (Grimmer *et al.*, 1983); coal-tar (Lang & Eigen, 1967); charcoal-broiled steaks (20 μg/kg) (Lijinsky & Shubik,

1964); surface water (4.7-6.5 ng/l) (Woidich *et al.*, 1976); tap water (2.6-132.6 ng/l) (Melchiorri *et al.*, 1973; Olufsen, 1980); rain water (5.6-1460 ng/l) (Commission of the European Communities, 1979); subterranean water (9.5-100.0 ng/l) (Borneff & Kunte, 1969); waste water (0.1-45 μg/l) and sludge (580-4090 μg/kg) (Borneff & Kunte, 1967); freeze-dried sewage sludge (610-5160 μg/kg) (Grimmer *et al.*, 1980); and dried sediment of lakes (13-5870 μg/kg) (Grimmer & Böhnke, 1975, 1977).

3. Biological Data Relevant to the Evaluation of Carcinogenic Risk to Humans

3.1 Carcinogenicity studies in animals[1]

(*a*) *Skin application*

Mouse: This compound was studied by skin application in the mouse by Barry *et al.* (1935). [The Working Group noted that no conclusion could be drawn from the data.]

A group of 20 female Swiss (Millerton) mice [age unspecified] were painted thrice weekly for life with a 0.1% solution of fluoranthene [purity unspecified] in acetone. All animals were still alive at six months, 12 at 12 months and none at 17 months. No papilloma or carcinoma was reported (Wynder & Hoffmann, 1959).

A group of 20 female Swiss albino (Ha/ICR/Mil) mice, aged seven to eight weeks, received thrice-weekly applications of a 1% solution of fluoranthene in acetone. The application was continued for 12 months with further observation for three months. At 12 months, no tumour was reported in the 20/20 surviving animals [no vehicle controls was included in this study] (Hoffmann *et al.*, 1972).

A group of 30 female Swiss albino (Ha/ICR/Mil) mice, aged seven to eight weeks, received 10 doses of 0.1 mg fluoranthene (purity, >99.9%) in 50 μl acetone every other day on the shaved back. Similarly treated positive controls received 10 doses of 5 μg benzo[*a*]pyrene in 50 μl acetone, and negative controls received 10 doses of 50 μl acetone. Ten days after the last application, all mice were painted thrice weekly with a 2.5% solution of croton oil in acetone (average dose, 3.8 mg) for 20 weeks. At the end of 20 weeks of promotion, one mouse in the fluoranthene-treated group (29/30 survivors) had developed a single skin tumour; 19 mice in the benzo[*a*]pyrene-treated group (29/30 survivors) developed a total of 67 skin tumours; and one mouse in the acetone-treated control group (30/30 survivors) developed a single skin tumour (Hoffmann *et al.*, 1972).

Groups of 15 male C3H mice, two months old, received twice-weekly applications for 82 weeks of 50 mg (60 μl) fluoranthene (purified by recrystallization) as a 0.5% solution in either decalin or a 1:1 mixture (by volume) of decalin:*n*-dodecane. None of the animals in the group

[1] The Working Group was aware of a study that has recently been completed in hamsters by intrapulmonary administration of pellets, and of a study in progress in mice by i.p. administration (IARC, 1983).

of mice receiving fluoranthene in decalin (13 alive at 52 weeks) nor any animal in the group receiving fluoranthene in the 1:1 decalin:*n*-dodecane solution (12 alive at 52 weeks) developed papillomas or carcinomas. No decalin control group was reported; but a group of 20 mice receiving 1:1 decalin:*n*-dodecane (13 of which survived at least 52 weeks) developed two papillomas by the end of the experiment, the mean latency of which was 75 weeks (Horton & Christian, 1974). [The Working Group noted that *n*-dodecane has been shown to have a cocarcinogenic effect when administered simultaneously with several polynuclear hydrocarbons (Bingham & Falk, 1969.]

Groups of 50 female ICR/Ha Swiss mice, six to eight weeks of age, received thrice-weekly applications by micropipette to the clipped back skin for life of fluoranthene (purified, melting-point 107-109°C), benzo[*a*]pyrene and solvents according to the following schedule: Group 1, 40 μg fluoranthene in 0.1 ml acetone; Group 2, 40 μg fluoranthene + 5 μg benzo[*a*]pyrene in 0.1 ml acetone; Group 3, 5 μg benzo[*a*]pyrene in 0.1 ml acetone; Group 4, 0.1 ml acetone alone. A further group of mice, receiving no treatment, served as untreated controls. No tumour was observed in Group 1 up to the end of the study (440 days). In Group 2, a total of 126 skin tumours developed in 39 mice by day 440, 37 of which were diagnosed as squamous-cell carcinomas, with the first tumour appearing after 99 days. In Group 3, 16 mice had developed 26 tumours, 12 of which were diagnosed as squamous-cell carcinomas, with the first tumour developing after 210 days. No skin tumour was seen in either the vehicle-control or untreated-control groups. Fluoranthene and benzo[*a*]pyrene applied together doubled the number of tumour-bearing mice over that produced by the same dose of benzo[*a*]pyrene alone (Van Duuren & Goldschmidt, 1976).

Groups of 50 female ICR/Ha Swiss mice [age unspecified] received a single application of 150 μg benzo[*a*]pyrene on the dorsal skin by micropipette, followed two weeks later by thrice-weekly applications of 40 μg fluoranthene in 0.1 ml acetone by micropipette to shaven back skin for 448 days. A control group received the initiating dose of benzo[*a*]pyrene, followed by thrice-weekly applications of 0.1 ml acetone for 450 days. No tumour was reported in the vehicle-control group; one squamous-cell carcinoma developed in one treated mouse after 401 days (Van Duuren & Goldschmidt, 1976).

(*b*) *Subcutaneous and/or intramuscular administration*

Mouse: A group of seven male and seven female strain A mice [age unspecified] received a total of five s.c. injections of 10 mg crystalline fluoranthene [purity unspecified] in a few drops of glycerol into the left flank [dose schedule unspecified]. The study was terminated at 19 months, with 14 animals alive at 12 months and six alive at 18 months; no tumour was reported (Shear, 1938).

3.2 Other relevant biological data[1]

(*a*) *Experimental systems*

Toxic effects

When fluoranthene was administered i.p. to mice for seven days in daily doses of 500 mg/kg bw, all of 10 animals survived (Gerarde, 1960).

---

[1] See also 'General Remarks on the Substances Considered', p. 53.

The LD_{50} for the rat (oral) is 2000 mg/kg bw (1270-3130) and for the rabbit (dermal) 3180 mg/kg bw (2350-4290) (Smyth *et al.*, 1962). Fluoranthene inhibited (38%) the growth rate of mouse ascites sarcoma cells in culture when added at a concentration of 1 μmol/ml in dimethyl sulphoxide (Pilotti *et al.*, 1975).

I.p. administration of 30 mg fluoranthene in sesame oil solution did not inhibit body growth of young rats as did administration of 10 mg benzo[*a*]pyrene (Haddow *et al.*, 1937).

Effects on reproduction and prenatal toxicity

Benzo[*a*]pyrene hydroxylase activity of the rat placenta can be induced by fluoranthene (Welch *et al.*, 1969).

Metabolism and activation

The 2,3-dihydrodiol has been detected as a metabolite of fluoranthene following incubation of this compound with a rat-liver preparation. The 2,3-dihydriol was found to be mutagenic to bacteria in the presence of a metabolic activation system (LaVoie *et al.*, 1982).

Mutagenicity and other short-term tests

Results from short-term tests are given in Table 1.

Table 1. Results from short-term tests: Fluoranthene

Test	Organism/assay[a]	Exogenous metabolic system[a]	Reported result	Comments	References
PROKARYOTES					
Mutation	*Salmonella typhimurium* (his⁻/his⁺)	Aro-R-PMS	Positive	At 5 μg/plate in strain TA98	Hermann *et al.* (1980)
		Aro-R-PMS	Positive	At 10 μg/plate in strain TA100 [Inconclusive data were reported in an earlier study (LaVoie *et al.* (1979).]	LaVoie *et al.* (1982)
		Aro-R-PMS	Negative	Tested at up to 1000 μg/plate in strains TA1535, TA1537, TA1538, TA98 and TA100	Salamone *et al.* (1979)
	Salmonella typhimurium ($8AG^S/8AG^R$)	Aro-R-PMS	Positive	At 5 nmol/ml in strain TM677	Kaden *et al.*, 1979
MAMMALIAN CELLS *IN VITRO*					
Mutation	Human lymphoblastoid HH-4 cells ($6TG^S/6TG^R$)	Aro-R-PMS	Positive	At 2.5 μg/ml	Thilly *et al.*, 1980

[a] For an explanation of the abbreviations see the Appendix, p. 452.

(*b*) *Humans*

No data were available to the Working Group.

3.3 Case reports and epidemiological studies of carcinogenicity to humans

No data were available to the Working Group.

4. Summary of Data Reported and Evaluation[1]

4.1 Experimental data

Fluoranthene was tested for carcinogenicity by skin application in mice in two studies, and no tumorigenic effect was observed. It was also tested in the mouse-skin initiation-promotion assay and was inactive as an initiator. A study in mice by subcutaneous administration was considered inadequate for evaluation. When fluoranthene was administered to mice by skin application together with benzo[*a*]pyrene, an excess of skin tumours was produced over that induced by the same dose of benzo[*a*]pyrene alone.

No data were available on the teratogenicity of fluoranthene.

Fluoranthene was mutagenic to *Salmonella typhimurium* and to cultured human lymphoblastoid cells in the presence of an exogenous metabolic system.

There is *limited evidence* that fluoranthene is active in short-term tests.

4.2 Human data[2]

Fluoranthene is present as a major component of the total content of polynuclear aromatic compounds in the environment. Human exposure to fluoranthene occurs primarily through the smoking of tobacco, inhalation of polluted air and by ingestion of food and water contaminated by combustion effluents (for details, see 'General Remarks on the Substances Considered', p. 35).

4.3 Evaluation

The available data provide no evidence that fluoranthene *per se* is carcinogenic to experimental animals.

---

[1] For definitions of the italicized terms, see Preamble, p. 19.

[2] Studies on occupational exposure to polynuclear aromatic compounds will be considered in future *IARC Monographs*.

5. References

Barry, G., Cook, J.W., Haslewood, G.A.D., Hewett, C.L., Hieger, I. & Kennaway, E.L. (1935) The production of cancer by pure hydrocarbons - Part III. *Proc. R. Soc. London Ser. B*, *117*, 318-351

Bingham, E. & Falk, H.L. (1969) Environmental carcinogens. The modifying effect of cocarcinogens on the threshold response. *Arch. environ. Health*, *19*, 779-783

Borneff, J. & Kunte, H. (1967) Carcinogenic substances in water and soil. XIX. Action of sewage purification on polycylic hydrocarbons. *Arch. Hyg. Bakt.*, *151*, 202-210

Borneff, J. & Kunte, H. (1969) Carcinogenic substances in water and soil. 25. A routine method for the determination of polynuclear aromatic hydrocarbons in water (Ger.). *Arch. Hyg. Bakt.*, *153*, 220-229

Brockhaus, A. & Tomingas, R. (1976) Emission of polycyclic hydrocarbons during burning processes in small heating installations and their concentration in the atmosphere (Ger.). *Staub-Reinhalt. Luft*, *36*, 96-101

Buckingham, J., ed. (1982) *Dictionary of Organic compounds*, 5th ed., Vol. 3, New York, Chapman & Hall, p. 2633

Butler, J.D. & Crossley, P. (1981) Reactivity of polycyclic aromatic hydrocarbons adsorbed on soot particles. *Atmos. Environ.*, *15*, 91-94

Clar, E., ed. (1964) *Polycyclic Hydrocarbons*, Vol. 2, New York, Academic Press, pp. 295-303

Commission of the European Communities (1979) *Concerted Action. Analysis of Organic Micropollutants in Water* (*COST 646 bis*), 3rd ed., Vol. II, Luxembourg, p. 17

Davis, W.W., Krahl, M.E. & Clowes, G.H.A. (1942) Solubility of carcinogenic and related hydrocarbons in water. *J. Am. chem. Soc.*, *64*, 108-110

Ellington, J.J., Schlotzhauer, P.F. & Schepartz, A.I. (1978) Quantitation of hexane-extractable lipids in serial samples of flue-cured tobaccos. *J. Food agric. Chem.*, *26*, 270-273

Gerarde, H.W. (1960) *Toxicology and biochemistry of aromatic hydrocarbons*. In: Browning, E., ed., *Elsevier Monographs on Toxic Agents*, Amsterdam, Elsevier Publishing, pp. 249-321

Grimmer, G. & Böhnke, H. (1975) Profile analysis of polycyclic aromatic hydrocarbons and metal content in sediment layers of a lake. *Cancer Lett.*, *1*, 75-84

Grimmer, G. & Böhnke, H. (1977) Investigation on drilling cores of sediments of Lake Constance. I. Profiles of the polycyclic aromatic hydrocarbons. *Z. Naturforsch.*, *32c*, 703-711

Grimmer, G., Böhnke, H. & Glaser, A. (1977a) Investigation on the carcinogenic burden by air pollution in man. XV. Polycyclic aromatic hydrocarbons in automobile exhaust gas - An inventory. *Zbl. Bakt. Hyg.*, *1 Abt.*, *Orig. B164*, 218-234

Grimmer, G., Böhnke, H. & Harke, H.-P. (1977b) Passive smoking: Intake of polycyclic aromatic hydrocarbons by breathing of cigarette smoke containing air (Ger.). *Int. Arch. occup. environ. Health*, *40*, 93-99

Grimmer, G., Böhnke, H. & Harke, H.-P. (1977c) Passive smoking: Measuring of concentrations of polycyclic aromatic hydrocarbons in rooms after machine smoking of cigarettes (Ger.). *Int. Arch. occup. environ. Health*, *40*, 83-92

Grimmer, G., Hilge, G. & Niemitz, W. (1980) Comparison of the profile of polycyclic aromatic hydrocarbons in sewage sludge samples from 25 sewage treatment works (Ger.). *Wasser*, *54*, 255-272

Grimmer, G., Jacob, J. & Naujack, K.-W. (1981a) Profile of the polycyclic aromatic compounds from lubricating oils. Inventory by GCGC/MS - PAH in environmental materials, Part 1. *Fresenius Z. anal. Chem.*, *306*, 347-355

Grimmer, G., Jacob, J., Naujack, K.-W. & Dettbarn, G. (1981b) Profile of the polycyclic aromatic compounds from used engine oil - Inventory by GCGC/MS - PAH in environmental materials, Part 2. *Fresenius Z. anal. Chem.*, *309*, 13-19

Grimmer, G., Jacob, J. & Naujack, K.-W. (1983) Profile of the polycyclic aromatic compounds from crude oils. Inventory by GCGC/MS - PAH in environmental materials. Part 3. *Fresenius Z. anal. Chem.*, *314*, 29-36

Hermann, M., Durand, J.P., Charpentier, J.M., Chaude, O., Hofnung, M., Pétroff, N., Vandercasteele, J.P. & Weill, N. (1980) *Correlations of mutagenic activity with polynuclear aromatic hydrocarbon content of various mineral oils*. In: Bjørseth, A. & Dennis, A.J., eds, *Polynuclear Aromatic Hydrocarbons: Chemistry and Biological Effects. 4th Int. Symposium*, Columbus, OH, Battelle Press, pp. 899-916

Hoffmann, D., Rathkamp, F., Nesnow, S. & Wynder, E.L. (1972) Fluoranthenes: Quantitative determination in cigarette smoke, formation by pyrolysis, and tumor-initiating activity. *J. natl Cancer Inst.*, *49*, 1165-1175

Horton, A.W. & Christian, G.M. (1974) Cocarcinogenic versus incomplete carcinogenic activity among aromatic hydrocarbons: Contrast between chrysene and benzo[*b*]triphenylene. *J. natl Cancer Inst.*, *53*, 1017-1020

IARC (1983) *Information Bulletin on the Survey of Chemicals Being Tested for Carcinogenicity*, No. 10, Lyon, pp. 11, 126

Kaden, D.A., Hites, R.A. & Thilly, W.G. (1979) Mutagenicity of soot and associated polycyclic aromatic hydrocarbons to *Salmonella typhimurium*. *Cancer Res.*, *39*, 4152-4159

Karcher, W., Fordham, R., Dubois, J. & Gloude, P. (1983) *Spectral Atlas of Polycyclic Aromatic Compounds*, Dordrecht, The Netherlands, D. Reidel (in press)

Kiryu, S. & Kuratsune, M. (1966) Polycyclic aromatic hydrocarbons in the cigarette tar produced by human smoking. *Gann*, *57*, 317-322

Korfmacher, W.A., Wehry, E.L., Manantov, G. & Natusch, D.F.S. (1980) Resistance to photochemical decomposition of polycyclic aromatic hydrocarbons vapor-adsorbed on coal fly ash. *Environ. Sci. Technol.*, *14*, 1094-1099

Kuratsune, M. & Hirohata, T. (1962) Decomposition of polycyclic aromatic hydrocarbons under laboratory illuminations. *Natl Cancer Inst. Monogr., 9,* 117-125

Lang, K.F. & Eigen, I. (1967) *Organic compounds in coal tar* (Ger.). In: Heilbronner, E., Hofmann, U., Schäfer, K. & Wittig, G., eds, *Fortschrifte der chemischer Forschung* (Advances in chemical research), Vol. 8, Berlin, Springer, pp. 91-170

LaVoie, E., Bedenko, V., Hirota, N., Hecht, S.S. & Hoffmann, D. (1979) *A comparison of the mutagenicity, tumor-initiating activity and complete carcinogenicity of polynuclear aromatic hydrocarbons*. In: Jones, P.W. & Leber, P., eds, *Polynuclear Aromatic Hydrocarbons*, Ann Arbor, MI, Ann Arbor Science Publishers, pp. 705-721

LaVoie, E., Hecht, S.S., Bedenko, V. & Hoffmann, D. (1982) Identification of the mutagenic metabolites of fluoranthene, 2-methylfluoranthene and 3-methylfluoranthene. *Carcinogenesis*, *3*, 841-846

Lee, M.L., Novotny, M. & Bartle, K.D. (1976) Gas chromatography/mass spectrometry and nuclear magnetic resonance spectrometric studies of carcinogenic polynuclear aromatic hydrocarbons in tobacco and marijuana smoke condensates. *Anal. Chem.*, *48*, 405-416

Lijinsky, W. & Shubik, P. (1964) Benzo(*a*)pyrene and other polynuclear hydrocarbons in charcoal-broiled meat. *Science*, *145*, 53-55

May, W.E., Wasik, S.P. & Freeman, D.H. (1978) Determination of the solubility behaviour of polycyclic aromatic hydrocarbons in water. *Anal. Chem.*, *50*, 997-1000

Melchiorri, C., Chiacchiarini, L., Grella, A. & D'Arca, S.U. (1973) Identification and determination of polycyclic aromatic hydrocarbons in some tap waters of Roma city (Ital.). *Nuovi Ann. Ig. Microbiol.*, *24*, 279-301

Müller, K. & Meyer, J.P. (1974) *Einfluss von Ottokraftstoffen auf die Emission von polynuklearen aromatischen Kohlenwasserstoffen in Automobilabgasen im Europa-Test* (Effect of gasoline components on emission of polynuclear aromatic hydrocarbons in car exhaust in the Europa test) (*Forschungsbericht 4568*), Hamburg, Deutsche Gesellschaft für Mineralölwissenschaft und Kohlechemie e.V.

NIH/EPA Chemical Information System (1982) *Mass Spectral Search System*, Washington DC, CIS Project, Information Services Corporation

Olufsen, B. (1980) *Polynuclear aromatic hydrocarbons in Norwegian drinking water resources*. In: Bjørseth, A. & Dennis, A.J., eds, *Polynuclear Aromatic Hydrocarbons: Chemistry and Biological Effects, 4th Int. Symposium*, Columbus, OH, Battelle Press, pp. 333-343

Pilotti, A., Ancker, K., Arrhenius, E. & Enzell, C. (1975) Effects of tobacco and tobacco smoke constituents on cell multiplication *in vitro*. *Toxicology*, *5*, 49-62

Salamone, M.F., Heddle, J.A. & Katz, M. (1979) The mutagenic activity of thirty polycyclic aromatic hydrocarbons (PAH) and oxides in urban airborne particulates. *Environ. Int.*, *2*, 37-43

Severson, R.F., Arrendale, R.F., Chaplin, J.F. & Williamson, R.E. (1979) Use of pale-yellow tobacco to reduce smoke polynuclear aromatic hydrocarbons. *J. agric. Food Chem.*, *27*, 896-900

Shear, M.J. (1938) Studies in carcinogenesis V. Methyl derivatives of 1:2-benzanthracene. *Am. J. Cancer*, *33*, 499-537

Smyth, H.F., Jr, Carpenter, C.P., Weil, C.S., Pozzani, U.C. & Striegel, J.A. (1962) Range-finding toxicity data: List VI. *Am. ind. Hyg. J.*, *23*, 95-107

Thilly, W.G., DeLuca, J.G., Furth, E.E., Hoppe IV, H., Kaden, D.A., Krolewski, J.J., Liber, H.L., Skopek, T.R., Slapikoff, S.A., Tizard, R.J. & Penman, B.W. (1980) *Gene-locus mutation assays in diploid human lymphoblast lines*. In: de Serres, F.J. & Hollaender, A., eds, *Chemical Mutagens. Principles and Methods for Their Detection*, Vol. 6, New York, Plenum Press, pp. 331-364

Tokiwa, H., Nakagawa, R. & Ohnishi, Y. (1981) Mutagenic assay of aromatic nitro compounds with *Salmonella typhimurium*. *Mutat. Res.*, *91*, 321-325

Van Duuren, B.L. & Goldschmidt, B.M. (1976) Cocarcinogenic and tumor-promoting agents in tobacco carcinogenesis. *J. natl Cancer Inst.*, *56*, 1237-1242

Weast, R.C., ed. (1975) *CRC Handbook of Chemistry and Physics*, 56th ed., Cleveland, OH, Chemical Rubber Co., p. C-300

Welch, R.M., Harrison, Y.E., Gommi, B.W., Poppers, P.J., Finster, M. & Conney, A.H. (1969) Stimulatory effect of cigarette smoking on the hydroxylation of 3,4-benzpyrene and the *N*-demethylation of 3-methyl-4-monomethylaminoazobenzene by enzymes in human placenta. *Clin. Pharmacol. Ther.*, *10*, 100-109

Woidich, H., Pfannhauser, W., Blaicher, G. & Tiefenbacher, K. (1976) Analysis of polycyclic aromatic hydrocarbons in drinking and industrial water (Ger.). *Lebensmittelchem. gerichtl. Chem.*, *30*, 141-146

Wynder, E.L. & Hoffmann, D. (1959) A study of tobacco carcinogenesis VII. The role of higher polycyclic hydrocarbons. *Cancer*, *12*, 1079-1086

FLUORENE

1. Chemical and Physical Data

1.1 Synonyms and trade names

Chem. Abstr. Services Reg. No.: 86-73-7

Chem. Abstr. Name: 9H-Fluorene

IUPAC Systematic Name: Fluorene

Synonyms: *ortho*-Biphenylenemethane; diphenylenemethane; 2,2′-methylenebiphenyl

1.2 Structural and molecular formulae and molecular weight

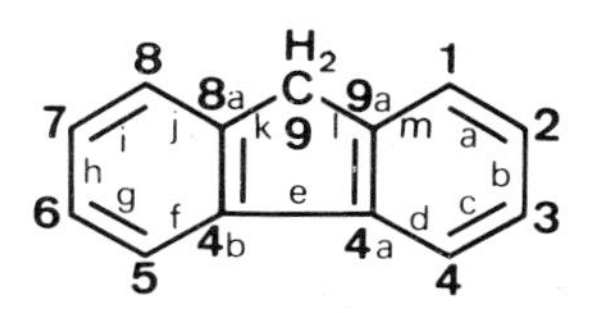

$C_{13}H_{10}$ Mol. wt: 166.2

1.3 Chemical and physical properties of the pure substances

From National Library of Medicine (1982), unless otherwise specified

(*a*) *Description*: White leaflets or flakes (recrystallized from ethanol)

(*b*) *Boiling-point*: 295°C

(*c*) *Melting-point*: 116-117°C

(*d*) *Spectroscopy data*: λ_{max} 208, 261, 289, 301 nm (in ethanol) (Clar, 1964). Mass spectra have been tabulated (NIH/EPA Chemical Information System, 1982).

(*e*) *Solubility*: Almost insoluble in water (1.68-1.98 mg/l) (May *et al*., 1978); soluble in hot acetic acid, acetone, benzene, carbon disulphide, carbon tetrachloride, diethyl ether, ethanol, pyrimidine solution and toluene

(*f*) *Volatility*: Vapour pressure, 10 mm at 146°C

(*g*) *Stability*: Does not undergo photo-oxidation in organic solvents under fluorescent light or indoor sunlight (Kuratsune & Hirohata, 1962)

(*h*) *Reactivity*: Hydrogenated by reaction with hydroiodic acid and red phosphorus. Undergoes Friedel-Crafts reactions and can be easily nitrated or oxidized. Dehydrogenation leads first to a colourless compound and then to the orange-red bisdiphenylene ethylene. Reacts with oxygen to form a hydroperoxide (Clar,1964). Reacts with NO_2 to give nitro derivatives (Tokiwa *et al*., 1981).

2. Production, Use, Occurrence and Analysis

2.1 Production and use

There is no commercial production or known use of this compound.

2.2 Occurrence and analysis

Data on occurrence and methods of analysis are summarized in the 'General Remarks on the Substances Considered', p. 35.

Fluorene occurs ubiquitously in products of incomplete combustion; it also occurs in fossil fuels. Levels of up to 1.6% have been found in coal-tar (Windholz, 1976). It has been detected in mainstream cigarette smoke (Grob & Voellmin, 1970; Snook *et al*., 1978); exhaust from gasoline engines (Grimmer *et al*., 1977); surface water (4.1-102.1 ng/l) (Grimmer *et al*., 1981); tap water (4-16 ng/l) (Thruston, 1978); and sewage sludge (0.61-51.60 mg/kg) (Grimmer *et al*., 1980).

3. Biological Data Relevant to Evaluation of Carcinogenic Risk to Humans

3.1 Carcinogenicity studies in animals[1]

(*a*) *Oral administration*

Rat: A group of 20 female Buffalo-strain rats, three months of age, were fed 0.05% fluorene ('highly purified') in a diet of 'natural' food with 3% corn oil for six months [average total dose,

---

[1] The Working Group was aware of a study in progress in mice by s.c. administration (IARC, 1983).

796 mg/rat]. The study was terminated at 10.7 months, at which time 11/20 animals were examined histologically; one squamous-cell carcinoma of the renal pelvis and one squamous-cell carcinoma of the ureter were reported (Morris *et al.*, 1960). [The Working Group noted the lack of concurrent controls and the short duration of the study.]

A group of 18 female Buffalo-strain rats, 0.9 months of age, were fed 0.05% fluorene in the diet [as described above] for 18 months [total average intake, 2553 mg/rat], and surviving animals were killed at 20.1 months; the average age at autopsy was 19 months. Tumours reported were one uterine carcinosarcoma, one uterine fibrosarcoma, one granulocytic leukaemia and four pituitary adenomas. In a control group of 18 rats, 3.5 months of age, fed basal diet for an average of 15.5 months, one uterine adenocarcinoma, two uterine fibro-epithelial polyps, five adrenal cortical adenomas, six pituitary adenomas and one inguinal region fibroma were reported (Morris *et al.*, 1960). [The Working Group noted the small number of animals used in the study.]

(*b*) *Skin application*

Mouse: Negative results were reported after nine months of repeated skin application of fluorene in 90% benzene in a group of 100 mice (Kennaway, 1924). [Full details were not reported.]

In an investigation to study the effect of fluorene on 3-methylcholanthrene-induced skin carcinogenesis, a group of 10 male and 10 female CF1 mice, 12 weeks of age, received applications of 0.30% fluorene in 0.02 ml acetone to the clipped interscapular region twice weekly for 31 weeks; at that time, 16 animals were alive, and no skin tumour was observed in this group. A similar group of 20 mice received skin applications of 0.15% fluorene + 0.15% 3-methylcholanthrene in 0.02 ml acetone twice weekly for 31 weeks. The first tumour appeared at nine weeks, at which time 17 mice were still alive. By the end of 31 weeks, 16/17 mice had skin tumours, with a mean latent period of 11.9 weeks. Of the 16 skin tumours, 13 were squamous-cell carcinomas and three were squamous-cell carcinomas that metastasized. A further group of 20 mice received 0.15% 3-methylcholanthrene in 0.02 ml acetone for 31 weeks; 19 surviving mice developed skin tumours within a mean latent period of 14.5 weeks (Riegel *et al.*, 1951). [The Group noted the short duration of treatment with fluorene alone in the assay for complete carcinogenicity.]

(*c*) *Subcutaneous and/or intramuscular administration*

Mouse: A group of 10 male strain A mice [age unspecified] received seven s.c. injections each of 10 mg fluorene in a few drops of glycerol into the left flank over a period of 16 months. No tumour was observed in seven mice still alive at termination of the study at 18 months (Shear, 1938). [The Working Group noted the small number of animals used and the low frequency of injections.]

3.2 Other relevant biological data[1]

(*a*) *Experimental systems*

No data were available to the Working Group on toxic effects, on effects on reproduction and prenatal toxicity.

Metabolism and activation

1-Hydroxy, 9-hydroxy and 9-ketofluorene have been detected as metabolites of fluorene following incubation of this compound with rat-liver preparations (LaVoie *et al.*, 1981a).

Mutagenicity and other short-term tests

Results from short-term tests are given in Table 1.

Table 1. Results from short-term tests: Fluorene

Test	Organism/assay[a]	Exogenous metabolic system[a]	Reported result	Comments	References
PROKARYOTES					
Mutation	*Salmonella typhimurium* (his-/his+)	Aro-R-PMS	Negative	Tested at up to 1000 μg/plate in strains TA1535, TA1537, TA98 and TA100	McCann *et al.* (1975); LaVoie *et al.* (1979, 1981b)
	Salmonella typhimurium (8AG[S]/8AG[R])	Aro-R-PMS, PB-R-PMS	Negative	Tested at up to 300 nmol/ml in strain TM677	Kaden *et al.* (1979)
MAMMALIAN CELLS *IN VITRO*					
DNA damage	Primary rat hepatocytes (unscheduled DNA synthesis)	--	Negative	Tested at up to 10 nmol/ml	Probst *et al.* (1981)

[a] For an explanation of the abbreviations see the Appendix, p. 452.

(*b*) *Humans*

No data were available to the Working Group.

3.3 Case reports and epidemiological studies of carcinogenicity to humans

No data were available to the Working Group.

[1] See also 'General Remarks on the Substances Considered', p. 53.

4. Summary of Data Reported and Evaluation[1]

4.1 Experimental data

Fluorene was tested for carcinogenicity in mice by skin application and by subcutaneous administration and in female rats by oral administration in the diet. The studies were considered inadequate for evaluation.

No data were available on the teratogenicity of fluorene.

Fluorene was not mutagenic to *Salmonella typhimurium*. In the one available study, it did not induce unscheduled DNA synthesis in primary rat hepatocyte cultures.

There is *inadequate evidence* that fluorene is active in short-term tests.

4.2 Human data[2]

Fluorene is present as a major component of the total content of polynuclear aromatic compounds in the environment. Human exposure to fluorene occurs primarily through the smoking of tobacco. inhalation of polluted air and by ingestion of food and water contaminated by combustion effluents (for details, see 'General Remarks on the Substances Considered', p. 35).

4.3 Evaluation

The available data are inadequate to permit an evaluation of the carcinogenicity of fluorene to experimental animals.

---

[1] For definitions of italicized terms, see Preamble, p. 19.

[2] Studies on occupational exposure to polynuclear aromatic compounds will be considered in future *IARC Monographs*.

5. References

Clar, E., ed. (1964) *Polycyclic Hydrocarbons*, Vol. 2, New York, Academic Press, pp. 6-11

Grimmer, G., Böhnke, H. & Glaser, A. (1977) Investigation on the carcinogenic burden by air pollution in man. XV. Polycyclic aromatic hydrocarbons in automobile exhaust gas - An inventory. *Zbl. Bakt. Hyg., 1 Abt., Orig. B164*, 218-234

Grimmer, G., Hilge, G. & Niemitz, W. (1980) Comparison of the profile of polycyclic aromatic hydrocarbons in sewage sludge samples from 25 sewage treatment works (Ger.). *Wasser*, *54*, 255-272

Grimmer, G., Schneider, D. & Dettbarn, G. (1981) The load of different rivers in the Federal Republic of Germany by PAH (PAH - profiles of surface water) (Ger.). *Wasser*, *56*, 131-144

Grob, K. & Voellmin, J.A. (1970) GC-MS analysis of 'semi-volatiles' of cigarette smoke. *J. chromatogr. Sci.*, *8*, 218-220

IARC (1983) *Information Bulletin on the Survey of Chemicals Being Tested for Carcinogenicity*, No. 10, Lyon, p. 85

Kaden, D.A., Hites, R.A. & Thilly, W.G. (1979) Mutagenicity of soot and associated polycyclic aromatic hydrocarbons to *Salmonella typhimurium*. *Cancer Res.*, *39*, 4152-4159

Kennaway, E.L. (1924) On cancer-producing tars and tar-fractions. *J. ind. Hyg.*, *5*, 462-490

Kuratsune, M. & Hirohata, T. (1962) Decomposition of polycyclic aromatic hydrocarbons under laboratory illuminations. *Natl Cancer Inst. Monogr.*, *9*, 117-125

LaVoie, E., Bedenko, V., Hirota, N., Hecht, S.S. & Hoffmann, D. (1979) *A comparison of the mutagenicity, tumor-initiating activity and complete carcinogenicity of polynuclear aromatic hydrocarbons*. In: Jones, P.W. & Leber, P., eds, *Polynuclear Aromatic Hydrocarbons*, Ann Arbor, MI, Ann Arbor Science Publishers Inc., pp. 705-721

LaVoie, E.J., Thulley-Freiler, L., Bedenko, V., Girach, Z. & Hoffmann, D. (1981a) *Comparative studies on the tumor initiating activity and metabolism of methylfluorenes and methylbenzofluorenes*. In: Cooke, M. & Dennis, A.J., eds, *Polynuclear Aromatic Hydrocarbons: Chemical Analysis and Biological Fate*, *5th Int. Symposium*, Columbus, OH, Battelle Press, pp. 417-427

LaVoie, E.J., Tulley, L., Bedenko, V. & Hoffmann, D. (1981b) Mutagenicity of methylated fluorenes and benzofluorenes. *Mutat. Res.*, *91*, 167-176

May, W.E., Wasik, S.P. & Freeman, D.H. (1978) Determination of the solubility behavior of some polycyclic aromatic hydrocarbons in water. *Anal. Chem.*, *50*, 997-1000

McCann, J., Choi, E., Yamasaki, E. & Ames, B.N. (1975) Detection of carcinogens as mutagens in the *Salmonella*/microsome test: Assay of 300 chemicals. *Proc. natl Acad. Sci. USA*, *72*, 5135-5139

Morris, H.P., Velat, C.A., Wagner, B.P., Dahlgard, M. & Ray, F.E. (1960) Studies of carcinogenicity in the rat of derivatives of aromatic amines related to *N*-2-fluorenylacetamide. *J. natl Cancer Inst.*, *24*, 149-180

National Library of Medicine (1982) *Toxicology Data Bank*, Bethesda, MD, National Library of Medicine Specialized Information Services, Toxicology Information Program

NIH/EPA Chemical Information System (1982) *Mass Spectral Search System*, Washington DC, CIS Project, Information Services Corporation

Probst, G.S., McMahon, R.E., Hill, L.E., Thompson, C.Z., Epp, J.K. & Neal, S.B. (1981) Chemically-induced unscheduled DNA synthesis in primary rat hepatocyte cultures: A comparison with bacterial mutagenicity using 218 compounds. *Environ. Mutagenesis*, *3*, 11-32

Riegel. B., Wartman, W.B., Hill, W.T., Reeb, B.B., Shubik, P. & Stanger, D.W. (1951) Delay of methylcholanthrene skin carcinogenesis in mice by 1,2,5,6-dibenzofluorene. *Cancer Res.*, *11*, 301-306

Shear, M.J. (1938) Studies in carcinogenesis V. Methyl derivatives of l:2-benzanthracene. *Am. J. Cancer*, *33*, 499-537

Snook, M.E., Severson, R.F., Arrendale, R.F., Higman, C.H. & Chortyk, O.T. (1978) Multi-alkylated polynuclear aromatic hydrocarbons of tobacco smoke: Separation and identification. *Beitr. Tabakforsch.*, *9*, 222-247

Thruston, A.D., Jr (1978) High pressure liquid chromatography techniques for the isolation and identification of organics in drinking water extracts. *J. chromatogr. Sci.*, *16*, 254-259

Tokiwa, H., Nakagawa, R., Morita, K. & Ohnishi, Y. (1981) Mutagenicity of nitro derivatives induced by exposure of aromatic compounds to nitrogen dioxide. *Mutat. Res.*, *85*, 195-205

Windholz, M., ed. (1976) *The Merck Index*, 9th ed., Rahway, NJ, Merck & Co., p. 537

INDENO[1,2,3-*cd*]PYRENE

This compound was considered by a previous working group, in December 1972 (IARC, 1973).

1. Chemical and Physical Data

1.1 Synonyms and trade names

Chem. Abstr. Services Reg. No.: 193-39-5

Chem. Abstr. Name: Indeno(1,2,3-cd)pyrene

IUPAC Systematic Name: Indeno[1,2,3-*cd*]pyrene

Synonyms: IP; *ortho*-phenylenepyrene; 1,10-(*ortho*-phenylene)pyrene; 1,10-(1,2-phenylene)pyrene; 2,3-*ortho*-phenylenepyrene

1.2 Structural and molecular formulae and molecular weight

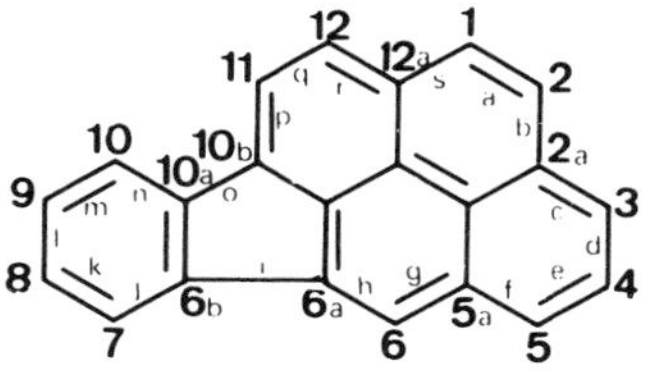

$C_{22}H_{12}$

Mol. wt: 276.3

1.3 Chemical and physical properties of the pure substance

From National Library of Medicine (1982), unless otherwise specified

(*a*) *Description*: Yellow plates or needles (recrystallized from light petroleum solution) showing a greenish-yellow fluorescence

(*b*) *Melting-point*: 163.6°C (Karcher *et al.*, 1983)

(*c*) *Spectroscopy data*: λ_{max} 242, 249, 274, 290, 301, 314, 340, 358, 376, 382, 407, 426, 446, 453, 460 nm (in cyclohexane) (Karcher *et al.*, 1983). Mass and nuclear magnetic resonance spectra have been tabulated (NIH/EPA Chemical Information System, 1982; Karcher *et al.*, 1983).

(*d*) *Solubility*: Soluble in organic solvents

(*e*) *Stability*: No data were available.

(*f*) *Reactivity*: No data were available.

2. Production, Use, Occurrence and Analysis

2.1 Production and use

There is no commercial production or known use of this compound. A reference material of certified high purity is available (Karcher *et al.*, 1980; Community Bureau of Reference, 1982).

2.2 Occurrence and analysis

Data on occurrence and methods of analysis are summarized in the 'General Remarks on the Substances Considered', p. 35.

Indeno[1,2,3-*cd*]pyrene occurs ubiquitously in products of incomplete combustion; it also occurs in fossil fuels. It has been identified in mainstream cigarette smoke (0.4-2.0 μg/100 cigarettes) (Wynder & Hoffmann, 1963; Ayres & Thornton, 1965); urban atmosphere (1.5-8.2 ng/m^3) (Hoffmann & Wynder, 1976); exhaust emissions from gasoline engines (32-86 μg/l fuel) (Grimmer *et al.*, 1977); exhaust from burnt coal (0.02-2.3 mg/kg) (Brockhaus & Tomingas, 1976); lubricating oils (0.00-0.02 mg/kg) (Grimmer *et al.*, 1981a) and used motor oils (0.06-12.5 mg/kg) (Grimmer *et al.*, 1981b); gasolines (0.04-0.38 mg/kg) (Müller & Meyer, 1974); surface water (0-2-0.5 ng/l) (Woidich *et al.*, 1976); tap water (0.3-4.8 ng/l) (Melchiorri *et al.*, 1973); rainfall (0.2-8.7 ng/l) (Woidich *et al.*, 1976); subterranean water (0.2-5.0 ng/l) (Borneff & Kunte, 1969); waste water (0.01-15 μg/l) (Borneff & Kunte, 1967); sludge (470-1200 μg/kg) (Borneff & Kunte, 1967); freeze-dried sewage sludge samples (300-7400 μg/kg) (Grimmer *et al.*, 1980); and dried sediment from lakes (1-2070 μg/kg) (Grimmer & Böhnke, 1975, 1977).

3. Biological Data Relevant to the Evaluation of Carcinogenic Risk to Humans

3.1 Carcinogenicity studies in animals

There is *sufficient evidence* that indeno[1,2,3-*cd*]pyrene is carcinogenic to experimental animals (see General Remarks on the Substances Considered', p. 33).

3.2 Other relevant biological data[1]

(*a*) *Experimental systems*

No data were available to the Working Group on toxic effects, on effects on reproduction and prenatal toxicity or on metabolism and activation.

Mutagenicity and other short-term tests

Indeno[1,2,3-*cd*]pyrene was reported to induce mutations in *Salmonella typhimurium* in strain TA100 (*his*$^-$/*his*$^+$) at a concentration of 20 μg/plate (LaVoie *et al.*, 1979) and in strain TA98 at a concentration of 2 μg/plate (Hermann *et al.*, 1980) in the presence of an exogenous metabolic system (postmitochondrial supernatant from Aroclor-induced rat liver).

(*b*) *Humans*

No data were available to the Working Group.

3.3 Case reports and epidemiological studies of carcinogenicity to humans

No data were available to the Working Group.

4. Summary of Data Reported and Evaluation[2]

4.1 Experimental data

Indeno[1,2,3-*cd*]pyrene has been shown to be carcinogenic to experimental animals (see 'General Remarks on the Substances Considered' p. 33 and IARC, 1973).

No data on the teratogenicity of this compound were available.

---

[1] See also 'General Remarks on the Substances Considered', p. 53.

[2] For definitions of the italicized terms, see Preamble, p. 19.

Indeno[1,2,3-*cd*]pyrene was mutagenic to *Salmonella typhimurium* in the presence of an exogenous metabolic system.

There is *inadequate evidence* that indeno[1,2,3-*cd*]pyrene is active in short-term tests.

4.2 Human data[1]

Indeno[1,2,3-*cd*]pyrene is present as a component of the polynuclear aromatic compound content in the environment. Human exposure to indeno[1,2,3-*cd*]pyrene occurs primarily through the smoking of tobacco, inhalation of polluted air and by ingestion of food and water contaminated by combustion effluents (for details, see 'General Remarks on the Substances Considered', p. 35).

4.3 Evaluation[2]

There is *sufficient evidence* that indeno[1,2,3-*cd*]pyrene is carcinogenic to experimental animals.

5. References

Ayres, C.I. & Thornton, R.E. (1965) Determination of benzo(*a*)pyrene and related compounds in cigarette smoke. *Beitr. Tabakforsch.*, *3*, 285-290

Borneff, J. & Kunte, H. (1967) Carcinogenic substances in water and soils. XIX. Action of sewage purification on polycyclic aromatics (Ger.). *Arch. Hyg. Bakt.*, *151*, 202-210

Borneff, J. & Kunte, H. (1969) Carcinogenic substances in water and soils. 25. A routine method for the determination of polynuclear aromatic hydrocarbons (Ger.). *Arch. Hyg. Bakt.*, *153*, 220-229

Brockhaus, A. & Tomingas, R. (1976) Emission of polycyclic hydrocarbons during burning processes in small heating installations and their concentration in the atmosphere (Ger.). *Staub-Reinhalt. Luft*, *36*, 96-101

Community Bureau of Reference (1982) *Polycyclic Aromatic Hydrocarbon Reference Materials of Certified Purity* (*Information Handout, No. 22*), Brussels, Commission of the European Communities

[1] Studies on occupational exposure to polynuclear aromatic compounds will be considered in future *IARC Monographs*.

[2] In the absence of adequate data on humans it is reasonable for practical purposes to regard chemicals for which there is *sufficient evidence* of carcinogenicity in animals as if they presented a carcinogenic risk to humans (see also Preamble, p. 20).

Grimmer, G. & Böhnke, H. (1975) Profile analysis of polycyclic aromatic hydrocarbons and metal content in sediment layers of a lake. *Cancer Lett.*, *1*, 75-84

Grimmer, G. & Böhnke, H. (1977) Investigation on drilling cores of sediments of Lake Constance. I. Profiles of the polycyclic aromatic hydrocarbons (Ger.). *Z. Naturforsch.*, *32c*, 703-711

Grimmer, G., Böhnke, H. & Glaser, A. (1977) Investigation on the carcinogenic burden by air pollution in man. XV. Polycyclic aromatic hydrocarbons in automobile exhaust gas - An inventory. *Zbl. Bakt. Hyg., 1 Abt., Orig. B164*, 218-234

Grimmer, G., Hilge, G. & Niemitz, W. (1980) Comparison of the profile of polycyclic aromatic hydrocarbons sewage sludge samples from 25 sewage treatment works (Ger.). *Wasser*, *54*, 255-272

Grimmer, G., Jacob, J. & Naujack, K.-W. (1981a) Profile of the polycyclic aromatic hydrocarbons from lubricating oils. Inventory by GCGC/MS-PAH in environmental materials, part 1. *Fresenius Z. anal. Chem.*, *306*, 347-355

Grimmer, G., Jacob, J., Naujack, K.-W. & Dettbarn, G. (1981b) Profile of the polycyclic aromatic hydrocarbons from used engine oil - Inventory by GCGC/MS-PAH in environmental materials, part 2. *Fresenius Z. anal. Chem.*, *309*, 13-19

Hermann, M., Durand, J.P., Charpentier, J.M., Chaudé, O., Hofnung, M., Pétroff, N., Vandercasteele, J.-P. & Weill, N. (1980) *Correlations of mutagenic activity with polynuclear aromatic hydrocarbon content of various mineral oils.* In: Bjørseth, A. & Dennis, A.J., eds, *Polynuclear Aromatic Hydrocarbons: Chemistry and Biological Effects*, *4th Int. Symposium*, Columbus, OH, Battelle Press, pp. 899-916

Hoffmann, D. & Wynder, E.L. (1976) *Environmental respiratory carcinogenesis.* In: Searle, C.E., ed., *Chemical Carcinogens* (*ACS Monograph 173*), Washington DC, American Chemical Society, p. 341

IARC (1973) *IARC Monographs on the Evaluation of Carcinogenic Risk of Chemicals to Man*, Vol. 3, *Certain polycyclic aromatic hydrocarbons and heterocyclic compounds*, Lyon, pp. 229-237

Karcher, W., Jacob, J. & Haemers, L. (1980) *The Certification of Eight Polycyclic Aromatic Hydrocarbon Materials (PAH) (BCR Reference Materials Nos. 46, 47, 48, 49, 50, 51, 52 and 53)* (*EUR 6967 EN*), Luxembourg, Commission of the European Communities

Karcher, W., Forham, R., Dubois, J. & Gloude, P. (1983) *Spectral Atlas of Polycyclic Aromatic Compounds*, Dordrecht, The Netherlands, D. Reidel (in press)

LaVoie, E., Bedenko, V., Hirota, N., Hecht, S.S. & Hoffmann, D. (1979) *A comparison of the mutagenicity, tumor-initiating activity and complete carcinogenicity of polynuclear aromatic hydrocarbons.* In: Jones, P.W. & Leber, P., eds, *Polynuclear Aromatic Hydrocarbons*, Ann Arbor, MI, Ann Arbor Science Publishers, pp. 705-721

Melchiorri, C., Chiacchiarini, L., Grella, A. & D'Arca, S.U. (1973) Research and determination of polycyclic aromatic hydrocarbons in some tap waters of the city of Roma (Ital.). *Nuov. Ann. Ig. Microbiol.*, *24*, 279-301

Müller, K. & Meyer, J.P. (1974) *Einfluss von Ottokraftstoffen auf die Emission von polynuklearen aromatischen Kohlenwasserstoffen in Automobilabgasen im Europa-Test* (Effect of gasoline components on emission of polynuclear aromatic hydrocarbons in car exhaust in the europa test) (*Forschungsbericht 4568*), Hamburg, Deutsche Gesellschaft für Mineralölwissenschaft und Kohlechemie e.V.

National Library of Medicine (1982) *Toxicology Data Bank*, Bethesda, MD, National Library of Medicine Specialized Information Services, Toxicology Information Program

NIH/EPA Chemical Information System (1982) *Mass Spectral Search System*, Washington DC, CIS Project, Information Services Corporation

Woidich, H., Pfannhauser, W., Blaicher, G. & Tiefenbacher, K. (1976) Analysis of polycyclic aromatic hydrocarbons in drinking and industrial water (Ger.). *Lebensmittelchem. gerichtl. Chem.*, *30*, 141-146

Wynder, E.L. & Hoffmann, D. (1963) Experimental contribution to tobacco smoke carcinogenesis (Ger.). *Dtsch. med. Wochenschr.*, *88*, 623-628

1-, 2-, 3-, 4-, 5- and 6-METHYLCHRYSENES

1. Chemical and Physical Data

1-Methylchrysene

1.1 Synonyms and trade names

Chem. Abstr. Services Reg. No.: 3351-28-8

Chem. Abstr. Name: Chrysene, 1-methyl-

IUPAC Systematic Name: 1-Methylchrysene

Synonym: 3-Methylchrysene (Clar, 1964)

1.2 Structural and molecular formulae and molecular weight

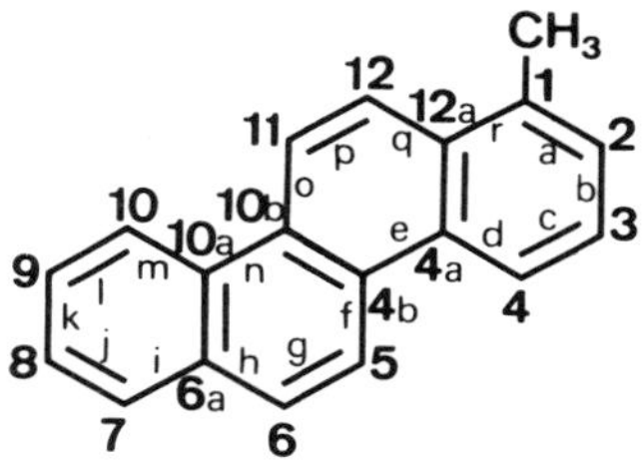

$C_{19}H_{14}$ Mol. wt: 242.3

1.3 Chemical and physical properties of the pure substance

(*a*) *Description*: Leaflets (recrystallized from benzene, hexane or toluene) (Buckingham, 1982)

(*b*) *Boiling-point*: Sublimes at 130-140°C (in vacuum) (Weast, 1975) (*c*) *Melting-point*: 253°C (Clar, 1964); 256.5-257°C (Weast, 1975); 247-249°C, 254-255°C (in vacuum) (Buckingham, 1982); 254.4°C (Karcher *et al.*, 1983)

(*d*) *Spectroscopy data*: λ_{max} 259, 281, 292, 302, 337, 352, 390 nm (in ethanol) (Weast, 1975). Mass spectra have been tabulated (NIH/EPA Chemical Information System, 1982).

(*e*) *Solubility*: Insoluble in water; soluble in ethanol (Weast, 1975); soluble in acetone (Hecht *et al.*, 1974)

(*f*) *Stability*: No data were available.

(*g*) *Reactivity*: No data were available.

2-Methylchrysene

1.1 Synonyms and trade names

Chem. Abstr. Services Reg. No.: 3351-32-4

Chem. Abstr. Name: Chrysene, 2-methyl-

IUPAC Systematic Name: 2-Methylchrysene

Synonym: 4-Methylchrysene (Clar, 1964)

1.2 Structural and molecular formulae and molecular weight

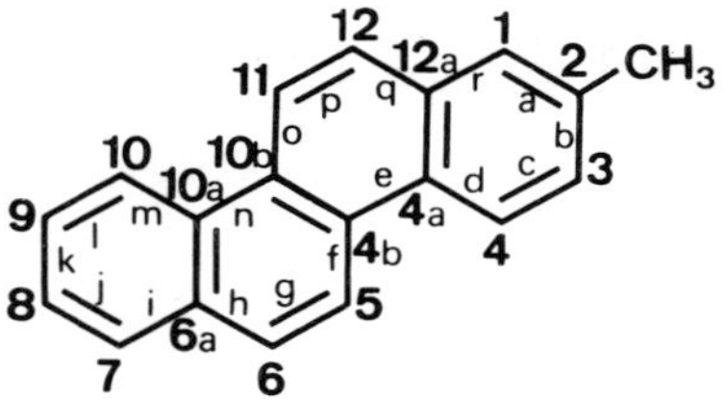

$C_{19}H_{14}$ Mol. wt: 242.3

1.3 Chemical and physical properties of the pure substance

(*a*) *Description*: Leaflets (recrystallized from benzene/ethanol) (Weast, 1975) (*b*) *Melting-point*: 229-230°C (Buckingham, 1982); 224.5-225.5°C (Rappoport, 1967); 224-225°C (Clar, 1964); 229.5-230°C (Weast, 1975); 230.2°C (Karcher *et al.*, 1983)

(*c*) *Spectroscopy data*: λ_{max}: 242, 260, 270, 295, 307, 320, 344, 361 nm (in cyclohexane); nuclear magnetic resonance spectra have been tabulated (Karcher *et al.*, 1983).

(*d*) *Solubility*: Insoluble in water; soluble in acetic acid and ethanol (Weast, 1975); soluble in acetone (Karcher *et al.*, 1983)

(*e*) *Stability*: No data were available.

(*f*) *Reactivity*: No data were available.

3-Methylchrysene

1.1 Synonyms and trade names

Chem. Abstr. Services Reg. No.: 3351-31-3

Chem. Abstr. Name: Chrysene, 3-methyl-

IUPAC Systematic Name: 3-Methylchrysene

Synonym: 5-Methylchrysene (Clar, 1964)

1.2 Structural and molecular formulae and molecular weight

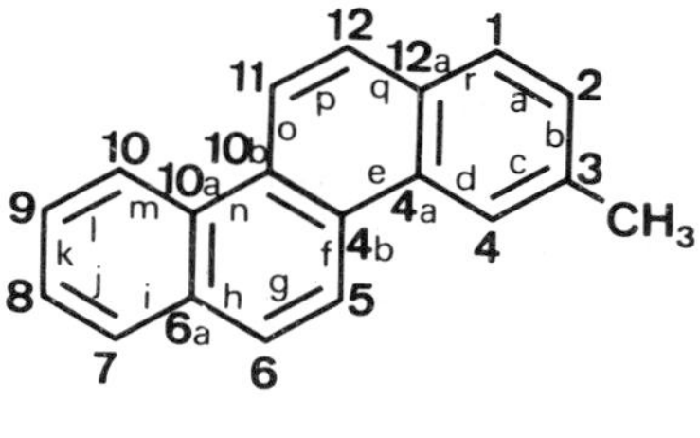

$C_{19}H_{14}$ Mol. wt: 242.3

1.3 Chemical and physical properties of the pure substance

(*a*) *Description*: Leaves (recrystallized from benzene/petroleum ether) (Weast, 1975) (*b*) *Melting-point*: 170°C (Clar, 1964); 171.8°C (Karcher *et al.*, 1983); 173-174°C (Buckingham, 1982); 172.5-173.5°C (Weast, 1975)

(*c*) *Spectroscopy data*: λ_{max} 268, 275, 297, 324, 355 nm (in cyclohexane) (Weast, 1975). Mass spectra have been tabulated (NIH/EPA Chemical Information System, 1982).

(*d*) *Solubility*: Soluble in ethanol (Weast, 1975); soluble in acetone (Hecht *et al.*, 1974)

(*e*) *Stability*: No data were available.

(*f*) *Reactivity*: No data were available.

4-Methylchrysene

1.1 Synonyms and trade names

Chem. Abstr. Services Reg. No.: 3351-30-2

Chem. Abstr. Name: Chrysene, 4-methyl-

IUPAC Systematic Name: 4-Methylchrysene

Synonym: 6-Methylchrysene (Clar, 1964)

1.2 Structural and molecular formulae and molecular weight

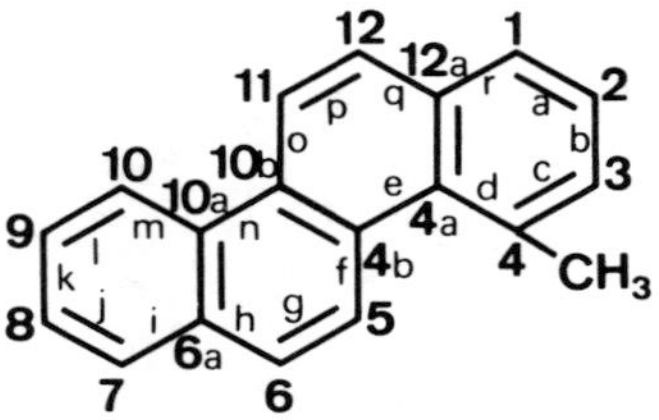

$C_{19}H_{14}$ Mol. wt: 242.3

1.3 Chemical and physical properties of the pure substance

(*a*) *Description*: Highly fluorescent plates (recrystallized from benzene/ethanol) (Buckingham, 1982) (*b*) *Melting-point*: 149°C (Clar, 1964); 151-152°C (Buckingham, 1982); 150.6°C (Karcher *et al.*, 1983)

(*c*) *Spectroscopy data*: Mass spectra have been tabulated (NIH/EPA Chemical Information System, 1982). Ultraviolet spectra have been reported (Peters, 1957).

(*d*) *Solubility*: Soluble in acetone (Hecht *et al.*, 1974)

(*e*) *Stability*: No data were available.

(*f*) *Reactivity*: No data were available.

5-Methylchrysene

1.1 Synonyms and trade names

Chem. Abstr. Services Reg. No.: 3697-24-3

Chem. Abstr. Name: Chrysene, 5-methyl-

IUPAC Systematic Name: 5-Methylchrysene

Synonym: 1-Methylchrysene (Clar, 1964)

1.2 Structural and molecular formulae and molecular weight

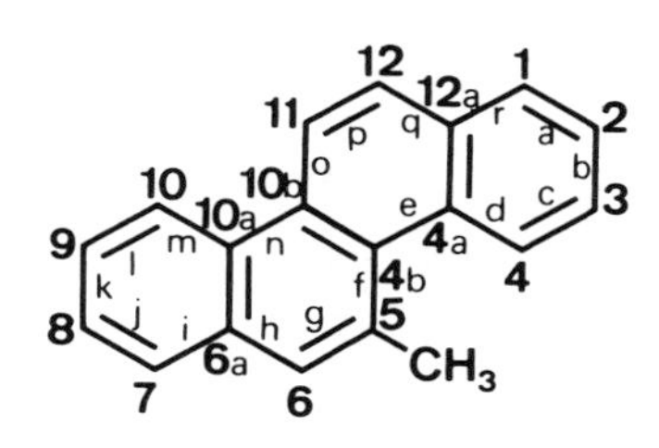

$C_{19}H_{14}$ Mol. wt: 242.3

1.3 Chemical and physical properties of the pure substance

(*a*) *Description*: Needles (recrystallized from benzene/ethanol) with a brilliant bluish-violet fluorescence in ultraviolet light (Buckingham, 1982)

(*b*) *Melting-point*: 118-119°C (Buckingham, 1982); 117-118°C (Clar, 1964); 117.2-117.8°C (Rappoport, 1967); 117.1°C (Karcher *et al.*, 1983) (*c*) *Spectroscopy data*: λ_{max} 217, 270, 300, 312, 326, 348, 366 nm (in cyclohexane) (Karcher *et al.*, 1983). Mass and nuclear magnetic resonance spectra have been tabulated (NIH/EPA Chemical Information System, 1982; Karcher *et al.*, 1983).

(*d*) *Solubility*: Soluble in acetone (Hecht *et al.*, 1974)

(*e*) *Stability*: No data were available.

(*f*) *Reactivity*: No data were available.

6-Methylchrysene

1.1 Synonyms and trade names

Chem. Abstr. Services Reg. No.: 1705-85-7

Chem. Abstr. Name: Chrysene, 6-methyl-

IUPAC Systematic Name: 6-Methylchrysene

Synonym: 2-Methylchrysene (Clar, 1964)

1.2 Structural and molecular formulae and molecular weight

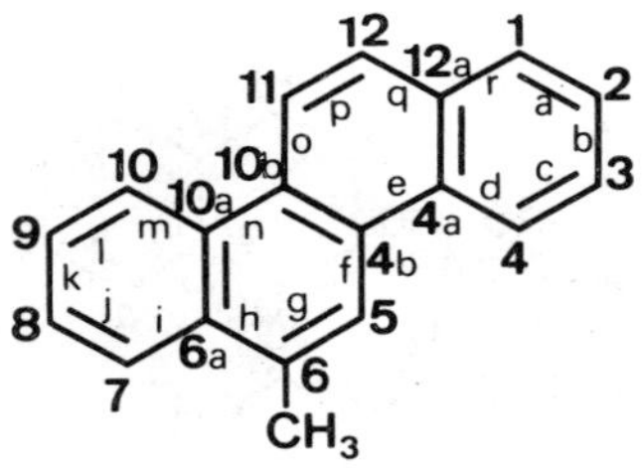

$C_{19}H_{14}$ Mol. wt: 242.3

1.3 Chemical and physical properties of the pure substance

(*a*) *Description*: Fluorescent needles (recrystallized from ethyl acetate/ethanol) (Buckingham, 1982)

(*b*) *Melting-point*: 161°C (Clar, 1964); 161-162°C (Buckingham, 1982); 161-161.4°C (Rappoport, 1967); 160.3°C (Karcher *et al.*, 1983) (*c*) *Spectroscopy data*: Mass spectra have been tabulated (NIH/EPA Chemical Information System, 1982). Ultraviolet spectra have been reported (Peters, 1957).

(*d*) *Solubility*: Soluble in acetone (Hecht *et al.*, 1974)

(*e*) *Stability*: No data were available.

(*f*) *Reactivity*: No data were available.

2. Production, Use, Occurrence and Analysis

2.1 Production and use

There is no commercial production or known use of these compounds. Reference samples of certified high purity are available for all six methylchrysenes (Karcher *et al.*, 1981).

2.2 Occurrence and analysis

Data on occurrence and methods of analysis are summarized in the 'General Remarks on the Substances Considered', p. 35.

1-Methylchrysene has been identified in tobacco smoke (3 ng/cigarette) (Hecht *et al.*, 1974); marijuana smoke (Lee *et al.*, 1976); lubricating oil (0.56 mg/kg) (Grimmer *et al.*, 1981a) and used motor oil (40.13 mg/kg) (Grimmer *et al.*, 1981b); crude oils (17.5 mg/kg) (Grimmer *et al.*, 1983); and coal-tar (Lang & Eigen, 1967).

2-Methylchrysene has been identified in tobacco smoke (1.2 ng/cigarette) (Hecht *et al.*, 1974); marijuana smoke (Lee *et al.*, 1976); gasoline engine exhaust gas (Grimmer *et al.*, 1977); used motor oil (73.15 ng/kg) (Grimmer *et al.*, 1981b); and vegetables (0.9-6.2 μg/kg) (Grimmer, 1979).

3-Methylchrysene has been identified in tobacco smoke (6.1 ng/cigarette) (Hecht *et al.*, 1974); marijuana smoke (Lee *et al.*, 1976); gasoline engine exhaust gas (Grimmer *et al.*, 1977); used motor oil (198.14 mg/kg) (Grimmer *et al.*, 1981b); lubricating oil (very little) (Grimmer *et al.*, 1981a); and vegetables (1.7-20.2 μg/kg) (Grimmer, 1979).

4-Methylchrysene has been identified in tobacco smoke (Snook *et al.*, 1978); gasoline engine exhaust gas (Grimmer *et al.*, 1977); and green kale (Grimmer, 1979).

5-Methylchrysene has been identified in tobacco smoke (0.6 ng/cigarette) (Hecht *et al.*, 1974); marijuana smoke (Lee *et al.*, 1976); and gasoline engine exhaust gas (Grimmer *et al.*, 1977).

6-Methylchrysene has been identified in tobacco smoke (7.2 ng/cigarette) (Hecht *et al.*, 1974); marijuana smoke (Lee *et al.*, 1976); gasoline engine exhaust gas (Grimmer *et al.*, 1977); coal-tar (Lang & Eigen, 1967); and vegetables (0.9-2.6 μg/kg) (Grimmer, 1979).

3. Biological Data Relevant to the Evaluation of Carcinogenic Risk to Humans

Reviews have been published on the comparative carcinogenicity of methylchrysenes and their metabolites in relation to their structure and biological activity (Hecht *et al.*, 1976a; Hoffmann *et al.*, 1982).

3.1 Carcinogenicity studies in animals[1]

1-Methylchrysene

Skin application

Mouse: A group of 20 female Swiss albino Ha/ICR/Mil mice, seven to eight weeks of age, received thrice-weekly applications of 100 μg 1-methylchrysene in 0.1 ml acetone on shaved back skin for 72 weeks. Further groups of mice received thrice-weekly applications of 5 μg benzo[*a*]pyrene in 0.2 ml acetone (positive controls) or 0.2 ml acetone alone (vehicle controls). The study was terminated at 72 weeks, at which time seven animals were still alive. No skin tumour was seen in the 1-methylchrysene-treated group. In the positive-control group, a total of 22 skin tumours (four of which were carcinomas found in four animals) were observed in 13 tumour-bearing mice; 7/20 mice survived until the end of the experiment. No skin tumour was reported in the vehicle-control group (Hecht *et al.*, 1974; Hoffmann *et al.*, 1974).

A group of 20 female Swiss albino Ha/ICR/Mil mice, seven to eight weeks of age, received applications of 100 μg 1-methylchrysene (purity, >99.9%) in 0.1 ml acetone every other day for 20 days (total dose, 1.0 mg), followed ten days later by thrice-weekly applications of 2.5 μg 12-*O*-tetradecanoylphorbol-13-acetate (TPA) in 0.1 ml acetone for 20 weeks (total dose, 150 μg), at which time the experiment was terminated. Further groups of 20 mice received 10 applications of 5 μg benzo[*a*]pyrene in 0.1 ml acetone (positive controls) or 10 applications of 0.1 ml acetone only (vehicle controls) according to the same dose schedule. At the end of the study there were 19 survivors in the 1-methylchrysene-treated group. In this group, six animals each had a skin tumour. In the positive-control group (20 survivors), six animals had ten skin tumours; no such tumour was observed among the vehicle control group (Hecht *et al.*, 1974).

[1] The Working Group was aware of a completed study in which 5-methylchrysene was given by skin application to mice (IARC, 1983).

2-Methylchrysene

Skin application

Mouse: A group of 20 female Ha/ICR/Mil mice, seven to eight weeks of age, received applications of 100 μg 2-methylchrysene in 0.1 ml acetone thrice weekly for 72 weeks. Groups in which benzo[*a*]pyrene was administered as a thrice-weekly application of 5 μg in 0.2 ml acetone, and 0.2 ml acetone administered alone were used as positive and vehicle controls, respectively. The first skin tumour was reported at 40 weeks, at which time 18 mice were still alive. At 72 weeks, 10/20 mice in the 2-methylchrysene-treated group were still alive; 21 skin tumours were observed in 11 tumour-bearing animals. Of these, seven carcinoma-bearing mice presented a total of seven carcinomas. In the benzo[*a*]pyrene group at 72 weeks, 22 skin tumours occurred in 13 tumour-bearing mice, four of which had a total of four carcinomas. No skin tumour was reported in the vehicle-control group (Hecht *et al*., 1974).

A group of 20 female Swiss albino (Ha/IRC/Mil) mice, seven to eight weeks of age, received ten applications of 100 μg 2-methylchrysene (purity, >99.9%) in 0.1 ml acetone (total dose, 1.0 mg) to the shaven back every other day. Promotion was begun ten days after the last dose, by application of 2.5 μg TPA in 0.1 ml acetone thrice weekly for 20 weeks (total dose, 150 μg); applications of 5 μg benzo[*a*]pyrene in 0.1 ml acetone and 0.1 ml acetone under the same dose schedule were given as positive and vehicle controls, respectively. At 20 weeks, no tumour had been observed in the acetone-treated controls. In the 2-methylchrysene-treated group (19 surviving mice), 13 skin tumours were present in eight tumour-bearing animals, as compared with ten skin tumours in six tumour-bearing mice in the benzo[*a*]pyrene-treated group (Hecht *et al*., 1974).

3-Methylchrysene

Skin application

Mouse: A group of 20 female Swiss albino Ha/ICR/Mil mice, aged seven to eight weeks, received thrice-weekly applications of 100 μg 3-methylchrysene in 0.1 ml acetone on shaved back skin for 72 weeks. Further groups of mice received thrice-weekly applications of 5 μg benzo[*a*]pyrene in 0.2 ml acetone (positive controls) or 0.2 ml acetone alone (vehicle controls). The first tumour was reported at 15 weeks at which time 20 mice were still alive. The study was terminated at 72 weeks, at which time eight mice were still alive. There were five tumour-bearing mice in this group, with a total of six skin tumours. Of these, four carcinomas were seen in four mice. In the positive-control group, a total of 22 skin tumours, four of which were carcinomas found in four animals, were observed in 13 tumour-bearing mice; seven mice survived until the end of the experiment. No skin tumour was reported in the vehicle-control group (Hecht *et al*., 1974; Hoffmann *et al*., 1974).

A group of 20 female Swiss albino Ha/ICR/Mil mice, aged seven to eight weeks, received applications of 100 μg 3-methylchrysene (purity, >99.9%) in 0.1 ml acetone every other day for 20 days (total dose, 1.0 mg), followed ten days later by thrice-weekly applications of 2.5 μg 12-*O*-tetradecanoylphorbol-13-acetate (TPA) in 0.1 ml acetone for 20 weeks (total dose, 150 μg), at which time the experiment was terminated. Further groups of 20 mice received 10 applications of 5 μg benzo[*a*]pyrene in 0.1 ml acetone (positive controls) or 10 applications of 0.1 ml acetone only (vehicle controls) according to the same dose schedule. At the end of the study, 20 animals were still alive in the 3-methylchrysene-treated group, 14 of which 14 had 26 skin tumours. In the positive-control group, 20 animals were still alive, of which six mice had 10 skin tumours; no such tumour was observed among the vehicle control group (Hecht *et al*., 1974).

In the same study, further groups of 20 mice received applications of 30 or 10 μg 3-methylchrysene in 0.1 ml acetone (total doses, 300 or 100 μg, respectively), 10 μg benzo[*a*]pyrene (positive controls) in 0.1 ml acetone or 0.1 ml acetone alone (vehicle controls) every other day for 20 days, followed by applications of 2.5 μg TPA for 20 weeks, at which time the study was terminated. Survival rates at the end of the study were 16/20, 17/20 and 19/20 in the high- and low-dose 3-methylchrysene-treated and positive-control groups, respectively. Skin tumour incidences in these groups were eight in four tumour-bearing mice, three in three tumour-bearing mice and 47 in 18 tumour-bearing mice, respectively (Hecht *et al.*, 1974).

4-Methylchrysene

(*a*) *Skin application*

Mouse: A group of 20 female Swiss albino Ha/ICR/Mil mice, seven to eight weeks of age, received thrice-weekly applications of 100 μg 4-methylchrysene in 0.1 ml acetone on shaved back skin for 72 weeks. Further groups of mice received thrice-weekly applications of 5 μg benzo[*a*]pyrene in 0.2 ml acetone (positive controls) or 0.2 ml acetone alone (vehicle controls). In the 4-methylchrysene-treated group, the first skin tumour was reported after 55 weeks, at which time 14 animals were still alive; at the end of the study (72 weeks) there were ten survivors. The three tumour-bearing mice had a total of five skin tumours, two of which were carcinomas found in two mice. In the positive-control group, a total of 22 skin tumours, four of which were carcinomas found in four animals, were observed in 13 tumour-bearing mice; seven mice survived until the end of the experiment. No skin tumour was reported in the vehicle-control group (Hecht *et al.*, 1974).

A group of 20 female Swiss albino Ha/ICR/Mil mice, seven to eight weeks of age, received applications of 100 μg 4-methylchrysene (purity, >99.9%) in 0.1 ml acetone every other day for 20 days (total dose, 1.0 mg), followed ten days later by thrice-weekly applications of 2.5 μg TPA in 0.1 ml acetone for 20 weeks (total dose, 150 μg), at which time the experiment was terminated. Further groups of 20 mice received 10 applications of 5 μg benzo[*a*]pyrene in 0.1 ml acetone (positive controls) or ten applications of 0.1 ml acetone only (vehicle controls) according to the same dose schedule. At the end of the study, seven tumour-bearing animals in the 4-methylchrysene-treated group (20 survivors) had nine skin tumours; six tumour-bearing animals in the positive-control group (20 survivors) had ten tumours; no such tumour was observed among the vehicle control group (Hecht *et al.*, 1974).

(*b*) *Subcutaneous and/or intramuscular administration*

Mouse: Groups of 5 male Swiss or 5 male C3H mice, two to three months old, received single s.c. injections of 4 mg 4-methylchrysene [purity unspecified] in 0.2 ml tricaprylin into the rump. No tumour at the injection site was reported in either group after six months [total duration of observation unspecified]. A further group of 20 male C3H mice, two to three months old, received a single injection of 2 mg 4-methylchrysene in 0.2 ml tricaprylin into the rump. After 170 days, one transplantable subcutaneous tumour appeared among the 17 animals still alive at that time (Dunlap & Warren, 1943). [The Working Group noted the small number of animals used in this study, the absence of details on purity and the incomplete reporting of study duration.]

5-Methylchrysene

(*a*) *Skin application*

Mouse: A group of 20 female Swiss albino Ha/ICR/Mil mice received applications of 100 μg 5-methylchrysene in 0.1 ml acetone thrice weekly on shaved back skin for 72 weeks. Further groups of mice receiving thrice-weekly applications of 5 μg benzo[*a*]pyrene in 0.2 acetone or 0.2 acetone alone served as positive and vehicle controls, respectively. All mice in the 5-methylchrysene-treated group had skin tumours (total number of tumours, 85) by 25 weeks; by 35 weeks all the animals in this group had died. At that time, the total number of tumours was 99, of which 37 were carcinomas found in 12 mice; in addition, three mice had multiple metastases to the lung, and one had metastases to the lung and spleen. In the positive control group, 13 tumour-bearing mice had a total of 22 skin tumours, four of which were skin carcinomas found in four animals; 7/20 mice survived until week 72. No skin tumours was reported in the vehicle-control group (Hecht *et al.*, 1974; Hoffmann *et al.*, 1974).

Groups of 20 female Swiss Ha/ICR mice received skin applications of 0.01% or 0.005% 5-methylchrysene in 0.1 ml acetone, 0.01% or 0.005% benzo[*a*]pyrene in 0.1 ml acetone or 0.1 ml acetone alone thrice weekly for 62 weeks. At 55 weeks, all mice in the high-dose groups (5-methylchrysene and benzo-[*a*]pyrene) were dead; 15 tumour-bearing mice treated with 5-methylchrysene had a total of 38 skin tumours, of which 12 were carcinomas found in ten mice; 18 tumour-bearing mice treated with benzo[*a*]pyrene had a total of 70 skin tumours, of which 16 were carcinomas found in 14 mice. At 62 weeks in the low-dose groups, seven and 12 mice survived; nine tumour-bearing mice treated with 5-methylchrysene had a total of 22 tumours, of which seven were carcinomas found in six mice; ten tumour-bearing mice treated with benzo[*a*]pyrene had 19 tumours, of which eight were carcinomas found in seven mice. Only skin tumours were reported. No skin tumour was found in the acetone-treated controls (Hecht *et al.*, 1976b).

A group of 20 female Swiss albino Ha/ICR/Mil mice, seven to eight weeks of age, received ten initiating applications of 100 μg 5-methylchrysene (purity, >99.9%) in 0.1 ml acetone (total dose, 1.0 mg) every other day. Ten days after the last dose, applications of 2.5 μg TPA in 0.1 ml acetone were given thrice weekly for 20 weeks (total dose, 150 μg), at which time the experiment was terminated. Further groups of 20 mice were given doses of 5 μg benzo[*a*]pyrene in 0.1 ml acetone (positive controls) or doses of 0.1 ml acetone only (vehicle controls) by the same dose schedule. At the end of the study, 17 tumour-bearing animals in the 5-methylchrysene-treated group had a total 96 skin tumours (18 animals survived). In the positive-control group, six tumour-bearing animals had ten tumours (20 animals survived). No tumour was reported in the vehicle-control group (Hecht *et al.*, 1974).

In the same study, similar groups of 20 mice received applications of either 30 μg or 10 μg 5-methylchrysene (total doses, 300 or 100 μg, respectively), 10 μg benzo[*a*]pyrene (positive controls) or 0.1 ml acetone (vehicle controls) according to the same dose schedule as above, followed by repeated applications of 2.5 μg TPA for 20 weeks. At the termination of the study, all 20 mice in the two 5-methylchrysene groups were still alive; 20 tumour-bearing mice in the high-dose 5-methylchrysene-treated group had 160 skin tumours, and 20 tumour-bearing mice in the low-dose 5-methylchrysene-treated group had 110 skin tumours. Among the 19 surviving positive controls, 18 tumour-bearing animals had 47 skin tumours. No such tumour was reported in the vehicle-control group (Hecht *et al.*, 1974).

Groups of 20 female Swiss Ha/ICR mice, seven to eight weeks of age, received a total of ten initiating applications of 3 μg or 1 μg highly purified 5-methylchrysene in 0.1 ml acetone.

Positive controls received applications of benzo[*a*]pyrene at the same dose levels and schedules, and vehicle controls received applications of acetone only. All groups received promoting applications of 2.5 μg TPA in 0.1 ml acetone thrice weekly for 20 weeks, at which time the study was terminated. By the end of the study period, 20 mice in the high-dose 5-methylchrysene-treated group had 45 skin tumours, one of which was a carcinoma, compared with six mice with tumours in the high-dose benzo[*a*]pyrene group. In the low-dose 5-methylchrysene-treated group, two tumour-bearing mice had one skin tumour each; no tumour was seen in the low-dose benzo[*a*]pyrene group. No skin tumour was reported in the vehicle-control group (Hecht *et al.*, 1976b).

Groups of 20 female Ha/ICR outbred Swiss albino mice, seven to eight weeks of age, received a total of ten applications of either 3 μg or 10 μg 5-methylchrysene (purity determined by melting-point, elemental analysis and gas chromatographic or high-performance liquid chromatographic analysis) in 0.1 ml acetone on shaved back skin every other day. A group of mice receiving applications of acetone only served as controls. Ten days after the final administration, mice received applications of 2.5 μg TPA thrice weekly for 20 weeks. In the low-dose group, one papilloma, 24 keratoacanthomas and two carcinomas were reported; in the high-dose group, there were three papillomas, 89 keratoacanthomas and two carcinomas. The incidences of skin tumour-bearing animals were 60% in the low-dose group and 80% in the high-dose group [precise numbers not specified]. No skin tumour was observed in the control group (Hecht *et al.*, 1978a).

Groups of 20 Swiss Ha/ICR/Mil albino mice received a total of 10 applications of 3 or 10 μg 5-methylchrysene (purity, >99.9%) in 0.1 ml acetone; controls received acetone alone. Ten days later, this treatment was followed by thrice-weekly applications of 2.5 μg TPA in 0.1 ml acetone for 20 weeks. The average incidence of skin tumour-bearing animals ranged from 55-95% in the low-dose groups (eight groups) and 80-90% in the high-dose groups (eight groups) (Hecht *et al.*, 1978b). Similar results were reported, with incidences of 75% and 85%, in groups of 20 female Ha(ICR) outbred Swiss albino mice receiving total doses of 30 μg and 10 μg 5-methylchrysene, respectively (Hecht *et al.*, 1979).

(*b*) *Subcutaneous and/or intramuscular administration*

Mouse: A group of 20 male Swiss mice and two groups of ten male C3H mice, two to three months old, received single s.c. injections of 2 mg 5-methylchrysene [purity unspecified] in 0.2 ml tricaprylin into the rump. No injection-site tumour was seen in the Swiss mice, but only 4/20 animals were alive six months after receiving the injection. In the first group of C3H mice, 7/10 animals developed a sarcoma at the injection site (nine mice were still alive at the appearance of the first tumour); the first tumour developed at 114 days, and average latency was 136 days. In the second group of C3H mice, 3/10 animals developed a sarcoma at the site of injection; the first tumour developed at 79 days, and average latency was 125 days (Dunlap & Warren, 1943).

A group of 25 male C57Bl mice received s.c. injections of 0.05 mg highly purified 5-methylchrysene in 0.1 ml trioctanoin once every two weeks for 20 weeks (total dose, 0.5 mg) and were observed for a further 12 weeks. A total of 22/25 mice had 24 fibrosarcomas, with an average latent period of 25 weeks. No such tumour was reported in a vehicle control group (Hecht *et al.*, 1976a).

6-Methylchrysene

(*a*) *Skin application*

Mouse: A group of 20 female Swiss albino Ha/ICR/Mil mice, seven to eight weeks old, received thrice-weekly applications of 100 μg 6-methylchrysene in 0.1 ml acetone on shaved back skin for 72 weeks. Further groups of mice received thrice-weekly applications of 5 μg benzo[*a*]pyrene in 0.2 ml acetone (positive controls) or 0.2 ml acetone alone (vehicle controls). In the 6-methylchrysene-treated group, the first skin tumour was reported after 20 weeks, at which time 19 animals were still alive. At the end of the study (72 weeks), 12 were alive; three tumour-bearing mice had a total of three skin tumours, one of which was a carcinoma. In the positive control group, a total of 22 skin tumours, four of which were carcinomas found in four animals, were observed in 13 mice; seven mice survived until the end of the experiment. No such tumour was reported in the vehicle-control group (Hecht *et al.*, 1974; Hoffmann *et al.*, 1974).

A group of 20 female Swiss albino Ha/ICR/Mil mice, seven to eight weeks of age, received applications of 100 μg 6-methylchrysene (purity, >99.9%) in 0.1 ml acetone every other day for 20 days (total dose, 1.0 mg), followed ten days later by thrice-weekly applications of 2.5 μg TPA in 0.1 ml acetone for 20 weeks (total dose, 150 μg), at which time the experiment was terminated. Further groups of 20 mice received 10 applications of 5 μg benzo[*a*]pyrene in 0.1 ml acetone (positive controls) or 10 applications of 0.1 ml acetone only (vehicle controls), according to the same dose schedule followed by the same promoting regime. At the end of the study, seven tumour-bearing animals in the 6-methylchrysene-treated group (19 surviving animals) had 11 skin tumours; six tumour-bearing animals still alive in the positive control group (20 surviving animals) had ten skin tumours; no such tumour was observed among the vehicle control group (Hecht *et al.*, 1974).

(*b*) *Subcutaneous and/or intramuscular application*

A group of 20 mice [strain unspecified] received s.c. injections of 2 mg 6-methylchrysene. A further 2-mg injection was given six months later. Nine mice were alive at one year, and the last animal died 16 months after the start of the study. At ten months, one mouse developed an injection-site sarcoma (Shear & Leiter, 1941). [The Working Group noted the absence of a control group.]

3.2 Other relevant biological data[1]

(*a*) *Experimental systems*

No data were available to the Working Group on toxic effects or on effects on reproduction and prenatal toxicity.

Metabolism and activation

1,2- and 7,8-Dihydrodiols were the major metabolites of *5-methylchrysene*, and small amounts of the 9,10-dihydrodiol were also formed following incubation of the compound with rat-liver preparations. Other metabolites detected include 1-, 7- and 9-hydroxy-5-methylchrysene and 5-hydroxymethylchrysene (Hecht *et al.*, 1978c).

---

[1] See also 'General Remarks on the Substances Considered', p. 53.

The 1,2-dihydrodiol and the 7,8-dihydrodiol of *5-methylchrysene* were mutagenic to bacteria in the presence of a metabolic activation system (Hecht *et al.*, 1978c) and were active as tumour initiators in mouse skin, while the 9,10-dihydrodiol was not (Hecht *et al*, 1980).

DNA adducts were detected following application of *5-methylchrysene* to mouse skin; they had chromatographic properties consistent with their formation by reaction of the 1,2-diol-3,4-epoxide and the 7,8-diol-9,10-epoxide with DNA (Melikian *et al.*, 1982).

Carcinogenicity of metabolites

Skin application: Groups of 20 female outbred Ha/ICR Swiss albino mice, seven to eight weeks of age, received applications of 3.0 μg of one of three diol metabolites of *5-methylchrysene* - 5-methylchrysene-9,10-diol, 5-methylchrysene-7,8-diol or 5-methychrysene-1,2-diol - or of the parent compound (all compounds were purified) in 0.1 ml acetone on shaved back skin every other day for 20 days (total doses, 30 μg). Controls received acetone only. Ten days after the final administration, mice received applications of 2.5 μg TPA in 0.1 ml acetone thrice weekly for 20 weeks. No skin tumour was seen in the vehicle-control group; 15 animals treated with 5-methylchrysene had papillomas, with an average of three tumours/animal; 19 animals given 5-methylchrysene-1,2-diol had papillomas, and seven had carcinomas (three of which had metastasized to lymph nodes and one to the lungs), with an average of 7.3 tumours/mouse; ten animals treated with 5-methylchrysene-7,8-diol had papillomas, with an average of 1.1 tumours/mouse; and no skin tumour was seen in the 5-methylchrysene-9,10-diol-treated group (Hecht *et al.*, 1980).

In three sets of experiments, groups of 20 female outbred Cr1:COBS CD-1(ICR)BR albino mice, seven to eight weeks of age, received applications of 30 μg (first experiment) or 1.0 μg or 3.0 μg (second experiment) *5-methylchrysene* or one of its major metabolites, 5-hydroxymethylchrysene, or 3.0 μg 5-hydroxymethylchrysene (third experiment) in 0.1 ml acetone every other day for 20 days (total doses, 10 or 30 μg). Control animals received acetone only. Ten days after the final dose, mice received thrice-weekly applications of 2.5 μg TPA in 0.1 ml acetone for 20 weeks. The incidence rates of skin tumours and average numbers of tumours/mouse were as follows: in the first experiment, 90% and 5.7 in the high-dose 5-methylchrysene-treated animals, and 95% and 8.8 in those receiving the high dose of 5-hydroxymethylchrysene, and 5% and 0.1 in controls; in the second experiment, 75% and 6.2 in the high-dose 5-methylchrysene-treated group, 55% and 5.6 in the low-dose 5-methylchrysene-treated animals, 90% and 9.5 in those receiving the high dose of 5-hydroxymethylchrysene and 45% and 2.6 in those receiving the low dose of 5-hydroxymethylchrysene; and, in the third experiment, 85% and 6.6 in those receiving 30 μg 5-hydroxymethylchrysene. These skin tumours were diagnosed as papillomas or keratoacanthomas (Amin *et al.*, 1981).

No data were available on the metabolites of other methylchrysenes.

Mutagenicity and other short-term tests

Results from short-term tests are given in Table 1.

Table 1. Results from short-term tests: 1-, 2-, 3-, 4-, 5-, and 6-Methylchrysenes

Test	Organism/assay[a]	Exogenous metabolic system[a]	Reported result	Comments	References
PROKARYOTES					
1-Methylchrysene					
Mutation	*Salmonella typhimurium* (his⁻/his⁺)	Aro-R-PMS	Positive	At 10 μg/plate in strain TA100	Coombs *et al.* (1976)
2-Methylchrysene					
Mutation	*Salmonella typhimurium* (his⁻/his⁺)	Aro-R-PMS	Positive	At 10 μg/plate in strain TA100	Coombs *et al.* (1976)
3-Methylchrysene					
Mutation	*Salmonella typhimurium* (his⁻/his⁺)	Aro-R-PMS	Positive	At 20 μg/plate in strain TA100	Coombs *et al.* (1976)
4-Methylchrysene					
Mutation	*Salmonella typhimurium* (his⁻/his⁺)	Aro-R-PMS	Positive	At 10 μg/plate in strain TA100	Coombs *et al.* (1976)
5-Methylchrysene					
Mutation	*Salmonella typhimurium* (his⁻/his⁺)	Aro-R-PMS	Positive	At 10 μg/plate in strain TA100	Coombs *et al.* (1976); Amin *et al.* (1979); Hecht *et al.* (1979)
6-Methylchrysene					
Mutation	*Salmonella typhimurium* (his⁻/his⁺)	Aro-R-PMS	Positive	At 10 μg/plate in strain TA100	Coombs *et al.* (1976)
MAMMALIAN CELLS *IN VITRO*					
5-Methylchrysene					
DNA damage	Primary rat hepatocytes (unscheduled DNA synthesis)	--	Positive	At 50 nmol/ml	Tong *et al.* (1981)

[a] For an explanation of the abbreviations see the Appendix, p. 452.

(*b*) *Humans*

No data were available to the Working Group.

3.3 Case reports and epidemiological studies of carcinogenicity to humans

No data were available to the Working Group.

4. Summary of Data Reported and Evaluation[1]

4.1 Experimental data

In comparative studies carried out in the same laboratory, 1-, 2-, 3-, 4-, 5- and 6-methylchrysenes were tested for carcinogenicity by skin application to female mice and in the mouse-skin initiation-promotion assay. 5-Methylchrysene induced the highest incidence of malignant skin tumours, when tested alone or together with a promoter. An intermediate response was observed with 2-, 3-, 4- and 6-methylchrysenes when tested as carcinogens; however, 1-methylchrysene was inactive. All the chrysene derivatives showed varying degrees of initiating activity.

5-Methylchrysene, when tested by subcutaneous injection in mice, produced a high incidence of sarcomas at the site of injection.

No data on the teratogenicity of these compounds were available.

1-, 2-, 3-, 4-, 5- and 6-Methylchrysenes were mutagenic to *Salmonella typhimurium* in the presence of an exogenous metabolic system. 5-Methylchrysene induced DNA damage in primary rat hepatocytes.

There is *inadequate evidence* that 1-, 2-, 3-, 4- and 6-methylchrysenes are active in short-term tests. There is *limited evidence* that 5-methylchrysene is active in short-term tests.

4.2 Human data[2]

1-, 2-, 3-, 4-, 5- and 6-Methychrysenes are present as minor components of the total content of polynuclear aromatic compounds in the environment. They occur primarily in products deriving from organic matter containing steroids, such as tobacco smoke and some petroleum-derived products.

4.3 Evaluation

There is *inadequate evidence* that 1-methylchrysene is carcinogenic to experimental animals.

There is *limited evidence* that 2-, 3-, 4- and 6-methylchrysenes are carcinogenic to experimental animals.

There is *sufficient evidence*[3] that 5-methylchrysene is carcinogenic to experimental animals.

---

[1] For definitions of the italicized terms, see Preamble, p. 19.

[2] Studies on occupational exposure to polynuclear aromatic compounds will be considered in future *IARC Monographs*.

[3] In the absence of adequate data on humans, it is reasonable, for practical purposes, to regard chemicals for which there is *sufficient evidence* of carcinogenicity in animals as if they presented a carcinogenic risk to humans (see also Preamble, p. 20).

5. References

Amin, S., Hecht, S.S., LaVoie, E. & Hoffmann, D. (1979) Synthesis and mutagenicity of 5,11-dimethylchrysene and some methyl-oxidized derivatives of 5-methylchrysene. *J. med. Chem.*, *22*, 1336-1340

Amin, S., Juchatz, A., Furuya, K. & Hecht, S.S. (1981) Effects of fluorine substitution on the tumor initiating activity and metabolism of 5-hydroxymethylchrysene, a tumorigenic metabolite of 5-methylchrysene. *Carcinogenesis*, *2*, 1027-1032

Buckingham, J., ed. (1982) *Dictionary of Organic Compounds*, 5th ed., Vol. 4, New York, Chapman & Hall, pp. 3791, 3792

Clar, E., ed. (1964) *Polycyclic Hydrocarbons*, Vol. 1, New York, Academic Press, p. 251

Coombs, M.M., Dixon, C. & Kissonerghis, A.-M. (1976) Evaluation of the mutagenicity of compounds of known carcinogenicity, belonging to the benz[*a*]anthracene, chrysene, and cyclopenta[*a*]phenanthrene series, using Ames's test. *Cancer Res.*, *36*, 4525-4529

Dunlap, C.E. & Warren, S. (1943) The carcinogenic activity of some new derivatives of aromatic hydrocarbons. I. Compounds related to chrysene. *Cancer Res.*, *3*, 606-607

Grimmer, G. (1979) *Sources and occurrence of polycyclic aromatic hydrocarbons*. In: Egan, H., Castegnaro, M., Bogovski, P., Kunte, H. & Walker, E.A., eds, *Environmental Carcinogens. Selected methods of Analysis* (*IARC Scientific Publications No. 29*), Lyon, International Agency for Research on Cancer, pp. 31-54

Grimmer, G., Böhnke, H. & Glaser, A. (1977) Investigation on the carcinogenic burden by air pollution in man. XV. Polycyclic aromatic hydrocarbons in automobile exhaust gas - An inventory. *Zbl. Bakt. Hyg., 1 Abt., Orig. B164*, 218-234

Grimmer, G., Jacob, J. & Naujack, K.-W. (1981a) Profile of the polycyclic aromatic hydrocarbons from lubricating oils. Inventory by GCGC/MS - PAH in environmental materials, Part 1. *Fresenius Z. anal. Chem.*, *306*, 347-355

Grimmer, G., Jacob, J., Naujack, K.-W. & Dettbarn, G. (1981b) Profile of the polycyclic aromatic hydrocarbons from used engine oil - Inventory by GCGC/MS - PAH in environmental materials, Part 2. *Fresenius Z. anal. Chem.*, *309*, 13-19

Grimmer, G., Jacob, J. & Naujack, K.-W. (1983) Profile of the polycyclic aromatic compounds from crude oils. Inventory by GCGC/MS - PAH in environmental materials, Part 3. *Fresenius Z. anal. Chem.*, *314*, 29-36

Hecht, S.S., Bondinell, W.E. & Hoffmann, D. (1974) Chrysene and methylchrysenes: Presence in tobacco smoke and carcinogenicity. *J. natl Cancer Inst.*, *53*, 1121-1133

Hecht, S.S., Loy, M. & Hoffmann, D. (1976a) *On the structure and carcinogenicity of methylchrysenes*. In: Freudenthal, R.I. & Jones, P.W., eds, *Carcinogenesis*, Vol. 1, *Polynuclear Aromatic Hydrocarbons: Chemistry, Metabolism, and Carcinogenesis*, New York, Raven Press, pp. 325-340

Hecht, S.S., Loy, M., Maronpot, R.R. & Hoffmann, D. (1976b) A study of chemical carcinogenesis: Comparative carcinogenicity of 5-methylchrysene, benzo(*a*)pyrene and modified chrysenes. *Cancer Lett.*, *1*, 147-154

Hecht, S.S., LaVoie, E., Mazzarese, R., Amin, S., Bedenko, V. & Hoffmann, D. (1978c) 1,2-Dihydro-1,2-dihydroxy-5-methylchrysene, a major activated metabolite of the environmental carcinogen 5-methylchrysene. *Cancer Res.*, *38*, 2191-2194

Hecht, S.S., Hirota, N., Loy, M. & Hoffmann, D. (1978a) Tumor-initiating activity of fluorinated 5-methylchrysenes. *Cancer Res.*, *38*, 1694-1698

Hecht, S.S., Loy, M., Mazzarese, R. & Hoffmann, D. (1978b) *On the carcinogenicity of 5-methylchrysene: Structure-activity studies and metabolism*. In: Gelboin, H.V. & Ts'O, P.O.P., eds, *Polycyclic Hydrocarbons and Cancer*, Vol. 1, *Environment, Chemistry and Metabolism*, New York, Academy Press, pp. 119-130

Hecht, S.S., Amin, S., Rivenson, A & Hoffmann, D. (1979) Tumor initiating activity of 5,11-dimethylchrysene and the structural requirements favoring carcinogenicity of methylated polynuclear aromatic hydrocarbons. *Cancer Lett.*, *8*, 65-70

Hecht, S.S., Rivenson, A. & Hoffmann, D. (1980) Tumor-initiating activity of dihydrodiols formed metabolically from 5-methylchrysene. *Cancer Res.*, *40*, 1396-1399

Hoffmann, D., Bondinell, W.E. & Wynder, E.L. (1974) Carcinogenicity of methylchrysenes. *Science*, *183*, 215-216

Hoffmann, D., LaVoie, E.J. & Hecht. S.S. (1982) *Polynuclear aromatic hydrocarbons: Effects of chemical structure on tumorigenicity*. In: Cooke, M., Dennis, A.J. & Fisher, G.L., eds, *Polynuclear Aromatic Hydrocarbons: Physical and Biological Chemistry*, *6th Int. Symposium*, Columbus, OH, Battelle Press, pp. 1-19

IARC (1983) *Information Bulletin on the Survey of Chemicals Being Tested for Carcinogenicity*, No. 10, Lyon, p. 175

Karcher, W., Jacob, J. & Fordham, R. (1981) *The Certification of Polycyclic Aromatic Hydrocarbon Materials (PAH)*, Part II, *The Six Methylchrysene Isomers (BRC Reference Materials No. 77, 78, 79, 80, 81 and 82)* (*EUR 7175 EN*), Luxembourg, Commission of the European Communities

Karcher, W., Fordham, R., Dubois, J. & Gloude, P. (1983) *Spectral Atlas of Polycyclic Aromatic Compounds*, Dordrecht, The Netherlands, D. Reidel (in press)

Lang, K.F. & Eigen, I. (1967) *Organic compounds in coal tar* (Ger.). In: Heilbronner, E., Hofmann, U., Schäfer, K. & Wittig, G., eds, *Fortschrifte der chemischer Forschung* (Advances in chemical research), Vol. 8, Berlin, Springer, pp. 91-170

Lee, M.L., Novotny, M. & Bartle, K.D. (1976) Gas chromatography/mass spectrometry and nuclear magnetic resonance spectrometric studies of carcinogenic polynuclear aromatic hydrocarbons in tobacco and marijuana smoke condensates. *Anal. Chem.*, *48*, 405-416

Melikian, A.A., LaVoie, E.J., Hecht, S.S. & Hoffmann, D. (1982) Influence of a bay-region methyl group on formation of 5-methylchrysene dihydrodiol epoxide: DNA adducts in mouse skin. *Cancer Res.*, *42*, 1239-1242

NIH/EPA Chemical Information System (1982) *Mass Spectral Search System*, Washington DC, CIS Project, Information Services Corporation

Peters, D. (1957) Colour and constitution. I. The effect of methyl substitution on the ultraviolet spectra of alternant hydrocarbons. *J. chem. Soc.*, *646*, 651

Rappoport, Z., ed. (1967) *CRC Handbook of Tables for Organic Compound Identification*, 3rd ed., Boca Raton, FL, Chemical Rubber Co., pp. 48, 50, 51

Shear, M.J. & Leiter, J. (1941) Studies in carcinogenesis. XVI. Production of subcutaneous tumors in mice by miscellaneous polycyclic compounds. *J. natl Cancer Inst.*, *11*, 241-258

Snook, M.E., Severson, R.F., Arrendale, R.F., Higman, H.C. & Chortyk, O.T. (1978) Multi-alkylated polynuclear aromatic hydrocarbons of tobacco smoke: Separation and identification. *Beitr. Tabakforsch. int.*, *9*, 222-247

Tong, C., Laspia, M.F., Telang, S. & Williams, G.M. (1981) The use of adult rat liver cultures in the detection of the genotoxicity of various polycyclic aromatic hydrocarbons. *Environ. Mutagenesis*, *3*, 477-487

Weast, R.C., ed. (1975) *CRC Handbook of Chemistry and Physics*, 56th ed., Cleveland, OH, Chemical Rubber Co., p. C-241

2- and 3-METHYLFLUORANTHENES[1]

1. Chemical and Physical Data

2-Methylfluoranthene

1.1 Synonyms and trade names

Chem. Abstr. Services Reg. No.: 33543-31-6

Chem. Abstr. Name: Fluoranthene, 2-methyl-

IUPAC Systematic Name: 2-Methylfluoranthene

Synonym: 3-Methylfluoranthrene (Clar, 1964)

1.2 Structural and molecular formulae and molecular weight

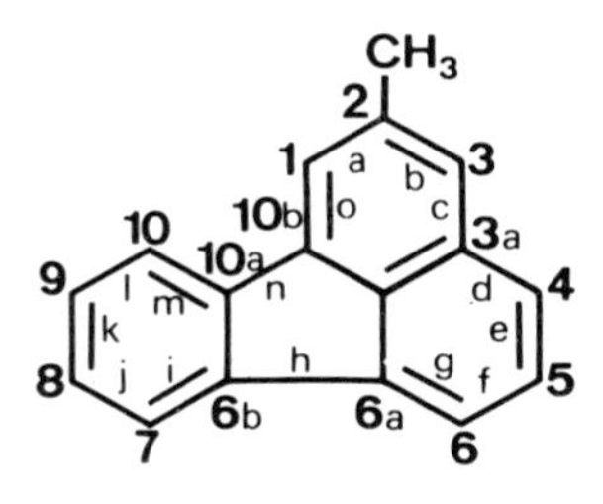

$C_{17}H_{12}$ Mol. wt: 216.3

[1] In earlier literature these compounds were referred to as 3- and 4-methylfluoranthene, respectively.

1.3 Chemical and physical properties of the pure substance

(*a*) *Description*: Pale-yellow needles [recrystallized from petroleum ether (a mixture of low-boiling hydrocarbons)]; vivid-blue fluorescence (Campbell & Wang, 1949)

(*b*) *Melting-point*: 78-80°C (Clar, 1964), 79-81°C (Tucker, 1952), 79-79.5°C (Hoffmann *et al.*, 1972)

(*c*) *Stability*: No data were available.

(*d*) *Reactivity*: No data were available.

3-Methylfluoranthene

1.1 Synonyms and trade names

Chem. Abstr. Services Reg. No.: 1706-01-0

Chem. Abstr. Name: Fluoranthene, 3-methyl-

IUPAC Systematic Name: 3-Methylfluoranthene

Synonym: 4-Methylfluoranthene (Clar, 1964)

1.2 Structural and molecular formulae and molecular weight

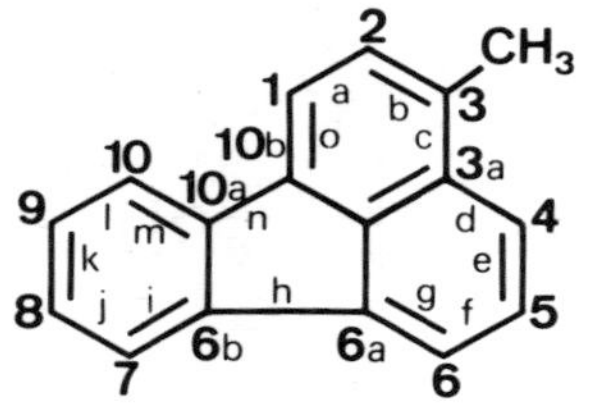

$C_{17}H_{12}$ Mol. wt: 216.3

1.3 Chemical and physical properties of the pure substance

(*a*) *Description*: Needles (recrystallized from methanol) (Stubbs & Tucker, 1950)

(*b*) *Melting-point*: 66°C (Clar, 1964), 65-66°C (Hoffmann *et al.*, 1972) (*c*) *Stability*: No data were available.

(*d*) *Reactivity*: No data were available.

2. Production, Use, Occurrence and Analysis

2.1 Production and use

There is no commercial production or known use of these compounds.

2.2 Occurrence and analysis

Data on occurrence and methods of analysis are summarized in the 'General Remarks on the Substances Considered', p. 35.

2- and 3-Methylfluoranthene have been reported sporadically to occur in products of incomplete combustion. They have been identified in mainstream cigarette smoke (Hoffmann *et al.*, 1972; Lee *et al.*, 1976); and smoke from marijuana cigarettes (Lee *et al.*, 1976).

3. Biological Data Relevant to the Evaluation of Carcinogenic Risk to Humans

3.1 Carcinogenicity studies in animals

2-Methylfluoranthene

Skin application

Mouse: A group of ten female Swiss albino Ha/ICR/Mil mice, seven to eight weeks old, in the second telogen stage of the hair cycle, received 50 μl of a 0.2% solution of 2-methylfluoranthene (purity checked by mass spectrometry) in acetone, applied thrice weekly on the shaved back for 12 months. The mice were observed for another three months. A total of 12 skin tumours (four carcinomas) were observed in six mice, four of which lived for more than one year. The number of pulmonary adenomas/mouse was increased in the experimental group (2.5 tumours/mouse *versus* 0.8 in controls) (Hoffmann *et al.*, 1972).

A group of 30 female Swiss albino Ha/ICR/Mil mice, seven to eight weeks of age, received 0.1 mg 2-methylfluoranthene (purity checked by mass spectroscopy) in 50 μl acetone by skin painting ten times every other day to the shaved backs. Ten days after the last application, promotion with 2.5% croton oil in acetone (average dose, 3.8 mg) was carried out for 20 weeks. Nine out of 30 animals surviving more than 20 weeks developed 14 skin papillomas. In a second, independent experiment with 30 animals under identical conditions, 29 animals lived to the end of treatment; ten animals developed 16 papillomas. Among 30 mice treated with 2.5% croton oil alone, one skin papilloma occurred (Hoffmann *et al.*, 1972).

3-Methylfluoranthene

Skin application

Mouse: A group of 30 Swiss albino Ha/ICR/Mil female mice, seven to eight weeks of age, received 0.1 mg 3-methylfluoranthene (purity checked by mass spectroscopy) in 50 μl acetone every other day ten times on the shaved backs. Ten days after the last application, promotion with 2.5% croton oil in acetone (average dose, 3.8 mg) was carried out for 20 weeks. In 9/28 animals surviving more than 20 weeks 14 skin papillomas developed. In 30 mice receiving 2.5% croton oil alone, one skin papilloma occurred (Hoffmann *et al.*, 1972).

3.2 Other relevant biological data[1]

(*a*) *Experimental systems*

No data were available to the Working Group on toxic effects or on effects on reproduction and prenatal toxicity.

Metabolism and activation

The 4,5-dihydrodiols of 2- and 3-methylfluoranthene, as well as their respective hydroxymethyl derivatives, have been detected as metabolites following incubation of these compounds with rat-liver preparations. All these metabolites were reported to be mutagenic to bacteria in the presence of an exogenous metabolic system (LaVoie *et al.*, 1982).

Mutagenicity and other short-term tests

Both 2- and 3-methylfluoranthenes were reported to induce mutations in *Salmonella typhimurium* strain TA100 (*his*$^-$/*his*$^+$) at, respectively, 20 μg/plate and 10 μg/plate in the presence of an exogenous metabolic system (postmitochondrial supernatant from Aroclor-induced rat liver) (LaVoie *et al.*, 1982).

[1] See also 'General Remarks on the Substances Considered', p. 53.

(*b*) *Humans*

No data were available to the Working Group.

3.3 Case reports and epidemiological studies of carcinogenicity to humans

No data were available to the Working Group.

4. Summary of Data Reported and Evaluation[1]

4.1 Experimental data

2-Methylfluoranthene was tested for carcinogenicity in one experiment in mice by skin application and produced benign and malignant skin tumours. It was also tested in the mouse-skin initiation-promotion assay in one experiment and was active as an initiator.

3-Methylfluoranthene was tested in only one experiment in the mouse-skin initiation-promotion assay, and benign skin tumours occurred.

No data on the teratogenicity of these compounds were available.

In the one available study, 2- and 3-methylfluoranthenes were mutagenic to *Salmonella typhimurium* in the presence of an exogenous metabolic system.

There is *inadequate evidence* that 2- and 3-methylfluoranthenes are active in short-term tests.

4.2 Human data[2]

2- and 3-Methylfluoranthenes are present as minor components of the total content of polynuclear aromatic compounds in tobacco smoke; however, they are also expected to occur in other combustion products contaminating the environment.

---

[1] For definitions of the italicized terms see Preamble, p. 19.

[2] Studies on occupational exposure to polynuclear aromatic compounds will be considered in future *IARC Monographs*.

4.3 Evaluation

There is *limited evidence* that 2-methylfluoranthene is carcinogenic to experimental animals.

The available data were inadequate to permit an evaluation of the carcinogenicity of 3-methylfluoranthene to experimental animals.

5. References

Campbell, N. & Wang, H. (1949) Syntheses in the fluoranthene series. *J. chem. Soc.*, 1513-1515

Clar, E., ed. (1964) *Polycyclic Hydrocarbons*, Vol. 2, New York, Academic Press, p. 301

Hoffmann, D., Rathkamp, G., Nesnow, S. & Wynder, E.L. (1972) Fluoranthenes: Quantitative determination in cigarette smoke, formation by pyrolysis, and tumor-initiating activity. *J. natl Cancer Inst.*, *49*, 1165-1175

LaVoie, E.J., Hecht, S.S., Bedenko, V. & Hoffmann, D. (1982) Identification of the mutagenic metabolites of fluoranthene, 2-methylfluoranthene and 3-methylfluoranthene. *Carcinogenesis*, *3*, 841-846

Lee, M.L., Novotny, M. & Bartle, K.D. (1976) Gas chromatography/mass spectrometric and nuclear magnetic resonance spectrometric studies of carcinogenic polynuclear aromatic hydrocarbons in tobacco and marijuana smoke condensates. *Anal. Chem.*, *48*, 405-416

Stubbs, H.W.D. & Tucker, S.H. (1950) Synthesis of fluoranthenes. VI. Utilization of the Mannich reaction. *J. chem. Soc.*, 3288-3292

Tucker, S.H. (1952) Synthesis of fluoranthenes, Part IX. 3-Methylfluoranthene. *J. chem. Soc.*, 803-807

1-METHYLPHENANTHRENE

1. Chemical and Physical Data

1.1 Synonyms and trade names

Chem. Abstr. Services Reg. No.: 832-69-9

Chem. Abstr. Name: Phenanthrene, 1-methyl-

IUPAC Systematic Name: 1-Methylphenanthrene

1.2 Structural and molecular formulae and molecular weight

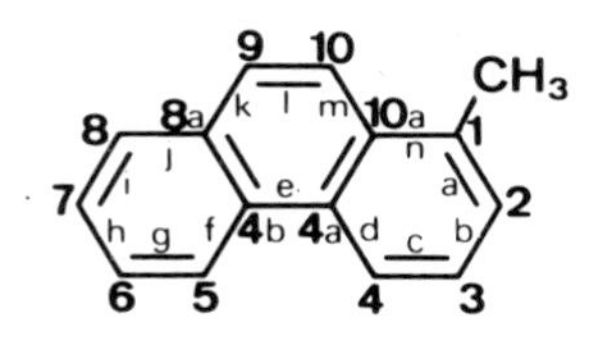

$C_{15}H_{12}$

Mol. wt: 192.3

1.3 Chemical and physical properties of the pure substance

From Weast (1975), unless otherwise specified

(*a*) *Description*: Leaves or plates (recrystallized from aqueous ethanol)

(*b*) *Melting-point*: 123°C

(*c*) *Spectroscopy data*: Infrared and nuclear magnetic resonance spectra have been reported (Brown & Raju, 1966). Mass spectra have been tabulated (NIH/EPA Chemical Information System, 1982).

(*d*) *Solubility*: Virtually insoluble in water (269 μg/l) (May *et al.*, 1978); soluble in ethanol

(*e*) *Stability*: No data were available.

(*f*) *Reactivity*: No data were available.

2. Production, Use, Occurrence and Analysis

2.1 Production and use

There is no commercial production or known use of this compound.

2.2 Occurrence and analysis

Data on occurrence and methods of analysis are summarized in the 'General Remarks on the Substances Considered', p. 35.

1-Methylphenanthrene occurs ubiquitously in products of incomplete combustion; it also occurs in fossil fuels. It has been identified in mainstream cigarette smoke (3.2 μg/100 cigarettes) and marijuana smoke (4.2 μg/100 cigarettes) (Lee *et al.*, 1976); gasoline engine exhaust (256-404 μg/l burned fuel) (Grimmer *et al.*, 1977); and tap water (0.37 μg/l) (Olufsen, 1980).

3. Biological Data Relevant to the Evaluation of Carcinogenic Risk to Humans

3.1 Carcinogenicity studies in animals

Skin application

Mouse: A group of 20 female Swiss albino Ha/ICR mice, seven to eight weeks of age, received applications of a 0.1% solution of 1-methylphenanthrene (purity, >99.5%) in 100 μl acetone, 10 times on alternative days to the shaved back skin (total dose, 1 mg). Ten days after the last dose, thrice-weekly applications of 2.5 μg 12-*O*-tetradecanoylphorbol-13-acetate (TPA) in 100 μl acetone were given to the shaved back for 20 weeks, at which time the study was terminated. A positive control group of 20 mice received a total dose of 0.3 mg benzo[*a*]pyrene [dose schedule unspecified] (the group was reduced to 15 mice after the fifth week of promotion); a vehicle-control group of 20 mice received acetone alone. At the end of the study period, no tumour was seen in the 1-methylphenanthrene-treated or vehicle-control groups. In contrast, 93% of the benzo[*a*]pyrene-treated animals had tumours (average number of tumours/animal, 3.7) (LaVoie *et al.*, 1981).

3.2 Other relevant biological data[1]

(a) Experimental systems

No data were available to the Working Group on toxic effects or on effects on reproduction and prenatal toxicity.

Metabolism and activation

1-Hydroxymethylphenanthrene, the 3,4- and 5,6-dihydrodiols of 1-hydroxymethyl-phenanthrene and 1-methylphenanthrene and certain unidentified hydroxylated derivatives were detected as metabolites following incubation of 1-methylphenanthrene with rat-liver preparations. When the dihydrodiols were assayed in the presence of a metabolic activation system, they were found to have mutagenic activity (LaVoie *et al.*, 1981).

Mutagenicity and other short-term tests

Results from short-term tests are given in Table 1.

Table 1. Results from short-term tests: 1-Methylphenanthrene

Test	Organism/assay[a]	Exogenous metabolic system[a]	Reported result	Comments	References
PROKARYOTES					
Mutation	*Salmonella typhimurium* (his^-/his^+)	Aro-R-PMS	Positive	At 25 μg/plate in strain TA100	LaVoie *et al.* (1981)
	Salmonella typhimurium ($8AG^S/8AG^R$)	Aro-R-PMS	Positive	At 80 nmol/ml in strain TM677	Kaden *et al.* (1979)
MAMMALIAN CELLS *IN VITRO*					
DNA damage	Primary rat hepatocytes (unscheduled DNA synthesis)	--	Positive	At 50 nmol/ml	Tong *et al.* (1981)
Mutation	Human lymphoblastoid TK6 cells (TFT^S/TFT^R)	Aro-R-PMS	Positive	At 3 nmol/ml	Barfknecht *et al.* (1981)

[a] For an explanation of the abbreviations, see the Appendix, p. 452.

(b) Humans

No data were available to the Working Group.

[1] See also 'General Remarks on the Substances Considered', p. 53.

3.3 Case reports and epidemiological studies of carcinogenicity to humans

No data were available to the Working Group.

4. Summary of Data Reported and Evaluation[1]

4.1 Experimental data

1-Methylphenanthrene was tested in the mouse-skin initiation-promotion assay in one study and was inactive as an initiator.

No data on the teratogenicity of this chemical were available.

1-Methylphenanthrene was mutagenic to *Salmonella typhimurium* in the presence of an exogenous metabolic system. It induced unscheduled DNA synthesis in cultured primary hepatocytes from rats and was mutagenic to human lymphoblasts *in vitro* in the presence of an exogenous metabolic system.

There is *sufficient* evidence that 1-methylphenanthrene is active in short-term tests.

4.2 Human data[2]

1-Methylphenanthrene is present as a major component of the total content of polynuclear aromatic compounds in the environment. Human exposure to 1-methylphenanthrene occurs primarily through the smoking of tobacco, inhalation of polluted air and by ingestion of food and water contaminated by combustion effluents (for details, see 'General Remarks on the Substances Considered', p. 35).

4.3 Evaluation

The available data are inadequate to permit an evaluation of the carcinogenicity of 1-methylphenanthrene to experimental animals.

---

[1] For definitions of the italicized terms, see Preamble, p. 19.

[2] Studies on occupational exposure to polynuclear aromatic compounds will be considered in future *IARC Monographs*.

5. References

Barfknecht, T.R., Andon, B.M., Thilly, W.G. & Hites R.A. (1981) *Soot and mutation in bacteria and human cells*. In: Cooke, M. & Dennis, A.J., eds, *Chemical Analysis and Biological Fate: Polynuclear Aromatic Hydrocarbons, 5th Int. Symposium*, Columbus, OH, Battelle Press, pp. 231-242

Brown, D.A. & Raju, J.R. (1966) Infrared and proton magnetic resonance spectra of II-complexes of substituted condensed hydrocarbons. *J. chem. Soc.*, 1617-1620

Grimmer, G., Böhnke, H. & Glaser, A. (1977) Investigation on the carcinogenic burden by air pollution in man. XV. Polycyclic aromatic hydrocarbons in automobile exhaust gas - An inventory. *Zbl. Bakt. Hyg., 1 Abt., Orig. B164*, 218-234

Kaden, D.A., Hites, R.A. & Thilly, W.G. (1979) Mutagenicity of soot and associated polycyclic aromatic hydrocarbons to *Salmonella typhimurium*. *Cancer Res.*, *39*, 4152-4159

LaVoie, E.J., Tulley-Freiler, L., Bedenko, V. & Hoffmann, D. (1981) Mutagenicity, tumor-initiating activity, and metabolism of methylphenanthrenes. *Cancer Res.*, *41*, 3441-3447

LaVoie, E.J., Tulley-Freiler, L., Bedenko, V., Girach, Z. & Hoffmann, D. (1981) *Comparative studies on the tumor initiating activity and metabolism of methylfluorenes and methylbenzofluorenes*. In: Cooke, M. & Dennis, A.J., eds, *Polynuclear Aromatic Hydrocarbons, Chemical Analysis and Biological Fate, 5th Int. Symposium*, Columbus, OH, Battelle Press, pp. 417-427

Lee, M.L., Novotny, M. & Bartle, K.D. (1976) Gas chromatography/mass spectrometry and nuclear magnetic resonance spectrometric studies of carcinogenic polynuclear aromatic hydrocarbons in tobacco and marijuana smoke condensates. *Anal. Chem.*, *48*, 405-416

May, W.E., Wasik, S.P. & Freeman, D.H. (1978) Determination of the solubility behavior of polycyclic aromatic hydrocarbons in water. *Anal. Chem.*, *50*, 997-1000

NIH/EPA Chemical Information System (1982) *Mass Spectral Search System*, Washington DC, CIS Project, Information Services Corporation

Olufsen, B. (1980) *Polynuclear aromatic hydrocarbons in Norwegian drinking water resources*. In: Bjørseth, A. & Dennis, A.J., eds, *Polynuclear Aromatic Hydrocarbons: Chemistry and Biological Effects*, *4th Int. Symposium*, Columbus, OH, Battelle Press, pp. 333-343

Tong, C., Laspia, M.F., Telang, S. & Williams, G.M. (1981) The use of adult rat liver cultures in the detection of the genotoxicity of various polycyclic aromatic hydrocarbons. *Environ. Mutagenesis*, *3*, 477-487

Weast, R.C., ed. (1975) *CRC Handbook of Chemistry and Physics*, 56th ed., Cleveland, OH, Chemical Rubber Co., p. C-419

PERYLENE

1. Chemical and Physical Data

1.1 Synonyms and trade names

Chem. Abstr. Services Reg. No.: 198-55-0

Chem. Abstr. Name: Perylene

IUPAC Systematic Name: Perylene

Synonyms: Dibenz(de,kl)anthracene; peri-dinaphthalene; perilene

1.2 Structural and molecular formulae and molecular weight

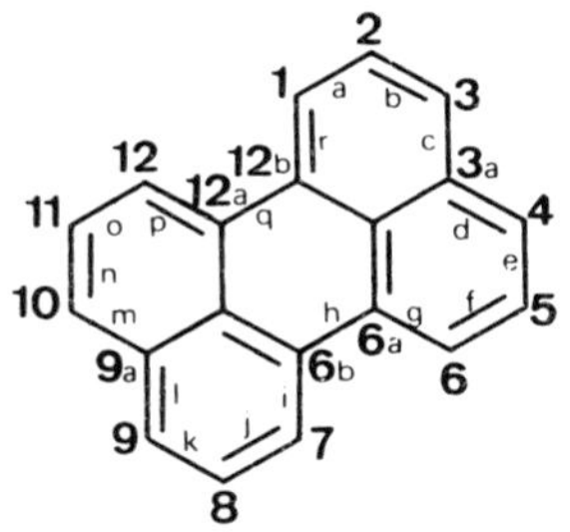

$C_{20}H_{12}$ Mol. wt: 252.3

1.3 Chemical and physical properties of the pure substance

From Windholz (1976), unless otherwise specified

(*a*) *Description*: Yellow to colourless crystals (recrystallized from toluene)

(*b*) *Melting point*: 273-274°C

(*c*) *Spectroscopy data*: λ_{max} 245, 251, 368, 387, 406, 434 nm (in methanol/ethanol) (Clar, 1964; Weast, 1975). Mass spectra have been tabulated (NIH/EPA Chemical Information System, 1982).

(*d*) *Solubility*: Virtually insoluble in water ($<$0.5 μg/l) (Davis *et al.*, 1942); very sparingly soluble in petroleum ether (a mixture of low-boiling hydrocarbons); slightly soluble in acetone, diethyl ether and ethanol; moderately soluble in benzene; freely soluble in carbon disulphide and chloroform

(*e*) *Stability*: Photo-oxidizable in the presence of acids (Clar, 1964)

(*f*) *Reactivity*: Forms deep-brown molecular compounds with halogens. Can be hydrogenated to tetradecahydroperylene. Disubstitution takes place at positions 3,9- and 3,10-, with the exception of the reaction with maleic anhydride. Perylene-ketones can be obtained by Friedel-Crafts reactions (Clar, 1964). Reacts with NO_2 to form nitro derivatives (Pitts *et al.*, 1978)

2. Production, Use, Occurrence and Analysis

2.1 Production and use

There is no commercial production or known use of this compound.

2.2 Occurrence and analysis

Data on occurrence and methods of analysis are summarized in the 'General Remarks on the Substances Considered', p. 35.

Perylene occurs ubiquitously in products of incomplete combustion; it also occurs in fossil fuels. It has also been found in peat (Aizenshtat, 1973); mainstream cigarette smoke (0.9 μg/cigarette) and sidestream cigarette smoke (3.9 μg/cigarette) (Grimmer *et al.*, 1977b); smoke-filled rooms (11 ng/m^3) (Grimmer *et al.*, 1977c); mainstream smoke from marijuana cigarettes (0.9 μg/100 cigarettes) (Lee *et al.*, 1976); urban air (0.7 ng/m^3) (Hoffmann & Wynder, 1976); gasoline engine exhausts (7-14 μg/l fuel) (Grimmer *et al.*, 1977a); emissions of burnt coal (0.09-1.9 mg/kg) (Brockhaus & Tomingas, 1976); lubricating oils (0.007-0.224 mg/kg) (Grimmer *et al.*, 1981a); various fresh and used motor oils (0.04-10.0 mg/kg) (Grimmer *et al.*, 1981b); gasolines (0.01-0.16 mg/kg) (Müller & Meyer, 1974); charcoal-broiled steaks (2 μg/kg) (Lijinsky & Shubik, 1964); surface water (0-520 ng/l) (Acheson *et al.*, 1976; Woidich *et al.*, 1976); tap water (0.1-1.4 ng/l) (Melchiorri *et al.*, 1973); rain water (0.0-1.0 ng/l) (Woidich *et al.*, 1976); subterranean water (0-0.2 ng/l) (Woidich *et al.*, 1976); waste water (0.03-3.0 μg/l) (Borneff & Kunte, 1967); freeze-dried sewage sludge (140-6400 μg/kg) (Grimmer *et al.*, 1980); and dried sediment from lakes (4-680 μg/kg) (Grimmer & Böhnke, 1975, 1977).

Analysis of sediments of a sewage sedimentation lagoon revealed the presence of 300 mg/kg perylene (Rose & Harshbarger, 1977). [One third of tiger salamanders inhabiting the lagoon had neoplastic skin lesions.]

3. Biological Data Relevant to the Evaluation of Carcinogenic Risk to Humans

3.1 Carcinogenicity studies in animals[1]

Skin application

Mouse: A group of 20 female ICR/Ha Swiss mice, six to eight weeks of age, received a single dose of 800 μg perylene (purified by recrystallization) in 200 μl benzene applied on the clipped dorsal skin. No skin tumour was seen after 58-60 weeks of observation. In a second experiment, the same dose of perylene was administered to 20 female ICR/Ha mice; two weeks later the animals received thrice-weekly applications of 2.5 μg 12-*O*-tetradecanoylphorbol-13-acetate (TPA) in 0.1 ml acetone for the duration of the experiment (58-60 weeks). A total of three skin papillomas were observed in three animals; in the TPA control group, one mouse developed two papillomas (Van Duuren *et al.*, 1970).

A group of 20 male C3H mice, two months of age, received 60 μl of a 0.15% solution of perylene (purified by recrystallization) in decalin applied twice weekly to the interscapular area for 82 weeks or until the mouse died or developed a tumour. At 52 weeks, 16 mice were still alive, and no skin tumour developed. In a second experiment, 20 male mice were treated with 0.15% perylene dissolved in a vehicle obtained by addition to decalin:*n*-dodecane (1:1 v/v); one papilloma, which regressed, was observed in 15 animals surviving more than 52 weeks. In 13 animals alive at 52 weeks and treated with the vehicle only, two papillomas occurred (Horton & Christian, 1974). [The Working Group noted that *n*-dodecane has been shown to have a cocarcinogenic effect when administered simultaneously with several polynuclear hydrocarbons (Bingham & Falk, 1969).]

3.2 Other relevant biological data[2]

(*a*) *Experimental systems*

Toxic effects

The growth rate of mouse ascites sarcoma cells in culture was not inhibited when perylene was added at a concentration of 1 μmol/ml in dimethyl sulphoxide (Pilotti *et al.*, 1975).

---

[1] The Working Group was aware of a study in progress in rats by oral administration as well as one in mice by skin application (IARC, 1983).

[2] See also 'General Remarks on the Substances Considered', p. 53.

Effects on reproduction and prenatal toxicity

Benzo[*a*]pyrene hydroxylase activity of the rat placenta can be induced by perylene (Welch *et al.*, 1969).

No data were available to the Working Group on metabolism and activation.

Mutagenicity and other short-term tests

Results from short-term tests are given in Table 1.

Table 1. Results from short-term tests: Perylene

Test	Organism/assay[a]	Exogenous metabolic system[a]	Reported	Comments	References
PROKARYOTES					
Mutation	*Salmonella typhimurium* (his⁻/his⁺)	Aro-R-PMS	Positive	At 10 μg/plate in strain TA100	LaVoie *et al.* (1979)
				At 5 μg/plate in strain TA98	Ho *et al.* (1981)
	Salmonella typhimurium ($8AG^S/8AG^R$)	Aro-R-PMS	Positive	At 1 nmol/ml in strain TM677	Kaden *et al.* (1979)
MAMMALIAN CELLS *IN VITRO*					
Mutation	Human lymphoblastoid TK6 cells (TFT^S/TFT^R; $6TG^S/6TG^R$)	Aro-R-PMS	Negative	Tested at up to 22 nmol/ml	Penman *et al.* (1980)
Chromosome effects	Chinese hamster V79 cells: aberrations	None	Positive	Tested at up to 10 μg/ml [results did not appear to be statistically significant]	Popescu *et al.* (1977)
	sister chromatid exchange	None	Negative	Tested at up to 10 μg/ml	

[a] For an explanation of the abbreviations see the Appendix, p. 452.

(*b*) *Humans*

No data were available to the Working Group.

3.3 Case reports and epidemiological studies of carcinogenicity to humans

No data were available to the Working Group.

4. Summary of Data Reported and Evaluation[1]

4.1 Experimental data

Perylene was tested for carcinogicity by skin application to mice in one experiment, and no carcinogenic effect was observed. It was also tested in the mouse-skin initiation-promotion assay in one study, with negative results.

No data on the teratogenicity of this compound were available.

Perylene was mutagenic to *Salmonella typhimurium* in the presence of an exogenous metabolic system. In one study, it did not induce mutations in cultured human lymphoblastoid cells. The data from one study on chromosomal effects were inadequate to make an evaluation.

There is *inadequate evidence* that perylene is active in short-term tests.

4.2 Human data[2]

Perylene is present as a minor component of the total content of polynuclear aromatic compounds of the environment. Human exposure to perylene occurs primarily through the smoking of tobacco, inhalation of polluted air and by ingestion of food and water contaminated with combustion products (for details, see 'General Remarks on the Substances Considered', p. 35).

4.3 Evaluation

The available data are inadequate to permit an evaluation of the carcinogenicity of perylene in experimental animals.

[1] For definitions of the italicized terms, see Preamble, p. 19.

[2] Studies on occupational exposure to polynuclear aromatic compounds will be considered in future *IARC Monographs*.

5. References

Acheson, M.A., Harrison, R.M., Perry, R. & Wellings, R.A. (1976) Factors affecting the extraction and analysis of polynuclear aromatic hydrocarbons in water. *Water Res.*, *10*, 207-212

Aizenshtat, Z. (1973) Perylene and its geochemical significance. *Geochim. cosmochim. Acta*, *37*, 559-567 [*Chem. Abstr.*, *78*, 127153t]

Bingham, E. & Falk, H.L. (1969) Environmental carcinogens. The modifying effect of cocarcinogens on the threshold response. *Arch. environ. Health*, *19*, 779-783

Borneff, J. & Kunte, H. (1967) Carcinogenic substances in water and soils. XIX. Action of sewage purification on polycyclic aromatics (Ger.). *Arch. Hyg. Bakt.*, *101*, 202-210

Brockhaus, A. & Tomingas, R. (1976) Emission of polycyclic hydrocarbons during burning processes in small heating installations and their concentration in the atmosphere (Ger.). *Staub-Reinhalt. Luft*, *36*, 96-101

Clar, E., ed. (1964) *Polycyclic Hydrocarbons*, Vol. 2, New York, Academic Press, pp. 24-36

Davis, W.W., Krahl, M.E. & Clowes, G.H.A. (1942) Solubility of carcinogenic and related hydrocarbons in water. *J. Am. chem. Soc.*, *64*, 108-110

Grimmer, G. & Böhnke, H. (1975) Profile analysis of polycyclic aromatic hydrocarbons and metal content in sediment layers of a lake. *Cancer Lett.*, *1*, 75-84

Grimmer, G. & Böhnke, H. (1977) Investigation on drilling cores of sediments of Lake Constance. I. Profiles of the polycyclic aromatic hydrocarbons (Ger.). *Z. Naturforsch.*, *32c*, 703-711

Grimmer, G., Böhnke, H. & Glaser, A. (1977a) Investigation on the carcinogenic burden of air pollution in man. XV. Polycyclic aromatic hydrocarbons in automobile exhaust gas - An inventory. *Zbl. Bakt. Hyg., 1 Abt., Orig. B164*, 218-234

Grimmer, G., Böhnke, H. & Harke, H.-P. (1977b) Passive smoking: Intake of polycyclic aromatic hydrocarbons by breathing of cigarette smoke containing air (Ger.). *Int. Arch. occup. environ. Health*, *40*, 93-99

Grimmer, G., Böhnke, H. & Harke, H.-P. (1977c) Passive smoking: Measuring of concentrations of polycyclic aromatic hydrocarbons in rooms after machine smoking of cigarettes (Ger.). *Int. Arch. occup. environ. Health*, *40*, 83-92

Grimmer, G., Hilge, G. & Niemitz, W. (1980) Comparison of the profile of polycyclic aromatic hydrocarbons in samples of sewage sludge from 25 sewage treatment works (Ger.). *Wasser*, *54*, 255-272

Grimmer, G. Jacob, J. & Naujack, K.-W. (1981a) Profile of the polycyclic aromatic hydrocarbons from lubricating oils. Inventory by GCGC/MS - PAH in environmental materials, part 1. *Fresenius Z. anal. Chem.*, *306*, 347-355

Grimmer, G., Jacob, J., Naujack, K.-W. & Dettbarn, G. (1981b) Profile of the polycyclic aromatic hydrocarbons from used engine oil. Inventory by GCGC/MS - PAH in environmental materials, part 2. *Fresenius Z. anal. Chem.*, *309*, 13-19

Ho, C.H., Clark, B.R., Guerin, M.R., Barkenbus, B.D., Rao, T.K. & Epler, J.L. (1981) Analytical and biological analyses of test materials from the synthetic fuel technologies. IV. Studies of chemical structure-mutagenic activity relationships of aromatic nitrogen compounds relevant to synfuels. *Mutat. Res.*, *85*, 335-345

Hoffmann, D. & Wynder, E.L. (1976) *Environmental respiratory carcinogenesis*. In: Searle, C.E., ed., *Chemical Carcinogenesis* (*ACS Monograph No. 173*), Washington DC, American Chemical Society, pp. 324-365

Horton, A.W. & Christian, G.M. (1974) Cocarcinogenic versus incomplete carcinogenic activity among aromatic hydrocarbons: Contrast between chrysene and benzo[*b*]triphenylene. *J. natl Cancer Inst.*, *53*, 1017-1020

IARC (1983) *Information Bulletin on the Survey of Chemicals Being Tested for Carcinogenicity*, No. 10, Lyon, pp. 22, 127

Kaden, D.A., Hites, R.A. & Thilly, W.G. (1979) Mutagenicity of soot and associated polycyclic aromatic hydrocarbons to *Salmonella typhimurium*. *Cancer Res.*, *39*, 4152-4159

LaVoie, E., Bedenko, V., Hirota, N., Hecht, S.S. & Hoffmann, D. (1979) *A comparison of the mutagenicity, tumor-initiating activity and complete carcinogenicity of polynuclear aromatic hydrocarbons*. In: Jones, P.W. & Leber, P., eds, *Polynuclear Aromatic Hydrocarbons*, Ann Arbor, MI, Ann Arbor Science Publishers, pp. 705-721

Lee, M.L., Novotny, M. & Bartle, K.D. (1976) Gas chromatography/mass spectrometric and nuclear magnetic resonance spectrometric studies of carcinogenic polynuclear aromatic hydrocarbons in tobacco and marijuana smoke condensates. *Anal. Chem.*, *48*, 405-416

Lijinsky, W. & Shubik, P. (1964) Benzo(*a*)pyrene and other polynuclear hydrocarbons in charcoal-broiled meat. *Science*, *145*, 53-55

Melchiorri, C., Chiacchiarini, L., Grella, A. & D'Arca, S.U. (1973) Research and determination of polycyclic aromatic hydrocarbons in some tap water in the city of Rome (Ital.). *Nuov. Ann. Ig. Microbiol.*, *24*, 279-301

Müller, K. & Meyer, J.P. (1974) *Einfluss von Ottokraftstoffen auf die Emission von polynuklearen aromatischen Kohlenwasserstoffen in Automobilabgasen im Europa-Test* (Effect of gasoline components on emission of polynuclear aromatic hydrocarbons in car exhaust in the Europa test) (*Forschungsbericht 4568*), Hamburg, Deutsche Gesellschaft für Mineralölwissenschaft und Kohlechemie e.V.

NIH/EPA Chemical Information System (1982) *Mass Spectral Search System*, Washington DC, CIS Project, Information Services Corporation

Penman, B.W., Kaden, D.A., Liber, H.L., Skopek, T.R. & Thilly, W.G. (1980) Perylene is a more potent mutagen than benzo[α]pyrene for *S. typhimurium*. *Mutat. Res.*, *77*, 271-277

Pilotti, A., Ancker, K., Arrhenius, E. & Enzell, C. (1975) Effects of tobacco and tobacco smoke constituents on cell multiplication *in vitro*. *Toxicology*, *5*, 49-62

Pitts, J.N., Jr, Van Cauwenberghe, K.W., Grosjean, D., Schmid, J.P., Fitz, D.R., Belser, W.L., Jr, Knudson, G.B. & Hynds, P.M. (1978) Atmospheric reactions of polycyclic aromatic hydrocarbons: Facile formation of mutagenic nitro derivatives. *Science*, *202*, 515-519

Popescu, N.C., Turnbull, D. & DiPaolo, J.A. (1977) Sister chromatid exchange and chromosome aberration analysis with the use of several carcinogens and noncarcinogens: Brief communication. *J. natl Cancer Inst.*, *59*, 289-293

Rose, F.L. & Harshbarger, J.C. (1977) Neoplastic and possibly related skin lesions in neotenic tiger salamanders from a sewage lagoon. *Science*, *196*, 315-317

Van Duuren, B.L., Sivak, A., Goldschmidt, B.M., Katz, C. & Melchionne, S. (1970) Initiating activity of aromatic hydrocarbons in two-stage carcinogenesis. *J. natl Cancer Inst.*, *44*, 1167-1173

Weast, R.C., ed. (1975) *CRC Handbook of Chemistry and Physics*, 56th ed., Cleveland, OH, Chemical Rubber Co., p. C-418

Welch, R.M., Harrison, Y.E., Gommi, B.W., Poppers, P.J., Finster, M. & Conney, A.H. (1969) Stimulatory effect of cigarette smoking on the hydroxylation of 3,4-benzpyrene and the *N*-demethylation of 3-methyl-4-monomethylaminoazobenzene by enzymes in human placenta. *Clin. Pharmacol. Ther.*, *10*, 100-109

Windholz, M., ed. (1976) *The Merck Index*, 9th ed., Rahway, NJ, Merck & Co., pp. 930, 931

Woidich, W., Pfannhauser, W., Blaicher, G. & Tiefenbacher, K. (1976) Analysis of polycyclic aromatic hydrocarbons in drinking and industrial water (Ger.). *Lebensmittelchem. gerichtl. Chem.*, *30*, 141-146

PHENANTHRENE

1. Chemical and Physical Data

1.1 Synonyms and trade names

Chem. Abstr. Services Reg. No.: 85-01-8

Chem. Abstr. Name: Phenanthrene

IUPAC Systematic Name: Phenanthrene

Synonyms: Phenanthren; phenantrin

1.2 Structural and molecular formulae and molecular weight

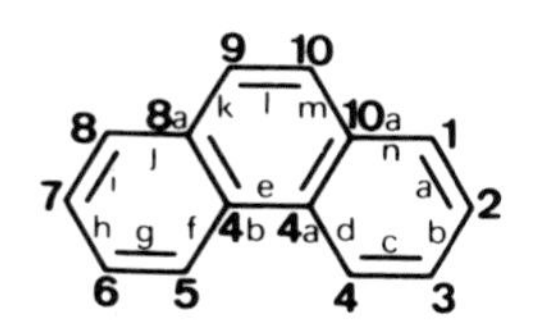

$C_{14}H_{10}$

Mol. wt: 178.2

1.3 Chemical and physical properties of the pure substance

From Windholz (1976), unless otherwise specified

(*a*) *Description:* Monoclinic plates (recrystallized from ethanol)

(*b*) *Boiling-point*: 340°C

(*c*) *Melting-point*: 100°C

(*d*) *Refractive index*: 1.59427

(*e*) *Spectroscopy data*: λ_{max} 210, 219, 242, 251, 273.5, 281, 292.5, 308.5, 314, 322.5, 329.5, 337, 345 nm (in methanol-ethanol) (Clar, 1964). Mass spectra and infrared spectra have been tabulated (NIH/EPA Chemical Information System, 1982).

(*f*) *Solubility*: Practically insoluble in water (1-1.6 mg/l) (Davis *et al.*, 1942; May *et al.*, 1978); soluble in glacial acetic acid; 1 g dissolves in the following volumes of organic solvents: 2 ml benzene, 1 ml carbon disulphide, 2.4 ml carbon tetrachloride, 3.3 ml anhydrous diethyl ether, 60 ml cold ethanol, 25 ml absolute ethanol, 10 ml boiling 95% ethanol and 2.4 ml toluene

(*g*) *Stability*: Does not undergo photo-oxidation in organic solvents under fluorescent light or indoor sunlight (Kuratsune & Hirohata, 1962); resistant to photodecomposition (Korfmacher *et al.*, 1980)

(*h*) *Reactivity*: Can be hydrogenated, nitrated, sulphonated and halogenated at various positions. Forms an 9,10-ozonide from which a 2,2′-diformyl diphenyl can be obtained. Lithium, sodium, and potassium add at the 9,10- position (Clar, 1964); reacts with NO and NO_2 to form nitro derivatives (Butler & Crossley, 1981; Tokiwa *et al*; 1981)

2. Production, Use, Occurrence and Analysis

2.1 Production and use

There is no commercial production or known use of this compound. Its derivative, cyclopentenophenanthrene, has been used as a starting material for synthesizing bile acids, cholesterol and other steroids (Clar, 1964).

2.2 Occurrence and analysis

Data on occurrence and methods of analysis are summarized in the 'General Remarks on the Substances Considered', p. 35.

Phenanthrene is present in products of incomplete combustion; it also occurs in fossil fuels. It is found in relatively high concentrations in coal-tar (Windholz, 1976). It has been identified in sidestream cigarette smoke (6.0-8.5 μg/100 cigarettes) (Lee *et al.*, 1976); mainstream smoke of marijuana cigarettes (8.9 μg/100 cigarettes) (Lee *et al.*, 1976); gasoline engine exhausts (2356-2930 μg/l burnt fuel) (Grimmer *et al.*, 1977); lubricating oils (> 7.09 mg/kg) (Grimmer *et al.*, 1981a); used motor oil (157.85 mg/kg) (Grimmer *et al.*, 1981b); crude oils (> 128.7 mg/kg) (Grimmer *et al.*, 1983); charcoal-broiled steaks (11 μg/kg) (Lijinsky & Shubik, 1964); surface water (0-1300 ng/l) (Commission of the European Communities, 1979); tap water (3.1-90 ng/l) (Thruston, 1978; Olufsen, 1980); waste water (70 μg/l) (Jungclaus *et al.*, 1976); and dried sediment (140-2740 μg/kg) (Grimmer & Böhnke, 1975).

3. Biological Data Relevant to the Evaluation of Carcinogenic Risk to Humans

3.1 Carcinogenicity studies in animals

(*a*) *Oral administration*

Rat: In a comparative study of mammary tumour induction, single oral doses of 200 mg phenanthrene [purity unspecified] dissolved in sesame oil were administered to 10 female Sprague-Dawley rats, 50 days old. No mammary tumour was produced within 60 days. In a positive-control group administered 20 mg 7,12-dimethylbenz[*a*]anthracene under the same conditions, mammary tumours occurred in 100% of 700 animals (Huggins & Yang, 1962).

(*b*) *Skin application*

Mouse: A solution of phenanthrene in 90% benzene was tested on 100 mice for a period of nine months. No effect upon the skin was observed (Kennaway, 1924).

A group of 20 'S' strain mice [sex and age unspecified] received 10 applications of 0.3 ml of an 18% solution of phenanthrene [purity unspecified] in acetone thrice weekly on the back (total dose, 0.54 g). Starting 25 days after the last phenanthrene application, 18 weekly treatments with 0.17% croton oil in acetone (0.3 ml) were given. At the end of the croton oil treatments, 12 benign papillomas were observed on 5/20 survivors compared with four papillomas on 4/19 surviving animals treated with croton oil (Salaman & Roe, 1956).

In a study reported as an abstract, thrice-weekly applications of a 5% solution of phenanthrene [solvent not indicated] applied for one year on mouse skin [number, strain and sex not indicated] failed to induce tumours or to promote tumour formation after an 'initiating dose' [unspecified] of benzo[*a*]pyrene. In other experiments, 12 applications of a 5% solution of phenanthrene failed to increase or decrease the initiating effect of a single dose of benzo[*a*]pyrene applied either before or afterwards. In the latter experiment the phenanthrene treatment was followed by weekly applications of 0.1% croton oil in acetone (Roe & Grant, 1964).

A group of 10 male and 10 female stock albino mice received four applications of 300 μg phenanthrene (high purity) dissolved in 0.25 ml acetone on the shaved back on days 0, 2, 6 and 8. A weekly application of 0.25 ml of a 0.1% croton oil solution in acetone was started from day 21 for 20 weeks. The number of papillomas was registered one week after the last application of croton oil. The incidences of skin papillomas were: four mice with papillomas among 19 survivors, compared to two mice with papillomas among 20 survivors in the control experiment, in which phenanthrene was replaced by acetone (Roe, 1962).

A group of 30 female CD1 mice, eight weeks of age, received 10 μmol phenanthrene (purified by thin-layer chromatography) in benzene applied to the shaved back. Starting one week later, treatment with 12-*O*-tetradecanoylphorbol-13-acetate (TPA) (5 μmol/dose) was given twice weekly for 34 weeks. Papillomas developed in 40% of the animals (0.6 papilloma/mouse); no papilloma developed in the control group treated with 10 μmol TPA alone (Scribner, 1973). [The Working Group noted that the dose of TPA was probably 5 μg (Scribner & Süss, 1978).]

A group of 20 female Swiss Ha/ICR mice, seven to eight weeks of age, received applications of 100 μl of a 0.1% solution of phenanthrene in acetone ten times on alternate days. Treatment was followed 10 days later by thrice weekly applications of 2.5 μg TPA in 100 μl acetone for 20 weeks. No skin tumour was found (LaVoie *et al.*, 1981).

A group of 30 female Charles River CD-1 mice, eight weeks of age, received single skin applications of 10 μmol phenanthrene (purity, >98%) in 200 μl acetone. A group of 30 females received acetone alone (vehicle controls). Starting one week later, the mice received twice-weekly applications of 16 nmol TPA in 200 μl acetone for 35 weeks. Skin papillomas developed in 5/30 mice (0.28 skin tumour/mouse) compared with 2/30 skin papillomas (0.1 tumour/mouse) in 30 controls ($p > 0.05$). Similar results were found in a second experiment (Wood *et al.*, 1979).

(*c*) *Subcutaneous and/or intramuscular administration*

Mouse: A group of 40 C57Bl mice (of both sexes) was injected s.c. in the interscapular region with 5 mg phenanthrene [purity unspecified] in 0.5 ml tricaprylin; 27 mice survived four months, and the experiment was terminated at 28 months. No local tumour was observed (Steiner, 1955). A group of 10 male and 10 female stock albino mice received 300 μg phenanthrene in 3% aqueous gelatine administered s.c. on days, 0, 2, 4, 6 and 8; starting at day 21, 0.25 ml 0.1% croton oil in acetone was applied for 20 weeks. One week after the last application of croton oil, papillomas were found in 3/17 survivors at the end of croton oil treatment and 2/20 surviving among controls receiving acetone instead of phenanthrene (Roe, 1962).

(*d*) *Perinatal exposure*

Mouse: A dose of 40 μg phenanthrene of highest purity in 1% aqueous gelatine was injected s.c. into neonatal stock albino mice; 49 animals survived more than 50 weeks. The incidence of tumours (pulmonary adenomas, hepatomas and skin papillomas) at 62 weeks was not higher than that seen in the two solvent control groups. In two other experiments, in which phenanthrene was mixed with benzo[*a*]pyrene (20 μg benzo[*a*]pyrene + 20 μg phenanthrene or 40 μg benzo[*a*]pyrene + 40 μg phenanthrene), the tumour yield was not different from that with treatment with benzo[*a*]pyrene alone (Grant & Roe, 1963).

Phenanthrene (purity, > 98%; homogeneous on reverse-phase high-pressure liquid chromatography) dissolved in anhydrous dimethyl sulphoxide, was injected i.p. into 100 newborn Blu-HA (ICR) Swiss-Webster mice, 0.2 μmol [35 μg] on the first day, 0.4 μmol [70 μg] on the eighth day and 0.8 μmol [140 μg] on the fifteenth day. The animals were killed when they were 38-42 weeks old. All tissues with suspected lesions were examined histologically. Of 35 animals still alive at 42 weeks, 17% displayed pulmonary adenomas (0.2/mouse) compared to 15% of 59 surviving animals (0.17 adenoma/mouse) in the dimethyl sulphoxide-control group (Buening *et al.*, 1979).

3.2 Other relevant biological data[1]

(*a*) *Experimental systems*

Toxic effects

The LD_{50} for the mouse (i.p) is 700 mg/kg bw (Simmon *et al.*, 1979).

The growth rate of mouse ascites sarcoma cells in culture was inhibited by 22% when phenanthrene was added at a concentration of 1 μmol/ml in dimethyl sulphoxide (Pilotti *et al.*, 1975).

Effects on reproduction and prenatal toxicity

No data were available to the Working Group.

Benzo[*a*]pyrene hydroxylase activity of the rat placenta can be induced by phenanthrene (Welch *et al.*, 1969).

Metabolism and activation

The 1,2-, 3,4- and 9,10-dihydrodiols are excreted either free or conjugated with glucuronic acid in the urine of rats and rabbits following i.p. administration of phenanthrene (Boyland & Wolf, 1950; Boyland & Sims, 1962). These metabolites have also been detected *in vitro* following incubation of phenanthrene with liver preparations from guinea-pigs, rats and mice (Sims, 1970; Chaturapit & Holder, 1978; Nordqvist *et al.*, 1981). Further oxidative metabolism of the 1,2-dihydrodiol by rat-liver preparations to the 1,2-diol-3,4-epoxide has also been reported (Norqvist *et al.*, 1981; Vyas *et al.*, 1982).

The 1,2-, 3,4- and 9,10-dihydrodiols of phenanthrene have little, if any tumour initiating activity on mouse skin (Wood *et al.*, 1979). The diastereoisomeric 1,2-diol-3,4-epoxides of phenanthrene did not induce pulmonary tumours in newborn mice (Buening *et al.*, 1979). These epoxides are mutagenic to bacteria and mammalian cells (Wood *et al.*,1979).

Mutagenicity and other short-term tests

Results from short-term tests are given in Table 1.

---

[1] See also 'General Remarks on the Substances Considered', p. 53.

Table 1. Results from short-term tests: Phenanthrene[a]

Test	Organism/assay[b]	Exogenous metabolic system[b]	Reported result	Comments	References
PROKARYOTES					
DNA damage	*Bacillus subtilis* (rec^+/rec^-)	Aro-R-PMS	Negative	Tested at 125 μg/well	McCarroll *et al*. (1981)
	Escherichia coli ($polA^+/polA^-$)	UI-R-PMS	Negative	Tested at up to 250 μg/ml [The Working Group noted that UI-PMS was used.]	Rosenkranz & Poirier (1979)
Mutation	*Salmonella typhimurium* (his^-/his^+)	Aro-R-PMS	Positive	At 12 μg/plate in strain TA100 [required a high concentration of PMS]	Oesch *et al*. (1981)
		Aro-R-Micr	Negative	Tested at up to 50 nmol/plate in strain TA100	Wood *et al*. (1979)
		Aro-R-PMS	Negative	Tested at up to 50 μg/plate in strains TA1535, TA1537, TA98 and TA100	McCann *et al*. (1975)
		Aro-R-PMS	Negative	Tested at up to 200 μg/plate in strain TA100	LaVoie *et al*. (1981)
	Salmonella typhimurium ($8AG^S/8AG^R$)	Aro-R-PMS, PB-R-PMS	Negative	Tested at up to 300 nmol/ml in strain TM677	Kaden *et al*. (1979)
FUNGI					
Mutation	*Saccharomyces cerevisiae* D3 (mitotic recombination)	Aro-R-PMS	Negative	Tested at up to 5% w/v [*sic*]	Simmon (1979)
MAMMALIAN CELLS *IN VITRO*					
DNA damage	Human foreskin epithelial cells (unscheduled DNA synthesis)	--	Negative	Tested at up to 400 μg/ml	Lake *et al*. (1978)
	Primary rat hepatocytes (unscheduled DNA synthesis)	--	Negative	Tested at up to 100 nmol/ml	Probst *et al*. (1981)
Mutation	Chinese hamster V79 cells ($8AG^S/8AG^R$; OUA^S/OUA^R)	SHE feeder layer	Negative	Tested at 1 μg/ml	Huberman & Sachs (1976)
	Human lymphoblastoid TK6 cells (TFT^S/TFT^R)	Aro-R-PMS	Positive	Tested at 50 nmol	Barfknecht *et al*. (1981)
Chromosome effects	Chinese hamster V79-4 cells (sister chromatid exchange, aberrations)	SHE feeder layer	Negative	Tested at up to 10 μg/ml	Popescu *et al*. (1977)
Cell transformation	Mouse prostate C3HG23 cells (morphological)	--	Negative	Tested at up to 10 μg/ml	Marquardt *et al*. (1972)
	Syrian hamster embryo cells (morphological)	--	Negative	Tested at up to 40 μg/ml	Pienta *et al*. (1977)
	Mouse BALB/3T3 cells (morphological)	--	Negative	Tested at up to 50 μg/ml	Kakunaga (1973)
	Guinea-pig foetal cells (morphological)	--	Negative	Tested at 5 μg/ml	Evans & DiPaolo (1975)

Test	Organism/assay[b]	Exogenous metabolic system[b]	Reported result	Comments	References
MAMMALIAN CELLS *IN VIVO*					
DNA damage	Chinese hamster bone-marrow cells (aberrations)	--	Negative	Treated i.p. with 100 mg/kg bw Treated i.p. with 2 x 450 mg/kg bw	Bayer (1978) Roszinsky-Kocher *et al.* (1979)
	Chinese hamster bone-marrow cells (sister chromatid exchange)	--	Positive	At 100 mg/kg bw i.p. At 2 x 450 mg/kg bw i.p.	Bayer (1978); Roszinsky-Kocher *et al.* (1979)
Chromosome effects	Chinese hamster bone-marrow cells (micronuclei)	--	Negative	Treated i.p. with 500 mg/kg bw	Bayer (1978)
Cell transformation	Hamster embryo cells (morphological)	--	Negative	Pregnant females treated i.p. with 30 mg/kg bw	Quarles *et al.* (1979)

[a] This table comprises selected assays and references and is not intended to be a complete review of the literature.
[b] For an explanation of the abbreviations see the Appendix, p. 452.

(*b*) *Humans*

No data were available to the Working Group.

3.3 Case reports and epidemiological studies of carcinogenicity to humans

No data were available to the Working Group.

4. Summary of Data Reported and Evaluation[1]

4.1 Experimental data

Phenanthrene was tested for carcinogenicity in two fragmentary studies in mice by skin painting and no skin tumour was reported. In the six studies in which phenanthrene was tested in the mouse-skin initiation-promotion assay, it was active as an initiator in one study, inactive as an initiator in four others, and inactive as a promoter in one study.

Phenanthrene administered by intraperitoneal or subcutaneous injection to neonatal mice did not increase the incidence of tumours over that in controls. Experiments involving a single oral administration to rats and a single subcutaneous injection to mice were inadequate for evaluation.

No data on the teratogenicity of this compound were available.

[1] For definitions of the italicized terms, see Preamble, p. 19.

Phenanthrene has generally been reported to be non-mutagenic to *Salmonella typhimurium*; however, in one study it was reported to be mutagenic to *Salmonella typhimurium* in the presence of a high concentration of an exogenous metabolic system. It gave negative results in an assay for differential survival using DNA-repair-proficient/-deficient strains of *Bacillus subtilis*. It did not induce DNA repair, chromosomal aberrations or sister chromatid exchange in cultured mammalian cells. It did induce mutation in one experiment in human cells in culture in the presence of an exogenous metabolic system, and induced sister chromatid exchange in Chinese hamster bone-marrow cells *in vivo*. The compound failed to induce morphological transformation.

There is *limited evidence* that phenanthrene is active in short-term tests.

4.2 Human data[1]

Phenanthrene occurs as a major component of the total content of polynuclear aromatic compounds in the environment. Human exposure to phenanthrene occurs primarily through the smoking of tobacco, inhalation of polluted air or by ingestion of food or water contaminated by combustion effluents (for details, see 'General Remarks on the Substances Considered', p. 35).

4.3 Evaluation

The available data are inadequate to permit an evaluation of the carcinogenicity of phenanthrene to experimental animals.

5. References

Barfknecht, T.R., Andon, B.M., Thilly, W.G. & Hites, R.A. (1981) *Soot and mutation in bacteria and human cells*. In: Cooke, M. & Dennis, A.J., eds, *Chemical Analysis and Biological Fate: Polynuclear Aromatic Hydrocarbons*, *5th Int. Symposium*, Columbus, OH, Battelle Press, pp. 231-242

Bayer, U. (1978) In vivo *induction of sister chromatid exchanges by three polyaromatic hydrocarbons*. In: Jones, P.W. & Freudenthal, R.I., eds, *Carcinogenesis*, Vol. 3, *Polynuclear Aromatic Hydrocarbons*, New York, Raven Press, pp. 423-428

Boyland, E. & Sims, P. (1962) Metabolism of polycyclic compounds. The metabolism of phenanthrene in rabbits and rats: Dihydro-dihydroxy compounds and related glucosiduronic acids. *Biochem. J.*, *84*, 571-582

[1] Studies on occupational exposure to polynuclear aromatic compounds will be considered in future *IARC Monographs*.

Boyland, E. & Wolf, G. (1950) Metabolism of polycyclic compounds. 6. Conversion of phenanthrene into dihydroxydihydrophenanthrenes. *Biochem. J.*, *47*, 64-69

Buening, M.K., Levin, W., Karle, J.M., Yagi, H., Jerina, D.M. & Conney, A.H. (1979) Tumorigenicity of bay-region epoxides and other derivatives of chrysene and phenanthrene in newborn mice. *Cancer Res.*, *39*, 5063-5068

Butler, J.D. & Crossley, P. (1981) Reactivity of polycyclic aromatic hydrocarbons adsorbed on soot particles. *Atmos. Environ.*, *15*, 91-94

Chaturapit, S. & Holder, G.M. (1978) Studies on the hepatic microsomal metabolism of [^{14}C]-phenanthrene. *Biochem. Pharmacol.*, *27*, 1865-1871

Clar, E., ed. (1964) *Polycyclic Hydrocarbons*, Vol. 1, New York, Academic Press, pp. 223-236

Commission of the European Communities (1979) *Concerted Action. Analysis of Organic Micropollutants in Water* (*COST 64 b bis*), 3rd ed., Vol. II, Luxemburg, p. 22

Davis, W.W., Krahl, M.E. & Clowes, G.H.A. (1942) Solubility of carcinogenic and related hydrocarbons in water. *J. Am. chem. Soc.*, *64*, 108-110

Evans, C.H. & DiPaolo, J.A. (1975) Neoplastic transformation of guinea pig fetal cells in culture induced by chemical carcinogens. *Cancer Res.*, 35, 1035-1044

Grant, G. & Roe, F.J.C. (1963) The effect of phenanthrene on tumour induction by 3,4-benzopyrene administered to newly born mice. *Br. J. Cancer*, *17*, 261-265

Grimmer, G. & Böhnke, H. (1975) Profile analysis of polycyclic aromatic hydrocarbons and metal content in sediment layers of a lake. *Cancer Lett.*, *1*, 75-84

Grimmer, G., Böhnke, H. & Glaser, A. (1977) Investigation on the carcinogenic burden by air pollution in man. XV. Polycyclic aromatic hydrocarbons in automobile exhaust gas - An inventory. *Zbl. Bakt. Hyg., 1 Abt., Orig. B164*, 218-234

Grimmer, G., Jacob, J. & Naujack, K.-W. (1981a) Profile of the polycyclic aromatic hydrocarbons from lubricating oils. Inventory by GCGC/MS - PAH in environmental materials, Part 1. *Fresenius Z. anal. Chem.*, *306*, 347-355

Grimmer, G., Jacob, J., Naujack, K.-W. & Dettbarn, G. (1981b) Profile of the polycyclic aromatic compounds from used engine oil - Inventory by GCGC/MS - PAH in environmental materials, Part 2. *Fresenius Z. anal. Chem.*, *309*, 13-19

Grimmer, G., Jacob, J. & Naujack, K.-W. (1983) Profile of the polycyclic aromatic compounds from crude oils. Inventory by GCGC/MS - PAH in environmental materials, Part 3. *Fresenius Z. anal. Chem.*, *314*, 29-36

Huberman, E. & Sachs, L. (1976) Mutability of different genetic loci in mammalian cells by metabolically activated carcinogenic polycyclic hydrocarbons. *Proc. natl Acad. Sci. USA*, *73*, 188-192

Huggins, C. & Yang, N.C. (1962) Induction and extinction of mammary cancer. *Science*, *137*, 257-262

Jungclaus, G.A., Games, L.M. & Hites, R.A. (1976) Identification of trace organic compounds in tire manufacturing plant waste waters. *Anal. Chem.*, *48*, 1894-1896

Kaden, D.A., Hites, R.A. & Thilly, W.G. (1979) Mutagenicity of soot and associated polycyclic aromatic hydrocarbons to *Salmonella typhimurium*. *Cancer Res.*, *39*, 4152-4159

Kakunaga, T. (1973) A quantitative system for assay of malignant transformation by chemical carcinogens using a clone derived from BALB/3T3. *Int. J. Cancer*, *12*, 463-473

Kennaway, E.L. (1924) On the cancer-producing factor in tar. *Br. med. J.*, *i*, 564-567

Korfimacher, W.A., Wehry, E.C., Mamantov, G. & Natusch, D.F.S. (1980) Resistance to photochemical decomposition of polycyclic aromatic hydrocarbons vapor-adsorbed on coal fly ash. *Environ. Sci. Technol.*, *14*, 1094-1098

Kuratsune, M. & Hirohata, T. (1962) Decomposition of polycyclic aromatic hydrocarbons under laboratory illuminations. *Natl Cancer Inst. Monogr.*, *9*, 117-125

Lake, R.S., Kropko, M.L., Prezzutti, M.R., Shoemaker, R.H. & Igel, H.J. (1978) Chemical induction of unscheduled DNA synthesis in human skin epithelial cell cultures. *Cancer Res.*, *38*, 2091-2098

LaVoie, E.J., Tulley-Freiler, L., Bedenko, V. & Hoffmann, D. (1981) Mutagenicity, tumor-initiating activity, and metabolism of methylphenanthrenes. *Cancer Res.*, *41*, 3441-3447

Lee, M.L., Novotny, M. & Bartle, K.D. (1976) Gas chromatography/mass spectrometric and nuclear magnetic resonance spectrometric studies of carcinogenic polynuclear aromatic hydrocarbons in tobacco and marijuana smoke condensates. *Anal. Chem.*, *48*, 405-416

Lijinsky, W. & Shubik, P. (1964) Benzo(*a*)pyrene and other polynuclear hydrocarbons in charcoal-broiled meat. *Science*, *145*, 53-55

Marquardt, H., Kuroki, T., Huberman, E., Selkirk, J.K., Heidelberger, C., Grover, P.L. & Sims, P. (1972) Malignant transformation of cells derived from mouse prostate by epoxides and other derivatives of polycyclic hydrocarbons. *Cancer Res.*, *32*, 716-720.

May, W.E., Wasik, S.P. & Freeman, D.H. (1978) Determination of the solubility of some polycyclic aromatic hydrocarbons in water. *Anal. Chem.*, *50*, 997-1000

McCann, J., Choi, E., Yamasaki, E. & Ames, B.N. (1975) Detection of carcinogens as mutagens in the *Salmonella*/microsome test: Assay of 300 chemicals. *Proc. natl Acad. Sci. USA*, *72*, 5135-5139

McCarroll, N.E., Keech, B.H. & Piper, C.E. (1981) A microsuspension adaptation of the *Bacillus subtilis* 'rec' assay. *Environ. Mutagenesis*, *3*, 607-616

NIH/EPA Chemical Information System (1982) *Mass Spectral Search System* and *Infrared Spectral Search System*, Washington DC, CIS Project, Information Services Corporation

Nordqvist, M., Thakker, D.R., Vyas, K.P., Yagi, H., Levin, W., Ryan, D.E., Thomas, P.E., Conney, A.H. & Jerina, D.M. (1981) Metabolism of chrysene and phenanthrene to bay-region diol epoxides by rat liver enzymes. *Mol. Pharmacol.*, *19*, 168-178

Oesch, F., Bücker, M. & Glatt, H.R. (1981) Activation of phenanthrene to mutagenic metabolites and evidence for at least two different activation pathways. *Mutat. Res., 8l*, 1-10

Olufsen, B. (1980) *Polynuclear aromatic hydrocarbons in Norwegian drinking water resources.* In: Bjørseth, A. & Dennis, A.J, eds, *Polynuclear Aromatic Hydrocarbons: Chemistry and Biological Effects*, *4th Int. Symposium*, Columbus, OH, Battelle Press, pp. 333-343

Pienta, R.J., Poiley, J.A. & Lebherz, W.B., III (1977) Morphological transformation of early passage golden Syrian hamster embryo cells derived from cryopreserved primary cultures as a reliable *in vitro* bioassay for identifying diverse carcinogens. *Int. J. Cancer*, *19*, 642-655

Pilotti, A., Ancker, K., Arrhenius, E. & Enzell, C. (1975) Effects of tobacco and tobacco smoke constituents on cell multiplication *in vitro*. *Toxicology*, *5*, 49-62

Popescu, N.C., Turnbull, D & DiPaolo, J.A. (1977) Sister chromatid exchange and chromosome aberration analysis with the use of several carcinogens and noncarcinogens. Brief communication. *J. natl Cancer Inst.*, *59*, 289-293

Probst, G.S., McMahon, R.E., Hill, L.E., Thompson, C.Z., Epp, J.K. & Neal, S.B. (1981) Chemically-induced unscheduled DNA synthesis in primary rat hepatocyte cultures: A comparison with bacterial mutagenicity using 218 compounds. *Environ. Mutagenesis*, *3*, 11-32

Quarles, J.M., Sega, M.W., Schenley, C.K. & Lijinsky, W. (1979) Transformation of hamster fetal cells by nitrosated pesticides in a transplacental assay. *Cancer Res.*, *39*, 4525-4533

Roe, F.J.C. (1962) Effect of phenanthrene on tumour-initiation by 3,4-benzopyrene. *Br. J. Cancer*, *16*, 503-506

Roe, F.J.C. & Grant, G.A. (1964) Tests of pyrene and phenanthrene for incomplete carcinogenic and anticarcinogenic activity (Abstract). *Br. Emp. Cancer Campaign*, *41*, 59-60

Rosenkranz, H.S. & Poirier, L.A. (1979) Evaluation of the mutagenicity and DNA-modifying activity of carcinogens and noncarcinogens in microbial systems. *J. natl Cancer Inst.*, *62*, 873-893

Rosenkranz, H.S. & Keifer, Z. (1980) *Determining the DNA-modifying activity of chemicals using DNA-polymerase-deficient* Escherichia coli. In: de Serres, F.J. & Hollaender, A., eds, *Chemical Mutagens: Principles and Methods for their Detection*, Vol. 6, New York, Plenum Press, pp. 109-147

Roszinsky-Köcher, G., Basler, A. & Röhrborn, G. (1979) Mutagenicity of polycyclic hydrocarbons. V. Induction of sister-chromatid exchanges *in vivo*. *Mutat. Res.*, *66*, 65-67

Salaman, M.H. & Roe, F.J.C. (1956) Further tests for tumour-initiating activity: *N,N*-Di(2-chloroethyl)-*p*-aminophenylbutyric acid (CB1348) as an initiator of skin tumour formation in the mouse. *Br. J. Cancer*, *10*, 363-378

Scribner, J.D. (1973) Brief communication: Tumor initiation by apparently noncarcinogenic polycyclic aromatic hydrocarbons. *J. natl Cancer Inst.*, *50*, 1717-1719

Scribner, J.D. & Süss, R. (1978) Tumour initiation and promotion. *Int. Rev. exp. Pathol.*, *18*, 137-198

Simmon, V.F. (1979) In vitro assays for recombinogenic activity of chemical carcinogens and related compounds with *Saccharomyces cerevisiae* D3. *J. natl cancer Inst.*, *62*, 901-909

Simmon, V.F., Rosenkranz, H.S., Zeiger, E. & Poirier, L.A. (1979) Mutagenic activity of chemical carcinogens and related compounds in the intraperitoneal host-mediated assay. *J. natl Cancer Inst.*, *62*, 911-918

Sims, P. (1970) Qualitative and quantitative studies on the metabolism of a series of aromatic hydrocarbons by rat-liver preparations. *Biochem. Pharmacol.*, *19*, 795-818

Steiner, P.E. (1955) Carcinogenicity of multiple chemicals simultaneously administered. *Cancer Res.*, *15*, 632-635

Thruston, A.D., Jr (1978) High pressure liquid chromatography techniques for the isolation and identification of organics in drinking water extracts. *J. chromatogr. Sci.*, *16*, 254-259

Tokiwa, H., Nagakawa, R., Morita, K. & Ohnishi, Y. (1981) Mutagenicity of nitroderivatives induced by exposure of aromatic compounds to nitrogen dioxide. *Mutat. Res.*, *85*, 195-205

Vyas, K.P., Thakker, D.R., Levin, W., Yagi, H., Conney, A.H. & Jerina D.M. (1982) Stereoselective metabolism of the optical isomers of *trans*-1,2-dihydroxy-1,2-dihydrophenanthrene to bay-region diol epoxides by rat liver microsomes. *Chem.-biol. Interactions*, *38*, 203-213

Welch, R.M., Harrison, Y.E., Gommi, B.W., Poppers, P.J., Finster, M. & Conney, A.H. (1969) Stimulatory effect of cigarette smoking on the hydroxylation of 3,4-benzpyrene and the *N*-demethylation of 3-methyl-4-monomethylaminoazobenzene by enzymes in human placenta. *Clin. Pharmacol. Ther.*, *10*, 100-109

Windholz, M., ed. (1976) *The Merck Index*, 9th ed., Rahway, NJ, Merck & Co., p. 934

Wood, A.W., Chang, R.L., Levin, W., Ryan, D.E., Thomas, P.E., Mah, H.D., Karle, J.M., Yagi, H., Jerina, D.M. & Conney, A.H. (1979) Mutagenicity and tumorigenicity of phenanthrene and chrysene epoxides and diol epoxides. *Cancer Res.*, *39*, 4069-4077

PYRENE

1. Chemical and Physical Data

1.1 Synonyms and trade names

Chem. Abstr. Services Reg. No.: 129-00-0

Chem. Abstr. Name: Pyrene

IUPAC Systematic Name: Pyrene

Synonyms: Benzo(def)phenanthrene; β-pyrene

1.2 Structural and molecular formulae and molecular weight

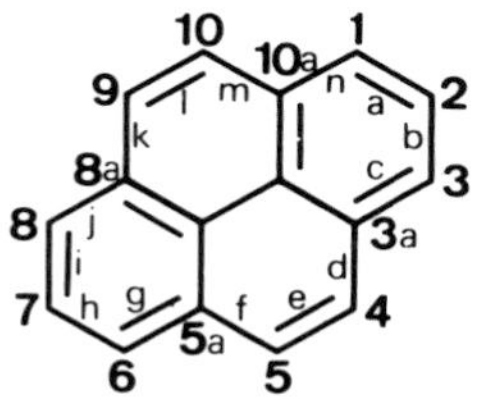

$C_{16}H_{10}$

Mol. wt: 202.3

1.3 Chemical and physical properties of the pure substance

From National Library of Medicine (1982), unless otherwise specified

(*a*) *Description*: Pale-yellow plates (recrystallized from toluene); monoclinic prismatic tablets with a slight blue fluorescence (recrystallized from ethanol or by sublimation)

(*b*) *Boiling-point*: 385°C (Rappoport, 1967); 404°C (Windholz, 1976)

(*c*) *Melting-point*: 149-150°C (Rappoport, 1967); 153-155°C (Hoffmann & Wynder, 1962); 156°C (Windholz, 1976)

(*d*) *Spectroscopy data*: λ_{max} 230.5, 241, 251, 261.5, 272, 292, 305, 318, 333.5, 351.5, 356, 362, 371.5 nm (in methanol/ethanol) (Clar, 1964). Mass spectra have been tabulated (NIH/EPA Chemical Information System, 1982).

(*e*) *Solubility*: Virtually insoluble in water (129-165 μg/l) (Davis *et al.*, 1942; May *et al.*, 1978); soluble in benzene, carbon disulphide, diethyl ether, ethanol, petroleum ether, toluene and acetone (Goldschmidt *et al.*, 1973)

(*f*) *Stability*: Does not undergo photo-oxidation in organic solvents under fluorescent light or indoor sunlight (Kuratsune & Hirohata, 1962)

(*g*) *Reactivity*: Reacts with nitrogen oxides to form nitro derivatives (Tokiwa *et al.*, 1981). Reacts with 70% nitric acid to produce 1-nitro, 1,3-/1,6- and 1,8-dinitro-, 1,3,6-trinitro- and 1,3,6,8-tetranitropyrenes (Rosenkranz *et al.*, 1980)

2. Production, Use, Occurrence and Analysis

2.1 Production and use

There is no commercial production or known use of this compound. Pyrene from coal-tar has been used as the starting material for the synthesis of benzo[*a*]pyrene (Clar, 1964; Buckingham 1982).

2.2 Occurrence and analysis

Data on occurrence and methods of analysis are summarized in the 'General Remarks on the Substances Considered', p. 35.

Pyrene occurs ubiquitously in products of incomplete combustion; it also occurs in fossil fuels. It is found in relatively high quantities in coal-tar (Windholz, 1976).

Pyrene has been identified in mainstream cigarette smoke (7 μg/100 cigarettes) (Mu"ller *et al.*, 1967), (28.3 mg/100 g burnt material) (Masuda & Karatsune, 1972), (6.8 μg/100 cigarettes) (Lee *et al.*, 1976), (10.1-11.4 μg/100 cigarettes) (Ellington *et al.*, 1978), (8.2-14 μg/100 cigarettes) (Severson *et al.*, 1979), (27 μg/cigarette) (Grimmer *et al.*, 1977b), (17-68 μg/1000 cigarettes) (Kiryu & Kuratsune, 1966); sidestream cigarette smoke (390 μg/cigarette) (Kotin & Falk, 1960), (101.1 μg/cigarette) (Grimmer *et al.*, 1977b); smoke-filled rooms (66 ng/m^3) (Grimmer *et al.*, 1977c); smoke from cigars (17.6 μg/100 g tobacco consumed) and pipes (75.5 μg/100 g tobacco consumed) (Campbell & Lindsey, 1957); mainstream marijuana smoke (6.6 μg/100 cigarettes) (Lee *et al.*, 1976); urban air (traces-35 ng/m^3) (Hoffmann & Wynder, 1976); gasoline engine exhaust (2150-2884 μg/l fuel) (Grimmer *et al.*, 1977a); exhaust of burnt coals (0.09-31.00 mg/kg) (Brockhaus & Tomingas, 1976); lubricating oils (1.83 mg/kg) (Grimmer *et al.*, 1981a); used motor oils (429.78 mg/kg) (Grimmer *et al.*, 1981b); crude oils (1.6-10.7 mg/kg) (Grimmer *et al.*, 1983); gasolines (1.48-21.70 mg/kg) (Mu"ller & Meyer, 1974); charcoal-broiled steaks (18 μg/kg) (Lijinsky & Shubik, 1964); surface water (2.0-3.7 ng/l) (Woidich *et al.*, 1976); rain water (5.8-27.8 ng/l) (Woidich *et al.*, 1976); subterranean water (1.6-2.5 ng/l) (Woidich *et al.*, 1976); tap water (1.1 ng/l) (Olufsen, 1980); waste water (0.00023-11.8 μg/l) (Wedgwood & Cooper, 1956; Borneff & Kunte, 1967); sludge (0.57-3.08 mg/kg) (Borneff & Kunte, 1967); freeze-dried sewage sludge (900-47200 μg/kg) (Grimmer *et al.*, I980); and dried sediment from lakes (7-3940 μg/kg) (Grimmer & Böhnke, 1975, 1977).

3. Biological Data Relevant to the Evaluation of Carcinogenic Risk to Humans

3.1 Carcinogenicity studies in animals[1]

(*a*) *Skin application*

Mouse: A 0.3% solution of pyrene obtained by synthesis was applied in benzene twice weekly on the back skin of 40 mice [strain, age and sex unspecified]. The longest observation time was 680 days. No skin lesion was reported (Badger *et al.*, 1940).

In a study reported as an abstract, a 5% solution of pyrene [purity unspecified; solvent not indicated] was applied three times weekly for one year to mouse skin [strain, age and sex unspecified]. No skin tumour was observed. A 5% solution of pyrene was also applied for one year on mouse skin previously treated with a tumour-initiating dose of benzo[*a*]pyrene. No skin tumour was observed (Roe & Grant, 1964).

---

[1] The Working Group was aware of three studies in progress in mice by i.p. administration (IARC, 1983)

A group of 20 male C3H mice, two months of age, received applications of 60 μl of a 0.5% solution pyrene [of 'high purity'] dissolved in decalin twice weekly on the interscapular area for 82 weeks; three papillomas were observed among 13 mice surviving 52 weeks. A group of 15 male C3H mice received 60 μl of 0.5% pyrene dissolved in a mixture of 50% decalin and 50% *n*-dodecane twice weekly. At the end of the treatment (82 weeks), two papillomas and two carcinomas were observed on 13 survivors. Among 13 surviving decalin/*n*-dodecane controls two papillomas occurred (Horton & Christian, 1974). [The Working Group noted that no vehicle control was used and that *n*-dodecane has been shown to have a cocarcinogenic effect when administered simultaneously with several polynuclear hydrocarbons (Bingham & Falk, 1969).]

Two groups of 50 female ICR/Ha Swiss mice (clipped two days before) received applications of 12 μg or 40 μg of pure recrystallized pyrene dissolved in 0.1 ml acetone thrice weekly on the back. After 368 and 440 days of treatment, respectively, no change was observed at the application site (Van Duuren & Goldschmidt, 1976).

Pyrene [purity unspecified] was tested for initiating activity on 20 'S' strain mice by 10 thrice-weekly applications of 8.3% pyrene solution in acetone (total dose, 0.25 g). Starting 25 days after the last treatment, the animals received 18 weekly applications of 0.17% croton oil in acetone (0.3 ml). At the end of the treatment with croton oil, nine papillomas were present on 6/20 surviving animals compared with four papillomas on 4/19 survivors in the croton oil-control group (Salaman & Roe, 1956). [The Working Group noted the short duration of the study.]

A group of 30 female CD-1 mice, eight weeks of age, received an application of 10 μmol pyrene (purified to homogeneity by thin-layer chromatography) dissolved in acetone to the shaved dorsal skin. Starting one week later, a twice-weekly treatment with 12-*O*-tetradecanoylphorbol-13-acetate (TPA) in acetone (5 μmol/dose) was given for 34 weeks. Papillomas developed in 5/29 surviving animals (0.21 papilloma/mouse); none were seen in a control group receiving 10 μmol TPA (Scribner, 1973). [The Working Group noted that the dose of TPA was probably 5 μg (Scribner & Süss, 1978).]

(*b*) *Subcutaneous and/or intramuscular administration*

Mouse: A group of 30 male and female Jackson A strain mice, three to four months old, received a s.c. injection of 10 mg pyrene crystals moistened with glycerol in the left flank. The injection was repeated four months later. There were 23 mice alive at the end of one year and nine at 18 months. No s.c. tumour was produced. The average number of pulmonary adenomas in the mice at 18 months was only 1.6 (Shear & Leiter, 1941). [The Working Group noted the small number of treatments.]

(*c*) *Intrabronchial or intratracheal administration*

Hamster: A chromatographically pure sample of pyrene (purity, >99%) was mixed in equal parts with haematite dust (Fe_2O_3, particle size 94%, <1.0 μ) and ground to form a finely aggregated dust mixture. The dust, suspended in 0.2 ml saline containing 3 mg pyrene, was instilled intratracheally weekly for 30 weeks into the respiratory tract of 48 male Syrian golden hamsters, nine to ten weeks old. There were 24 animals alive after 50 weeks and seven after

90 weeks. All animals were examined histologically. One tumour of the trachea and two malignant lymphomas were observed. A group receiving dibenz[*a,i*]pyrene developed a high incidence of carcinomas. In a group of 82 effective controls, no respiratory-tract tumour was found (Sellakumar & Shubik, 1974).

(*d*) *Administration with other compounds*

Mouse: In order to investigate the cocarcinogenicity of pyrene, three groups of 50 female ICR/Ha Swiss mice, six to eight weeks of age, were given thrice-weekly applications on the shaved skin of the back of 0.1 ml acetone containing 4, 12 or 40 μg pure recrystallized pyrene + 5 μg benzo[*a*]pyrene. Two positive-control groups of 50 mice received acetone alone; and groups of 100 mice were untreated. The low- and mid-dose groups receiving pyrene + benzo[*a*]pyrene, one of the untreated groups and one of the positive control groups were followed for 368 days; the other groups were followed for 440 days. There were 12 papilloma-bearing mice (total, 14 papillomas) and six with squamous-cell carcinomas in the low-dose group; 26 papilloma-bearing mice (total, 42 papillomas) and 20 with squamous-cell carcinomas in the mid-dose group; and 35 papilloma-bearing mice (total, 66 papillomas) and 26 with squamous-cell carcinomas in the high-dose group. In the positive-control group treated for 368 days, there were 14 papilloma-bearing mice (total, 16 papillomas) and 10 with squamous-cell carcinomas. In the other positive-control group, treated for 440 days, there were 16 papilloma-bearing mice (total, 26 papillomas) and 12 with squamous-cell carcinomas. No skin tumour occurred in the solvent or untreated controls (Van Duuren & Goldschmidt, 1976).

In a study reported as an abstract, 5 μg benzo[*a*]pyrene [of 'high purity'] dissolved in 0.1 ml acetone were applied thrice weekly on the skin of 50 female ICR/Ha mice simultaneously with pyrene (of high purity) at the molar ratios: pyrene:benzo[*a*]pyrene, 1:1 or 3:1. At the 3:1 molar ratio, the effect of benzo[*a*]pyrene was enhanced by pyrene: 13/50 mice had papillomas and 5/50, carcinomas after 33 weeks of treatment. When benzo[*a*]pyrene was applied alone, 6/50 mice had papillomas; application of pyrene alone did not induce tumours (Goldschmidt *et al.*, 1973).

3.2 Other relevant biological data[1]

(*a*) *Experimental system*

Toxic effects

The $LD_{50(7)}$ (dose that killed half the animals in seven days) for mice (i.p.) is 514 and the $LD_{50(4)}$ (dose that killed half the animals in four days), 678 mg/kg bw (Salamone, 1981).

The growth rate of mouse ascites sarcoma cells in culture was inhibited (33%) when pyrene was added at a concentration of 1 μmol/ml in dimethyl sulphoxide (Pilotti *et al.*, 1975).

I.p. administration of 20 mg pyrene in sesame oil produced no immediate, long-lasting reduction in the growth rate of young rats, as occurred with administration of 10 mg benzo[*a*]pyrene (Haddow *et al.*, 1937). When pyrene was fed at a concentration of 2000 mg/kg of diet to young rats for 100 days, inhibition of growth was observed. The livers were enlarged and of fatty appearance, with no any other pathological condition (White & White, 1939).

---

[1] See also 'General Remarks on the Substances Considered', p. 53.

Goblet-cell hyperplasia and cases of so-called 'transitional hyperplasia' were observed when pyrene was implanted (in a beeswax pellet) into isogenically transplanted rat tracheas (Topping *et al.*, 1978).

Effects on reproduction and prenatal toxicity

When pyrene, 4 mg/mouse was injected in sunflower oil i.m. daily during the last week of gestation (into BALB/c, C3H/a, C57Bl x CBA F_1 hybrids), increased survival and hyperplastic changes were seen in explants of embryonic mouse kidney in organ culture, compared with controls. The effects were similar but less marked than those produced by benzo[*a*]pyrene (Shabad *et al.*, 1972). Administration of two s.c. injections of 6 mg pyrene on days 18 and 19 of pregnancy to strain A mice did not induce an increase in tumour incidence in the offspring in studies in which benzo[*a*]pyrene injections did have that effect (Nikonova, 1977).

Metabolism and activation

The 1-hydroxy, 1,6-, 1,8-dihydroxy, and the 4,5-dihydrodiol have been identified as metabolites of pyrene in rats and rabbits (Boyland & Sims, 1964). In addition, two trihydroxy derivatives were isolated following incubation of pyrene with rat-liver preparations (Jacob *et al.*, 1982).

Mutagenicity and other short-term tests

Results from short-term tests are given in Table 1.

Table 1. Results from short-term tests: Pyrene[a]

Test	Organism/assay[b]	Exogenous metabolic system[b]	Reported result	Comments	References
PROKARYOTES					
DNA damage	*Escherichia coli* (rec^+/rec^-; $polA^+/polA^-$); $polA^+$ $uvrA^+/polA^-$ $uvrA^-$)	Various	Negative	International collaborative programme (concensus view of participants)	Ashby & Kilbey (1981)
	Bacillus subtilis (rec^+/rec^-)	Various	Negative	International collaborative programme (concensus view of participants)	Ashby & Kilbey (1981)
Mutation	*Salmonella typhimurium* (his^-/his^+)	Aro-R-PMS	Negative	Tested at up to 1000 μg/plate in strains TA1535, TA1537, TA98 and TA100	McCann *et al.* (1975); LaVoie *et al.* (1979); Ho *et al.* (1981)
		Various	Positive	At 25 μg/plate in strain TA1537; international collaborative programme (concensus view of participants)	Bridges *et al.* (1981)
	Salmonella typhimurium ($8AG^S/8AG^R$)	Aro-R-PMS	Positive	At 140 nmol/ml in strain TM677	Kaden *et al.* (1979)

Test	Organism/assay[b]	Exogenous metabolic system[b]	Reported result	Comments	References
FUNGI					
Mutation	*Saccharomyces cerevisiae*; *Schizosaccharomyces pombe* (different genetic end-points)	Various	Negative	International collaborative programme (concensus view of participants)	de Serres & Hoffman (1981)
INSECTS					
Mutation	*Drosophila melanogaster* (sex-linked recessive lethals)	--	Negative	Fed at 800 mg/kg in the diet for up to 72 h	Valencia & Houtchens (1981)
MAMMALIAN CELLS *IN VITRO*					
DNA damage	Human foreskin epithelial cells (unscheduled DNA synthesis)	--	Negative	Tested at up to 400 μg/ml	Lake *et al*. (1978)
	Primary rat hepatocytes (unscheduled DNA synthesis)	--	Negative	Tested at up to 500 nmol/ml	Probst *et al*. (1981)
	HeLa cells (unscheduled DNA synthesis)	3MC-R-PMS	Negative	[Doses unspecified]	Martin *et al*. (1978)
	Human fibroblast cell line WI38 (unscheduled DNA synthesis)	Aro-R-PMS	Positive	At 7.2 μg/ml	Robinson & Mitchell (1981)
Mutation	Mouse lymphoma L5178Y cells (TFT^S/TFT^R)	Aro-R-PMS	Positive	At 10 μg/ml	Jotz & Mitchell (1981)
Chromosome effects	Rat liver epithelial ARL 18 cells (sister chromatid exchange)	--	Negative	Tested at up to 1 μmol/ml	Tong *et al*. (1981)
	Chinese hamster ovary cells (sister chromatid exchange)	Aro-R-PMS	Positive Negative	At between 19-300 μg/ml Tested at up to 100 μg/ml	Evans & Mitchell (1981) Perry & Thomson (1981)
	Chinese hamster V79 cells (sister chromatid exchange; aberrations)	SHE feeder layer	Positive	At 10 μg/ml	Popescu *et al*. (1977)
	Rat liver RL_1 cells (aberrations)	--	Negative	Tested at up to 100 μg/ml	Dean (1981)
Cell transformation	Syrian hamster embryo cells (morphological)	--	Negative	Tested at up to 20 μg/ml	DiPaolo *et al*. (1969); Pienta *et al*. (1977); Casto (1979)
	Mouse prostate C3H cells (morphological)	MEF feeder layer	Negative	Tested at 1 μg/ml	Chen & Heidelberger (1969)
	Mouse BALB/C-3T3 cells (morphological)	--	Negative	Tested at 20 μg/ml Tested at 50 μg/ml	DiPaolo *et al*. (1972) Kakunaga (1973)
	Guinea-pig fetal cells (morphological)	--	Negative	Tested at 10 μg/ml	Evans & DiPaolo (1975)
MAMMALIAN CELLS *IN VIVO*					
Chromosome effects	Mouse bone-marrow cells (sister chromatid exchange)	--	Negative	Treated i.p. with up to 400 mg/kg bw	Paika *et al*. (1981)
	Mouse bone-marrow cells (micronucleus)	--	Negative	Treated i.p. with 2 x 400 mg/kg bw	Salamone *et al*. (1981)
		--	Negative	Treated i.p. with 2 x 260 mg/kg bw	Tsuchimoto & Matter (1981)

[a] This table comprises selected assays and references and is not intended to be a complete review of the literature.
[b] For an explanation of the abbreviations see the Appendix, p. 452.

(*b*) *Humans*

No data were available to the Working Group.

3.3 Case reports and epidemiological studies of carcinogenicity to humans

No data were available to the Working Group.

4. Summary of Data Reported and Evaluation[1]

4.1 Experimental data

Pyrene was tested for carcinogenicity in several experiments by skin application to mice, and no skin tumour was observed. It was also tested in several studies in the mouse-skin initiation-promotion assay, with inconclusive results. When tested on mice skin simultaneously with benzo[*a*]pyrene it enhanced the carcinogenic effects of benzo[*a*]pyrene.

A study in mice by subcutaneous injection was inadequate for evaluation of carcinogenicity.

Intratracheal administration to hamsters of pyrene attached to haematite did not produce tumours.

No data on the teratogenicity of this compound were available.

Pyrene has been tested extensively in both in-vitro and in-vivo short-term tests. It was negative in assays for differential survival in DNA-repair-proficient/-deficient strains of bacteria and was mutagenic in some assays in *Salmonella typhimurium* in the presence of an exogenous metabolic system. Tests for genetic activity in yeast were negative. It was not mutagenic to *Drosophila melanogaster*. It did induce mutations and unscheduled DNA synthesis in some in-vitro assays in mammalian cells. Pyrene did not induce morphological transformation. In tests in mammals *in vivo* it did not induce sister chromatid exchange or micronuclei.

There is *limited evidence* that pyrene is active in short-term tests.

---

[1] For definitions of the italicized terms, see Preamble, p. 19.

4.2 Human data[1]

Pyrene is present as a major component of the total content of polynuclear aromatic compounds in the environment. Human exposure to pyrene occurs primarily through the smoking of tobacco, inhalation of polluted air and by ingestion of food and water contaminated by combustion effluents (for details, see 'General Remarks on the Substances Considered', p. 35).

4.3 Evaluation

The available data provide no evidence that pyrene *per se* is carcinogenic to experimental animals.

5. References

Ashby, J. & Kilby, B. (1981) *Summary report on the performance of bacterial repair, phage induction, degranulation, and nuclear enlargement assays*. In: de Serres, F.J. & Ashby, J., eds, *Evaluation of Short-Term Tests for Carcinogens. Report of the International Collaborative Program. Progress in Mutation Research*, Vol. 1, New York, Elsevier/North Holland, pp. 33-48

Badger, G.M., Cook, J.W., Hewett, C.L., Kennaway, E.L., Kennaway, N.M., Martin, R.H. & Robinson, A.M. (1940) The production of cancer by pure hydrocarbons. V. *Proc. R. Soc. London Ser. B*, *129*, 439-467

Bingham, E. & Falk, H.L. (1969) Environmental carcinogens: The modifying effect of cocarcinogens on the threshold response. *Arch. environ. Health*, *19*, 779-783

Borneff, J. & Kunte, H. (1967) Carcinogenic substances in water and soils. XIX. Action of sewage purification on polycyclic aromatics (Ger.). *Arch. Hyg. Bakt.*, *151*, 202-210

Boyland, E. & Sims, P. (1964) Metabolism of polycyclic compounds. The metabolism of pyrene in rats and rabbits. *Biochem. J.*, *90*, 391-398

Bridges, B.A., Zeiger, E. & McGregor, D.B. (1981) *Summary report on the performance of bacterial mutation assays*. In: de Serres, F.J. & Ashby, J., eds, *Evaluation of Short-Term Tests for Carcinogens. Report of the International Collaborative Program. Progress in Mutation Research*, Vol. 1, New York, Elsevier/North Holland, pp. 49-67

Brockhaus, A. & Tomingas, R. (1976) Emission of polycyclic hydrocarbons during burning processes in small heating installations and their concentration in the atmosphere (Ger.). *Staub-Reinhalt. Luft*, *36*, 96-101

[1] Studies on occupational exposure to polynuclear aromatic compounds will be considered in future *IARC Monographs*.

Buckingham, J., ed. (1982) *Dictionary of Organic Compounds*, 5th ed., Vol. 5, New York, Chapman & Hall, p. 4840

Campbell, J.M. & Lindsey, A.J. (1957) Polycyclic hydrocarbons in cigar smoke. *Br. J. Cancer*, *11*, 192-195

Casto, B.C. (1979) *Polycyclic hydrocarbons and Syrian hamster embryo cells: Cell transformation, enhancement of viral transformation and analysis of DNA damage*. In: Jones, P.W. & Leber, P., eds, *Polynuclear Aromatic Hydrocarbons*, Ann Arbor, MI, Ann Arbor Science Publishers, pp. 51-66

Chen, T.T. & Heidelberger, C. (1969) Quantitative studies on the malignant transformation of mouse prostate cells by carcinogenic hydrocarbons *in vitro*. *Int. J. Cancer*, *4*, 166-178

Clar, E., ed. (1964) *Polycyclic Hydrocarbons*, Vol. 1, New York, Academic Press, pp. 236-243

Davis, W.W., Krahl, M.E. & Clowes, G.H.A. (1942) Solubility of carcinogenic and related hydrocarbons in water. *J. Am. chem. Soc.*, *64*, 108-110

Dean, B.J. (1981) *Activity of 27 coded compounds in the RL_1 chromosome assay*. In: de Serres, F.J. & Ashby, J., eds, *Evaluation of Short-Term Tests for Carcinogens. Report of the International Collaborative Program. Progress in Mutation Research*, Vol. 1, New York, Elsevier/North-Holland, pp. 570-579

DiPaolo, J.A., Donovan, P. & Nelson, R. (1969) Quantitative studies of *in vitro* transformation by chemical carcinogens. *J. natl Cancer Inst.*, *42*, 867-876

DiPaolo, J.A., Takano, K. & Popescu, N.C. (1972) Quantitation of chemically induced neoplastic transformation of BALB/3T3 cloned cell lines. *Cancer Res.*, *32*, 2686-2695

Ellington, J.J., Schlotzhauer, P.F. & Schepartz, A.I. (1978) Quantitation of hexane-extractable lipids in serial samples of flue-cured tobaccos. *J. Food agric. Chem.*, *26*, 270-273

Evans, C.H. & DiPaolo, J.A. (1975) Neoplastic transformation of guinea pig fetal cells in culture induced by chemical carcinogens. *Cancer Res.*, *35*, 1035-1044

Evans, E.L. & Mitchell, A.D. (1981) *Effect of 20 coded chemicals on sister chromatid exchange frequencies in cultured Chinese hamster cells*. In: de Serres F.J. & Ashby J., eds, *Evaluation of Short-Term Tests for Carcinogens. Report of the International Collaborative Program. Progress in Mutation Research*, Vol. 1, New York, Elsevier/North-Holland, pp. 538-550

Goldschmidt, B.M., Katz, C. & Van Duuren, B.L. (1973) The cocarcinogenic activity of non-carcinogenic aromatic hydrocarbons (Abstract No. 334). *Proc. Am. Assoc. Cancer Res.*, *17*, 84

Grimmer, G. & Böhnke, H. (1975) Profile analysis of polycyclic aromatic hydrocarbons and metal content in sediment layers of a lake. *Cancer Lett.*, *1*, 75-84

Grimmer, G. & Böhnke, H. (1977) Investigation on drilling cores of sediments of Lake Constance. I. Profiles of the polycyclic aromatic hydrocarbons. *Z. Naturforsch.*, *32c*, 703-711

Grimmer, G., Böhnke, H. & Glaser, A. (1977a) Investigation on the carcinogenic burden by air pollution in man. XV. Polycyclic aromatic hydrocarbons in automobile exhaust gas - An inventory. *Zbl. Bakt. Hyg., 1. Abt., Orig. B164*, 218-234

Grimmer, G., Böhnke, H. & Harke, H.-P. (1977b) Passive smoking: Intake of polycyclic aromatic hydrocarbons by breathing of cigarette smoke containing air (Ger.). *Int. Arch. occup. environ. Health*, *40*, 93-99

Grimmer, G., Böhnke, H. & Harke, H.-P. (1977c) Passive smoking: Measurement of concentrations of polycyclic aromatic hydrocarbons in rooms after machine smoking of cigarettes (Ger.). *Int. Arch. occup. environ. Health*, *40*, 83-92

Grimmer, G., Hilge, G. & Niemitz, W. (1980) Comparison of the profile of polycyclic aromatic hydrocarbons in sewage sludge samples from 25 sewage treatment works (Ger.). *Wasser*, *54*, 255-272

Grimmer, G., Jacob, J., & Naujack, K.-W. (1981a) Profile of the polycyclic aromatic hydrocarbons from lubricating oils. Inventory by GCGC/MS - PAH in environmental materials. Part 1. *Fresenius Z. anal. Chem.*, *306*, 347-355

Grimmer, G., Jacob, J., Naujack, K.-W. & Dettbarn, G. (1981b) Profile of the polycyclic aromatic compounds from used engine oil - Inventory by GCGC/MS - PAH in environmental materials, Part 2. *Fresenius Z. anal. Chem.*, *309*, 13-19

Grimmer, G., Jacob, J. & Naujack, K.-W. (1983) Profile of the polycyclic aromatic compounds from crude oils. Inventory by GCGC/MS - PAH in environmental materials, Part 3. *Fresenius Z. anal. Chem.*, *314*, 29-36

Haddow, A., Scott, C.M. & Scott, J.D. (1937) The influence of certain carcinogenic and other hydrocarbons on body growth in the rat. *Proc. R. Soc. London Ser. B*, *122*, 477-507

Ho, C.-H., Clark, B.R., Guerin, M.R., Barkenbus, B.D., Rao, T.K. & Epler, J.L. (1981) Analytical and biological analyses of test materials from the synthetic fuel technologies. IV. Studies of chemical structure-mutagenic activity relationships of aromatic nitrogen compounds relevant to synfuels. *Mutat. Res.*, *85*, 335-345

Hoffmann, D. & Wynder, E.L. (1962) Analytical and biological studies on gasoline engine exhaust. *Natl Cancer Inst. Monogr.*, *9*, 91-116

Hoffmann, D. & Wynder, E. (1976) *Environmental respiratory carcinogenesis*. In: Searle, C.E., ed., *Chemical Carcinogens* (*ACS Monograph 173*), Washington DC, American Chemical Society, p. 341

Horton, A.W. & Christian, G.M. (1974) Cocarcinogenic versus incomplete carcinogenic activity among aromatic hydrocarbons: Contrast between chrysene and benzo[*b*]triphenylene. *J. natl Cancer Inst.*, *53*, 1017-1020

IARC (1983) *Information Bulletin on the Survey of Chemicals Being Tested for Carcinogenicity*, No. 10, Lyon, p. 43

Jacob, J., Grimmer, G., Raab, G. & Schmoldt, A. (1982) The metabolism of pyrene by rat liver microsomes and the influence of various mono-oxygenase inducers. *Xenobiotica*, *12*, 45-53

Jotz, M.M. & Mitchell, A.D. (1981) *Effects of 20 coded chemicals on the forward mutation frequency at the thymidine kinase locus in L5178Y mouse lymphoma cells.* In: de Serres, F.J. & Ashby, J., eds, *Evaluation of Short-Term Tests for Carcinogens. Report of the International Collaborative Program. Progress in Mutation Research*, Vol. 1, New York, Elsevier/North-Holland, pp. 580-593

Kaden, D.A., Hites, R.A. & Thilly, W.G. (1979) Mutagenicity of soot and associated polycyclic aromatic hydrocarbons to *Salmonella typhimurium. Cancer Res.*, *39*, 4152-4159

Kakunaga, T. (1973) A quantitative system for assay of malignant transformation by chemical carcinogens using a clone derived from BALB/3T3. *Int. J. Cancer*, *12*, 463-473

Kiryu, S. & Kuratsune, M. (1966) Polycyclic aromatic hydrocarbons in the cigarette tar produced by human smoking. *Gann*, *57*, 317-322

Kotin, P. & Falk, H.L. (1960) The role and action of environmental agents in the pathogenesis of lung cancer. II. Corneff, J. & Kunteigarette smoke. *Cancer*, *13*, 250-262

Kuratsune, M. & Hirohata, T. (1962) Decomposition of polycyclic aromatic hydrocarbons under laboratory illuminations. *Natl Cancer Inst. Monogr.*, *9*, 117-125

Lake, R.S., Kropko, M.L., Pezzutti, M.R., Shoemaker, R.H. & Igel, H.J. (1978) Chemical induction of unscheduled DNA synthesis in human skin epithelial cell cultures. *Cancer Res.*, *38*, 2091-2098

LaVoie, E., Bedenko, V., Hirota, N., Hecht, S.S. & Hoffmann, D. (1979) *A comparison of the mutagenicity, tumor-initiating activity and complete carcinogenicity of polynuclear aromatic hydrocarbons*. In: Jones, P.W. & Leber, P., eds, *Polynuclear Aromatic Hydrocarbons*, Ann Arbor, MI, Ann Arbor Science Publishers, pp. 705-721

Lee, M.L., Novotny, M. & Bartle, M.D. (1976) Gas chromatography/mass spectrometric and nuclear magnetic resonance spectrometric studies of carcinogenic polynuclear aromatic hydrocarbons in tobacco and marijuana smoke condensates. *Anal. Chem.*, *48*, 405-416

Lijinsky, W. & Shubik, P. (1964) Benzo[*a*]pyrene and other polynuclear hydrocarbons in charcoal-broiled meat. *Science*, *145*, 53-55

Martin, C.N., McDermid, A.C. & Garner, R.C. (1978) Testing of known carcinogens and noncarcinogens for their ability to induce unscheduled DNA synthesis in Hela cells. *Cancer Res.*, *38*, 2621-2627

Masuda, Y. & Kuratsune, M. (1972) Comparison of the yield of polycyclic aromatic hydrocarbons in smoke from Japanese tobacco. *Jpn. J. Hyg.*, *27*, 339-341

May, W.E., Wasik, S.P. & Freeman, D.H. (1978) Determination of the solubility behavior of some polycyclic aromatic hydrocarbons in water. *Anal. Chem.*, *50*, 997-1000

McCann, J., Choi, E., Yamasaki, E. & Ames, B.N. (1975) Detection of carcinogens as mutagens in the *Salmonella*/microsome test: Assay of 300 chemicals. *Proc. natl Acad. Sci. USA*, *72*, 5135-5139

Müller, K. & Meyer, J.P. (1974) *Einfluss von Ottokraftstoffen auf die Emission von polynuklearen aromatischen Kohlenwasserstoffen in Automobilabgasen im Europa-Test* (Effect of gasoline components on emission of polynuclear aromatic hydrocarbons in car exhaust in the Europa test) (*Forschungsbericht 4568*), Hamburg, Deutsche Gesellschaft für Mineralölwissenschaft und Kohlechemie e.V.

Müller, R., Moldenhauer, W. & Schlemmer, P. (1967) Experiences with the quantitative determination of polycyclic hydrocarbons in tobacco smoke (Ger.). *Ber. Inst. Tabakforsch. Dresden*, *14*, 159-173

National Library of Medicine (1982) *Toxicology Data Bank*, Bethesda, MD, National Library of Medicine Specialized Information Services, Toxicology Information Program

NIH/EPA Chemical Information System (1982) *Mass Spectral Search System*, Washington DC, CIS Project, Information Services Corporation

Nikonova, T.V. (1977) Transplacental action of benzo[*a*]pyrene and pyrene. *Bull. exp. Biol. Med.*, *84*, 1025-1027

Olufsen, B. (1980) *Polynuclear aromatic hydrocarbons in Norwegian drinking water resources.* In: Bjørseth, A. & Dennis, A.J., eds, *Polynuclear Aromatic Hydrocarbons: Chemistry and Biological Effects*, *4th Int. Symposium*, Columbus, OH, Battelle Press, pp. 333-343

Paika, I.J., Beauchesne, M.T., Randall, M., Schreck, R. & Latt, S.A. (1981) In vivo *SCE analysis of 20 coded compounds.* In: de Serres, F.J. & Ashby, J., eds, *Evaluation of Short-Term Tests for Carcinogens. Report of the International Collaborative Program. Progress in Mutation Research*, Vol. 1, New York, Elsevier/North-Holland, pp. 672-681

Perry, P.E. & Thomson, E.J. (1981) *Evaluation of the sister chromatid exchange method in mammalian cells as a screening system for carcinogens.* In: de Serres, F.J. & Ashby, J., eds, *Evaluation of Short-Term Tests for Carcinogens. Report of the International Collaborative Program. Progress in Mutation Research*, Vol. 1, New York, Elsevier/North-Holland, pp. 560-569

Pienta, R.J., Poiley, J.A. & Lebherz, W.B., III (1977) Morphological transformation of early passage golden Syrian hamster embryo cells derived from cryopreserved primary cultures as a reliable *in vitro* bioassay for identifying diverse carcinogens. *Int. J. Cancer*, *19*, 642-655

Pilotti, A., Ancker, K., Arrhenius, E. & Enzell, C. (1975) Effects of tobacco and tobacco smoke constituents on cell multiplication *in vitro*. *Toxicology*, *5*, 49-62

Popescu, N.C., Turnbull, D. & DiPaolo, J.A. (1977) Sister chromatid exchange and chromosome aberration analysis with the use of several carcinogens and noncarcinogens: Brief communication. *J. natl Cancer Inst.*, *59*, 289-293

Probst, G.S., McMahon, R.E., Hill, L.E., Thompson, C.Z., Epp, J.K. & Neal, S.B. (1981) Chemically-induced unscheduled DNA synthesis in primary rat hepatocytes: a comparison with bacterial mutagenicity using 218 compounds. *Environ. Mutagenesis*, *3*, 11-32

Rappoport, Z., ed. (1967) *CRC Handbook of Tables for Organic Compound Identification*, 3rd ed., Boca Raton, FL, Chemical Rubber Co., p. 49

Robinson, D.E. & Mitchell, A.D. (1981) *Unscheduled DNA synthesis response of human fibroblasts, WI-38 cells, to 20 coded chemicals*. In: de Serres, F.J. & Ashby, J., eds, *Evaluation of Short-Term Tests for Carcinogens. Report of the International Collaborative Program. Progress in Mutation Research*, Vol. 1, New York, Elsevier/North Holland, pp. 517-527

Roe, F.J.C. & Grant, G.A. (1964) Tests of pyrene and phenanthrene for incomplete carcinogenic and anticarcinogenic activity (Abstract). *Br. Emp. Cancer Campaign*, *41*, 59-60

Rosenkranz, H.S., McCoy, E.C., Sanders, D.R., Butler, M., Kiriazides, D.K. & Mermelstein, R. (1980) Nitropyrenes: Isolation, identification, and reduction of mutagenic impurities in carbon black and toners. *Science*, *209*, 1039-1043

Salaman, M.H. & Roe, F.J.C. (1956) Further tests for tumour-initiating activity: *N,N*-Di(2-chloroethyl)-*p*-aminophenylbutyric acid (CB1348) as an initiator of skin tumour formation in the mouse. *Br. J. Cancer*, *10*, 363-378

Salamone, M.F. (1981) *Toxicity of 41 carcinogens and noncarcinogenic analogs*. In: de Serres, F.J. & Ashby, J., eds, *Evaluation of Short-Term Tests for Carcinogens. Report of the International Collaborative Program. Progress in Mutation Research*, Vol. 1, New York, Elsevier/North Holland, pp. 682-685

Salamone, M.F., Heddle, J.A. & Katz, M. (1981) *Mutagenic activity of 41 compounds in the* in vivo *micronucleus assay*. In: de Serres, F.J. & Ashby, J., eds, *Evaluation of Short Term Tests for Carcinogens. Report of the International Collaborative Program. Progress in Mutation Research*, Vol. 1, New York, Elsevier/North Holland, pp. 686-697

Scribner, J.D. (1973) Brief communication: Tumor initiation by apparently noncarcinogenic polycyclic aromatic hydrocarbons. *J. natl Cancer Inst.*, *50*, 1717-1719

Scribner, J.D. & Süss, R. (1978) Tumour initiation and promotion. *Int. Rev. exp. Pathol.*, *18*, 137-198

Sellakumar, A. & Shubik, P. (1974) Carcinogenicity of different polycyclic hydrocarbons in the respiratory tract of hamsters. *J. natl Cancer Inst.*, *53*, 1713-1719

de Serres, F.J. & Hoffman, G.R. (1981) *Summary report on the performance of yeast assays*. In: de Serres, F.J. & Ashby, J., eds, *Evaluation of Short-Term Tests for Carcinogens. Report of the International Collaborative Program. Progress in Mutation Research*, Vol. 1, New York, Elsevier/North Holland, pp. 68-76

Severson, R.F., Arrendale, R.F., Chaplin, J.F. & Williamson, R.E. (1979) Use of pale-yellow tobacco to reduce smoke polynuclear aromatic hydrocarbons. *J. agric. Food chem.*, *27*, 896-900

Shabad, L.M., Sorokina, J.D., Golub, N.I. & Bogovski, S.P. (1972) Transplacental effect of some chemical compounds on organ cultures of embryonic kidney tissue. *Cancer Res.*, *32*, 617-627

Shear, M.J. & Leiter, J. (1941) Studies in carcinogenesis. XVI. Production of subcutaneous tumors on mice by miscellaneous polycyclic compounds. *J. natl Cancer Inst.*, *11*, 241-258

Tokiwa, H., Nakagawa, R., Morita, K. & Ohnishi, Y. (1981) Mutagenicity of nitro derivatives induced by exposure of aromatic compounds to nitrogen dioxide. *Mutat. Res.*, *85*, 195-205

Tong, C., Brat, S.V. & Williams, G.M. (1981) Sister-chromatid exchange induction by polycyclic aromatic hydrocarbons in an intact cell system of adult rat-liver epithelial cells. *Mutat. Res.*, *91*, 467-473

Topping, D.C., Pal, B.C., Martin, D.H., Nelson, F.R. & Nettesheim, P. (1978) Pathologic changes induced in respiratory tract mucosa by polycyclic hydrocarbons of differing carcinogenic activity. *Am. J. Pathol.*, *93*, 311-324

Tsuchimoto, T. & Matter, B.E. (1981) *Activity of coded compounds in the micronucleus test.* In: de Serres, F.J. & Ashby, J., eds, *Evaluation of Short-Term Tests for Carcinogens. Report of the International Collaborative Program. Progress in Mutation Research*, Vol. 1, New York, Elsevier/North-Holland, pp. 705-711

Valencia, R. & Houtchens, K. (1981) *Mutagenic activity of 10 coded compounds in the* Drosophila *sex-linked recessive lethal test.* In: de Serres, F.J. & Ashby, J., eds, *Evaluation of Short-Term Tests for Carcinogens. Report of the International Collaborative Program. Progress in Mutation Research*, Vol. 1, New York, Elsevier/North-Holland, pp. 651-659

Van Duuren, B.L. & Goldschmidt, B.M. (1976) Cocarcinogenic and tumor-promoting agents in tobacco carcinogenesis. *J. natl Cancer Inst.*, *56*, 1237-1242

Van Duuren, B.L., Katz, C. & Goldschmidt, B.M. (1973) Cocarcinogenic agents in tobacco carcinogenesis. *J. natl Cancer Inst.*, *51*, 703-705

Wedgwood, P. & Cooper, R.L. (1956) The detection and determination of traces of polynuclear hydrocarbons in industrial effluents and sewage - Part IV. The quantitative examination of effluents. *Analyst*, *81*, 42-44

White, J. & White, A. (1939) Inhibition of growth of the rat by oral administration of methylcholanthrene, benzpyrene, or pyrene and the effects of various dietary supplements. *J. biol. Chem.*, *131*, 149-161

Windholz, M., ed. (1976) *The Merck Index*, 9th ed., Rahway, NJ, Merck & Co., p. 1032

Woidich, H., Pfannhauser, W., Blaicher, G. & Tiefenbacher, K. (1976) Analysis of polycyclic aromatic hydrocarbons in drinking and industrial water (Ger.). *Lebensmittelchem. gerichtl. Chem.*, *30*, 141-160

TRIPHENYLENE

1. Chemical and Physical Data

1.1 Synonyms and trade names

Chem. Abstr. Services Reg. No.: 217-59-4

Chem. Abstr. Name: Triphenylene

IUPAC Systematic Name: Triphenylene

Synonyms: Benzo(1)phenanthrene; 9,10-benzophenanthrene; 9,10-benzphenanthrene; 1,2,3,4-dibenznaphthalene; isochrysene

1.2 Structural and molecular formulae and molecular weight

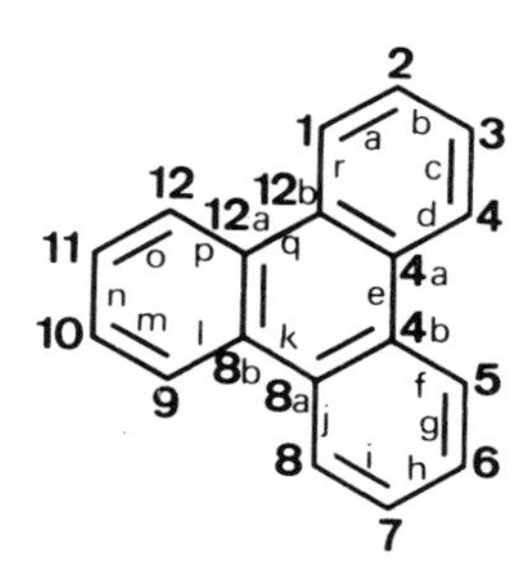

$C_{18}H_{12}$ Mol. wt: 228.3

1.3 Chemical and physical properties of the pure substance

From Windholz (1976), unless otherwise specified

(*a*) *Description*: Long needles (recrystallized from chloroform or ethanol)

(*b*) *Boiling-point*: 425°C

(*c*) *Melting-point*: 199°C; 196.5°C (Clar, 1964)

(*d*) *Spectroscopy data*: λ_{max} 248.5, 257, 273, 284, 302, 321, 327, 333.5, 340 nm (in methanol/ethanol) (Clar, 1964). Mass spectra have been tabulated (NIH/EPA Chemical Information System, 1982).

(*e*) *Solubility*: Virtually insoluble in water (38 μg/l) (Davis *et al.*, 1942); soluble in acetic acid and ethanol; very soluble in benzene and chloroform (Weast, 1975)

(*f*) *Stability*: No data were available.

(*g*) *Reactivity*: Oxidation with chromic acid gives a quinone. Reacts with bromine. Nitration yields 1- and 2-nitrotriphenylene (Clar, 1964).

2. Production, Use, Occurrence and Analysis

2.1 Production and use

There is no commercial production or known use of this compound.

2.2 Occurrence and analysis

Data on occurrence and methods of analysis are summarized in the 'General Remarks on the Substances Considered', p. 35.

Triphenylene occurs ubiquitously in products of incomplete combustion; it also occurs in fossil fuels. It has been identified in mainstream cigarette smoke (Snook *et al.*, 1977); gasoline engine exhaust (40-60 μg/l burned fuel) (Grimmer *et al.*, 1977) and exhaust tar (440 mg/kg tar) (Hoffmann & Wynder, 1962).

3. Biological Data Relevant to the Evaluation of Carcinogenic Risk to Humans

3.1 Carcinogenicity studies in animals[1]

Skin application

Mouse: A group of l0 mice was painted twice weekly with a 0.3% solution of triphenylene in benzene on the dorsal skin. No skin lesion was observed after 548 days (Barry *et al.*, 1935).

---

[1] The Working Group was aware of a study in progress in mice by skin application (IARC, 1983).

A group of 20 male C3H mice, two months of age, received skin applications of 60 μl of a 0.5% solution of triphenylene (purified by recrystallization) in decalin twice weekly on the interscapular area for 82 weeks. Among 14 mice surviving 52 weeks, no skin tumour was observed. In a second, similar experiment, 15 mice were treated with 0.5% triphenylene dissolved in a mixture of 1:1 decalin:*n*-dodecane (v/v). Among 11 mice that survived 52 weeks one carcinoma and four papillomas were observed compared to two papillomas in 13 control animals treated with the vehicle only. Median times of appearance of tumours in the two groups were 70 and 75 weeks (Horton & Christian, 1974)). [The Working Group noted that *n*-dodecane has been shown to have a cocarcinogenic effect when tested simultaneously with several polynuclear hydrocarbons (Bingham & Falk, 1969).]

3.2 Other relevant biological data[1]

(*a*) *Experimental systems*

No data were available to the Working Group on toxic effects, on effects on reproduction and prenatal toxicity or on metabolism and activation.

Mutagenicity and other short-term tests

Results from short-term tests are given in Table 1.

Table 1. Results from short-term tests: Triphenylene

Test	Organism/assay[a]	Exogenous metabolic system[a]	Reported result	Comments	References
PROKARYOTES					
Mutation	*Salmonella typhimurium* (his^-/his^+)	Aro-R-PMS	Positive	At 20 μg/plate in strain TA100	Mossanda *et al.* (1979)
		Aro-R-Micr	Positive	At 20 nmol/plate in strain TA100; dose-response seen with increasing concentrations of Micr	Wood *et al.* (1980)
	Salmonella typhimurium ($8AG^S/8AG^R$)	Aro-R-PMS	Positive	At 44 nmol/ml in strain TM677	Kaden *et al.* (1979)

[a] For an explanation of the abbreviations, see the Appendix, p. 452.

(*b*) *Humans*

No data were available to the Working Group.

[1] See also 'General Remarks on the Substances Considered', p. 53.

3.3 Case reports and epidemiological studies of carcinogenicity to humans

No data were available to the Working Group.

4. Summary of Data Reported and Evaluation[1]

4.1 Experimental data

Triphenylene was tested for carcinogenicity in one study by skin application to male mice. No increase in the incidence of skin tumours was observed.

No data on the teratogenicity of this compound were available.

Triphenylene was mutagenic to *Salmonella typhimurium* in the presence of an exogenous metabolic system.

There is *inadequate evidence* that triphenylene is active in short-term tests.

4.2 Human data[2]

Triphenylene is present as a major component of the total content of polynuclear aromatic compounds in the environment. Human exposure to triphenylene occurs primarily through the smoking of tobacco, inhalation of polluted air and by ingestion of food and water contaminated by combustion effluents (for details, see 'General Remarks on the Substances Considered', p. 35).

4.3 Evaluation

The available data are inadequate to permit an evaluation of the carcinogenicity of triphenylene to experimental animals.

---

[1] For definitions of the italicized terms, see Preamble, p. 19.

[2] Studies on occupational exposure to polynuclear aromatic compounds will be considered in future *IARC Monographs*.

5. References

Barry, G., Cook, J.W., Haslewood, G.A.D., Hewett, C.L., Hieger, I. & Kennaway, E.L. (1935) The production of cancer by pure hydrocarbons - Part III. *Proc. R. Soc. London Ser. B*, *117*, 318-351

Bingham, E. & Falk, H.L. (1969) Environmental carcinogens. The modifying effect of cocarcinogens on the threshold response. *Arch. environ. Health*, *19*, 779-783

Clar, E., ed. (1964) *Polycyclic Hydrocarbons*, Vol. 1, New York, Academic Press, pp. 236-243

Davis, W.W., Krahl, M.E. & Clowes, G.H.A. (1942) Solubility of carcinogenic and related hydrocarbons in water. *J. Am. chem. Soc.*, *64*, 108-110

Grimmer, G., Böhnke, H. & Glaser, A. (1977) Investigation on the carcinogenic burden by air pollution in man. XV. Polycyclic aromatic hydrocarbons in automobile exhaust gas - An inventory. *Zbl. Bakt. Hyg., 1 Abt., Orig. B164*, 218-234

Hoffmann, D. & Wynder, E.L. (1962) A study of air pollution carcinogenesis. II. The isolation and identification of polynuclear aromatic hydrocarbons from gasoline engine exhaust condensate. *Cancer*, *15*, 93-102

Horton, A.W. & Christian, G.M. (1974) Cocarcinogenic versus incomplete carcinogenic activity among aromatic hydrocarbons: Contrast between chrysene and benzo(*b*)triphenylene. *J. natl Cancer Inst.*, *53*, 1017-1020

IARC (1983) *Information Bulletin on the Survey of Chemicals Being Tested for Carcinogenicity*, No. 10, Lyon, p. 23

Kaden, D.A., Hites, R.A. & Thilly, W.G. (1979) Mutagenicity of soot and associated polycyclic aromatic hydrocarbons to *Salmonella typhimurium*. *Cancer Res.*, *39*, 4152-4159

Mossanda, K., Poncelet, F., Fouassin, A. & Mercier, M. (1979) Detection of mutagenic polycyclic aromatic hydrocarbons in African smoked fish. *Food Cosmet. Toxicol.*, *17*, 141-143

NIH/EPA Chemical Information System (1982) *Mass Spectral Search System*, Washington DC, CIS Project, Information Services Corporation

Snook, M.E., Severson, R.F., Arrendale, R.F., Higman, H.C. & Chortyk, O.T. (1977) The identification of high molecular weight polynuclear aromatic hydrocarbons in a biologically active reaction of cigarette smoke condensate. *Beitr. Tabakforsch.*, *9*, 79-101

Weast, R.C., ed. (1975) *CRC Handbook of Chemistry and Physics*, 56th ed., Cleveland, OH, Chemical Rubber Co., p. C-194

Windholz, M., ed. (1976) *The Merck Index*, 9th ed., Rahway, NJ, Merck & Co., p. 1250

Wood, A.W., Chang, R.L. Huang, M.-T., Levin, W., Lehr, R.E., Kumar, S., Thakker, D.R., Yagi, H., Jerina, D.M. & Conney, A.H. (1980) Mutagenicity of benzo(*e*)pyrene and triphenylene tetrahydroepoxides and diol-epoxides in bacterial and mammalian cells. *Cancer Res.*, *40*, 1985-1989

APPENDIX : Abbreviations used in tables of results of short-term tests

8AG	8-Azaguanine
Aro	Aroclor-induced
CytP450	Purified cytochrome P-450
GP	Guinea-pig
his	Histidine
3MC	3-Methylcholanthrene-induced
MEF	Mouse embryo fibroblast
Micr	Microsomes
None	None used (tests using these cells normally require an exogenous metabolic system)
OUA	Ouabain
PB	Phenobarbital-induced
PCB	Polychlorinated biphenyl-induced
PMS	Postmitochondrial supernatant
pol	Polymerase
R	Rat
R	Resistant
Rec	Recombination
S	Sensitive
SHE	Syrian hamster embryo
TFT	Trifluorothymidine
6TG	6-Thioguanine
UI	Uninduced
uvr A	Excision repair
	Not required

SUPPLEMENTARY CORRIGENDA TO VOLUMES 1-31

Corrigenda covering Volumes 1-6 appeared in Volume 7 ; others appeared in Volumes 8, 10-13, 15-31

Volume 20

p. 195	α-*isomer,* line 6	*delete ;* α-lindane
	β-*isomer,* line 5	*delete ;* β-lindane
p. 196	γ-*isomer,* line 5	*delete ;* γ-lindane
	δ-*isomer,* line 5	*delete ;* δ-lindane
	ε-*isomer,* line 4	*delete ;* ε-lindane
p. 197	ξ-*isomer,* line 3	*delete ;* ξ-lindane
	η-*isomer,* line 3	*delete ;* η-lindane
	θ-isomer, line 3	*delete ;* θ-lindane

Supplement 3

p. 117	*delete from* α-Lindane *see* Hexachlorocyclohexane (α-isomer) *20* 195 *through* θ-Lindane *see* Hexachlorocyclohexane θ-isomer) *20* 197 *inclusive*	

Volume 29

p. 193	*(k) Conversion factor*	*replace* ppm = 0.782 × mg/m³ *by* 1 ppm = 12.8 mg/m³
p. 269	*Chem. Abstr. Services Reg. No.*	*replace* 117-82-7 *by* 117-81-7
p. 270	*(k) Conversion factor*	*replace* ppm = 0.626 × mg/m³ *by* 1 ppm = 16 mg/m³
p. 273	*(b) Occupational exposure* 2nd para, 1st line	*replace* [0.25 – 2 ppm] *by* [0.025 – 0.2 ppm]
	(c) Air	*replace* [0.25 ppt] *by* [0.025 ppt]
	2nd line	*replace* [1.8 ppt] *by* [0.18 ppt]
p. 274	2nd line	*replace* [0.73 ppt] *by* [0.073 ppt]
p. 276	Limit of detection lines 1 and 2	*replace* [1.27 – 6.82 ppt] *by* [0.13 – 0.68 ppt]
p. 284	*Mutagenicity and other chromosomal effects*	*replace* [0.001 – 0.016 mg/m³] *by* [0.01 – 0.16 mg/m³]
p. 296	line 21	*replace* Tetrodirect Black E ; Tetrodirect Black EFD *by* Tertrodirect Black E ; Tertrodirect Black EFG

CUMULATIVE INDEX TO IARC MONOGRAPHS ON THE EVALUATION OF THE CARCINOGENIC RISK OF CHEMICALS TO HUMANS

Numbers in italics indicate volume, and other numbers indicate page. References to corrigenda are given in parentheses. Compounds marked with an asterisk(*) were considered by the working groups, but monographs were not prepared becaused adequate data on carcinogenicity were not available.

E

IARC MONOGRAPHS ON THE EVALUATION OF THE CARCINOGENIC RISK OF CHEMICALS TO HUMANS

Some Inorganic Substances, Chlorinated Hydrocarbons, Aromatic Amines, *N*-Nitroso Compounds, and Natural Products	Volume 1, 1972; 184 pages (out of print)
Some Inorganic and Organometallic Compounds	Volume 2, 1973; 181 pages US$ 3.60; Sw. fr. 12.-- (out of print)
Certain Polycyclic Aromatic Hydrocarbons and Heterocyclic Compounds	Volume 3, 1973; 271 pages (out of print)
Some Aromatic Amines, Hydrazine and Related Substances, *N*-Nitroso Compounds and Miscellaneous Alkylating Agents	Volume 4, 1974; 286 pages US$ 7.20; Sw. fr. 18.--
Some Organochlorine Pesticides	Volume 5, 1974; 241 pages US$ 7.20; Sw. fr. 18.-- (out of print)
Sex Hormones	Volume 6, 1974; 243 pages US$ 7.20; Sw. fr. 18.--
Some Anti-thyroid and Related Substances, Nitrofurans and Industrial Chemicals	Volume 7, 1974; 326 pages US$ 12.80; Sw. fr. 32.--
Some Aromatic Azo Compounds	Volume 8, 1975; 357 pages US$ 14.40; Sw. fr. 36.--
Some Aziridines, *N*-, *S*- and *O*-Mustards and Selenium	Volume 9, 1975; 268 pages US$ 10.80; Sw. fr. 27.--
Some Naturally Occurring Substances	Volume 10, 1976; 353 pages US$ 15.00; Sw. fr. 38.--
Cadmium, Nickel, Some Epoxides, Miscellaneous Industrial Chemicals and General Considerations on Volatile Anaesthetics	Volume 11, 1976; 306 pages US$ 14.00; Sw. fr. 34.--
Some Carbamates, Thiocarbamates and Carbazides	Volume 12, 1976; 282 pages US$ 14.00; Sw. fr. 34.--
Some Miscellaneous Pharmaceutical Substances	Volume 13, 1977; 255 pages US$ 12.00; Sw. fr. 30.--
Asbestos	Volume 14, 1977; 106 pages US$ 6.00; Sw. fr. 14.--
Some Fumigants, the Herbicides 2,4-D and 2,4,5-T, Chlorinated Dibenzodioxins and Miscellaneous Industrial Chemicals	Volume 15, 1977; 354 pages US$ 20.00; Sw. fr. 50.--
Some Aromatic Amines and Related Nitro Compounds - Hair Dyes, Colouring Agents and Miscellaneous Industrial Chemicals	Volume 16, 1978; 400 pages US$ 20.00; Sw. fr. 50.--
Some *N*-Nitroso Compounds	Volume 17, 1978; 365 pages US$ 25.00; Sw. fr. 50.--

Polychlorinated Biphenyls and Polybrominated Biphenyls	Volume 18, 1978; 140 pages US$ 13.00; Sw. fr. 20.--
Some Monomers, Plastics and Synthetic Elastomers, and Acrolein	Volume 19, 1979; 513 pages US$ 35.00; Sw. fr. 60.--
Some Halogenated Hydrocarbons	Volume 20, 1979; 609 pages US$ 35.00; Sw. fr. 60.--
Sex Hormones (II)	Volume 21, 1979; 583 pages US$ 35.00; Sw. fr. 60.--
Some Non-nutritive Sweetening Agents	Volume 22, 1980; 208 pages US$ 15.00; Sw. fr. 25.--
Some Metals and Metallic Compounds	Volume 23, 1980; 438 pages US$ 30.00; Sw. fr. 50.--
Some Pharmaceutical Drugs	Volume 24, 1980; 337 pages US$ 25.00; Sw. fr. 40.--
Wood, Leather and Some Associated Industries	Volume 25, 1980; 412 pages US$ 30.00; Sw. fr. 60.--
Some Anticancer and Immunosuppressive Drugs	Volume 26, 1981; 411 pages US$ 30.00; Sw. fr. 62.--
Some Aromatic Amines, Anthraquinones and Nitroso Compounds and Inorganic Fluorides Used in Drinking-Water and Dental Preparations	Volume 27, 1982; 341 pages US$ 25.00; Sw. fr. 40.--
The Rubber Industry	Volume 28, 1982; 486 pages US$ 35.00; Sw. fr. 70.--
Some Industrial Chemicals and Dyestuffs	Volume 29, 1982; 416 pages US$ 30.00; Sw. fr. 60.--
Miscellaneous Pesticides	Volume 30, 1983; 424 pages US$ 30.00; Sw. fr. 60.--
Some Feed Additives, Food Additives and Naturally Occurring Substances	Volume 31, 1983; 314 pages US$ 30.00; Sw. fr. 60. --
Chemicals and Industrial Processes Associated with Cancer in Humans (IARC Monographs 1-20)	Supplement 1, 1979; 71 pages (out of print)
Long-term and Short-term Screening Assays for Carcinogens: A Critical Appraisal	Supplement 2, 1980; 426 pages US$ 25.00; Sw. fr. 40.--
Cross Index of Synonyms and Trade Names in Volumes 1 to 26	Supplement 3, 1982; 199 pages US$ 30.00; Sw. fr. 60.--
Chemicals, Industrial Processes and Industries Associated with Cancer in Humans (IARC Monographs Volumes 1 to 29)	Supplement 4, 1982; 292 pages US$ 30.00; Sw. fr. 60.--
Polynuclear Aromatic Compounds, Part 1, Chemical, Environmental and Experimental Data	Volume 32, 1983 ; 477 pages US $ 35.00 ; Sw.fr. 70

Liver Cancer	No. 1, 1971; 176 pages US$ 10.000; Sw. fr. 30.—
Oncogenesis and Herpesviruses	No. 2, 1972; 515 pages US$ 25.00; Sw. fr. 100.—
N-Nitroso Compounds, Analysis and Formation	No. 3, 1972; 140 pages US$ 6.25; Sw. fr. 25.—
Transplacental Carcinogenesis	No. 4, 1973; 181 pages US$ 12.00; Sw. fr. 40.—
Pathology of Tumours in Laboratory Animals—Volume I—Tumours of the Rat, Part 1	No. 5, 1973; 214 pages US$ 15.00; Sw. fr. 50.—
Pathology of Tumours in Laboratory Animals—volume I—Tumours of the Rat, Part 2	No. 6, 1976; 319 pages US$ 35.00; Sw. fr. 90.— (OUT OF PRINT)
Host Environment Interactions in the Etiology of Cancer in Man	No. 7, 1973; 464 pages US$ 40.00; Sw. fr. 100.—
Biological Effects of Asbestos	No. 8, 1973; 346 pages US$ 32.00; Sw. fr. 80.—
N-Nitroso Compounds in the Environment	No. 9, 1974; 243 pages US$ 20.00; Sw. fr. 50.—
Chemical Carcinogenesis Essays	No. 10, 1974; 230 pages US$ 20.00; Sw. fr. 50.—
Oncogenesis and Herpesviruses II	No. 11, 1975; Part 1, 511 pages US$ 38.00; Sw. fr. 100.— Part 2, 403 pages US$ 30.00; Sw. fr. 80.—
Screening Tests in Chemical Carcinogenesis	No. 12, 1976; 666 pages US$ 48.00; Sw. fr. 120.—
Environmental Pollution and Carcinogenic Risks	No. 13, 1976; 454 pages US$ 20.00; Sw. fr. 50.—
Environmental *N*-Nitroso Compounds—Analysis and Formation	No. 14, 1976; 512 pages US$ 45.00; Sw. fr. 110.—
Cancer Incidence in Five Continents—Volume III	No. 15, 1976; 584 pages US$ 40.00; Sw. fr. 100.—
Air Pollution and Cancer in Man	No. 16, 1977; 331 pages US$ 35.00; Sw. fr. 90.—
Directory of On-going Research in Cancer Epidemiology 1977	No. 17, 1977; 599 pages US$ 10.00; Sw. fr. 25.— (OUT OF PRINT)
Environmental Carcinogens—Selected Methods of Analysis, Vol. 1: Analysis of Volatile Nitrosamines in Food	No. 18, 1978; 212 pages US$ 45.00; Sw. fr. 90.—
Environmental Aspects of *N*-Nitroso Compounds	No. 19, 1978; 566 pages US$ 50.00; Sw. fr. 100.—
Nasopharyngeal Carcinoma: Etiology and Control	No. 20, 1978; 610 pages US$ 60.00; Sw. fr. 100.—
Cancer Registration and Its Techniques	No. 21, 1978; 235 pages US$ 25.00; Sw. fr. 40.—
Environmental Carcinogens—Selected Methods of Analysis, Vol. 2: Methods for the Measurement of Vinyl Chloride in Poly (vinyl chloride), Air, Water and Foodstuffs	No. 22, 1978; 142 pages US$ 45.00; Sw. fr. 75.—
Pathology of Tumours in Laboratory Animals—Volume II—Tumours of the Mouse	No. 23, 1979; 669 pages US$ 60.00; Sw. fr. 100.—
Oncogenesis and Herpesviruses III	No. 24, 1978; Part 1, 580 pages US$ 30.00; Sw. fr. 50.— Part 2, 522 pages US$ 30.00; Sw. fr. 50.—
Carcinogenic Risks—Strategies for Intervention	No. 25, 1979; 283 pages US$ 30.00; Sw. fr. 50.—
Directory of On-Going Research in Cancer Epidemiology 1978	No. 26, 1978; 550 pages Sw. fr. 30.—
Molecular and Cellular Aspects of Carcinogen Screening Tests	No. 27, 1980; 371 pages US$ 40.00; Sw. fr. 60.—
Directory of On-Going Research in Cancer Epidemiology 1979	No. 28, 1979; 672 pages Sw. fr. 30.— (OUT OF PRINT)
Environmental Carcinogens—Selected Methods of Analysis, Vol. 3: Analysis of Polycyclic Aromatic Hydrocarbons in Environmental Samples	No. 29, 1979; 240 pages US$ 30.00; Sw. fr. 50.—
Biological Effects of Mineral Fibres	No. 30, 1980; Volume 1, 494 pages US$ 35.00; Sw. fr. 60.— Volume 2, 513 pages US$ 35.00; Sw. fr. 60.—
N-Nitroso Compounds: Analysis, Formation and Occurrence	No. 31, 1980; 841 pages US$ 40.00; Sw. fr. 70.—
Statistical Methods in Cancer Research, Vol. 1: The Analysis of Case-Control Studies	No. 32, 1980; 338 pages US$ 30.00; Sw. fr. 50.—
Handling Chemical Carcinogens in the Laboratory—Problems of Safety	No. 33, 1979; 32 pages US$ 8.00; Sw. fr. 12.—
Pathology of Tumours in Laboratory Animals—Volume III—Tumours of the Hamster	No. 34, 1982; 461 pages US$ 40.00; Sw. fr. 80.—
Directory of On-Going Research in Cancer Epidemiology 1980	No. 35, 1980; 660 pages Sw. fr. 35.—
Cancer Mortality by Occupation and Social Class 1851-1971	No. 36, 1982; 253 pages US$ 30.00; Sw. fr. 60.—
Laboratory Decontamination and Destruction of Aflatoxins B_1, B_2, G_1, G_2 in Laboratory Wastes	No. 37, 1980; 59 pages US$ 10.00; Sw. fr. 18.—
Directory of On-Going Research in Cancer Epidemiology 1981	No. 38, 1981; 696 pages Sw. fr. 40.—
Host Factors in Human Carcinogenesis	No. 39, 1982; 583 pages US$ 50.00; Sw. fr. 100.—
Environmental Carcinogens—Selected Methods of Analysis, Vol. 4: Some Aromatic Amines and Azo Dyes in the General and Industrial Environment	No. 40, 1981; 347 pages US$ 30.00; Sw. fr. 60.—
N-Nitroso Compounds: Occurrence and Biological Effects	No. 41, 1982; 755 pages US$ 55.00; Sw. fr. 110.—
Cancer Incidence in Five Continents—Volume IV	No. 42, 1982; 811 pages US$ 50.00; Sw. fr. 100.—
Laboratory Decontamination and Destruction of Carcinogens in Laboratory Wastes: Some *N*-Nitrosamines	No. 43, 1982; 73 pages US$ 10.00; Sw. fr. 18.—

Environmental Carcinogens—Selected Methods of Analysis, Vol. 5: Mycotoxins	No. 44, 1983; 455 pages US$ 30.00; Sw. fr. 60.—
Environmental Carcinogens—Selected Methods of Analysis, Vol. 6: *N*-Nitroso Compounds	No. 45, 1983 ; 508 pages US $ 40.00; Sw.fr. 80.—
Directory of On-Going Research in Cancer Epidemiology 1982	No. 46, 1982; 722 pages Sw. fr. 40.—
Cancer Incidence in Singapore	No. 47, 1982; 174 pages US$ 15.00; Sw. fr. 30.—
Cancer Incidence in the USSR Second Revised Edition	No. 48, 1982; 75 pages US$ 15.00; Sw. fr. 30.—
Laboratory Decontamination and Destruction of Carcinogens in Laboratory Wastes: Some Polycyclic Aromatic Hydrocarbons	No. 49, 1983; 81 pages US$ 10.00; Sw. fr. 20.-
Directory of On-Going Research in Cancer Epidemiology 1983	No. 50, 1983; 740 pages Sw. fr. 50.-

NON-SERIAL PUBLICATIONS

Alcool et Cancer	1978; 42 pages Fr. fr. 35-; Sw. fr. 14.-
Information Bulletin on the Survey of Chemicals Being Tested for Carcinogenicity No. 8	1979, 604 pages US$ 20.00; Sw.fr. 40.-
Cancer Morbidity and Causes of Death Among Danish Brewery Workers	1980, 145 pages US$ 25.00; Sw.fr. 45.-
Information Bulletin on the Survey of Chemicals Being Tested for Carcinogenicity No. 9	1981, 294 pages US$ 20.00; Sw.fr. 41.-
Information Bulletin on the Survey of Chemicals Being Tested for Carcinogenicity No. 10	1982, 326 pages US$ 20.00; Sw.fr. 42.-

Composition, impression et façonnage
Groupe MCP-Mame
Dépôt légal : Février 1984

00461855
3 1378 00461 8552